Nonlinear Fractional Differential Equations of Variable Order

Mohammed Said Souid
Department of Economic Sciences
University of Tiaret, Tiaret, Algeria

Mustafa Inc
Science Faculty, Department of Mathematics
Firat University, Elazig, Turkey

Amar Benkerrouche
Faculty of Exact Science and Computer Science
University of Djelfa, Djelfa, Algeria

CRC Press
Taylor & Francis Group
Boca Raton London New York

CRC Press is an imprint of the
Taylor & Francis Group, an **informa** business
A SCIENCE PUBLISHERS BOOK

First edition published 2026
by CRC Press
2385 NW Executive Center Drive, Suite 320, Boca Raton FL 33431

and by CRC Press
4 Park Square, Milton Park, Abingdon, Oxon, OX14 4RN

CRC Press is an imprint of Taylor & Francis Group, LLC

Library of Congress Cataloging-in-Publication Data (applied for)

ISBN: 978-1-041-12067-4 (hbk)
ISBN: 978-1-041-17093-8 (pbk)
ISBN: 978-1-003-68790-0 (ebk)

DOI: 10.1201/9781003687900

Typeset in Times New Roman
by Prime Publishing Services

To all those who tirelessly pursue knowledge, pushing the boundaries of what is known, and to the scientific community that fosters innovation and collaboration.
We dedicate this work to our families, whose unwavering support and patience made this endeavor possible, and to our mentors and colleagues, who inspired and guided us throughout our academic journeys.
May this book serve as a valuable resource for researchers and students alike, contributing to the ongoing exploration and understanding of nonlinear fractional differential equations of variable order. It is our hope that the insights presented herein will ignite new ideas and pave the way for future advancements in this fascinating field, inspiring generations to come to delve deeper into the intricate beauty and profound applications of mathematics.

Preface

The field of fractional calculus has seen remarkable growth in recent years, driven by its ability to model complex phenomena in physics, biology, engineering, and beyond, where traditional integer-order differential equations fall short. Nonlinear Fractional Differential Equations of Variable Order addresses the pressing need for a comprehensive resource on variable-order fractional differential equations, an emerging area that captures dynamic systems with time- or space-dependent fractional orders. This book fills a critical gap by providing a rigorous mathematical framework, practical applications, and advanced analytical techniques for tackling these sophisticated equations. This book offers an in-depth exploration of nonlinear fractional differential equations with variable order, covering Riemann-Liouville, Caputo, and Hadamard fractional derivatives. It begins with foundational notations and definitions, progresses through initial and terminal value problems, and delves into advanced topics like Ulam-Hyers stability and fixed point theorem techniques. The text emphasizes real-world applications, such as anomalous diffusion, bone remodeling, and drug distribution models, illustrating the power of variable-order fractional calculus in addressing intricate systems. Each chapter builds systematically, blending theoretical rigor with practical examples, and includes a detailed bibliography for further research. The target audience comprises graduate students, researchers, and professionals in applied mathematics, physics, engineering, and related fields. It is intended for those with a solid background in calculus and differential equations who wish to deepen their understanding of fractional calculus and its applications. The book serves as both a textbook for advanced courses and a reference for researchers exploring cutting-edge problems in variable-order fractional systems. By presenting clear derivations, illustrative examples, and open problems, it aims to inspire further innovation and application in this dynamic field. This work is dedicated to those who tirelessly pursue knowledge, pushing the boundaries of mathematics to uncover its profound beauty and utility. We hope it becomes a valuable resource for readers, authors, and researchers advancing the study of fractional differential equations.

Contents

1

Introduction

1.1 Introduction

Variable-order fractional operators are derivatives and integrals whose order depends on specific variables, forming a more complex class. They generalize constant order fractional operators to accommodate variable-order behavior.

In 1993, Samko and Ross [92] introduced fractional integration and differentiation with variable order, where the order $\ltimes$ varies with time as $\ltimes(s)$. They extended the R-Liouville and Fourier definitions [76, 87, 91]. Various definitions of variable-order fractional operators have since emerged, including Riemann–Liouville, Caputo, Hadamard, and Caputo–Hadamard; see [8, 10, 77, 91, 92].

Subsequent research has focused on variable-order fractional operators, their applications and interpretations; see [7, 37, 39]. Samko's formulation has proven especially useful in mechanics and viscous flow theory [37, 39, 41, 75, 78, 79], where many physical processes exhibit fractional-order behavior that varies with time or space [67]. For instance, [39] studies a variable-order fractional differential equation relevant to viscoelasticity, while [41] analyzes the dynamics and control of a nonlinear viscoelastic oscillator, proposing two controllers for tracking arbitrary reference functions. The work in [75] examines drag force on a particle in oscillatory viscous flow, using variable-order calculus to model flow-dependent derivative order. In [79] variable-order equation is developed for a particle in a quiescent viscous liquid. For a broader overview of dynamic system modeling using variable-order operators, see the recent review [78].

In recent years, variable-order fractional differential equations and operators have gained traction in engineering applications; see [48, 98] and references therein. Despite their broad utility, foundational research on existence, uniqueness, and stability of solutions—particularly for initial and terminal *v*alue problems—remains limited. Current studies, such as [94, 101], address only basic cases with restricted methodologies, highlighting the need for deeper exploration in this emerging field.

DOI: 10.1201/9781003687900-1

Complex competitive interactions—such as trophic dynamics, nutrient diffusion, and cellular competition—are common in ecological and biological systems. These systems often exhibit long-range memory and spatial effects that vary over time and space. Variable-order fractional operators offer an effective framework for modeling such evolving dynamics.

Ghanbari et al. [43] modeled nutrient–phytoplankton–zooplankton dynamics using variable-order operators, showing that memory effects depend on both population levels and order variation. A variable-order growth model in [9] accurately captured population histories across countries, outperforming constant-order models. Similarly, Sweilam et al. [104] applied variable-order fractional equations with Grünwald-Letnikov operators to simulate interactions among individuals affected by three tuberculosis strains—drug-sensitive, multidrug-resistant, and extensively drug-resistant—and those uninfected.

Using WHO data, Sweilam et al. [104] estimated the treatment rate needed to control tuberculosis spread in Egypt. Another biological application involves bone cell dynamics: Neto et al. [69, 70] demonstrated that variable-order fractional models capture tumor–healthy cell interactions with fewer parameters than integer-order models. These models incorporate non-local memory effects, where the order varies with time and spatial location, reflecting tumor influence on healthy bone tissue.

The authors suggest that aligning variable-order models with experimental data could yield deeper insights into tumor growth and inform targeted therapies. An intriguing application involves modeling Twitter's influence on alcoholism spread using time-delayed variable-order operators, which outperformed constant-order models [44]. Across these studies, variable-order fractional calculus effectively captures transitions between dynamic regimes in biological systems. It has also been extended to random-order models [100, 102, 116], though research remains limited. A rigorous mathematical framework for random-order operators is still lacking, yet their potential in modeling chaotic phenomena—such as financial volatility, turbulence, and vibration control—makes them promising for advanced risk analysis and control strategies.

Fractional calculus has become a widely used and influential tool across disciplines such as physics, chemistry, and dynamical systems control.

Recent interest has grown in the existence and uniqueness of solutions to fractional differential equations with initial and terminal *v*alue conditions. While fixed-order cases are well-studied, variable-order formulations remain underexplored.

This absence complicates the analysis of variable-order fractional operators, both integral and differential. Without key properties like the semigroup law, nonlinear functional analysis offers limited tools for proving existence of solutions to variable-order differential equations. This prevents transforming such equations into equivalent integral forms. Addressing initial and terminal value problems in this context remains a significant challenge—but one that must be overcome due to the importance of variable-order models.

Between 2021 and 2023, Souid et al. published over thirty papers advancing the theory of variable-order fractional differential equations. Their work focused on existence, uniqueness, and stability of solutions using fixed point methods; examples include [24, 25, 26, 27, 28, 29, 30, 31, 32, 33, 34, 62, 85, 86, 97, 106, 117].

Inspired by recent work, this book explores the development of fractional calculus of variations with variable order. It focuses on the existence, uniqueness, and stability of approximate solutions to nonlinear variable-order fractional differential equations—covering implicit, multiterm, and thermostat models—with various derivatives (Riemann–Liouville, Caputo, Hadamard, Caputo–Hadamard, Atangana–Baleanu) under local and nonlocal conditions (initial, terminal, impulsive, and delay). The analysis employs several techniques, including upper–lower solution methods, Banach and Schauder fixed point theorems, Schaefer's and Krasnoselskii's fixed point results, and Darbo's theorem via measures of noncompactness.

Our main approach introduces an abstract variable-order problem for fractional differential equations, where the system's hierarchical variation is modeled by a function $\ltimes(s) : J \longrightarrow (k-1, k], k = 1, 2$. The domain $J := [0, \mathsf{T}]$ or $[1, \mathsf{T}]$ is partitioned into subintervals $\mathcal{P} := \{J_1 = [\mathsf{T}_0, \mathsf{T}_1], J_2 = (\mathsf{T}_1, \mathsf{T}_2], J_3 = (\mathsf{T}_2, \mathsf{T}_3], \ldots, J_n = (\mathsf{T}_{n-1}, \mathsf{T}_n]\}$, where n is a given natural number, in which $\mathsf{T}_0 = 0$ or 1, and $\mathsf{T}_n = \mathsf{T}$. We define a piecewise constant function $\ltimes(s) : J \to (k-1, k]$ over $\mathcal{P}$, as:

$$\ltimes(s) = \sum_{i=1}^{n} \ltimes_i I_i(s), \ s \in J,$$

where $k - 1 < \ltimes_i \leq k, i = 1, 2, \ldots, n$ are constants and $X_{Ji}(s)$ is the indicator function for J_i. This formulation transforms the variable-order fractional problem into an equivalent constant-order problem on each subinterval.

The stability theory of functional equations began with Ulam's question [109, 110] on the stability of group homomorphisms: "*Under what conditions does an additive mapping exist near an approximately additive one?*" Hyers [51] provided the first partial answer for ßanach spaces. This was extended by Aoki [13] for additive mappings and by Th.M. ℜassias [83] for linear mappings with unbounded Cauchy differences. Gavruta [42] further generalized the ℜassias result. Since then, numerous studies have explored Hyers–Ulam stability across various functional equations [1, 11, 22, 53, 56, 58, 59, 61, 64, 113, 114, 115], with comprehensive treatments in [38, 81, 82]. Our work examines Ulam–Hyers and Ulam-Hyers-ℜassias stability for variable-order fractional differential equations with initial and terminal $v$alue conditions, highlighting their growing role in mathematical analysis.

Finally, we used Maple and MATLAB® with Euler's discretization over the interval J to compute and illustrate approximate solutions for selected examples.

Overall, our findings highlight the broad applicability of transdisciplinary sciences and offer a foundation for future research on this open challenge.

The book opens with Chapter 1, introducing key notations, definitions, and foundational concepts used throughout.

Section 1.1 introduces key terms like partition, piecewise constant functions, and generalized intervals, along with Banach space notations. Section 1.2 reviews gamma function properties and foundational concepts in constant- and variable-order fractional calculus, including two lemmas on the Hadamard fractional integral. Section 1.3 covers the Kuratowski measure of noncompactness, while Section 1.4 presents fixed point theorems used later. Section 1.5 revisits definitions related to Ulam-type stability.

Chapter 2 focuses on the existence, uniqueness, and stability of continuous solutions found in the Ulam-Hyers and Ulam-Hyers-$\Re$assias sense for four initial and terminal value problems (ITVPs) involving variable-order Riemann–Liouville fractional differential equations. The operators $\mathcal{F}_{0+}^{\ltimes(s)}$, $I_{0+}^{\ltimes(s)}$ denote the variable-order Riemann–Liouville derivative and integral, respectively, where $\lambda(s) : J := [0, \mathsf{T}]$ to $(1, 2]$ with $0 < \mathsf{T} < \infty$. Section 2.1 examines the multiterm ITVP:

$$\begin{cases} \mathcal{F}_{0+}^{\ltimes(s)} \varkappa(s) = f(s, \varkappa(s), I_{0+}^{\ltimes(s)} \varkappa(s)), & s \in J, \\ \varkappa(0) = 0, \quad \varkappa(\mathsf{T}) = 0, \end{cases} \tag{1.1}$$

Let X be a real or complex Banach space and J a given set. Consider a continuous function $f : J \times X \times X \to X$. This section employs the Kuratowski measure of noncompactness and Darbo's fixed point theorem to establish key results. We also analyze the Ulam–Hyers–Rassias stability of the ITVP (1.1) solution. The final section presents an example validating the theoretical findings. Section 2.2 focuses on the implicit ITVP.

$$\begin{cases} \mathcal{F}_{0+}^{\ltimes(s)} y(s) = f(s, y(s), \mathcal{F}_{0+}^{\ltimes(s)} y(s)), & s \in J, \\ y(0) = 0, \quad y(\mathsf{T}) = 0, \end{cases} \tag{1.2}$$

Let J be a set and consider a continuous $f : J \times \mathbb{R} \times \mathbb{R} \to \mathbb{R}$ investigate Ulam-Hyers-$\Re$assias stability for the given problem using Krasnoselskii's fixed point theorem, supported by a numerical example. Section 2.3 examines the following variable-order fractional ITVP:

$$\begin{cases} \mathcal{F}^{\ltimes(s)} y(s) = g(s, y(s)), & s \in J, \\ I^{2-\ltimes(s)} y(0) = a I^{2-\ltimes(s)} y(a), & \mathcal{F}^{\ltimes(s)-1} y(0) = b \mathcal{F}^{\ltimes(s)-1} y(a), \end{cases} \tag{1.3}$$

given a function $g : J \times \mathbb{R} \to \mathbb{R}$, and a $=$ b $= 1$. Two results are presented: one via Schaefer's theorem in ßanach spaces and another using a fixed point theorem fixed point approach. We conclude with an example illustrating the Ulam-Hyers-$\Re$assias stability of the solutions. Section 2.4 investigates the existence and Ulam–Hyers stability of solutions to the following initial-terminal value problem (ITVP) with finite delay:

$$\begin{cases} F_{0+}^{\ltimes(s)}\varkappa(s) = f(s,\varkappa_s), \;\; s \in J, \\ \varkappa(s) = \chi(s), s \in [-\gamma, 0], \gamma > 0, \end{cases} \tag{1.4}$$

where $f : J \times C([-\gamma, 0], \mathbb{R}) \to \mathbb{R}$, $\chi \in C([-\gamma, 0], \mathbb{R})$ with $\chi(0) = 0$, and the delay function $\varkappa_s \in C([-\gamma, 0], \mathbb{R})$ is defined by

$$\varkappa_s(t) := \varkappa(s+t), \;\; t \in [-\gamma, 0], \tag{1.5}$$

for any $\varkappa$ defined on $[-\gamma, \mathsf{T}]$ and $s \in J$. The results are established using Green's function, Darbo's fixed point theorem, and Kuratowski's measure of noncompactness. In conclusion, two examples are presented to validate the results. Section 2.5 addresses the following implicit ITVP:

$$\begin{cases} F_{0+}^{\ltimes(s)}\varkappa(s) + f(s, \varkappa(s), F_{0+}^{\ltimes(s)}\varkappa(s)) = 0, \;\; s \in J := [0, \mathsf{T}], \\ \varkappa(0) = 0, \;\; \varkappa(\mathsf{T}) = 0. \end{cases} \tag{1.6}$$

The function $f : J \times X \times X \to X$ is continuous for $1 < \ltimes(s) \leq 2$. The R-Liouville fractional derivative of the order variable $\ltimes(s)$ is $F_{0+}^{\ltimes(s)}$. Let X be a real (or complex) ßanach space. The novel results gained are based on Kuratowski's metric of noncompactness and Darbo's fixed point theorem. Furthermore, sufficient conditions for Ulam-Hyers $\Re$assias stability are established, confirming the applicability of the core results. Finally, numerical examples illustrate the validity and effectiveness of the findings.

Chapter 3 focuses on variable-order Caputo fractional differential equations. Section 3.1 studies the ITVP

$$\begin{cases} {}^{c}F_{0+}^{\ltimes(s)}\varkappa(s) = f(s, \varkappa(s)), \;\; s \in J, \;\; 1 < \ltimes(s) \leq 2, \\ \varkappa(0) = 0, \;\; \varkappa(\mathsf{T}) = 0, \end{cases} \tag{1.7}$$

where $f : J \times \mathbb{R} \to \mathbb{R}$ is continuous and ${}^{c}F_{0+}^{\ltimes(s)}$ denotes the variable-order Caputo fractional derivative $\ltimes(s)$. Using Kuratowski's measure of noncompactness and Darbo's fixed point theorem, we establish existence and Ulam–Hyers stability. Section 3.2 addresses the impulsive IVP:

$${}^{c}F_{0+}^{\ltimes(s)}\varkappa(s) = f(s, \varkappa), \;\text{for}\; s \in J, \;\; s \neq \mathsf{T}_i, \;\; i = 1, \ldots, n, \tag{1.8}$$

$$\Delta\varkappa|_{s=\mathsf{T}_i} = \Phi_i(\varkappa(\mathsf{T}_i^-)), \;\; i = 1, \ldots, n, \tag{1.9}$$

$$\varkappa(0) = \varkappa_0, \tag{1.10}$$

where $0 < \ltimes(s) \leq 1$, $f : J \times \mathbb{R} \to \mathbb{R}$, $\Phi_i : \mathbb{R} \to \mathbb{R}$, continuous functions. The jump condition is defined by $\Delta\varkappa|_{s=\mathsf{T}_i} = \varkappa(\mathsf{T}_i^+) - \varkappa(\mathsf{T}_i^-)$. Two results are presented: one via Schauder's fixed point theorem and another using Banach's contraction principle. Ulam–Hyers stability is analyzed, followed by a validating example. Section 3.3 examines the multiterm ITVP:

$$\begin{cases} {}^{c}\digamma_{0+}^{\ltimes(s)}\varkappa(s) = f(s, \varkappa(s), I_{0+}^{\ltimes(s)}\varkappa(s)), & s \in J,\ 1 < \ltimes(s) \leq 2, \\ \varkappa(0) = 0,\ \varkappa(\mathsf{T}) = 0, \end{cases} \tag{1.11}$$

where the continuous function $f : J \times \mathbb{R} \times \mathbb{R} \to \mathbb{R}$ and ${}^{c}\digamma_{0+}^{\ltimes(s)}$, $I_{0+}^{\ltimes(s)}$ are the fractional derivatives of Caputo and the R-Liouville fractional integral varying in sequence $\ltimes(s)$, respectively. Using the Schauder and ßanach fixed point theorems, our goal is to examine if a solution exists for the ITVP (1.11). We also investigate the derived solution's stability in the Ulam-Hyers-$\mathfrak{R}$assias sense. A numerical example is presented to validate the theoretical results. Section 3.4 investigates the existence, uniqueness, and Ulam-Hyers stability of positive solutions for the fractional thermostat model:

$$\begin{cases} -{}^{c}\digamma^{\ltimes(s)}y(s) = f(s, y(s)), & s \in J,\ 1 < \ltimes(s) \leq 2, \\ y'(0) = 0, \quad {}^{c}\digamma^{\ltimes(s)-1}y(\mathsf{T}) + y(0) = 0, \end{cases} \tag{1.12}$$

where the function $f : J \times \mathbb{R} \to \mathbb{R}$ is given. The results are established via the Banach contraction principle and the Guo-Krasnoselskii fixed point theorem, with a final example illustrating their applicability.

In Chapter 4 focuses on two boundary value problems for Hadamard-type fractional differential equations with variable order. $\ltimes(t)$, where $\mathcal{J} := [1, \mathsf{T}] \to (1, 2]$, $1 < \mathsf{T} < \infty$. Section 4.1 explores qualitative aspects of the initial terminal value problem:

$$\begin{cases} ({}^{H}\digamma_{1+}^{\ltimes(s)}\varkappa)(s) = f(s, \varkappa(s)), \ s \in \mathcal{J}, \\ \varkappa(1) = 0, \quad \varkappa(\mathsf{T}) = 0, \end{cases} \tag{1.13}$$

where $f : \mathcal{J} \times \mathbb{R} \to \mathbb{R}$ is smooth, and $H\phi_{1+}^{(s)}$ denotes the Hadamard derivative of variable order $\phi(s)$. Using Schauder and ßanach fixed point theorems, we establish conditions for existence and uniqueness of solutions to (1.13), and analyze their Ulam-Hyers-$\mathfrak{R}$assias stability. A detailed example supports the theoretical findings. Section 4.2 examines the nonlinear initial terminal value problem:

$$\begin{cases} {}^{H}\digamma_{1+}^{\ltimes(s)}\varkappa(s) + f(s, \varkappa(s)) = 0, & s \in \mathcal{J}, \\ \varkappa(1) = \varkappa(\mathsf{T}) = 0, \end{cases} \tag{1.14}$$

where $f : \mathcal{J} \times X \to X$ is continuous and (X) is a Banach space. The operator ${}^{H}\digamma_{1+}^{\ltimes(s)}$ denotes the Hadamard derivative of variable order $\ltimes(s)$. Existence of solutions is established using Kuratowski's measure of non-compactness and Darbo's fixed point theorem. Ulam-Hyers-$\mathfrak{R}$assias stability is also analyzed, supported by a numerical example.

The second part of Chapter 4 investigates multiterm initial-terminal value problems. Section 4.3 considers:

$$\begin{cases} {}^{H}\digamma_{1^+}^{\ltimes(s)}\varkappa(s) = f(s,\varkappa(s),{}^{H}I_{1^+}^{\ltimes(s)}\varkappa(s)),\ s \in \mathcal{J}, \\ \varkappa(1) = \varkappa(\mathsf{T}) = 0, \end{cases} \tag{1.15}$$

where $f : \mathcal{J} \times X \times X \to X$ is smooth, X is a Banach space, and ${}^{H}\digamma_{1^+}^{\ltimes(s)}$, ${}^{H}I_{1^+}^{\ltimes(s)}$ denote the Hadamard fractional derivative and integral of variable-order $\ltimes(s)$. Existence and Ulam-Hyers-$\mathfrak{R}$assias stability are established using Kuratowski's measure of noncompactness. A numerical example illustrates the results.

$$\begin{cases} {}^{H}\digamma_{1^+}^{\ltimes(s)}\varkappa(s) = f(s,\varkappa(s),{}^{H}I_{1^+}^{\ltimes(s)}\varkappa(s)),\ s \in \mathcal{J}, \\ \varkappa(1) = \varkappa(\mathsf{T}) = 0, \end{cases} \tag{1.16}$$

with $1 < \mathsf{T} < +\infty$, $1 < \ltimes(s) \le 2$, $f : \mathcal{J} \times \mathbb{R} \times \mathbb{R} \to \mathbb{R}$ is a smooth function. Existence and uniqueness are proved via Krasnoselskii's fixed point theorem and Banach's contraction principle. Ulam-Hyers stability is analyzed, supported by a validating example.

Chapter 5 focuses on nonlinear fractional differential equations involving the variable-order Caputo–Hadamard derivative. Section 5.1 examines existence, uniqueness, and Ulam–Hyers stability for the impulsive ITVP:

$${}^{CH}\digamma_{\mathsf{T}_i^+}^{\ltimes(s)}\varkappa(s) = f(s,\varkappa(s),{}^{H}I_{\mathsf{T}_i^+}^{\ltimes(s)}\varkappa(s)),\ for\, s \in \mathcal{J},\ s \neq \mathsf{T}_i,\ i = 1,...,n, \tag{1.17}$$

$$\Delta\varkappa|_{s=\mathsf{T}_i} = L_i(\varkappa(\mathsf{T}_i^-)),\ i = 1,...,n, \tag{1.18}$$

$$a\varkappa(1) + b\varkappa(\mathsf{T}) = c, \tag{1.19}$$

Here, $\mathsf{T} > e$, $0 < \ltimes(s) < 1$, $f : \mathcal{J} \times \mathbb{R} \times \mathbb{R} \longrightarrow \mathbb{R}$, $L_i : \mathbb{R} \longrightarrow \mathbb{R}$, $i = 1,...,n$ are continuous. The operators ${}^{CH}\digamma_{\mathsf{T}_i^+}^{\ltimes(s)}$ and ${}^{H}I_{\mathsf{T}_i^+}^{\ltimes(s)}$ denote the Caputo–Hadamard derivative and Hadamard integral of variable order $\ltimes(s)$, respectively, a, b, and c are real constants with $a + b \neq 0$, $1 = \mathsf{T}_0 < \mathsf{T}_1 < ... < \mathsf{T}_n < \mathsf{T}_{n+1} = \mathsf{T}$, $\Delta\varkappa|_{s=\mathsf{T}_i} = \varkappa(\mathsf{T}_i^+) - \varkappa(\mathsf{T}_i^-)$, $\varkappa(\mathsf{T}_i^+) = \lim_{\mathsf{h}\longrightarrow 0^+}\varkappa(\mathsf{T}_i + \mathsf{h})$ and $\varkappa(\mathsf{T}_i^-) = \lim_{\mathsf{h}\longrightarrow 0^-}\varkappa(\mathsf{T}_i + \mathsf{h})$ represent the right-hand side and left-hand side limits of $\varkappa(s)$. The jump condition is defined by $\Delta\varkappa|_{s=\mathsf{T}_i} = \varkappa(\mathsf{T}_i^+) - \varkappa(\mathsf{T}_i^-)$, with $\mathrm{a} + \mathrm{b} = 0$. Existence and uniqueness results are derived using Schauder and Banach fixed point theorems. Section 5.2 investigates the applicability of the upper–lower solution method to the IVP:

$$\begin{cases} {}^{CH}\digamma_{p^+}^{\ltimes(s)}\varkappa(s) = f(s,\varkappa(s)),\ s \in \mathbf{J} := [p,\mathsf{T}], \\ \varkappa(p) = \varkappa_p, \end{cases} \tag{1.20}$$

where $0 < p < \mathsf{T} < \infty$, $\varkappa_p \in \mathbb{R}$, and $0 < \ltimes(s) \le 1$ and $f : \mathbf{J} \times \mathbb{R} \to \mathbb{R}$ is given. The operator ${}^{CH}\digamma_{p^+}^{\ltimes(s)}$ denotes the variable-order Caputo–Hadamard derivative. Section 5.3, investigates the existence and Ulam–Hyers stability of solutions to a non-local multiterm impulsive terminal value problem (ITVP) involving the variable-order Atangana–Baleanu fractional derivative in the Riemann–Liouville sense:

$$\begin{cases} {}^{ABR}\mathcal{F}_{0+}^{\ltimes(s)}y(s) = f_y(s), \quad s \in J, \\ y(0) = 0, y(\mathsf{T}) = \sum_{k=1}^{m} a_k y(s_k), \quad s_k \in]0, \mathsf{T}], \end{cases} \tag{1.21}$$

where $f_y(s) := f(s, y(s), {}^{AB}I_{0+}^{\ltimes(s)}y(s))$, $2 < \ltimes(s) \leq 3$. Here, $f : J \times \mathbb{R}^2 \to \mathbb{R}$ is a given function, and ${}^{ABR}\mathcal{F}_{0+}^{\ltimes(s)}$, ${}^{AB}I_{0+}^{\ltimes(s)}$ denote the Atangana–Baleanu fractional derivative and integral of variable order $\phi(s)$ in the R-Liouville sense. Existence results are derived using Krasnoselskii's fixed point theorem. An illustrative example confirms the applicability of the theoretical findings.

2

Preliminary

In this chapter definitions and some auxiliary results are given regarding the main objects of the monograph: some notations and definitions of some terms related to fractional calculus theory and proprieties of noncompactness measure, some fixed point theorems, some Ulam type stability theorems.

2.1 Notations and Definitions

Definition 1 *We indicate using $C(J, X)$, the continuous function Banach the norm in space $x : J \to X$*

$$\|x\| = Sup\{\|x(s)\| : s \in J\},$$

where X represents a Banach space that can be real or complex, and by $C(J,\mathbb{R})$, we denote the Banach space of all continuous functions $x : J \to \mathbb{R}$ with the usual norm

$$\|x\| = Sup\{|x(s)| : s \in J\}.$$

Definition 2 *The set of functions $PC(J,\mathbb{R})$ is defined as follows*

$PC(J,\mathbb{R}) = \{\varkappa : J \to \mathbb{R},\ \varkappa \in C((\mathsf{T}_i, \mathsf{T}_{i+1}],\mathbb{R}),$ *and there exist* $\varkappa(\mathsf{T}_i^-)$ *and* $\varkappa(\mathsf{T}_i^+),\ i = 1,\ldots,n$ *with* $\varkappa(\mathsf{T}_i^-) = \varkappa(\mathsf{T}_i)\}$.

$PC(J,\mathbb{R})$ is a Banach space with norm

$$\|\varkappa\| = \sup\{|\varkappa(s)| : s \in J\}.$$

Definition 3 *The set of functions $PC^1(J,\mathbb{R})$ is given*

$PC^1(J,\mathbb{R}) = \{\varkappa \in PC(J,\mathbb{R}),\ \varkappa \in C^1((\mathsf{T}_i, \mathsf{T}_{i+1}],\mathbb{R}),$ *and there exist* $\varkappa'(\mathsf{T}_i^-)$ *and* $\varkappa'(\mathsf{T}_i^+),\ i = 1,\ldots,n$ *with* $\varkappa'(\mathsf{T}_i^-) = \varkappa'(\mathsf{T}_i)\}$.

Definition 4 *[63] For $k \in \mathbb{N}$ define the set $AC^k(J)$ by*

$$AC^k(J) = \{\varkappa : J \to \mathbb{R},\ \varkappa, \varkappa', ..., \varkappa^{(k-1)} \in C(J,\mathbb{R}),\ \varkappa^{(k-1)} \text{ is absolutely continuous}\},$$

DOI: 10.1201/9781003687900-2

Definition 5 *[63] For* $0 < a < b < \infty$, *we define the set* $AC^k_{s\frac{d}{ds}}([a,b])$, *by*

$$AC^k_{s\frac{d}{ds}}([a,b]) = \left\{ \varkappa : [a,b] \to \mathbb{R}, \left(s\frac{d}{ds}\right)^{(k-1)} \varkappa(s) \in AC([a,b]) \right\}.$$

Definition 6 *[5, 63] For* $-\infty < a < b < +\infty$, $1 \leq p \leq \infty$, *we denote* $L^p((a,b),\mathbb{R})$, *the ßanach space of measurable functions* $x : (a,b) \to \mathbb{R}$ *for which* $\|x\|_p < \infty$, *where*

$$\|x\|_p = \left(\int_a^b |x(s)|^p ds \right)^{\frac{1}{p}} \; if \; 1 \leq p < \infty,$$

and

$$\|x\|_\infty = esssup_{a<s<b} |x(s)|.$$

Definition 7 *[4, 5] For* $-\infty < a < b < +\infty$, *the Sobolev space* $H^1((a,b),\mathbb{R})$ *is defined as follows*

$$H^1((a,b),\mathbb{R}) := \left\{ x : x \in L^2((a,b),\mathbb{R})), x' \in L^2((a,b),\mathbb{R}) \right\}.$$

Definition 8 *[55, 121, 123] A subset* I *of* $\mathbb{R}$ *that represents a generalized interval is either an interval (i.e., a set of the form* $[a,b]$, (a,b), $[a,b)$, *or* (a,b)*), a singleton* $\{a\}$, *or the null set* $\emptyset$.

Definition 9 *[55, 121, 123] Assuming a generalized interval* I. *When* $\varkappa$ *in* I *lies in exactly one of the generalized intervals* E *in* $\mathcal{P}$, *then* I *is partitioned. This is due to the fact that there exists a finite collection* $\mathcal{P}$ *of generalized intervals contained within* I.

Definition 10 *[55, 121, 123] For any* $E \in \mathcal{P}$, f *is constant on* E. *When* f *is piecewise constant, it means that* $\mathcal{P}$. *Consider* $f : I \to \mathbb{R}$ *as a function,* I *as a generalized interval, and* $\mathcal{P}$ *as a subdivision of* I.

2.2 Fractional Calculus

2.2.1 Gamma Function

The Gamma function $\Gamma(\ltimes)$, a cornerstone of fractional calculus, was introduced by Leonhard Euler (1707–1783) to extend the factorial to non-integer values. Its significance attracted further study by leading mathematicians including Legendre, Gauss, Gudermann, Liouville, Weierstrass, Hermite, and others.

Definition 11 *[3, 54] The Gamma function* $\Gamma(\ltimes)$ *is defined by the integral*

$$\Gamma(\ltimes) = \int_0^\infty s^{\ltimes - 1} e^{-s} ds, \ltimes > 0.$$

Proposition 1 *[3, 54] The Gamma function generalizes the factorial, satisfying*

$$\Gamma(n+1) = n\Gamma(n) = n!.$$

Proposition 2 *[54, 73] For $\ltimes \in [0,1]$, the Gamma function satisfies*

$$\frac{\ltimes^2 + 1}{\ltimes + 1} \leq \Gamma(\ltimes + 1) \leq \frac{\ltimes^2 + 2}{\ltimes + 2}. \tag{2.1}$$

Remark 1 *For $\ltimes \in [0,1]$, Proposition 2 yields*

$$\frac{1}{\Gamma(\ltimes + 1)} \leq \frac{1}{2(\sqrt{2} - 1)}. \tag{2.2}$$

2.2.2 Fractional Calculus of Constant-order

Fractional calculus extends derivative orders from integers to rationals—a concept that, while simple, yields profound results across bioengineering, dynamics, modeling, control theory, and physical sciences. Its growing relevance is reflected in numerous recent studies [44, 91, 101, 105, 111, 112, 118, 120, 119, 124], and references therein.

Definition 12 *[63] For $0 \leq a < b < \infty$, the left-sided Riemann–Liouville fractional integral of a function $h \in L^1([a,b], \mathbb{R}_+)$ of order $\ltimes \in \mathbb{R}_+$ is defined as*

$$I_a^{\ltimes} h(s) = \frac{1}{\Gamma(\ltimes)} \int_a^s (s-t)^{\ltimes - 1} h(t)dt,$$

where $\Gamma(.)$ denotes the gamma function.

Definition 13 *[63] For $0 \leq a < b < \infty$, the left-sided Riemann–Liouville fractional derivative of order $\ltimes > 0$ for a function $h \in L^1([a,b], \mathbb{R}_+)$ is defined as*

$$(\digamma_{a+}^{\ltimes} h)(s) = \frac{1}{\Gamma(n - \ltimes)} \left(\frac{d}{ds}\right)^n \int_a^s (s-t)^{n - \ltimes - 1} h(t)dt,$$

where $n = [\ltimes]$ and $\Gamma(.)$ is the Gamma function.

Definition 14 *[63] For $0 \leq a < b < \infty$, the left Caputo fractional derivative of order $\ltimes > 0$ for function $h(s)$ is defined as*

$${}^{c}\digamma_{a+}^{\ltimes} h(s) = \int_a^s \frac{(s-t)^{n - \ltimes - 1}}{\Gamma(n - \ltimes)} h^{(n)}(t)dt, \quad s > a,$$

where $n = [\ltimes]$ and $\Gamma(.)$ is the Gamma function.

Definition 15 *[63] For $1 \leq a < b < \infty$, the left-sided Hadamard fractional integral of a function $h \in L^1([a,b], \mathbb{R}_+)$ of order $\ltimes \in \mathbb{R}_+$ is defined as*

$${}^{H}I_{a+}^{\ltimes} h(s) = \frac{1}{\Gamma(\ltimes)} \int_a^s \left(\ell og \frac{s}{t}\right)^{\ltimes - 1} h(t) \frac{dt}{t}, \quad s > a. \tag{2.3}$$

Definition 16 *[63] For $1 \leq a < b < \infty$, the left-sided Hadamard fractional derivative of order $\ltimes \in \mathbb{R}_+$ for function $h(s)$ is defined as*

$$({}^{H}\digamma^{\ltimes}_{a+}h)(s) = \frac{1}{\Gamma(n-\ltimes)}(s\frac{d}{ds})^n \int_a^s (\ell og\frac{s}{t})^{n-\ltimes-1}\frac{h(t)}{t}dt, \quad s > a, \tag{2.4}$$

where $n = [\ltimes]$ and $\Gamma(.)$ is the Gamma function.

Definition 17 *[8] For $1 \leq a < b < \infty$, the Caputo-type Hadamard fractional derivative of order $\ltimes \in (0,1)$ for a function $h(s) \in AC^1([a,b])$ is defined as*

$${}^{CH}\digamma^{\ltimes}_{a}\, h(s) = \frac{1}{\Gamma(1-\ltimes)}\int_a^s \left(\ell og\frac{s}{t}\right)^{-\ltimes} h'(t)\, dt, \quad s > a, \tag{2.5}$$

where $\Gamma(.)$ denotes the Gamma function and $AC^1([a(.)b])$ is the space of absolutely continuous functions with integrable first derivatives.

Definition 18 *[2, 14] For $-\infty < a < b < +\infty$, the Riemann–Liouville fractional integral in the sense of Atangana-Baleanu of order $\ltimes \in (0,1)$ for a function $y(.) \in H^1((a,b),\mathbb{R})$ is expressed as*

$${}^{AB}I^{\ltimes}_{a+}y(s) = \frac{1-\ltimes}{N(\ltimes)}y(s) + \frac{\ltimes}{N(\ltimes)}\int_a^s \frac{(s-\tau)^{\ltimes-1}}{\Gamma(\ltimes)}y(\tau)d\tau, \quad s > a, \tag{2.6}$$

where

$$N(\ltimes) = 1 - \ltimes + \frac{\ltimes}{\Gamma(\ltimes)}.$$

Definition 19 *[2, 14] For $-\infty < a < b < +\infty$, the Riemann–Liouville fractional differential operator in the sense of Atangana-Baleanu of order $\text{ß} \in (0,1)$ for a function $y \in H^1((a,b),\mathbb{R})$ is expressed as*

$${}^{ABR}\digamma^{\text{ß}}_{a+}y(s) = \frac{N(\text{ß})}{1-\text{ß}}\frac{d}{ds}\int_a^s \mathcal{E}_{\text{ß}}\left(-\text{ß}\frac{(s-\tau)^{\text{ß}}}{(1-\text{ß})}\right)y(\tau)d\tau, \quad s > a, \tag{2.7}$$

where $\mathcal{E}_{\text{ß}}(.)$ denotes the Mittag-Leffler function.

We now outline key properties of constant-order fractional integrals and derivatives.

Lemma 1 *[63] Assume $\ltimes,\ \text{ß} > 0$, $a > 0$, $h \in L^1(a,b)$, and $\digamma^{\ltimes}_{a+}h \in L^1(a,b)$. Then, the fractional differential equation*

$$\digamma^{\ltimes}_{a+}h = 0$$

has a unique solution of the form

$$h(s) = \omega_1(s-a)^{\ltimes-1} + \omega_2(s-a)^{\ltimes-2} + \cdots + \omega_n(s-a)^{\ltimes-n}$$

and

$$I^{\ltimes}_{a+}\digamma^{\ltimes}_{a+}h(s) = h(s) + \omega_1(s-a)^{\ltimes-1} + \omega_2(s-a)^{\ltimes-2} + \cdots + \omega_n(s-a)^{\ltimes-n}.$$

where $n-1 < \ltimes \leq n$, $\omega_i \in \mathbb{R}$, $i = 1, 2, ..., n$.

Moreover, the following identities hold:

$$\digamma^{\ltimes}_{a+}I^{\ltimes}_{a+}h(s) = h(s).$$

and

$$I^{\ltimes}_{a+}I^{\ss}_{a+}h(s) = I^{\ss}_{a+}I^{\ltimes}_{a+}h(s) = I^{\ltimes+\ss}_{a+}h(s).$$

Lemma 2 *[63] Let* $\ltimes$, $\ss > 0$, $a > 0$, $h \in L^1(a,b)$, ${}^{c}\digamma^{\ltimes}_{a+}h \in L^1(a,b)$. *Then the Caputo fractional differential equation*

$${}^{c}\digamma^{\ltimes}_{a+}h = 0$$

has a unique solution given by

$$h(s) = \omega_0 + \omega_1(s-a) + \omega_2(s-a)^2 + ... + \omega_{n-1}(s-a)^{n-1},$$

and

$$I^{\ltimes}_{a+}\ {}^{c}\digamma^{\ltimes}_{a+}h(s) = h(s) + \omega_0 + \omega_1(s-a) + \omega_2(s-a)^2 + ... + \omega_{n-1}(s-a)^{n-1},$$

where $n-1 < \ltimes \leq n$, $\omega_i \in \mathbb{R}$, $i = 0, 1, ..., n-1$.

Moreover, the following identity holds:

$${}^{c}\digamma^{\ltimes}_{a+}I^{\ltimes}_{a+}h(s) = h(s).$$

Lemma 3 *[63, 90] Let* $\ltimes$, $\ss > 0$, $a > 1$, $h \in L^1(a,b)$, *and* ${}^{H}\digamma^{\ltimes}_{a+}h \in L^1(a,b)$. *Then the Hadamard-type fractional differential equation*

$${}^{H}\digamma^{\ltimes}_{a+}h = 0$$

has a unique solution given by

$$h(s) = \omega_1\left(\ell og\frac{s}{a}\right)^{\ltimes-1} + \omega_2\left(\ell og\frac{s}{a}\right)^{\ltimes-2} + \cdots + \omega_n\left(\ell og\frac{s}{a}\right)^{\ltimes-n},$$

and

$${}^{H}I^{\ltimes}_{a+}\left({}^{H}\digamma^{\ltimes}_{a+}h(s)\right) = h(s) + \omega_1\left(\ell og\frac{s}{a}\right)^{\ltimes-1} + \omega_2\left(\ell og\frac{s}{a}\right)^{\ltimes-2} + \cdots + \omega_n\left(\ell og\frac{s}{a}\right)^{\ltimes-n},$$

where $n = \lfloor \ltimes \rfloor + 1$ *and* $\omega_i \in \mathbb{R}$, $i = 1, 2, \ldots, n$, *and*

$${}^{H}\digamma^{\ltimes}_{a+}\left({}^{H}I^{\ltimes}_{a+}\right)h(s) = h(s),$$

$${}^{H}I^{\ltimes}_{a+}\left({}^{H}I^{\ss}_{a+}\right)h(s) = {}^{H}I^{\ss}_{a+}\left({}^{H}I^{\ltimes}_{a+}\right)h(s) = {}^{H}I^{\ltimes+\ss}_{a+}h(s).$$

Lemma 4 *[63]. Let* $\ltimes, \text{ß} > 0$, $a, b > 0$, $h \in AC^k_{s\frac{d}{ds}}([a,b])$. *Then,*

$$^{H}I^{\ltimes}_{a+}\, {}^{CH}\digamma^{\ltimes}_{a+}\, h(s) = h(s) - \sum_{i=0}^{k-1} \frac{(s\frac{d}{ds})^i\, h(t)}{i!} \left(\ell og\frac{s}{t}\right)^i,$$

We have the following special cases:

$$^{H}I^{\ltimes}_{a+}\left(\ell og\frac{s}{a}\right)^{\text{ß}-1} = \frac{\Gamma(\text{ß})}{\Gamma(\text{ß}+\ltimes)}\left(\ell og\frac{s}{a}\right)^{\text{ß}+\ltimes-1}, \quad \digamma^{\ltimes}_{a+}\left(\ell og\frac{s}{a}\right)^{\text{ß}-1} = \frac{\Gamma(\text{ß})}{\Gamma(\text{ß}-\ltimes)}\left(\ell og\frac{s}{a}\right)^{\text{ß}-\ltimes-1},$$

for each $s \in [a,b]$.

Lemma 5 *[92] Let* $n = [\ltimes] + 1$ *for* $\ltimes > 0$.

(i) *If* $\varkappa \in C([a,b],\mathbb{R})$, *then*

$$^{CH}\digamma^{\ltimes}_{a+}(^{H}I^{\ltimes}_{a+}\varkappa(s)) = \varkappa(s) \quad \text{for } s \in [a,b].$$

(ii) *If* $\varkappa \in AC([a,b])$, *then*

$$^{H}I^{\ltimes}_{a+}(^{CH}\digamma^{\ltimes}_{a+}\varkappa(s)) = \varkappa(s) - \varkappa(a) \quad \text{for } s \in [a,b].$$

Lemma 6 *[2] For* $y^{(n)}(.) \in H^1((a,b),\mathbb{R})$ *and* $\ltimes \in (n, n+1]$, *we have*

(i) $^{ABR}\digamma^{\ltimes}_{a+}\ {}^{AB}I^{\ltimes}_{a+}y(s) = y(s)$,

(ii) $^{AB}I^{\ltimes}_{a+}\ {}^{ABR}\digamma^{\ltimes}_{a+}y(s) = y(s) - \sum_{k=0}^{n-1}\frac{y^{(k)}(a)}{k!}(s-a)^k$.

2.2.3 Fractional Calculus of Variable-order

Variable-order fractional calculus generalizes constant-order operators by allowing the order to vary as a function. While constant-order problems have been extensively studied, including solution existence, variable-order cases remain less explored—particularly regarding solution stability and existence; see [24, 26, 27, 30, 32, 101, 105, 118]. This gap underscores the originality and significance of the findings presented in this book.

Definition 20 *[91, 92, 111] Let* $-\infty < a < b < +\infty$, *and consider functions* $\ltimes(s) : [a,b] \to (0,+\infty)$ *and* $\text{ß}(s) : [a,b] \to (n-1,n)$. *Then, the left-sided Riemann–Liouville Fractional Integral (RLFI) of variable order* $\ltimes(s)$ *for a function* $h(s)$ *is defined by*

$$I^{\ltimes(s)}_{a+}h(s) = \int_a^s \frac{(s-t)^{\ltimes(s)-1}}{\Gamma(\ltimes(s))}h(t)dt, \quad s > a, \tag{2.8}$$

and the Left-Sided Riemann–Liouville Fractional Integral (RLFI) of Variable Order for a function $h(s)$, *the RLFI of variable order* $\text{ß}(s)$ *is defined by*

$$\digamma^{\text{ß}(s)}_{a+}h(s) = \left(\frac{d}{ds}\right)^n I^{n-\text{ß}(s)}_{a+}h(s) = \left(\frac{d}{ds}\right)^n \int_a^s \frac{(s-t)^{n-\text{ß}(s)-1}}{\Gamma(n-\text{ß}(s))}h(t)dt, \quad s > a. \tag{2.9}$$

and the Left-Sided Caputo Fractional Derivative (CFD) of $h(s)$ with variable order $ß(s)$ is defined as:

$$ {}^{c}\digamma_{a+}^{ß(s)} h(s) = \int_a^s \frac{(s-t)^{n-ß(s)-1}}{\Gamma(n-ß(s))} h^{(n)}(t)dt, \quad s > a. \tag{2.10}$$

Definition 21 *[77, 10] For $1 \leq a < b < +\infty$, consider the mappings $\ltimes(s) : [a,b] \to (0,+\infty)$ and $ß(s) : [a,b] \to (n-1,n)$. Then, the left Hadamard fractional integral (HFI) of variable-order $\ltimes(s)$ for a function $h(s)$ is defined by:*

$$ {}^{H}I_{a+}^{\ltimes(s)} h(s) = \frac{1}{\Gamma(\ltimes(s))} \int_a^s (\ell og\frac{s}{t})^{\ltimes(s)-1} \frac{h(t)}{t} dt, \quad s > a. \tag{2.11}$$

Similarly, the left Hadamard fractional derivative (HFD) of a function $h(s)$ with respect to a variable-order $ß(s)$ is given by:

$$ ({}^{H}\digamma_{a+}^{ß(s)} h)(s) = \frac{1}{\Gamma(n-ß(s))} (s\frac{d}{ds})^n \int_a^s (\ell og\frac{s}{t})^{n-ß(s)-1} \frac{h(t)}{t} dt, \quad s > a, \tag{2.12}$$

Here, $\Gamma(.)$ denotes the gamma function.

Definition 22 *[8, 77] For $1 \leq a < b < \infty$, consider the mapping $\ltimes(s) : [a,b] \to (0,1)$. The Caputo-type Hadamard fractional derivative (CHFD) of variable order $\ltimes(s)$ for a function $h(s) \in AC^1([a,b])$ is defined by*

$$ \digamma_{a+}^{\ltimes(s)} h(s) = \frac{1}{\Gamma(1-\ltimes(s))} \int_a^s \frac{h'(\tau)}{(s-\tau)^{\ltimes(s)}} d\tau,$$

where $\Gamma(\cdot)$ denotes the Gamma function and $h'(s)$ is the derivative of $h(s)$. An alternative representation of the Caputo-type Hadamard fractional derivative is given by

$$ {}^{c}\digamma_{a}^{\ltimes(s)} h(s) = \frac{\ltimes'(s)}{\Gamma(2-\ltimes(s))} \int_a^s \left(\ell og\frac{s}{t}\right)^{1-\ltimes(s)} h'(t) \left[\frac{1}{1-\ltimes(s)} - \ell og\left(\ell og\frac{s}{t}\right)\right] dt$$

$$ + \frac{1}{\Gamma(1-\ltimes(s))} \int_a^s \left(\ell og\frac{s}{t}\right)^{-\ltimes(s)} h'(t)\, dt, \quad s > a. \tag{2.13}$$

Definition 23 *[2, 14] Let $-\infty < a < b < +\infty$, and consider the mapping $\ltimes : (a,b) \to [0,1]$. Then, the R-Liouville fractional integral in the sense of Atangana-Baleanu (RLFIAB) of variable-order $\ltimes(\cdot)$ for a function $y(\cdot) \in H^1((a,b),\mathbb{R})$ is defined by*

$$ I_{a+}^{\ltimes(s)} y(s) = \frac{1}{\Gamma(\ltimes(s))} \int_a^s \frac{y(\tau)}{(s-\tau)^{\ltimes(s)-1}} e^{-\frac{(s-\tau)^{\nu}}{\eta}} d\tau,$$

where $\Gamma(\cdot)$ *denotes the Gamma function, and* ν *and* η *are parameters associated with Atangana-Baleanu fractional integral. The Atangana–Baleanu integral in the Caputo sense (ABI) of variable order* $\Gamma(\cdot)$ *is given by*

$$ {}^{AB}I_{a+}^{\ltimes(s)}y(s) = \frac{1-\ltimes(s)}{N(\ltimes(s))}y(s) + \frac{\ltimes(s)}{N(\ltimes(s))}\int_a^s \frac{(s-\tau)^{\ltimes(\tau)-1}}{\Gamma(\ltimes(\tau))}y(\tau)d\tau, \quad s > a, \tag{2.14}$$

where the normalization function $N(\ltimes(s))$ *is defined as*

$$N(\ltimes(s)) = 1 - \ltimes(s) + \frac{\ltimes(s)}{\Gamma(\ltimes(s))}.$$

Definition 24 *[2, 14] Let* $-\infty < a < b < +\infty$*, and consider the mapping* $\text{ß} : [a,b] \to (0,1)$*. Then, the Riemann–Liouville fractional derivative of variable-order* $\text{ß}(s)$ *for a function* $h(s)$ *is defined as:*

$$\digamma_{a+}^{\text{ß}(s)}h(s) = \frac{1}{\Gamma(n-\text{ß}(s))}\frac{d^n}{ds^n}\int_a^s \frac{h(\tau)}{(s-\tau)^{\text{ß}(s)-n+1}}\,d\tau,$$

where $\Gamma(\cdot)$ *denotes the Gamma function, and* $n \in N$ *is such that* $n-1 < \text{ß}(s) < n$*. In the sense of Atangana-Baleanu (RLFDAB), the Riemann–Liouville fractional derivative of variable-order* $\text{ß}(.)$ *for a function* $y \in H^1((a,b),\mathbb{R})$ *is expressed by*

$$ {}^{ABR}\digamma_{a+}^{\text{ß}(s)}y(s) = \frac{N(\text{ß}(s))}{1-\text{ß}(s)}\frac{d}{dt}\int_a^s \mathcal{E}_{\text{ß}(\tau)}\left(-\text{ß}(\tau)\frac{(s-\tau)^{\text{ß}(\tau)}}{(1-\text{ß}(\tau))}\right)y(\tau)d\tau, \quad s > a, \tag{2.15}$$

where $\mathcal{E}_{\text{ß}(\tau)}(.)$ *denotes the Mittag-Leffler function.*

We now generalize Definition (24) to an arbitrary mapping $\ltimes : (a,b) \to (n, n+1]$.

Definition 25 *[2, 14] For the same interval* $a < b$ *introduced above, consider a mapping* $\ltimes : (a,b) \to (n, n+1]$ *associated with the function* $y(.) \in H^1((a,b),\mathbb{R})$*. Define* $\nu = \ltimes - n : (a,b) \to (0,1]$*. Then, the Riemann–Liouville fractional integral in the sense of Atangana-Baleanu of variable-order* $\ltimes(.)$ *is given by*

$$ {}^{AB}I_{a+}^{\ltimes(s)}y(s) = I_{a+}^{n}\ {}^{AB}I_{a+}^{\nu(s)}\ y(s), \quad s > a. \tag{2.16}$$

Definition 26 *[2, 14] For the same interval* $a < b$ *introduced above, consider a mapping* $\text{ß} : (a,b) \to (n, n+1]$ *and let* $y(.)$ *be such that* $y^{(n)}(.) \in H^1((a,b),\mathbb{R})$*. Define* $\nu = \text{ß} - n : (a,b) \to (0,1]$*. Then, the Riemann–Liouville fractional derivative in the sense of Atangana-Baleanu of variable-order* $\text{ß}(.)$ *is given by*

$$ {}^{ABR}\digamma_{a+}^{\text{ß}(s)}y(s) = {}^{ABR}\digamma_{a+}^{\nu(s)}y^{(n)}(s), \quad s > a. \tag{2.17}$$

Remark 2 *For the case where* $0 < \ltimes(s) \leq 1, 0 < \text{ß}(s) \leq 1$, *we have:*

(i) $\nu = \ltimes$

(ii) ${}^{AB}I_{a+}^{\ltimes(s)} y(s) = {}^{AB} I_{a+}^{\ltimes(s)} y(s)$

(iii) ${}^{ABR}F_{a+}^{\text{ß}(s)} y(s) = {}^{ABR} F_{a+}^{\text{ß}(s)} y(s)$

The classical operators—Riemann–Liouville fractional integral and derivative, Caputo fractional derivative (RLFI, RLFD), Caputo derivative (CFD), Hadamard integral and derivative (HFI, HFD), Caputo-type Hadamard derivative (CHFD), and their Atangana–Baleanu counterparts (RLFIAB, RLFDAB)—all coincide when the variable orders $\ltimes(s)$ and $\beta(s)$ are constant; see [2, 8, 10, 14, 63, 77, 91, 92].

Remark 3 *As noted in [119, 120, 122], the semigroup property does not generally hold for arbitrary functions* $\ltimes(s)$, $\text{ß}(s)$, *that is,*

$$I_{a+}^{\ltimes(s)} I_{a+}^{\text{ß}(s)} h(s) \neq I_{a+}^{\ltimes(s)+\text{ß}(s)} h(s).$$

Example 1 *Let*

$$\ltimes(s) = s, \quad s \in [0,4], \quad \text{ß}(s) = \begin{cases} 2, & s \in [0,1] \\ 3, & s \in]1,4]. \end{cases} \qquad h(s) = 2, \quad s \in [0,4]$$

$$\begin{aligned} I_{0+}^{\ltimes(s)} I_{0+}^{\text{ß}(s)} h(s) &= \int_0^s \frac{(s-t)^{\ltimes(s)-1}}{\Gamma(\ltimes(s))} \int_0^t \frac{(t-\tau)^{\text{ß}(t)-1}}{\Gamma(\text{ß}(t))} h(\tau) d\tau dt \\ &= \int_0^s \frac{(s-t)^{s-1}}{\Gamma(s)} [\int_0^1 \frac{(t-\tau)}{\Gamma(2)} 2d\tau + \int_1^t \frac{(t-\tau)^2}{\Gamma(3)} 2d\tau] dt \\ &= \int_0^s \frac{(s-t)^{s-1}}{\Gamma(s)} [2t - 1 + \frac{(t-1)^3}{3}] dt, \end{aligned}$$

and

$$I_{0+}^{\ltimes(s)+\text{ß}(s)} h(s)| = \int_0^s \frac{(s-t)^{\ltimes(s)+\text{ß}(s)-1}}{\Gamma(\ltimes(s)+\text{ß}(s))} h(t) dt.$$

So, we get

$$\begin{aligned} I_{0+}^{\ltimes(s)} I_{0+}^{\text{ß}(s)} h(s)|_{s=3} &= \int_0^3 \frac{(3-t)^2}{\Gamma(3)} [2t - 1 + \frac{(t-1)^3}{3}] dt \\ &= \frac{21}{10}, \end{aligned}$$

$$
\begin{aligned}
I_{0+}^{\ltimes(s)+\text{ß}(s)}h(s)|_{s=3} &= \int_0^3 \frac{(3-t)^{\ltimes(s)+\text{ß}(s)-1}}{\Gamma(\ltimes(s)+\text{ß}(s))}h(t)dt \\
&= \int_0^1 \frac{(3-t)^4}{\Gamma(5)}2dt + \int_1^3 \frac{(3-t)^5}{\Gamma(6)}2dt \\
&= \frac{1}{12}\int_0^1 (t^4 - 12t^3 + 54t^2 - 108t + 81)dt \\
&+ \frac{1}{60}\int_1^3 (-t^5 + 15t^4 - 90t^3 + 270t^2 - 405t + 243)dt \\
&= \frac{665}{180}.
\end{aligned}
$$

Therefore, we obtain

$$I_{0+}^{\ltimes(s)} I_{0+}^{\text{ß}(s)} h(s)|_{s=3} \neq I_{0+}^{\ltimes(s)+\text{ß}(s)} h(s)|_{s=3}.$$

Remark 4 *Note that the semigroup property is not fulfilled for general functions* $\ltimes(s)$, $\text{ß}(s)$, *i.e.:*

$${}^{H}I_{1+}^{\ltimes(s)}({}^{H}I_{1+}^{\text{ß}(s)})h(s) \neq^{H} I_{1+}^{\ltimes(s)+\text{ß}(s)}h(s).$$

Example 2 *Let*

$$\text{ß}(s) = \begin{cases} 1, & s \in [1,2] \\ 2, & s \in]2,4], \end{cases} \qquad \ltimes(s) = \begin{cases} 2, & s \in [1,2] \\ 1, & s \in]2,4], \end{cases} \qquad h(s) = 1, \ \ s \in [1,4],$$

$$
\begin{aligned}
{}^{H}I_{1+}^{\ltimes(s)}({}^{H}I_{1+}^{\text{ß}(s)})h(s) &= \frac{1}{\Gamma(\ltimes(s))}\int_1^s \frac{1}{t}(\ell og\frac{s}{t})^{\ltimes(s)-1} \\
&\quad \Big[\frac{1}{\Gamma(\text{ß}(t))}\int_1^t \frac{1}{\tau}(\ell og\frac{t}{\tau})^{\text{ß}(t)-1}h(\tau)d\tau\Big]dt \\
&= \frac{1}{\Gamma(2)}\int_1^2 \frac{1}{t}(\ell og\frac{s}{t})^1\Big[\frac{1}{\Gamma(1)}\int_1^2 \frac{1}{\tau}(\ell og\frac{t}{\tau})^{1-1}d\tau \\
&+ \frac{1}{\Gamma(2)}\int_2^t \frac{1}{\tau}(\ell og\frac{t}{\tau})^{2-1}d\tau\Big]dt \\
&+ \frac{1}{\Gamma(1)}\int_2^s \frac{1}{t}(\ell og\frac{s}{t})^0\Big[\frac{1}{\Gamma(1)}\int_1^2 \frac{1}{\tau}(\ell og\frac{t}{\tau})^{1-1}d\tau \\
&+ \frac{1}{\Gamma(2)}\int_2^t \frac{1}{\tau}(\ell og\frac{t}{\tau})^{2-1}d\tau\Big]dt \\
&= \frac{1}{\Gamma(2)}\int_1^2 \frac{1}{t}(\ell og\frac{s}{t})^1\Big[\ell og2 + \frac{1}{2}(\ell og\frac{t}{2})^2\Big]dt \\
&+ \frac{1}{\Gamma(1)}\int_2^s \frac{1}{t}(\ell og\frac{s}{t})^0\Big[\ell og2 + \frac{1}{2}(\ell og\frac{t}{2})^2\Big]dt,
\end{aligned}
$$

and

$$^{H}I_{1^+}^{\ltimes(s)+\text{ß}(s)}h(s) = \frac{1}{\Gamma(\ltimes(s)+\text{ß}(s))}\int_1^s \frac{1}{t}(\ell og\frac{s}{t})^{\ltimes(s)+\text{ß}(s)-1}h(t)dt,$$

we find that

$$\begin{aligned}
^{H}I_{1^+}^{\ltimes(s)}(^{H}I_{1^+}^{\text{ß}(s)})h(s)|_{s=3} &= \frac{1}{\Gamma(2)}\int_1^2 \frac{1}{t}(\ell og\frac{3}{t})^1\Big[\ell og2+\frac{1}{2}(\ell og\frac{t}{2})^2)\Big]dt\\
&+ \frac{1}{\Gamma(1)}\int_2^3 \frac{1}{t}(\ell og\frac{3}{t})^0\Big[\ell og2+\frac{1}{2}(\ell og\frac{t}{2})^2\Big]dt\\
&= \ell og2\int_1^2 \frac{1}{t}(\ell og\frac{3}{t})dt+\frac{1}{2}\int_1^2 \frac{1}{t}(\ell og\frac{3}{t})(\ell og\frac{t}{2})^2dt\\
&+ \ell og2\int_2^3 \frac{1}{t}dt+\frac{1}{2}\int_2^3 \frac{1}{t}(\ell og\frac{t}{2})^2dt\\
&= \frac{\ell og2}{2}[(\ell og3)^2-(\ell og\frac{3}{2})^2]+\frac{1}{24}[(\ell og3)^4-(\ell og\frac{3}{2})^4]\\
&+ \frac{1}{4}(\ell og3)^2(\ell og2)^2\\
&- \frac{1}{6}(\ell og2)(\ell og3)^3+\ell og2[(\ell og3)-(\ell og2)]+\frac{1}{6}(\ell og\frac{3}{2})^3\\
&\simeq 0.9013.
\end{aligned}$$

$$\begin{aligned}
^{H}I_{1^+}^{\ltimes(t)+\text{ß}(t)}h(s)|_{s=3} &= \frac{1}{\Gamma(\ltimes(t)+\text{ß}(t))}\int_1^s \frac{1}{t}(\ell og\frac{s}{t})^{\ltimes(s)+\text{ß}(s)-1}h(t)dt\\
&= \frac{1}{\Gamma(3)}\int_1^2 \frac{1}{t}(\ell og\frac{3}{t})^2dt+\frac{1}{\Gamma(3)}\int_2^3 \frac{1}{t}(\ell og\frac{3}{t})^2dt\\
&= \frac{1}{\Gamma(4)}[(\ell og3)^3-(\ell og\frac{3}{2})^3]+\frac{1}{\Gamma(4)}(\ell og\frac{3}{2})^3\\
&\simeq 0.2209.
\end{aligned}$$

Therefore, we obtain

$$^{H}I_{1^+}^{\ltimes(s)}(^{H}I_{1^+}^{\text{ß}(s)})h(s)|_{s=3} \neq {}^{H}I_{1^+}^{\ltimes(s)+\text{ß}(s)}h(s)|_{s=3}.$$

Remark 5 *Typically, for variable-order functions $\ltimes(s)$ and $\text{ß}(s)$, the semi-group property is does not hold; that is,*

$$^{AB}I_{a^+}^{\ltimes(s)}\ {}^{AB}I_{a^+}^{\text{ß}(s)}y(s) \neq {}^{AB}\ I_{a^+}^{\ltimes(s)+\text{ß}(s)}y(s).$$

Lemma 7 *[124] Let $\ltimes : J \to (1,2]$ be a continuous function. Then, for $h \in C_\delta(J,X) = \{h(s) \in C(J,X) \mid s^\delta h(s) \in C(J,X)\}$, where $0 \leq \delta \leq \min_{s\in J}|\ltimes(s)|$, the variable-order fractional integral $I_{0^+}^{\ltimes(s)}h(s)$ is well-defined at every point in J.*

Lemma 8 *[124] Let* $\ltimes : J \to (1,2]$ *be a continuous function. Then* $I_{0+}^{\ltimes(s)}h(s) \in C(J,X)$ *for* $h \in C(J,X)$*, the variable-order fractional integral* $I_{0+}^{\ltimes(s)}h(s)$ *belongs to* $C(J,X)$*; that is,* $I_{0+}^{\ltimes(s)}h(s) \in C(J,X)$.

Lemma 9 *[29, 50, 84] Let* $\ltimes : J \to (1,2]$ *be a continuous function. Then, for* $h \in C_\delta(J,\mathbb{R}) = \{h(s) \in C(J,\mathbb{R}) \mid (\ell ogs)^\delta h(s) \in C(J,\mathbb{R})\}$*, with* $0 \le \delta \le 1$*, the Hilfer-type variable-order integral* ${}^HI_{1+}^{\ltimes(s)}h(s)$ *well-defined at every point* $s \in J$.

Proof
Considering the continuity of $\Gamma(\ltimes(s))$, we assert that $M_\ltimes = \max_{s\in J}\left|\frac{1}{\Gamma(\ltimes(s))}\right|$ exists.

Let $\ltimes^* = \max_{s\in J}|(\ltimes(s))|$. Then, for $1 \le t \le s \le \mathsf{T}$, we have

$$(\ell og\frac{s}{t})^{\ltimes(s)-1} \le 1, \qquad if \;\; 1 \le \frac{s}{t} \le e,$$

$$(\ell og\frac{s}{t})^{\ltimes(s)-1} \le (\ell og\frac{s}{t})^{\ltimes^*-1}, \; if \;\; \frac{s}{t} > e.$$

Then, for $1 \le \frac{s}{t} < +\infty$, we obtain

$$(\ell og\frac{s}{t})^{\ltimes(s)-1} \le \max\left\{1, (\ell og\frac{s}{t})^{\ltimes^*-1}\right\} = M^*.$$

Now, for $h \in C_\delta(J,\mathbb{R})$, by the definition of (2.11),

we deduce:

$$\begin{aligned}
|{}^HI_{1+}^{\ltimes(s)}h(s)| &= \frac{1}{\Gamma(\ltimes(s))}\int_1^s (\ell og\frac{s}{t})^{\ltimes(s)-1}\frac{|h(t)|}{t}dt \\
&\le M_\ltimes \int_1^s (\ell og\frac{s}{t})^{\ltimes(s)-1}(\ell ogt)^{-\delta}(\ell ogt)^{\delta}\frac{|h(t)|}{t}dt \\
&\le M_\ltimes M^* \int_1^s \frac{1}{t}(\ell ogt)^{-\delta}\max_{t\in J}(\ell ogt)^{\delta}|h(t)|dt \\
&\le M_\ltimes M^* \max_{t\in J}(\ell ogt)^{\delta}h^\star \int_1^s \frac{1}{t}(\ell ogt)^{-\delta}dt \\
&\le M_\ltimes M^* \max_{t\in J}(\ell ogt)^{\delta}h^\star \frac{(\ell og\mathsf{T})^{1-\delta}}{1-\delta} < \infty,
\end{aligned}$$

where $h^\star = \max_{s\in J}|h(s)|$. This implies that the variable-order fractional integral ${}^HI_{1+}^{\ltimes(s)}h(s)$ is well-defined at every point in J.

Lemma 10 *[29, 50, 84] Proposition Let* $\ltimes : J \to (1,2]$ *be a continuous function. Then*

$${}^HI_{1+}^{\ltimes(s)}h(s) \in C(J,\mathbb{R}) \text{ for } h \in C(J,\mathbb{R}).$$

Proof

For $s, s_0 \in J$, $s_0 \leq s$, and let $h \in C(J,\mathbb{R})$. We aim to show that

$$
\begin{aligned}
&\Big|{}^{H}I_{1^+}^{\ltimes(s)}h(s) - {}^{H}I_{1^+}^{\ltimes(s_0)}h(s_0)\Big| = \Big|\int_1^s \frac{1}{\Gamma(\ltimes(s))}(\ell og\frac{s}{t})^{\ltimes(s)-1}\frac{h(t)}{t}dt \\
- &\int_1^{s_0}\frac{1}{\Gamma(\ltimes(s_0))}(\ell og\frac{s_0}{t})^{\ltimes(s_0)-1}\frac{h(t)}{t}dt\Big| \\
= &\Big|\int_0^1 \frac{1}{\Gamma(\ltimes(s))}\frac{(s-1)}{r(s-1)+1}(\ell og\frac{s}{r(s-1)+1})^{\ltimes(s)-1}h(r(s-1)+1)dr \\
- &\int_0^1 \frac{1}{\Gamma(\ltimes(s_0))}\frac{(s_0-1)}{r(s_0-1)+1}(\ell og\frac{s_0}{r(s_0-1)+1})^{\ltimes(s_0)-1}h(r(s_0-1)+1)dr\Big| \\
\\
= &\Big|\int_0^1 \Big[\frac{1}{\Gamma(\ltimes(s))}\frac{(s-1)}{r(s-1)+1}(\ell og\frac{s}{r(s-1)+1})^{\ltimes(s)-1}h(r(s-1)+1) \\
- &\frac{1}{\Gamma(\ltimes(s))}\frac{(s_0-1)}{r(s_0-1)+1}(\ell og\frac{s}{r(s-1)+1})^{\ltimes(s)-1}h(r(s-1)+1)\Big]dr \\
+ &\int_0^1 \Big[\frac{1}{\Gamma(\ltimes(s))}\frac{(s_0-1)}{r(s_0-1)+1}(\ell og\frac{s}{r(s-1)+1})^{\ltimes(s)-1}h(r(s-1)+1) \\
- &\frac{1}{\Gamma(\ltimes(s))}\frac{(s_0-1)}{r(s_0-1)+1}(\ell og\frac{s_0}{r(s_0-1)+1})^{\ltimes(s_0)-1}h(r(s-1)+1)\Big]dr \\
+ &\int_0^1 \Big[\frac{1}{\Gamma(\ltimes(s))}\frac{(s_0-1)}{r(s_0-1)+1}(\ell og\frac{s_0}{r(s_0-1)+1})^{\ltimes(s_0)-1}h(r(s-1)+1) \\
- &\frac{1}{\Gamma(\ltimes(s_0))}\frac{(s_0-1)}{r(s_0-1)+1}(\ell og\frac{s_0}{r(s_0-1)+1})^{\ltimes(s_0)-1}h(r(s-1)+1)\Big]dr \\
+ &\int_0^1 \Big[\frac{1}{\Gamma(\ltimes(s_0))}\frac{(s_0-1)}{r(s_0-1)+1}(\ell og\frac{s_0}{r(s_0-1)+1})^{\ltimes(s_0)-1}h(r(s-1)+1) \\
- &\frac{1}{\Gamma(\ltimes(s_0))}\frac{(s_0-1)}{r(s_0-1)+1}(\ell og\frac{s_0}{r(s_0-1)+1})^{\ltimes(s_0)-1}h(r(s_0-1)+1)\Big]dr\Big| \\
\leq &\; h^\star\int_0^1 \frac{1}{\Gamma(\ltimes(s))}(\ell og\frac{s}{r(s-1)+1})^{\ltimes(s)-1}\Big|\frac{(s-1)}{r(s-1)+1} - \frac{(s_0-1)}{r(s_0-1)+1}\Big|dr \\
+ &\; h^\star\int_0^1 \frac{1}{\Gamma(\ltimes(s))}\frac{(s_0-1)}{r(s_0-1)+1}\Big|(\ell og\frac{s}{r(s-1)+1})^{\ltimes(s)-1} \\
- &(\ell og\frac{s_0}{r(s_0-1)+1})^{\ltimes(s)-1}\Big|dr \\
+ &\; h^\star\int_0^1 \frac{(s_0-1)}{r(s_0-1)+1}(\ell og\frac{s_0}{r(s_0-1)+1})^{\ltimes(s_0)-1}\Big|\frac{1}{\Gamma(\ltimes(s))} - \frac{1}{\Gamma(\ltimes(s_0))}\Big|dr \\
+ &\int_0^1 \frac{1}{\Gamma(\ltimes(s_0))}\frac{(s_0-1)}{r(s_0-1)+1}(\ell og\frac{s_0}{r(s_0-1)+1})^{\ltimes(s_0)-1}\Big|h(r(s-1)+1) \\
- &h(r(s_0-1)+1)\Big|dr,
\end{aligned}
$$

where $h^\star = \max_{s\in J}|h(s)|$. Since $h \in C(J,\mathbb{R})$ is continuous, the integrands are continuous in s, and the difference tends to zero as $s \to s_0$. Hence, ${}^{H}I_{1+}^{\ltimes(s)}h(s) \in C(J,\mathbb{R})$.

2.2.4 Application of Variable Order Fractional Calculus

The next section highlights select applications of variable-order fractional calculus, extending those of constant-order models. Although the field remains underexplored, growing interest is driving active research and development.

First Application: Existing Models For Tumoruos Bone Remodeling

Biochemical models for bone remodeling aim to enhance understanding of bone dynamics and support early-stage clinical decision systems for targeted therapies. Simplified models simulating tumor-affected bone behavior are essential for personalized treatment. In this context, new approaches to existing models (Ayati et al., 2010; Komarova et al., 2003) were proposed, introducing variable-order derivatives for the first time in local biochemical modeling.

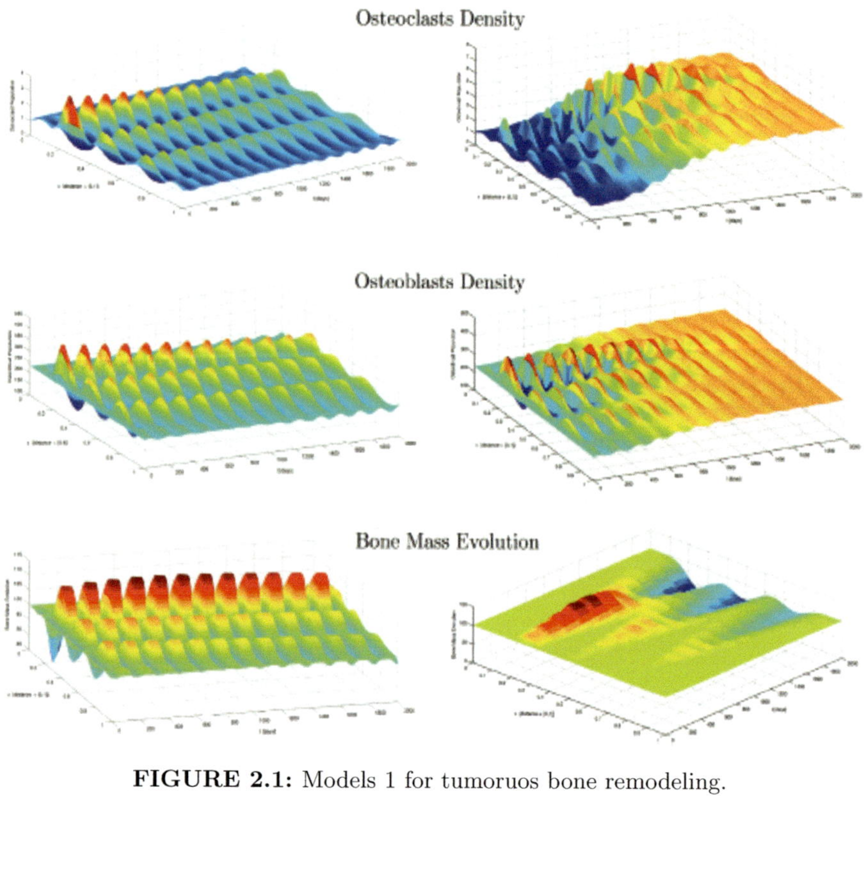

FIGURE 2.1: Models 1 for tumoruos bone remodeling.

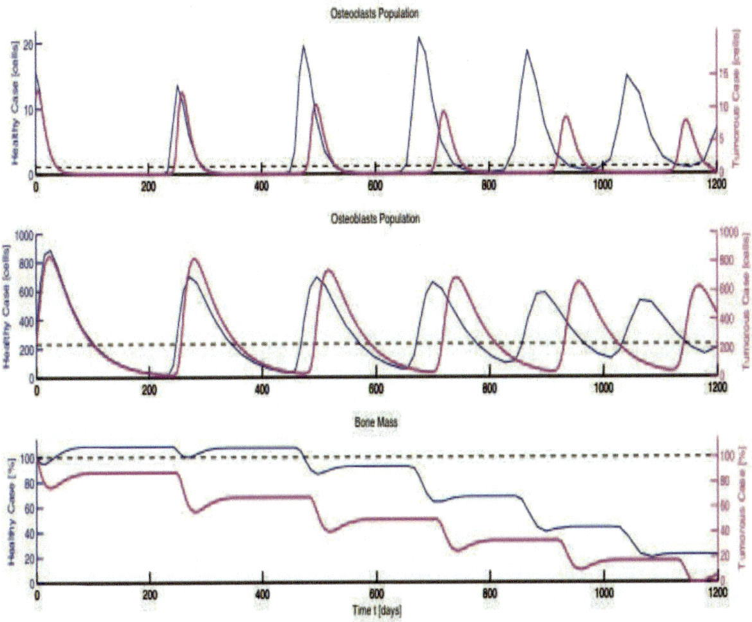

FIGURE 2.2: Model 2: Tumorous bone remodeling.

Simulation Results: The model incorporating variable-order derivatives simulates tumorous effects on the bone micro-environment—specifically osteoclasts, osteoblasts, and bone mass (see Figure 2.1). The blue curves represent the evolution of tumor-disrupted dynamics based on Equation 2 in [69], as presented by Ayati et al. (2010). In contrast, the magenta bold curves depict the same populations modeled using variable-order derivatives, as in Equation 3 in [69]. For this latter case, the resorption rate applied was $R = 0.0044$, with parameters detailed in Table 2.1.

TABLE 2.1: Additional variables and parameters introduced in the variable-order models (Figure 2.2).

variables/parameters	Descreption	*values*	Units
$\ltimes(t)$	Time dependent variable order	–	–
θ	variable order gradient	8.33×10^{-9}	–
κ_C	Bonne respetion rate	0.1548	$\%\text{day}^{-1}$
κ_B	Bonne respetion rate	6.8176×10^{-4}	$\%\text{day}^{-1}$
γ_T	Bonne metastates growth rate	0.004	$\%\text{day}^{-1}$

Enable simplified bone remodeling simulations that reproduce tumorous behavior with fewer inputs than traditional formulations. This leads to a more compact model that effectively captures key bone–tumor interactions and yields qualitatively consistent dynamics.

Second Application: Viscoelasticity

Viscoelastic media exhibit time-dependent behavior, motivating the use of variable-order fractional calculus in their modeling. Fractional time derivatives of displacement or strain effectively capture energy dissipation due to the inherent damping properties of fractional operators, enabling accurate representation of viscoelastic dynamics.

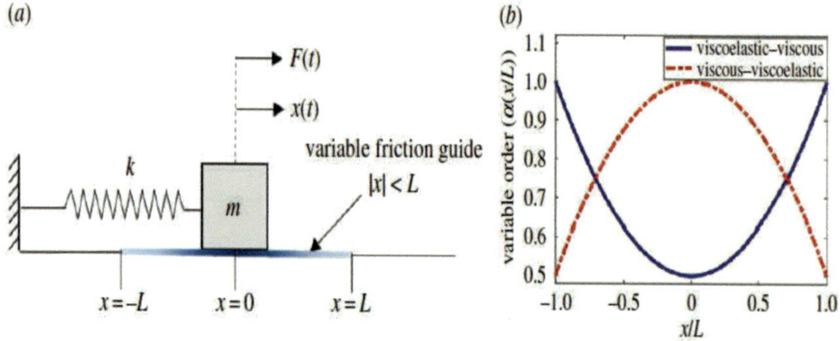

FIGURE 2.3: Viscoelasticity.

Figure 2.3(a): A mass-spring system moves along a guide [39] with spatially varying friction. In the 'viscoelastic-viscous' model, the guide is viscoelastic at $\varkappa = 0$ and becomes viscous at $|\varkappa| = L$; the reverse defines the 'viscous-viscoelastic' model. Shading illustrates the smooth transition in damping behavior.

Figure 2.3(b): Variable-order damping profiles are plotted for both models, showing a continuous shift between viscous and viscoelastic regions along the guide.

Third Application: Anomalous Diffusion

In classical diffusion, the mean square displacement (MSD) scales linearly with time, while in Caputo–order (CO) fractional models, it follows a fractional power law. Sun et al.[100] derived an MSD expression from a variable-order diffusion equation:

$$ {}_{0}^{C}F_{t}^{\ltimes(t)} c(t,\varkappa) = K\frac{\partial^2 c(t,\varkappa)}{\partial \varkappa^2}, \tag{2.18}$$

$0 < \ltimes(t) < 1$ yielding the MSD:

$$ < \varkappa^2(t) > = \frac{2Kt^{\ltimes(t)}}{\Gamma(\ltimes(t)+1)}, \tag{2.19}$$

Here, the Hurst exponent varies with time, a feature used in [16, 15, 71, 99] to study multi-fractional Brownian motion. Anomalous diffusion in fractals has also been explored in [89, 93], where MSD depends on both time and fractal dimension. Notably [93] modeled ion diffusion in Purkinje cell dendrites by treating the dendritic structure as a space- and time-dependent fractal (see Figure 2.4(a)).

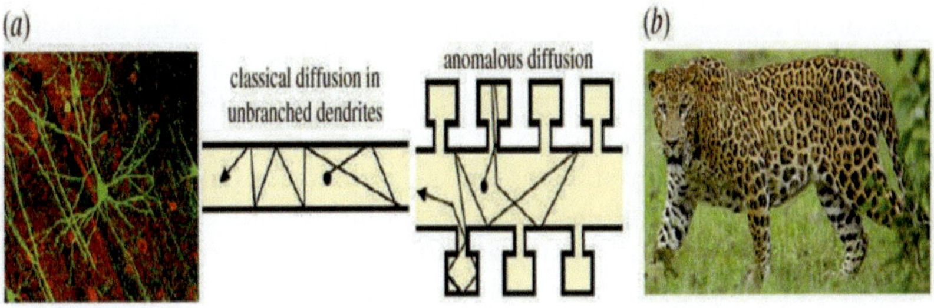

FIGURE 2.4: Anomalous diffusion.

Figure 2.4(a) Ion diffusion in neuronal dendrites is effectively captured by variable-order models, where the Hurst exponent depends on the fractal dimension, varying spatiotemporally [93]. Time-dependent Hurst exponents, $\ltimes = \ltimes(t)$, enhance modeling accuracy. Variable-order $\ltimes(\kappa)$ has also been applied to heterogeneous media like porous materials [103].

Figure 2.4(b) The variable-order Gray–Scott model captures natural spatial patterns, attributing anomalous diffusion to medium heterogeneity [40]. Dendrite and leopard images are from Wikipedia; the particle diffusion image is adapted from [93].

Fourth Application: Control Theory Proportional-integral-derivative (PID)

Extensions of the $PI^{\lambda}\digamma^{\mu}$ controller to variable order emerged in the early 2010s. In [94], Sheng et al. modeled temperature-dependent behavior of a circuit element ("fractor") using a variable-order system, showing that its transfer function order varied with temperature over time [94]. Figure 2.5 Block diagram illustrating the feedback variable order $PI^{\lambda}\digamma^{\mu}$ controller based on [94]. The orders of the variable order integral and derivative are both $0 \leq \lambda, \mu \leq 1$. Propose a variable order $PI^{\lambda}\digamma^{\mu}$ controller for the fractor system described by

$$u(t) = kpe(t) + k_I \digamma_t^{-\lambda(t)} e(t) + k_{\digamma} \digamma_t^{\mu(t)} e(t),$$

where $u(t)$ is the input voltage and $e(t)$ is the error between the output voltage $y(t)$ and the reference signal $r(t)$. The controller is depicted in Figure 2.5.

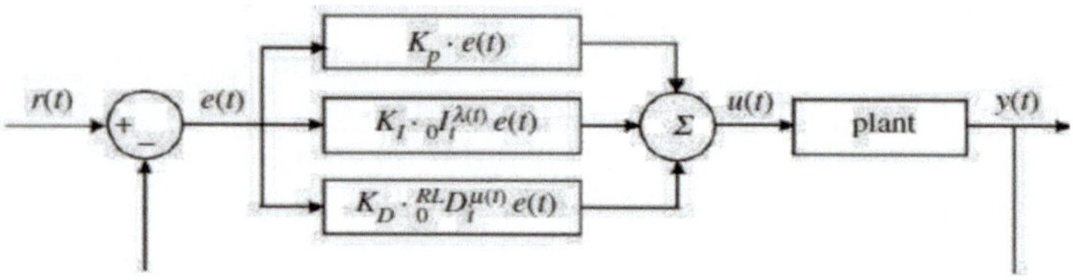

FIGURE 2.5: Control theory (PID).

Variable-order digital filters are widely used in signal processing for their flexibility in adjusting frequency characteristics without redesign. Conventional constant-order fractional differentiators have been extended to variable-order finite impulse response designs [108]. Taylor series [107] and iterative methods [95] have also been employed. Charef et al. proposed a polynomial interpolation technique for analog filter design [36]. Adaptive-order filters for parameter estimation were developed using a gradient-based algorithm that identifies the variable order from real-time input–output data, constrained by an upper bound [80].

Fifth application: Tumor growth or drug diffusion in biological tissue

In medicine, variable order fractional derivatives offer a powerful framework for modeling complex biological processes like tumor growth and drug diffusion.

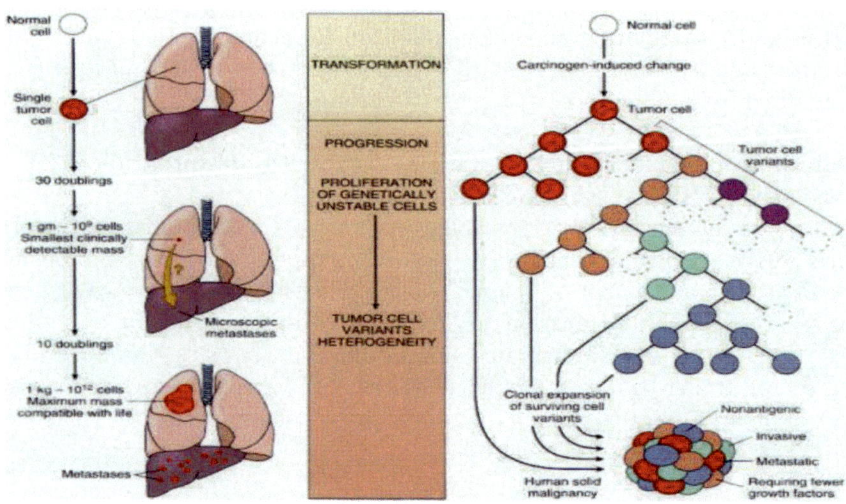

FIGURE 2.6: Tumor growth or drug diffusion in biological tissue.

For instance, chemotherapy drug diffusion. For instance, chemotherapy drug distribution can be described using diffusion equations with time- or space-varying parameters, improving upon traditional linear or constant-parameter models.

Drug diffusion in biological tissue is influenced by factors like uneven vascularization and variable cell density, which deviate from classical diffusion assumptions. These complexities make variable-order models more suitable for capturing realistic physiological behavior.

Variable-order fractional derivatives enable more accurate modeling of drug diffusion in tissues with spatial or temporal variability. For instance, changes in permeability due to cell density or vascularization can be captured effectively using these models.

These models can more accurately predict drug behavior in tissue, helping tailor treatments to local conditions—improving efficacy and reducing side effects.

Assuming drug concentration $C(\varkappa, t)$ in tissue evolves via variable-order fractional diffusion, it can be modeled by:

$$\frac{\partial^{\ltimes(t)} C(\varkappa, t)}{\partial t^{\ltimes(t)}} = \digamma(t) \cdot \frac{\partial^2 C(\varkappa, t)}{\partial \varkappa^2}$$

In this model [74],

- $C(\varkappa, t)$ denotes drug concentration at position $\varkappa$ and time t.

- $\frac{\partial^{\ltimes(t)} C(\varkappa,t)}{\partial t^{\ltimes(t)}}$ is a variable order fractional time derivative, where $\ltimes(t)$ represents the variable order of differentiation. This order could vary depending on local conditions of the biological tissue.

- $\digamma(t)$ captures time-varying permeability.

- $\frac{\partial^2 C(\varkappa,t)}{\partial \varkappa^2}$ represents spatial diffusion. This formulation integrates both spatial and temporal heterogeneity beyond classical models.

For more details on this refer to [74].

2.3 Measure of Noncompactness

The measure of noncompactness is a widely used tool in nonlinear analysis, especially for integral equations (see Akhmerov et al. [6], Alvarez [12], Banas et al. [17, 18, 19, 20, 21], Guo et al. [46]). This section outlines key concepts based on Kuratowski's foundational work.

Definition 27 *[65] Let (X, d) be a complete metric space and let $\mathcal{D}$ denote the family of bounded subsets of X. For any $D \in \mathcal{D}$, the Kuratowski measure of noncompactness $\varsigma(\digamma)$ $d > 0$ such that $\digamma$ can be covered by a finite collection of subsets with diameter less than d.*

Remark 6 *The diameter of a set $\digamma$ is defined as* $\sup\{d(\varkappa, y) : \varkappa \in \digamma, y \in \digamma\}$ *denoted by diam($\digamma$), with the convention that diam($\emptyset$) $= 0$.*

It is clear that $0 \leq \varsigma(F) \leq diam(F) < +\infty$ *for each nonempty bounded subset* F *of* X *and that* $diam(F) = 0$ *if and only if* F *is an empty set or consists of exactly one point.*

Definition 28 *[18] Let* X *be a Banach space, and let* Ω_X *represent the bounded subsets of* X. *The Kuratowski measure of noncompactness is a mapping* $\varsigma : \Omega_X \to [0, \infty]$ *which is constructed as follows:*

$$\varsigma(F) = \inf\{\epsilon > 0 : F(\in \Omega_X) \subseteq \cup_{i=1}^{n} F_i,\ diam(F_i) \leq \epsilon\},$$

where

$$diam(F_i) = sup\{||\varkappa - y|| :\ \varkappa,\ y \in F_i\}.$$

The following properties are valid for the Kuratowski measure of noncompactness:

Proposition 3 *[18, 19]. Let* X *be a Banach space,* F, F_1, F_2 *are bounded subsets of* X, *then*

1. $\varsigma(F) = 0 \Longleftrightarrow F$ *is relatively compact.*
2. $\varsigma(\phi) = 0$.
3. $\varsigma(F) = \varsigma(\overline{F}) = \varsigma(convF)$.
4. $F_1 \subset F_2 \Longrightarrow \varsigma(F_1) \leq \varsigma(F_2)$.
5. $\varsigma(F_1 + F_2) \leq \varsigma(F_1) + \varsigma(F_2)$.
6. $\varsigma(\lambda F) = |\lambda|\varsigma(F), \lambda \in \mathbb{R}$.
7. $\varsigma(F_1 \cup F_2) = Max\{\varsigma(F_1), \varsigma(F_2)\}$.
8. $\varsigma(F_1 \cap F_2) = Min\{\varsigma(F_1), \varsigma(F_2)\}$.
9. $\varsigma(F + \varkappa_0) = \varsigma(F)$ *for any* $\varkappa_0 \in X$.

Remark 7 *The a-measure of noncompactness was introduced by Kuratowski in order to generalize the Cantor intersection theorem*

Theorem 1 *[65] Let* (X, d) *be a complete metric space, and let* $\{F_n\}$ *be a decreasing sequence of nonempty, closed and bounded subsets of* X *and* $\lim_{n\to\infty} \varsigma(F_n) = 0$. *Then the intersection* F_∞ *of all* F_n *is nonempty and compact.*

In [49], Horvath has proved the following generalization of the Kuratowski theorem:

Theorem 2 *[65] Let* (X, d) *be a complete metric space and* $\{F_i\}_{i\in I}$ *be a family of nonempty closed and bounded subsets of* X *having the finite intersection property. If* $\inf_{i\in I} \varsigma(F_i) = 0$ *then the intersection* F_∞ *of all* F_i *is nonempty and compact.*

Lemma 11 *[46] If $U \subset C(J, X)$ is a equicontinuous and bounded set, then*

(i) *the function $\varsigma(\ltimes(s))$ is continuous for $s \in J$, and*

$$\widehat{\varsigma}(U) = \sup_{s \in J} \varsigma(\ltimes(s)).$$

(ii) $\varsigma\left(\int_0^{\mathsf{T}} \varkappa(\theta) d\theta : \varkappa \in U\right) \leq \int_0^{\mathsf{T}} \varsigma(U(\theta)) d\theta,$

where

$$U(t) = \{\varkappa(t) : \varkappa \in U\}, \ t \in J.$$

Theorem 3 *[65] Let $B(0,1)$ be the unit ball in a Banach space X. Then*

$$\begin{cases} \varsigma(B(0,1)) = 0, & if \ X \ is \ finite \ dimensional, \\ \varsigma(B(0,1)) = 2, & otherwise. \end{cases}$$

2.4 Some Fixed Point Theorems

Theorem 4 *(The fixed point theorem of Banach) [45]. If C is a closed, non-empty subset of the Banach space X, then each contraction mapping T on C has a unique fixed point.*

Theorem 5 *(Schauder fixed point theorem) [63] Let Q be a convex subset of E, $F : Q \longrightarrow Q$ be a continuous, compact map, and let E be a Banach space. Then, at least one fixed point for F exists in Q.*

Theorem 6 *(Krasnoselskii fixed point theorem)[35] Given a real Banach space E, let Q be a closed, bounded, and convex subset. Let W_1 and W_2 be operators on Q that meet the given constraints:*

(i) $W_1(Q) + W_2(Q) \subset Q$,
(ii) W_1 is continuous on Q and $W_1(Q)$ is a relatively compact portion of E ,
(iii) If W_2 is a strict contraction on Q, there exists $k \in [0,1)$ such that

$$\|W_2(\varkappa) - W_2(y)\| \leq k\|\varkappa - y\| \quad for \ all \ \varkappa, y \in Q,$$

Then there exists a point $\varkappa \in Q$ such that $W_1(\varkappa) + W_2(\varkappa) = \varkappa$.

Theorem 7 *(Darbo's fixed point theorem) [18] Let Λ be a continuous operator satisfying $D : \Lambda \longrightarrow \Lambda$ be a nonempty, closed, bounded, and convex subset of a Banach space X.*

$$\varsigma(D(\mathcal{S})) \leq k\varsigma(\mathcal{S}), \ for \ any \ (\mathcal{S} \neq \emptyset) \subset \Lambda, \ k \in [0,1),$$

i.e., D is $k-$ set contractions.
Then D has at least one fixed point in Λ.

Theorem 8 *(Schaefer's fixed point theorem) [45, 96] Suppose E is a $\mathfrak{B}$anach space. If $D : E \to E$ is a completely continuous operator and $\Omega = \{\varkappa \in E : \varkappa = \eta D\varkappa, 0 < \eta < 1\}$ is bounded, then D has a fixed point in E.*

Theorem 9 *(Guo-Krasnosel'skii fixed point theorem) [47]*
Let E be a $\mathfrak{B}$anach space and let $P \subset E$ be a cone. Assume Ω_1, Ω_2 are open bounded subsets of E such that $0 \in \Omega_1 \subset \overline{\Omega_1} \subset \Omega_2$, and let $\mathsf{T} \cap (\overline{\Omega_2} \setminus \Omega_1) \to P$ be a completely continuous operator such that

(i) $\| \mathsf{T}\, y\| \geq \|y\|, y \in P \cap \partial\Omega_1$ *and* $\| \mathsf{T}\, y\| \leq \|y\|, y \in P \cap \partial\Omega_2$*; or*

(ii) $\| \mathsf{T}\, y\| \leq \|y\|, y \in P \cap \partial\Omega_1$ *and* $\| \mathsf{T}\, y\| \geq \|y\|, y \in P \cap \partial\Omega_2$.

Then the operator P has a fixed point in $P \cap (\overline{\Omega_2} \setminus \Omega_1)$.

2.5 The Ulam Type Stability

Stability theory for functional equations has advanced rapidly in recent decades. Ulam first posed the problem in 1940 [110], and Hyers [51] provided its initial solution for Banach spaces, leading to the concept of Ulam–Hyers stability. Later, in 1978, $\mathfrak{R}$assias [83] introduced a broader formulation of this stability type.

In differential equations, two key stability concepts are widely studied: Ulam-Hyers stability and Ulam-Hyers-$\mathfrak{R}$assias stability. For detailed discussions, see [52, 57, 60, 68, 72, 88, 113].

Definition 29 *[88] Let $\vartheta \in C(J, X)$. Then, if a positive constant c_f can be found such that for all $z \in C(J, X)$ satisfying*

$$\|z(s) - \vartheta(s)\| \leq c_f \text{ for every } s \in J$$

and $\epsilon > 0$, the initial and terminal condition problem (ITVP) (1.1) *exhibits Ulam-Hyers-$\mathfrak{R}$assias stability with respect to ϑ.*

$$\|F_{0+}^{\ltimes(s)} z(s) - f(s, z(s), I_{0+}^{\ltimes(s)} z(s))\| \leq \epsilon\vartheta(s), \ s \in J \tag{2.20}$$

there exists a solution $\varkappa \in C(J, X)$ of ITVP (1.1) *with*

$$\|z(s) - \varkappa(s)\| \leq c_f \epsilon \vartheta(s), \ s \in J$$

Definition 30 *[88] Let $\varphi \in C(J, \mathbb{R}^+)$. The initial value problem (1.2) is said to be Ulam-Hyers-$\mathfrak{R}$assias stable if $c_f > 0$ exists, meaning that for any $\epsilon > 0$ and for each solution $z \in C(J, \mathbb{R})$, the following condition holds. For*

$$|F_{0+}^{\ltimes(s)} z(s) - f(s, z(s), F_{0+}^{\ltimes(s)} z(s))| \leq \epsilon\varphi(s), \ s \in J, \tag{2.21}$$

there exists $\varkappa \in C(J, \mathbb{R})$ of ITVP (1.2) with

$$|z(s) - \varkappa(s)| \leq c_f \epsilon \varphi(s), \ s \in J.$$

Definition 31 *[88] Let $\vartheta \in C(J, \mathbb{R}_+)$. The initial value problem (1.3) is said to be Hyers-Ulam-$\Re$assias stable if there exists a constant $c_g > 0$ such that satisfying*

$$|F_{0+}^{\ltimes(s)} z(s) - g(s, z(s))| \leq \epsilon\vartheta(s), \;\; s \in J \tag{2.22}$$

there exists a solution $y \in C(J, \mathbb{R})$ of ITVP (1.3) such that

$$|z(s) - y(s)| \leq c_g \epsilon \vartheta(s), \;\; s \in J$$

Definition 32 *([88]) An equation (E) is said to be Ulam-Hyers stable if there exists a constant $c_f > 0$ such that, for every $\epsilon > 0$ and any function $y \in C([-\gamma, \mathsf{T}], \mathbb{R})$ satisfying*

$$|F(s, y(s), y'(s))| \leq \epsilon,$$

there exists a solution $\tilde{y}$ to equation (E) such that

$$\|y - \tilde{y}\|_{C([-\gamma, \mathsf{T}])} \leq c_f \epsilon.$$

$$\begin{cases} |F_{0+}^{\ltimes(s)} y(s) - f(s, y_s)| \leq \epsilon, & s \in J := [0, \mathsf{T}], \\ y(s) = \chi(s), & s \in [-\gamma, 0], \end{cases} \tag{2.23}$$

then there exists a solution $\varkappa \in C([-\gamma, \mathsf{T}], \mathbb{R})$ of equation (1.4) such that

$$|y(s) - \varkappa(s)| \leq c_f \epsilon, \;\; s \in [-\gamma, \mathsf{T}]. \tag{2.24}$$

Remark 8 *An expression for $y \in C([-\gamma, \mathsf{T}], \mathbb{R})$ is said to be an inequality solution of (2.23) if and only if there exists a function $h \in C([-\gamma, \mathsf{T}], \mathbb{R})$, depending on y, such that*

(i) $|h(s)| \leq \epsilon$, *for all $s \in [-\gamma, \mathsf{T}]$.*

(ii) $F_{0+}^{\ltimes(s)} y(s) = f(s, y_s) + h(s)$ *for all $s \in J$.*

Definition 33 *[88] The equation (1.6) is said to be Ulam-Hyers $\Re$assias stable with respect to ψ if there exists a constant $c_f > 0$ exists such that, for every $\epsilon > 0$ and for every function $z \in C(J, X)$ satisfying the inequality:*

$$\|F_{0+}^{\ltimes(s)} z(s) + f(s, z(s), F_{0+}^{\ltimes(s)} z(s))\| \leq \epsilon\psi(s), \;\; s \in J,$$

there exists of (1.6) a solution $\varkappa \in C(J, X)$ such that

$$\|z(s) - \varkappa(s)\| \leq c_f \epsilon \psi(s), \;\; s \in J.$$

Definition 34 *[88] The initial terminal value problem (ITVP) (1.7) is said to be Ulam-Hyers stable if there exists a constant $c_f > 0$ such that, for any $\epsilon > 0$ and for every function $z \in C(J, \mathbb{R})$ satisfying the inequality*

$$|{}^c F_{0+}^{\ltimes(s)} z(s) - f(s, z(s))| \leq \epsilon, \;\; s \in J, \tag{2.25}$$

there exists a solution $\varkappa \in C(J, \mathbb{R})$ of ITVP (1.7) such that

$$|z(s) - \varkappa(s)| \leq c_f \epsilon, \;\; s \in J.$$

Definition 35 *[88] The initial value problem (IVP)* (1.8 − 1.10) *is said to be Ulam-Hyers stable if there exists a constant* $c_{f,n} > 0$ *such that, for any* $\epsilon > 0$, *and for each function* $z \in PC^1(J, \mathbb{R})$ *satisfying*

$$\begin{cases} |{}^{c}F_{0+}^{\ltimes(s)} z(s) - f(s, z(s))| \leq \epsilon, \ s \in J, \ s \neq \mathsf{T}_i, & i = 1, \ldots, n, \\ |\Delta \varkappa|_{s=\mathsf{T}_i} - \Phi_i(\varkappa(\mathsf{T}_i^-))| \leq \epsilon, & i = 1, \ldots, n, \end{cases} \tag{2.26}$$

there exists a solution $\varkappa \in PC^1(J, \mathbb{R})$ *of IVP* (1.8 − 1.10), *such that*

$$|z(s) - \varkappa(s)| \leq c_{f,n}\epsilon, \ \ s \in J.$$

Definition 36 *[88] Let* $\vartheta \in C(J, \mathbb{R})$. *The initial terminal value problem (ITVP)* (1.11) *is said to be Ulam-Hyers-*$\mathfrak{R}$*assias stable with respect to* ϑ *if there exists a constant* $c_f > 0$, *such that for any* $\epsilon > 0$ *and for every function* $z \in C(J, \mathbb{R})$ *satisfying the inequality*

$$|{}^{c}F_{0+}^{\ltimes(s)} z(s) - f(s, z(s), I_{0+}^{\ltimes(s)} z(s))| \leq \epsilon\vartheta(s), \ \ s \in J \tag{2.27}$$

there exists a solution $\varkappa \in C(J, \mathbb{R})$ *of ITVP* (1.11) *with*

$$|z(s) - \varkappa(s)| \leq c_f \epsilon \vartheta(s), \ \ s \in J$$

Definition 37 *([88]). The initial and terminal value problem of Caputo fractional thermostat model (1.12) is said to be Ulam-Hyers stable if there exists a constant* $c_f > 0$, *such that, for any* $\epsilon > 0$ *and for every function* $z \in C(J, \mathbb{R})$ *satisfying the inequality*

$$|-{}^{c}F_{0+}^{\ltimes(s)} z(s) - f(s, z(s))| \leq \epsilon, \ \ s \in J, \tag{2.28}$$

there exists a solution $\varkappa \in C(J, \mathbb{R})$ *of Caputo fractional thermostat model (1.12) such that*

$$|z(s) - \varkappa(s)| \leq c_f \epsilon, \ \ s \in J.$$

Definition 38 *[88] The variable-order Hadamard initial terminal value problem (ITVP)* (1.13) *is said to be Ulam-Hyers-*$\mathfrak{R}$*assias stable with respect to the function* $\phi \in C(\mathcal{J}, \mathbb{R}_+)$ *if there exists a constant* $0 < c_f \in \mathbb{R}$ *such that* $\forall \epsilon > 0$ *and for every function* $\forall \mathrm{r} \in C(\mathcal{J}, \mathbb{R})$ *satisfying*

$$|{}^{H}F_{1+}^{\ltimes} \mathrm{r}(s) - f(s, \mathrm{r}(s))| \leq \epsilon\phi(s), \ \ s \in \mathcal{J},$$

there exists a solution $\varkappa \in C(\mathcal{J}, \mathbb{R})$ *of the variable-order Hadamard ITVP* (1.13) *such that*

$$|\mathrm{r}(s) - \varkappa(s)| \leq c_f \epsilon \phi(s), \ \ s \in \mathcal{J}.$$

Definition 39 *[88] Let $\vartheta \in C(\mathcal{J}, X)$. The initial terminal value problem (ITVP)* (1.14) *is said to be Ulam-Hyers-$\mathfrak{R}$assias stable with respect to ϑ if there exists a constant $c_f > 0$ such that, for any $\epsilon > 0$ and for every function $z \in C(\mathcal{J}, X)$ satisfying*

$$\|{}^{H}F_{1+}^{\ltimes(s)} z(s) + f(s, z(s))\| \leq \epsilon\vartheta(s), s \in \mathcal{J}, \tag{2.29}$$

there exists a solution $\varkappa \in C(\mathcal{J}, X)$ of ITVP (1.14) *such that*

$$\|z(s) - \varkappa(s)\| \leq c_f \epsilon \vartheta(s), \ \ s \in \mathcal{J}$$

Definition 40 *[88] Let $\vartheta \in C(\mathcal{J}, X)$. The initial terminal value problem (ITVP)* (1.15) *is said to be Ulam-Hyers-$\mathfrak{R}$assias stable with respect to ϑ if there exists a constant $c_f > 0$ such that, for any $\epsilon > 0$ and for every function $z \in C(\mathcal{J}, X)$ satisfying*

$$\|{}^{H}F_{\mathsf{T}_{i-1}+}^{\ltimes_i} z(s) - f(s, z(s), {}^{H}I_{\mathsf{T}_{i-1}^{+}}^{\ltimes_i} z(s))\| \leq \epsilon\vartheta(s), s \in \mathcal{J}_i, \tag{2.30}$$

there exists a solution $\varkappa \in C(\mathcal{J}, X)$ of ITVP (1.15) *such that*

$$\|z(s) - \varkappa(s)\| \leq c_f \epsilon \vartheta(s), \ \ s \in \mathcal{J}$$

Definition 41 *[88] The initial terminal value problem (ITVP)* (1.16) *is said to be Ulam-Hyers stable if there exists a constant $c_f > 0$ such that, for any $\epsilon > 0$ and for every function $z \in C(\mathcal{J}, \mathbb{R})$ satisfying*

$$|{}^{H}F_{1+}^{\ltimes(s)} z(s) - f(s, z(s), {}^{H}I_{1+}^{\ltimes(s)} z(s))| \leq \epsilon, s \in \mathcal{J}, \tag{2.31}$$

there exists a solution $\varkappa \in C(\mathcal{J}, \mathbb{R})$ of ITVP (1.16) *satisfying*

$$|z(s) - \varkappa(s)| \leq c_f \epsilon, \ \ s \in \mathcal{J}.$$

Definition 42 *[88] The initial terminal value problem (ITVP) (1.17)–(1.19) is said to be Ulam–Hyers stable if there exists a constant $c_g > 0$ such that, for any $\epsilon > 0$ and for every function $z \in PC^1(\mathcal{J}, \mathbb{R})$ satisfying*

$$\begin{cases} |{}^{CH}F_{\mathsf{T}_i^{+}}^{\ltimes(s)} z(s) - f(s, z(s), {}^{H}I_{\mathsf{T}_i^{+}}^{\ltimes(s)} z(s))| \leq \epsilon, \ \ s \in \mathcal{J}, \\ |\Delta\varkappa|_{s=\mathsf{T}_i} - L_i(\varkappa(\mathsf{T}_i^{-}))| \leq \varepsilon, \, i = 1, ..., n, \end{cases} \tag{2.32}$$

there exists a solution $\varkappa \in PC^1(\mathcal{J}, \mathbb{R})$ of ITVP (1.17)–(1.19), such that

$$|z(s) - \varkappa(s)| \leq c_g \, \epsilon, \ \ s \in \mathcal{J}.$$

Definition 43 *[88] The R-Liouville fractional derivative of variable order in the sense of Atangana–Baleanu, is said to be Ulam–Hyers stable if, for all* $k \in \{1, 2, ..., n\}$, *there exists a constant* $c_f > 0$ *such that, for any* $\epsilon > 0$ *and for every function* $z \in C([\mathsf{T}_{k-1}, \mathsf{T}_k], \mathbb{R})$ *satisfying*

$$|F_{\mathsf{T}_{k-1}^+}^{\ltimes(s)} z(s) - f_z(s)| \leq \epsilon, \quad s \in J_k := (\mathsf{T}_{k-1}, \mathsf{T}_k], \tag{2.33}$$

there exists a solution $y \in C([\mathsf{T}_{k-1}, \mathsf{T}_k], \mathbb{R})$ *of* (1.21) *such that*

$$|z(s) - y(s)| \leq c_f \epsilon, \quad s \in [\mathsf{T}_{k-1}, \mathsf{T}_k].$$

3

Initial and Terminal Value Problems of R-Liouville Fractional Differential Equations of Variable Order

This chapter explores various types of variable-order Riemann–Liouville initial and terminal value problems—including multiterm and implicit forms—under diverse conditions such as fractional-order boundary constraints and finite delays. We examine existence, uniqueness, and stability of solutions using tools like Green's functions and fixed point theorems (Krasnoselskii, ßanach, Schaefer, and Darbo), alongside Kuratowski's measure of noncompactness.

3.1 Initial and Terminal Value Issues for Variable-order Nonlinear Fractional Differential Equations using Kuratowski MNC Method

We consider the following border *v*alue problem (ITVP):

$$\begin{cases} F_{0+}^{\ltimes(s)}\varkappa(s) = f(s, \varkappa(s), I_{0+}^{\ltimes(s)}\varkappa(s)), \quad s \in J := [0, \mathsf{T}] \\ \varkappa(0) = 0, \quad \varkappa(\mathsf{T}) = 0, \end{cases} \tag{3.1}$$

where $1 < \ltimes(s) \leq 2$, $f : J \times X \times X \to X$ is a continuous function, and $F_{0+}^{\ltimes(s)}$, $I_{0+}^{\ltimes(s)}$ denote the Riemann–Liouville fractional derivative and integral of variable order $\ltimes(s)$, respectively.

This section is devoted to solving problem (3.1)and analyzing the Ulam-Hyers-$\mathfrak{R}$assias stability of its solutions.

3.1.1 Existence Criteria

Let's state the underlying presumptions:

(H1) Let $n \in \mathbb{N}$, and let $\mathcal{P} = \{J_1 := [0, \mathsf{T}_1], J_2 := (\mathsf{T}_1, \mathsf{T}_2], J_3 := (\mathsf{T}_2, \mathsf{T}_3], \ldots, J_n := (\mathsf{T}_{n-1}, \mathsf{T}]\}$ be a partition of the interval J. Consider a

DOI: 10.1201/9781003687900-3

function $\ltimes : J \to (1,2]$ that is piecewise constant with respect to the partition $\mathcal{P}$; that is, $\ltimes(s)$ is constant on each subinterval J_i. Specifically, there exist constants $\ltimes_i \in (1,2]$ such that for all $s \in J_i$, $\ltimes(s) = \ltimes_i$.

$$\ltimes(s) = \sum_{i=1}^{n} \ltimes_i I_i(s) = \begin{cases} \ltimes_1, & if\ s \in J_1, \\ \ltimes_2, & if\ s \in J_2, \\ \cdot \\ \cdot \\ \cdot \\ \ltimes_n, & if\ s \in J_n, \end{cases}$$

where $1 < \ltimes_i \leq 2$ are unchanging, where I_i is the indicator function of the interval $J_i := (\mathsf{T}_{i-1}, \mathsf{T}_i]$, $i = 1, 2, ..., n$, (with $\mathsf{T}_0 = 0$, $\mathsf{T}_n = \mathsf{T}$), defined by

$$I_i(s) = \begin{cases} 1, & for\ s \in J_i, \\ 0, & otherwise, with \mathsf{T}_0 = 0,\ \mathsf{T}_n = \mathsf{T}. \end{cases}$$

(H2) Let $f : J \times X \times X \to X$ be a given function. Suppose there exists a constant δ satisfying $(0 \leq \delta \leq \min_{s \in J} |(\ltimes(s))|)$ such that $s^{\delta} f \in C(J \times X \times X, X)$. Moreover, assume there exist constants $K,\ L > 0$ such that $s^{\delta} \|f(s, y_1, z_1) - f(s, y_2, z_2)\| \leq K\|y_1 - y_2\| + L\|z_1 - z_2\|$, for all $y_1,\ y_2,\ z_1,\ z_2 \in X$ and $s \in J_i$.

Let $E_i = C(J_i, X)$ denote the for ßanach space of continuous functions on J_i, for each $i \in \{1, 2, ..., n\}$. For any function$\kappa : J_i \to X$, the norm in E_i is defined by

$$\|\varkappa\|_{E_i} = \sup_{s \in J_i} \|\varkappa(s)\|.$$

We now aim to demonstrate the existence of a solution to the initial-type value problem (ITVP) described in equation (3.1), using the Darbo fixed point theorem and the Kuratowski measure of noncompactness. First, we examine the ITVP as defined in (3.1).

For each $i = 1, 2, ..., n$, the left R-Liouville fractional derivative (RLFD), as defined in equation (2.9), can be expressed as a sum of left fractional derivatives of constant orders $\ltimes_k$, where $k = 1, \ldots, i$. Specifically,

$$F_{0+}^{\ltimes(s)} \varkappa(s) = \frac{d^2}{ds^2}\Big(\int_0^{\mathsf{T}_1} \frac{(s-t)^{1-\ltimes_1}}{\Gamma(2-\ltimes_1)} \varkappa(t)dt\Big) + ... + \frac{d^2}{ds^2}\Big(\int_{\mathsf{T}_{i-1}}^{s} \frac{(s-t)^{1-\ltimes_i}}{\Gamma(2-\ltimes_i)} \varkappa(t)dt\Big) \tag{3.2}$$

Taking assumption $(H1)$ into account, equation (3.2) on the interval J_i, for $i = 1, 2, ..., n$, becomes

$$\frac{d^2}{ds^2}\Big(\int_0^{\mathsf{T}_1} \frac{(s-t)^{1-\ltimes_1}}{\Gamma(2-\ltimes_1)} \varkappa(t)dt\Big) + ... + \frac{d^2}{ds^2}\Big(\int_{\mathsf{T}_{i-1}}^{s} \frac{(s-t)^{1-\ltimes_i}}{\Gamma(2-\ltimes_i)} \varkappa(t)dt\Big) = f(s, \varkappa(s), I_{0+}^{\ltimes_i} \varkappa(s)), \quad s \in J_i. \tag{3.3}$$

What follows is an overview of the definition and existence of a solution to the ITVP described in equation (3.1).

Definition 44 *ITVP (3.1) is said to a solution if there exist functions* $\varkappa_i \in C([0, \mathsf{T}_i], X)$*, for* $i = 1, 2, \ldots, n$*, satisfying equation (3.3) and the boundary conditions* $\varkappa_i(0) = 0 = \varkappa_i(\mathsf{T}_i)$*.*

Let $\widetilde{\varkappa}(s) \in C(J_i, X)$ be such that $\widetilde{\varkappa}(s) \equiv 0$ for $s \in [0, \mathsf{T}_{i-1}]$, and suppose it satisfies the integral equation (3.3). Then (3.3)reduces to

$$F^{\ltimes_i}_{\mathsf{T}^+_{i-1}} \varkappa(s) = f(s, \widetilde{\varkappa}(s), I^{\ltimes_i}_{\mathsf{T}^+_{i-1}} \widetilde{\varkappa}(s)), \quad s \in J_i.$$

Taking this into account, for any $i = 1, 2, ..., n$, we consider the following auxiliary ITVP with boundary conditions $\kappa(\mathsf{T}_i - 1) = 0, \kappa(\mathsf{T}_i) = 0$.

$$\begin{cases} F^{\ltimes_i}_{\mathsf{T}^+_{i-1}} \varkappa(s) = f(s, \varkappa(s), I^{\ltimes_i}_{\mathsf{T}^+_{i-1}} \varkappa(s)), \quad s \in J_i \\ \varkappa(\mathsf{T}_{i-1}) = 0, \varkappa(\mathsf{T}_i) = 0. \end{cases} \tag{3.4}$$

The next lemma will serve as the foundation for establishing the existence of a solution to equation (3.4).

Lemma 12 *For* $i \in \{1, 2, \ldots, n\}$*, the function* $\varkappa \in E_i$ *is a solution to equation (3.4) if and only if* $\varkappa$ *satisfies the integral equation*

$$\varkappa(s) = -(\mathsf{T}_i - \mathsf{T}_{i-1})^{1-\ltimes_i}(s - \mathsf{T}_{i-1})^{\ltimes_i - 1} I^{\ltimes_i}_{\mathsf{T}^+_{i-1}} f(\mathsf{T}_i, \varkappa(\mathsf{T}_i), I^{\ltimes_i}_{\mathsf{T}^+_{i-1}} \varkappa(\mathsf{T}_i)) + I^{\ltimes_i}_{\mathsf{T}^+_{i-1}} f(s, \varkappa(s), I^{\ltimes_i}_{\mathsf{T}^+_{i-1}} \varkappa(s)). \tag{3.5}$$

Proof Assume that $\varkappa \in E_i$ is a solution to the problem (3.4). By applying the operator $I^{\ltimes_i}_{\mathsf{T}^+_{i-1}}$ to both sides of equation (3.4) and invoking Lemma 1, we obtain

$$\varkappa(s) = \omega_1 (s - \mathsf{T}_{i-1})^{\ltimes_i - 1} + \omega_2 (s - \mathsf{T}_{i-1})^{\ltimes_i - 2} + \frac{1}{\Gamma(\ltimes_i)} \int_{\mathsf{T}_{i-1}}^{s} (s - t)^{\ltimes_i - 1} f(t, \varkappa(t), I^{\ltimes_i}_{\mathsf{T}^+_{i-1}} \varkappa(t)) dt, \quad s \in J_i.$$

Due to the assumption on the function f and the condition $\varkappa(\mathsf{T}_{i-1}) = 0$, we conclude that $\omega_2 = 0$.

Now, let $\varkappa(s)$ satisfy $\varkappa(\mathsf{T}_i) = 0$. Then we observe that

$$\omega_1 = -(\mathsf{T}_i - \mathsf{T}_{i-1})^{1-\ltimes_i} I^{\ltimes_i}_{\mathsf{T}^+_{i-1}} f(\mathsf{T}_i, \varkappa(\mathsf{T}_i), I^{\ltimes_i}_{\mathsf{T}^+_{i-1}} \varkappa(\mathsf{T}_i))$$

Substituting this into the expression for $\kappa(s)$, we obtain

$$\begin{aligned} \varkappa(s) &= -(\mathsf{T}_i - \mathsf{T}_{i-1})^{1-\ltimes_i}(s - \mathsf{T}_{i-1})^{\ltimes_i - 1} I^{\ltimes_i}_{\mathsf{T}^+_{i-1}} f(\mathsf{T}_i, \varkappa(\mathsf{T}_i), I^{\ltimes_i}_{\mathsf{T}^+_{i-1}} \varkappa(\mathsf{T}_i)) \\ &+ I^{\ltimes_i}_{\mathsf{T}^+_{i-1}} f(s, \varkappa(s), I^{\ltimes_i}_{\mathsf{T}^+_{i-1}} \varkappa(s)), \quad s \in J_i. \end{aligned}$$

Conversely, assume that $\varkappa \in E_i$ is a solution to integral equation (3.5). Then, by Lemma 1 and the continuity of function $s^\delta f$, we infer that κ is also a solution to problem (3.4).

Our first existence result is based on the Darbo's fixed point theorem.

Theorem 10 *Let conditions (H1), (H2) be satisfied, and suppose the inequality*

$$\frac{2(\mathrm{T}_i - \mathrm{T}_{i-1})^{\ltimes_i - 1}(\mathrm{T}_i^{1-\delta} - \mathrm{T}_{i-1}^{1-\delta})}{(1-\delta)\Gamma(\ltimes_i)}(K + L\frac{(\mathrm{T}_i - \mathrm{T}_{i-1})^{\ltimes_i}}{\Gamma(\ltimes_i + 1)}) < 1. \tag{3.6}$$

holds. Then, the problem (3.4) admits at least one solution in E_i.

Remark 9 *As stated on page 20 of [23], it is straightforward to demonstrate with ease that the following inequality, together with condition (H2)*

$$\varsigma(s^\delta \|f(s, B_1, B_2)\|) \leq K\varsigma(B_1) + L\varsigma(B_2),$$

is equivalent for any bounded sets $B_1, B_2 \subset X$ and for each $s \in J_i$.

Proof We define the operator

$$W : E_i \to E_i$$

as follows:

$$W\varkappa(s) = -(\mathrm{T}_i - \mathrm{T}_{i-1})^{1-\ltimes_i}(s - \mathrm{T}_{i-1})^{\ltimes_i - 1} I_{\mathrm{T}_{i-1}^+}^{\ltimes_i} f(\mathrm{T}_i, \varkappa(\mathrm{T}_i), I_{\mathrm{T}_{i-1}^+}^{\ltimes_i} \varkappa(\mathrm{T}_i))$$

$$+\frac{1}{\Gamma(\ltimes_i)}\int_{\mathrm{T}_{i-1}}^{s}(s-t)^{\ltimes_i - 1} f(t, \varkappa(t), I_{\mathrm{T}_{i-1}^+}^{\ltimes_i}\varkappa(t))dt, \quad s \in J_i. \tag{3.7}$$

The operator $W : E_i \to E_i$, defined in (3.7), is well-posed, as can be inferred from the continuity of the function f and the properties of fractional integrals.

Let

$$R_i \geq \frac{\frac{2f^\star(\mathrm{T}_i - \mathrm{T}_{i-1})^{\ltimes_i}}{\Gamma(\ltimes_i)}}{1 - \frac{2(\mathrm{T}_i - \mathrm{T}_{i-1})^{\ltimes_i - 1}(\mathrm{T}_i^{1-\delta} - \mathrm{T}_{i-1}^{1-\delta})}{(1-\delta)\Gamma(\ltimes_i)}(K + L\frac{(\mathrm{T}_i - \mathrm{T}_{i-1})^{\ltimes_i}}{\Gamma(\ltimes_i + 1)})},$$

where

$$f^\star = \sup_{s \in J_i} \|f(s, 0, 0)\|.$$

We consider the set

$$B_{R_i} = \{\varkappa \in E_i, \ \|\varkappa\|_{E_i} \leq R_i\}.$$

It is evident that B_{R_i} is bounded, closed, convex, and nonempty.

We now show that the fixed point theorem of Darbo applies to the operator W. The proof proceeds in four stages.

STEP 1: Claim: $W(B_{R_i}) \subseteq (B_{R_i})$.
For $\varkappa \in B_{R_i}$ and by (H2), we get

$$\begin{aligned}
\|W\varkappa(s)\| &\leq \tfrac{(\mathsf{T}_i-\mathsf{T}_{i-1})^{1-\ltimes_i}(s-\mathsf{T}_{i-1})^{\ltimes_i-1}}{\Gamma(\ltimes_i)} \int_{\mathsf{T}_{i-1}}^{\mathsf{T}_i} (\mathsf{T}_i - t)^{\ltimes_i-1} \|f(t,\varkappa(t), I^{\ltimes_i}_{\mathsf{T}_{i-1}^+}\varkappa(t))\|dt \\
&+ \tfrac{1}{\Gamma(\ltimes_i)} \int_{\mathsf{T}_{i-1}}^{s} (s-t)^{\ltimes_i-1} \|f(t,\varkappa(t), I^{\ltimes_i}_{\mathsf{T}_{i-1}^+}\varkappa(t))\|dt \\
&\leq \tfrac{2}{\Gamma(\ltimes_i)} \int_{\mathsf{T}_{i-1}}^{\mathsf{T}_i} (\mathsf{T}_i - t)^{\ltimes_i-1} \|f(t,\varkappa(t), I^{\ltimes_i}_{\mathsf{T}_{i-1}^+}\varkappa(t))\|dt \\
&\leq \tfrac{2}{\Gamma(\ltimes_i)} \int_{\mathsf{T}_{i-1}}^{\mathsf{T}_i} (\mathsf{T}_i - t)^{\ltimes_i-1} \|f(t,\varkappa(t), I^{\ltimes_i}_{\mathsf{T}_{i-1}^+}\varkappa(t)) - f(t,0,0)\|dt \\
&+ \tfrac{2}{\Gamma(\ltimes_i)} \int_{\mathsf{T}_{i-1}}^{\mathsf{T}_i} (\mathsf{T}_i - t)^{\ltimes_i-1} \|f(t,0,0)\|dt \\
&\leq \tfrac{2}{\Gamma(\ltimes_i)} \int_{\mathsf{T}_{i-1}}^{\mathsf{T}_i} (\mathsf{T}_i - t)^{\ltimes_i-1} t^{-\delta}(K\|\varkappa(t)\| + L\|I^{\ltimes_i}_{\mathsf{T}_{i-1}^+}\varkappa(t)\|)dt + \tfrac{2f^\star(\mathsf{T}_i-\mathsf{T}_{i-1})^{\ltimes_i}}{\Gamma(\ltimes_i)} \\
&\leq \tfrac{2(\mathsf{T}_i-\mathsf{T}_{i-1})^{\ltimes_i-1}}{\Gamma(\ltimes_i)} \int_{\mathsf{T}_{i-1}}^{\mathsf{T}_i} t^{-\delta}(K + L\tfrac{(\mathsf{T}_i-\mathsf{T}_{i-1})^{\ltimes_i}}{\Gamma(\ltimes_i+1)})\|\varkappa(t)\|dt + \tfrac{2f^\star(\mathsf{T}_i-\mathsf{T}_{i-1})^{\ltimes_i}}{\Gamma(\ltimes_i)} \\
&\leq \tfrac{2(\mathsf{T}_i-\mathsf{T}_{i-1})^{\ltimes_i-1}(\mathsf{T}_i^{1-\delta}-\mathsf{T}_{i-1}^{1-\delta})}{(1-\delta)\Gamma(\ltimes_i)}(K + L\tfrac{(\mathsf{T}_i-\mathsf{T}_{i-1})^{\ltimes_i}}{\Gamma(\ltimes_i+1)})R_i + \tfrac{2f^\star(\mathsf{T}_i-\mathsf{T}_{i-1})^{\ltimes_i}}{\Gamma(\ltimes_i)} \\
&\leq R_i,
\end{aligned}$$

which means that $W(B_{R_i}) \subseteq B_{R_i}$.

STEP 2: Claim: W is continuous.
Assumed to converge to $\varkappa$ in E_i and $s \in J_i$ is the sequence $(\varkappa_n)$. Then,

$$\|(W\varkappa_n)(s) - (W\varkappa)(s)\|$$

$$\begin{aligned}
&\leq \tfrac{(\mathsf{T}_i-\mathsf{T}_{i-1})^{1-\ltimes_i}(s-\mathsf{T}_{i-1})^{\ltimes_i-1}}{\Gamma(\ltimes_i)} \int_{\mathsf{T}_{i-1}}^{\mathsf{T}_i} (\mathsf{T}_i - t)^{\ltimes_i-1} \|f(t,\varkappa_n(t), I^{\ltimes_i}_{\mathsf{T}_{i-1}^+}\varkappa_n(t)) - f(t,\varkappa(t), I^{\ltimes_i}_{\mathsf{T}_{i-1}^+}\varkappa(t))\|dt \\
&+ \tfrac{1}{\Gamma(\ltimes_i)} \int_{\mathsf{T}_{i-1}}^{s} (s-t)^{\ltimes_i-1} \|f(t,\varkappa_n(t), I^{\ltimes_i}_{\mathsf{T}_{i-1}^+}\varkappa_n(t)) - f(t,\varkappa(t), I^{\ltimes_i}_{\mathsf{T}_{i-1}^+}\varkappa(t))\|dt \\
&\leq \tfrac{(\mathsf{T}_i-\mathsf{T}_{i-1})^{1-\ltimes_i}(\mathsf{T}_i-\mathsf{T}_{i-1})^{\ltimes_i-1}}{\Gamma(\ltimes_i)} \int_{\mathsf{T}_{i-1}}^{\mathsf{T}_i} (\mathsf{T}_i - t)^{\ltimes_i-1} \|f(t,\varkappa_n(t), I^{\ltimes_i}_{\mathsf{T}_{i-1}^+}\varkappa_n(t)) - f(t,\varkappa(t), I^{\ltimes_i}_{\mathsf{T}_{i-1}^+}\varkappa(t))\|dt \\
&+ \tfrac{1}{\Gamma(\ltimes_i)} \int_{\mathsf{T}_{i-1}}^{\mathsf{T}_i} (s-t)^{\ltimes_i-1} \|f(t,\varkappa_n(t), I^{\ltimes_i}_{\mathsf{T}_{i-1}^+}\varkappa_n(t)) - f(t,\varkappa(t), I^{\ltimes_i}_{\mathsf{T}_{i-1}^+}\varkappa(t))\|dt \\
&\leq \tfrac{2}{\Gamma(\ltimes_i)} \int_{\mathsf{T}_{i-1}}^{\mathsf{T}_i} (\mathsf{T}_i - t)^{\ltimes_i-1} \|f(t,\varkappa_n(t), I^{\ltimes_i}_{\mathsf{T}_{i-1}^+}\varkappa_n(t)) - f(t,\varkappa(t), I^{\ltimes_i}_{\mathsf{T}_{i-1}^+}\varkappa(t))\|dt \\
&\leq \tfrac{2}{\Gamma(\ltimes_i)} \int_{\mathsf{T}_{i-1}}^{\mathsf{T}_i} t^{-\delta}(\mathsf{T}_i - t)^{\ltimes_i-1} (K\|\varkappa_n(t) - \varkappa(t)\| + LI^{\ltimes_i}_{\mathsf{T}_{i-1}^+}\|\varkappa_n(t) - \varkappa(t))\|)dt \\
&\leq \tfrac{2K}{\Gamma(\ltimes_i)}\|\varkappa_n - \varkappa\|_{E_i} \int_{\mathsf{T}_{i-1}}^{\mathsf{T}_i} t^{-\delta}(\mathsf{T}_i - t)^{\ltimes_i-1}dt + \tfrac{2L}{\Gamma(\ltimes_i)}\|I^{\ltimes_i}_{\mathsf{T}_{i-1}^+}(\varkappa_n - \varkappa)\|_{E_i} \int_{\mathsf{T}_{i-1}}^{\mathsf{T}_i} t^{-\delta}(\mathsf{T}_i - t)^{\ltimes_i-1}dt \\
&\leq \tfrac{2K}{\Gamma(\ltimes_i)}\|\varkappa_n - \varkappa\|_{E_i} \int_{\mathsf{T}_{i-1}}^{\mathsf{T}_i} t^{-\delta}(\mathsf{T}_i - t)^{\ltimes_i-1}dt + \tfrac{2L(\mathsf{T}_i-\mathsf{T}_{i-1})^{\ltimes_i}}{\Gamma(\ltimes_i)\Gamma(\ltimes_i+1)}\|\varkappa_n - \varkappa\|_{E_i} \int_{\mathsf{T}_{i-1}}^{\mathsf{T}_i} t^{-\delta}(\mathsf{T}_i - t)^{\ltimes_i-1}dt \\
&\leq \left(\tfrac{2K}{\Gamma(\ltimes_i)} + \tfrac{2L(\mathsf{T}_i-\mathsf{T}_{i-1})^{\ltimes_i}}{\Gamma(\ltimes_i)\Gamma(\ltimes_i+1)}\right)\|\varkappa_n - \varkappa\|_{E_i} \int_{\mathsf{T}_{i-1}}^{\mathsf{T}_i} t^{-\delta}(\mathsf{T}_i - t)^{\ltimes_i-1}dt \\
&\leq \tfrac{(\mathsf{T}_i-\mathsf{T}_{i-1})^{\ltimes_i-1}({\mathsf{T}_i}^{1-\delta}-{\mathsf{T}_{i-1}}^{1-\delta})}{(1-\delta)\Gamma(\ltimes_i)}\left(2K + \tfrac{2L(\mathsf{T}_i-\mathsf{T}_{i-1})^{\ltimes_i}}{\Gamma(\ltimes_i+1)}\right)\|\varkappa_n - \varkappa\|_{E_i}
\end{aligned}$$

Hence we obtain

$$\|(W\varkappa_n) - (W\varkappa)\|_{E_i} \to 0 \quad as \quad n \to \infty.$$

Ergo, the operator W is continuous on E_i.

STEP 3: Claim — W is bounded and equicontinuous.
For every $\varkappa \in B_{R_i}$, we have $\|W(\varkappa)\|_{E_i} \leq R_i$ which implies that $W(B_{R_i})$ is bounded. This follows from Step 1, where we established that $W(B_{R_i}) = \{W(\varkappa) : \varkappa \in B_{R_i}\} \subset B_{R_i}$. The equicontinuity of $W(B_{R_i})$ remains to be shown.

For $s_1, s_2 \in J_i,\ s_1 < s_2$ and $\varkappa \in B_{R_i}$, we consider

$$
\begin{aligned}
&\|(W\varkappa)(s_2) - (W\varkappa)(s_1)\| \\
&= \| - \tfrac{(\mathsf{T}_i-\mathsf{T}_{i-1})^{1-\Bbbk_i}(s_2-\mathsf{T}_{i-1})^{\Bbbk_i-1}}{\Gamma(\Bbbk_i)} \int_{\mathsf{T}_{i-1}}^{\mathsf{T}_i} (\mathsf{T}_i - t)^{\Bbbk_i-1} f(t, \varkappa(t), I_{\mathsf{T}_{i-1}^+}^{\Bbbk_i} \varkappa(t))dt \\
&+ \tfrac{1}{\Gamma(\Bbbk_i)} \int_{\mathsf{T}_{i-1}}^{s_2} (s_2 - t)^{\Bbbk_i-1} f(t, \varkappa(t), I_{\mathsf{T}_{i-1}^+}^{\Bbbk_i} \varkappa(t))dt + \tfrac{(\mathsf{T}_i-\mathsf{T}_{i-1})^{1-\Bbbk_i}(s_1-\mathsf{T}_{i-1})^{\Bbbk_i-1}}{\Gamma(\Bbbk_i)} \\
&\quad \int_{\mathsf{T}_{i-1}}^{\mathsf{T}_i} (\mathsf{T}_i - t)^{\Bbbk_i-1} f(t, \varkappa(t), I_{\mathsf{T}_{i-1}^+}^{\Bbbk_i} \varkappa(t))dt - \tfrac{1}{\Gamma(\Bbbk_i)} \int_{\mathsf{T}_{i-1}}^{s_1} (s_1 - t)^{\Bbbk_i-1} f(t, \varkappa(t), I_{\mathsf{T}_{i-1}^+}^{\Bbbk_i} \varkappa(t))dt\| \\
&\leq \tfrac{(\mathsf{T}_i-\mathsf{T}_{i-1})^{1-\Bbbk_i}}{\Gamma(\Bbbk_i)} \Big((s_2 - \mathsf{T}_{i-1})^{\Bbbk_i-1} - (s_1 - \mathsf{T}_{i-1})^{\Bbbk_i-1}\Big) \int_{\mathsf{T}_{i-1}}^{\mathsf{T}_i} (\mathsf{T}_i - t)^{\Bbbk_i-1} \|f(t, \varkappa(t), I_{\mathsf{T}_{i-1}^+}^{\Bbbk_i} \varkappa(t))\|dt \\
&+ \tfrac{1}{\Gamma(\Bbbk_i)} \int_{\mathsf{T}_{i-1}}^{s_1} \Big((s_2 - t)^{\Bbbk_i-1} - (s_1 - s)^{\Bbbk_i-1}\Big) \|f(t, \varkappa(t), I_{\mathsf{T}_{i-1}^+}^{\Bbbk_i} \varkappa(t))\|dt \\
&+ \tfrac{1}{\Gamma(\Bbbk_i)} \int_{s_1}^{s_2} (s_2 - t)^{\Bbbk_i-1} \|f(t, \varkappa(t), I_{\mathsf{T}_{i-1}^+}^{\Bbbk_i} \varkappa(t))\|dt \\
&\leq \tfrac{(\mathsf{T}_i-\mathsf{T}_{i-1})^{1-\Bbbk_i}}{\Gamma(\Bbbk_i)} \Big((s_2 - \mathsf{T}_{i-1})^{\Bbbk_i-1} - (s_1 - \mathsf{T}_{i-1})^{\Bbbk_i-1}\Big) \\
&\quad \int_{\mathsf{T}_{i-1}}^{\mathsf{T}_i} (\mathsf{T}_i - t)^{\Bbbk_i-1} \|f(t, \varkappa(t), I_{\mathsf{T}_{i-1}^+}^{\Bbbk_i} \varkappa(t)) - f(t,0,0)\|dt \\
&+ \tfrac{(\mathsf{T}_i-\mathsf{T}_{i-1})^{1-\Bbbk_i}}{\Gamma(\Bbbk_i)} \Big((s_2 - \mathsf{T}_{i-1})^{\Bbbk_i-1} - (s_1 - \mathsf{T}_{i-1})^{\Bbbk_i-1}\Big) \int_{\mathsf{T}_{i-1}}^{\mathsf{T}_i} (\mathsf{T}_i - t)^{\Bbbk_i-1} \|f(t,0,0)\|dt \\
&+ \tfrac{1}{\Gamma(\Bbbk_i)} \int_{\mathsf{T}_{i-1}}^{s_1} \Big((s_2 - t)^{\Bbbk_i-1} - (s_1 - t)^{\Bbbk_i-1}\Big) \|f(t, \varkappa(t), I_{\mathsf{T}_{i-1}^+}^{\Bbbk_i} \varkappa(t)) - f(t,0,0)\|dt \\
&+ \tfrac{1}{\Gamma(\Bbbk_i)} \int_{\mathsf{T}_{i-1}}^{s_1} \Big((s_2 - t)^{\Bbbk_i-1} - (s_1 - t)^{\Bbbk_i-1}\Big) \|f(t,0,0)\|dt \\
&+ \tfrac{1}{\Gamma(\Bbbk_i)} \int_{s_1}^{s_2} (s_2 - t)^{\Bbbk_i-1} \|f(t, \varkappa(t), I_{\mathsf{T}_{i-1}^+}^{\Bbbk_i} \varkappa(t)) - f(t,0,0)\|dt \\
&+ \tfrac{1}{\Gamma(\Bbbk_i)} \int_{s_1}^{s_2} (s_2 - t)^{\Bbbk_i-1} \|f(t,0,0)\|dt. \\
&\leq \tfrac{(\mathsf{T}_i-\mathsf{T}_{i-1})^{1-\Bbbk_i}}{\Gamma(\Bbbk_i)} \Big((s_2 - \mathsf{T}_{i-1})^{\Bbbk_i-1} - (s_1 - \mathsf{T}_{i-1})^{\Bbbk_i-1}\Big) \\
&\quad \int_{\mathsf{T}_{i-1}}^{\mathsf{T}_i} (\mathsf{T}_i - t)^{\Bbbk_i-1} t^{-\delta} (K\|\varkappa(t)\| + L\|I_{\mathsf{T}_{i-1}^+}^{\Bbbk_i} \varkappa(t)\|)dt \\
&+ \tfrac{f^\star(\mathsf{T}_i-\mathsf{T}_{i-1})^{1-\Bbbk_i}}{\Gamma(\Bbbk_i)} \Big((s_2 - \mathsf{T}_{i-1})^{\Bbbk_i-1} - (s_1 - \mathsf{T}_{i-1})^{\Bbbk_i-1}\Big) \int_{\mathsf{T}_{i-1}}^{\mathsf{T}_i} (\mathsf{T}_i - t)^{\Bbbk_i-1} dt \\
&+ \tfrac{1}{\Gamma(\Bbbk_i)} \int_{\mathsf{T}_{i-1}}^{s_1} t^{-\delta} \Big((s_2 - t)^{\Bbbk_i-1} - (s_1 - t)^{\Bbbk_i-1}\Big) (K\|\varkappa(t)\| + L\|I_{\mathsf{T}_{i-1}^+}^{\Bbbk_i} \varkappa(t)\|)dt \\
&+ \tfrac{f^\star}{\Gamma(\Bbbk_i)} \int_{\mathsf{T}_{i-1}}^{s_1} \Big((s_2 - t)^{\Bbbk_i-1} - (s_1 - t)^{\Bbbk_i-1}\Big) dt \\
&+ \tfrac{1}{\Gamma(\Bbbk_i)} \int_{s_1}^{s_2} t^{-\delta} (s_2 - t)^{\Bbbk_i-1} (K\|\varkappa(t)\| + L\|I_{\mathsf{T}_{i-1}^+}^{\Bbbk_i} \varkappa(t)\|)dt + \tfrac{f^\star}{\Gamma(\Bbbk_i)} \int_{s_1}^{s_2} (s_2 - t)^{\Bbbk_i-1} dt \\
&\leq \tfrac{1}{\Gamma(\Bbbk_i)} \Big((s_2 - \mathsf{T}_{i-1})^{\Bbbk_i-1} - (s_1 - \mathsf{T}_{i-1})^{\Bbbk_i-1}\Big) (K\|\varkappa\|_{E_i} + L\|I_{\mathsf{T}_{i-1}^+}^{\Bbbk_i} \varkappa\|_{E_i}) \int_{\mathsf{T}_{i-1}}^{\mathsf{T}_i} t^{-\delta} dt \\
&+ \tfrac{f^\star(\mathsf{T}_i-\mathsf{T}_{i-1})}{\Gamma(\Bbbk_i+1)} \Big((s_2 - \mathsf{T}_{i-1})^{\Bbbk_i-1} - (s_1 - \mathsf{T}_{i-1})^{\Bbbk_i-1}\Big) \\
&+ \tfrac{1}{\Gamma(\Bbbk_i)} (K\|\varkappa\|_{E_i} + L\|I_{\mathsf{T}_{i-1}^+}^{\Bbbk_i} \varkappa\|_{E_i}) \int_{\mathsf{T}_{i-1}}^{s_1} t^{-\delta} ((s_2 - s_1)^{\Bbbk_i-1})dt \\
&+ \tfrac{f^\star}{\Gamma(\Bbbk_i)} \Big(\tfrac{(s_2-\mathsf{T}_{i-1})^{\Bbbk_i}}{\Bbbk_i} - \tfrac{(s_2-s_1)^{\Bbbk_i}}{\Bbbk_i} - \tfrac{(s_1-\mathsf{T}_{i-1})^{\Bbbk_i}}{\Bbbk_i}\Big) \\
&+ \tfrac{(s_2-s_1)^{\Bbbk_i-1}}{\Gamma(\Bbbk_i)} (K\|\varkappa\|_{E_i} + L\|I_{\mathsf{T}_{i-1}^+}^{\Bbbk_i} \varkappa\|_{E_i}) \int_{s_1}^{s_2} t^{-\delta} dt + \tfrac{f^\star}{\Gamma(\Bbbk_i)} \tfrac{(s_2-s_1)^{\Bbbk_i}}{\Bbbk_i}
\end{aligned}
$$

$$\begin{aligned}
&\leq \tfrac{\mathsf{T}_i{}^{1-\delta}-\mathsf{T}_{i-1}{}^{1-\delta}}{(1-\delta)\Gamma(\ltimes_i)}\Big((s_2-\mathsf{T}_{i-1})^{\ltimes_i-1}-(s_1-\mathsf{T}_{i-1})^{\ltimes_i-1}\Big)(K\|\varkappa\|_{E_i}+L\tfrac{(\mathsf{T}_i-\mathsf{T}_{i-1})^{\ltimes_i}}{\Gamma(\ltimes_i+1)}\|\varkappa\|_{E_i})\\
&+ \tfrac{f^\star(\mathsf{T}_i-\mathsf{T}_{i-1})}{\Gamma(\ltimes_i+1)}\Big((s_2-\mathsf{T}_{i-1})^{\ltimes_i-1}-(s_1-\mathsf{T}_{i-1})^{\ltimes_i-1}\Big)\\
&+ \Big(\tfrac{(s_1{}^{1-\delta}-\mathsf{T}_{i-1}{}^{1-\delta})(s_2-s_1)^{\ltimes_i-1}}{(1-\delta)\Gamma(\ltimes_i)}\Big)(K\|\varkappa\|_{E_i}+L\tfrac{(\mathsf{T}_i-\mathsf{T}_{i-1})^{\ltimes_i}}{\Gamma(\ltimes_i+1)}\|\varkappa\|_{E_i})\\
&+ \tfrac{f^\star}{\Gamma(\ltimes_i+1)}\Big((s_2-\mathsf{T}_{i-1})^{\ltimes_i}-(s_2-s_1)^{\ltimes_i}-(s_1-\mathsf{T}_{i-1})^{\ltimes_i}\Big)\\
&+ \tfrac{(s_2{}^{1-\delta}-s_1{}^{1-\delta})(s_2-s_1)^{\ltimes_i-1}}{(1-\delta)\Gamma(\ltimes_i)}(K\|\varkappa\|_{E_i}+L\tfrac{(\mathsf{T}_i-\mathsf{T}_{i-1})^{\ltimes_i}}{\Gamma(\ltimes_i+1)}\|\varkappa\|_{E_i})+\tfrac{f^\star(s_2-s_1)^{\ltimes_i}}{\Gamma(\ltimes_i+1)}\\
&\leq \Big(\tfrac{\mathsf{T}_i{}^{1-\delta}-\mathsf{T}_{i-1}{}^{1-\delta}}{(1-\delta)\Gamma(\ltimes_i)}(K+L\tfrac{(\mathsf{T}_i-\mathsf{T}_{i-1})^{\ltimes_i}}{\Gamma(\ltimes_i+1)})\|\varkappa\|_{E_i}+\tfrac{f^\star(\mathsf{T}_i-\mathsf{T}_{i-1})}{\Gamma(\ltimes_i+1)}\Big)\Big((s_2-\mathsf{T}_{i-1})^{\ltimes_i-1}-(s_1-\mathsf{T}_{i-1})^{\ltimes_i-1}\Big)\\
&+ \Big(\tfrac{s_2{}^{1-\delta}-\mathsf{T}_{i-1}{}^{1-\delta}}{(1-\delta)\Gamma(\ltimes_i)}(K+L\tfrac{(\mathsf{T}_i-\mathsf{T}_{i-1})^{\ltimes_i}}{\Gamma(\ltimes_i+1)})\|\varkappa\|_{E_i}\Big)(s_2-s_1)^{\ltimes_i-1}\\
&+ \tfrac{f^\star}{\Gamma(\ltimes_i+1)}\Big((s_2-\mathsf{T}_{i-1})^{\ltimes_i}-(s_1-\mathsf{T}_{i-1})^{\ltimes_i}\Big)
\end{aligned}$$

Hence $\|(W\varkappa)(s_2)-(W\varkappa)(s_1)\|_{E_i}\to 0$ as $|s_2-s_1|\to 0$. It implies that $\mathsf{T}(B_{R_i})$ is equicontinuous.

STEP 4: Claim — W is a set contraction.
Let $U\subset B_{R_i}$, and $s\in J_i$. Then,

$$\begin{aligned}
\varsigma(W(U)(s)) &= \varsigma((W\varkappa)(s),\varkappa\in U)\\
&\leq \Big\{\tfrac{(\mathsf{T}_i-\mathsf{T}_{i-1})^{1-\ltimes_i}(s-\mathsf{T}_{i-1})^{\ltimes_i-1}}{\Gamma(\ltimes_i)}\int_{\mathsf{T}_{i-1}}^{\mathsf{T}_i}(\mathsf{T}_i-t)^{\ltimes_i-1}\varsigma f(t,\varkappa(t),I^{\ltimes_i}_{\mathsf{T}^+_{i-1}}\varkappa(t))dt\\
&+ \tfrac{1}{\Gamma(\ltimes_i)}\int_{\mathsf{T}_{i-1}}^{s}(s-t)^{\ltimes_i-1}\varsigma f(t,\varkappa(t),I^{\ltimes_i}_{\mathsf{T}^+_{i-1}}\varkappa(t))dt,\varkappa\in U\Big\}.
\end{aligned}$$

Then Remark 9 implies that, for each $t\in J_i$,

$$\begin{aligned}
\varsigma(W(U)(s)) &\leq \Big\{\tfrac{(\mathsf{T}_i-\mathsf{T}_{i-1})^{1-\ltimes_i}(s-\mathsf{T}_{i-1})^{\ltimes_i-1}}{\Gamma(\ltimes_i)}\\
&\quad \int_{\mathsf{T}_{i-1}}^{\mathsf{T}_i}(\mathsf{T}_i-t)^{\ltimes_i-1}\Big[K\widehat{\varsigma}(U)\int_{\mathsf{T}_{i-1}}^{\mathsf{T}_i}t^{-\delta}dt+L\tfrac{(\mathsf{T}_i-\mathsf{T}_{i-1})^{\ltimes_i}}{\Gamma(\ltimes_i+1)}\widehat{\varsigma}(U)\int_{\mathsf{T}_{i-1}}^{\mathsf{T}_i}t^{-\delta}dt\Big]\\
&+ \tfrac{1}{\Gamma(\ltimes_i)}\int_{\mathsf{T}_{i-1}}^{s}(s-t)^{\ltimes_i-1}\Big[K\widehat{\varsigma}(U)\int_{\mathsf{T}_{i-1}}^{s}t^{-\delta}dt+L\tfrac{(\mathsf{T}_i-\mathsf{T}_{i-1})^{\ltimes_i}}{\Gamma(\ltimes_i+1)}\widehat{\varsigma}(U)\int_{\mathsf{T}_{i-1}}^{s}t^{-\delta}dt\Big],\varkappa\in U\Big\}.\\
&\leq \Big\{\tfrac{(s-\mathsf{T}_{i-1})^{\ltimes_i-1}}{\Gamma(\ltimes_i)}\int_{\mathsf{T}_{i-1}}^{\mathsf{T}_i}\Big[K\widehat{\varsigma}(U)\int_{\mathsf{T}_{i-1}}^{\mathsf{T}_i}t^{-\delta}dt+L\tfrac{(\mathsf{T}_i-\mathsf{T}_{i-1})^{\ltimes_i}}{\Gamma(\ltimes_i+1)}\widehat{\varsigma}(U)\int_{\mathsf{T}_{i-1}}^{\mathsf{T}_i}t^{-\delta}dt\Big]\\
&+ \tfrac{(s-\mathsf{T}_{i-1})^{\ltimes_i-1}}{\Gamma(\ltimes_i)}\int_{\mathsf{T}_{i-1}}^{s}\Big[K\widehat{\varsigma}(U)\int_{\mathsf{T}_{i-1}}^{s}t^{-\delta}dt+L\tfrac{(\mathsf{T}_i-\mathsf{T}_{i-1})^{\ltimes_i}}{\Gamma(\ltimes_i+1)}\widehat{\varsigma}(U)\int_{\mathsf{T}_{i-1}}^{s}t^{-\delta}dt\Big],\varkappa\in U\Big\}.\\
&\leq \tfrac{[(\mathsf{T}_i{}^{1-\delta}-\mathsf{T}_{i-1}{}^{1-\delta})+(s^{1-\delta}-\mathsf{T}_{i-1}{}^{1-\delta})](s-\mathsf{T}_{i-1})^{\ltimes_i-1}}{(1-\delta)\Gamma(\ltimes_i)}\Big(K+L\tfrac{(\mathsf{T}_i-\mathsf{T}_{i-1})^{\ltimes_i}}{\Gamma(\ltimes_i+1)}\Big)\widehat{\varsigma}(U)\\
&\leq \tfrac{2(\mathsf{T}_i{}^{1-\delta}-\mathsf{T}_{i-1}{}^{1-\delta})(\mathsf{T}_i-\mathsf{T}_{i-1})^{\ltimes_i-1}}{(1-\delta)\Gamma(\ltimes_i)}\Big(K+L\tfrac{(\mathsf{T}_i-\mathsf{T}_{i-1})^{\ltimes_i}}{\Gamma(\ltimes_i+1)}\Big)\widehat{\varsigma}(U)
\end{aligned}$$

Therefore

$$\widehat{\varsigma}(WU)\leq \frac{2(\mathsf{T}_i{}^{1-\delta}-\mathsf{T}_{i-1}{}^{1-\delta})(\mathsf{T}_i-\mathsf{T}_{i-1})^{\ltimes_i-1}}{(1-\delta)\Gamma(\ltimes_i)}\Big(K+L\frac{(\mathsf{T}_i-\mathsf{T}_{i-1})^{\ltimes_i}}{\Gamma(\ltimes_i+1)}\Big)\widehat{\varsigma}(U)$$

where $\mu<1$ under the condition given in inequality (3.6). Hence, W is a set contraction. By Darbo's fixed point theorem, problem (3.4) admits at least one solution $\widetilde{\varkappa}_i$ in B_{R_i}. Since $B_{R_i}\in E_i$, this completes the proof of Theorem 10.

We now turn to proving the existence of solutions for the ITVP (3.1). The following premise will be established.

(H3) Let $s^{\delta}f : J \times X \times X \to X$ be a continuous function. Suppose there exist constants $K, L > 0$ such that,
$s^{\delta}\|f(s, y_1, z_1) - f(s, y_2, z_2)\| \leq K\|y_1 - y_2\| + L\|z_1 - z_2\|$, for any $y_1,\ y_2,\ z_1,\ z_2 \in X$ and $s \in J$.

Theorem 11 *Suppose that conditions (H1) and (H3) hold, and that inequality (3.6) is satisfied for each $i \in \{1, 2, \ldots, n\}$. Then the boundary value problem (ITVP) (3.1) admits at least one solution in $C(J, X)$.*

Proof
According to Theorem 10, the initial and terminal value problem (ITVP) (3.4) admits a solution $\widetilde{\varkappa}_i \in E_i$ for all $i \in \{1, 2, \ldots, n\}$.

Define the function κ_i

$$\kappa_i(s) = \begin{cases} 0, & s \in [0, \mathsf{T}_{i-1}], \\ \widetilde{\varkappa}_i, & s \in J_i, \end{cases} \qquad i \in \{1, 2, ..., n\}.$$

Thus, for $s \in J_i$ the integral equation (3.3) has a solution $\varkappa_i \in C([0, \mathsf{T}_i], X)$ satisfying $\varkappa_i(0) = 0$ and $\varkappa_i(\mathsf{T}_i) = \widetilde{\varkappa}_i(\mathsf{T}_i) = 0$.

Consequently, define the function $\kappa : J \to X$ by

$$\varkappa(s) = \begin{cases} \varkappa_1(s), \quad s \in J_1, \\ \varkappa_2(s) = \begin{cases} 0, & s \in J_1, \\ \widetilde{\varkappa}_2, & s \in J_2 \end{cases} \\ \cdot \\ \cdot \\ \cdot \\ \varkappa_n(s) = \begin{cases} 0, & s \in [0, \mathsf{T}_{i-1}], \\ \widetilde{\varkappa}_i, & s \in J_i \end{cases} \end{cases}.$$

Then $\kappa \subset C(J, X)$ is a solution of the initial and terminal value problem (3.1).

3.1.2 Criterion of Stability

Theorem 12 *Let all the conditions of Theorem 11 be satisfied. Suppose that condition (H3) holds, and let $\vartheta \in C(J_i, X)$ be an increasing function for which there exists a constant $\lambda_\vartheta > 0$ such that*

$$I^{\ltimes_i}_{\mathsf{T}_{i-1}^+}\vartheta(s) \leq \lambda_{\vartheta(s)}\vartheta(s), \ \ for\ any\ s \in J_i,$$

holds. Then the initial and terminal value problem (ITVP) (3.1) is Ulam-Hyers-$\mathfrak{R}$assias stable with respect to ϑ

Proof Let $\epsilon > 0$ be an arbitrary positive number, and assume that the function $z(s) \in C(J, X)$ satisfies the inequality (2.20).

For any $i \in \{1,2,\ldots,n\}$ define the functions $z_1(s) \equiv z(s), for\ s \in [0, \mathsf{T}_1]$ and for $i = 2,3,\ldots,n$:

$$z_i(s) = \begin{cases} 0, & s \in [0, \mathsf{T}_{i-1}], \\ z(s), & s \in J_i, \end{cases}$$

For $s \in J_i$ and $i \in \{1,2,...,n\}$, using equality (3.2), we obtain:

$$F_{0^+}^{-\ltimes(s)} z_i(s) = \frac{d^2}{ds^2}\Big(\int_{\mathsf{T}_{i-1}}^{s} \frac{(s-t)^{1-\ltimes_i}}{\Gamma(2-\ltimes_i)} z(t)dt\Big)$$

Applying the Riemann–Liouville fractional integral $I^{\ltimes_i}_{\mathsf{T}^+_{i-1}}$ to both sides of inequality (2.20), we derive

$$\|z(s) + \frac{(\mathsf{T}_i - \mathsf{T}_{i-1})^{1-\ltimes_i}(s-\mathsf{T}_{i-1})^{\ltimes_i - 1}}{\Gamma(\ltimes_i)} \int_{\mathsf{T}_{i-1}}^{\mathsf{T}_i} (\mathsf{T}_i - t)^{\ltimes_{i-1}} f(t, z(t), I^{\ltimes_i}_{\mathsf{T}^+_{i-1}} z(t))dt$$
$$-\frac{1}{\Gamma(\ltimes_i)} \int_{\mathsf{T}_{i-1}}^{s} (s-t)^{\ltimes_{i-1}} f(t, z(t), I^{\ltimes_i}_{\mathsf{T}^+_{i-1}} z(t))dt\|$$
$$\leq \epsilon \int_{\mathsf{T}_{i-1}}^{s} \frac{(s-t)^{\ltimes(i)-1}}{\Gamma(u(i))} \vartheta(t)dt$$
$$\leq \epsilon \lambda_{\vartheta(s)} \vartheta(s)$$

By Theorem 11, the ınitial and terminal υalue problem (ITVP) (3.1) admits a solution $\varkappa \in C(J, X)$, defined by

$$\varkappa(s) = \varkappa_i(s) \text{ for } s \in J_i,\ i = 1,2,\ldots,n,$$

where

$$\varkappa_i = \begin{cases} 0, & s \in [0, \mathsf{T}_{i-1}], \\ \widetilde{\varkappa}_i, & s \in J_i, \end{cases} \tag{3.8}$$

and $\widetilde{\varkappa}_i$ is the solution of ITVP(3.4), given by Lemma 12:

$$\begin{aligned} \widetilde{\varkappa}_i(s) &= -\frac{(\mathsf{T}_i - \mathsf{T}_{i-1})^{1-\ltimes_i}(s-\mathsf{T}_{i-1})^{\ltimes_i - 1}}{\Gamma(\ltimes_i)} \int_{\mathsf{T}_{i-1}}^{\mathsf{T}_i} (\mathsf{T}_i - t)^{\ltimes_{i-1}} f(t, \widetilde{\varkappa}_i(t), I^{\ltimes_i}_{\mathsf{T}^+_{i-1}} \widetilde{\varkappa}_i(t))dt \\ &+ \frac{1}{\Gamma(\ltimes_i)} \int_{\mathsf{T}_{i-1}}^{s} (s-t)^{\ltimes_{i-1}} f(t, \widetilde{\varkappa}_i(t), I^{\ltimes_i}_{\mathsf{T}^+_{i-1}} \widetilde{\varkappa}_i(t))dt. \end{aligned}$$

Then, for $s \in J_i$, we have

$$|z(s) - \varkappa(s)| = |z(s) - \varkappa_i(s)| = |z_i(s) - \widetilde{\varkappa}_i(s)|$$

$$\begin{aligned}
= & \ \|z_i(s) + \frac{(\mathsf{T}_i - \mathsf{T}_{i-1})^{1-\ltimes_i}(s - \mathsf{T}_{i-1})^{\ltimes_i - 1}}{\Gamma(\ltimes_i)} \int_{\mathsf{T}_{i-1}}^{\mathsf{T}_i} (\mathsf{T}_i - t)^{\ltimes_i - 1} f(t, \widetilde{\varkappa}_i(t), I^{\ltimes_i}_{\mathsf{T}^+_{i-1}} \widetilde{\varkappa}_i(t))dt \\
- & \ \frac{1}{\Gamma(\ltimes_i)} \int_{\mathsf{T}_{i-1}}^{s} (s - t)^{\ltimes_i - 1} f(t, \widetilde{\varkappa}_i(t), I^{\ltimes_i}_{\mathsf{T}^+_{i-1}} \widetilde{\varkappa}_i(t))dt\| \\
= & \ \|z_i(s) + \frac{(\mathsf{T}_i - \mathsf{T}_{i-1})^{1-\ltimes_i}(s - \mathsf{T}_{i-1})^{\ltimes_i - 1}}{\Gamma(\ltimes_i)} \int_{\mathsf{T}_{i-1}}^{\mathsf{T}_i} (\mathsf{T}_i - t)^{\ltimes_i - 1} f(t, z_i(t), I^{\ltimes_i}_{\mathsf{T}^+_{i-1}} z_i(t))dt \\
- & \ \frac{1}{\Gamma(\ltimes_i)} \int_{\mathsf{T}_{i-1}}^{s} (s - t)^{\ltimes_i - 1} f(t, z_i(t), I^{\ltimes_i}_{\mathsf{T}^+_{i-1}} z_i(t))dt\| \\
+ & \ \frac{(\mathsf{T}_i - \mathsf{T}_{i-1})^{1-\ltimes_i}(s - \mathsf{T}_{i-1})^{\ltimes_i - 1}}{\Gamma(\ltimes_i)} \int_{\mathsf{T}_{i-1}}^{\mathsf{T}_i} (\mathsf{T}_i - t)^{\ltimes_i - 1} \\
& \ \|f(t, z_i(t), I^{\ltimes_i}_{\mathsf{T}^+_{i-1}} z_i) - f(t, \widetilde{\varkappa}_i(t), I^{\ltimes_i}_{\mathsf{T}^+_{i-1}} \widetilde{\varkappa}_i)\|dt \\
+ & \ \frac{1}{\Gamma(\ltimes_i)} \int_{\mathsf{T}_{i-1}}^{s} (s - t)^{\ltimes_i - 1} \|f(t, z_i(t), I^{\ltimes_i}_{\mathsf{T}^+_{i-1}} z_i) - f(t, \widetilde{\varkappa}_i(t), I^{\ltimes_i}_{\mathsf{T}^+_{i-1}} \widetilde{\varkappa}_i)\|dt \\
\leq & \ \lambda_{\vartheta(s)} \epsilon \vartheta(s) + \frac{(\mathsf{T}_i - \mathsf{T}_{i-1})^{1-\ltimes_i}(s - \mathsf{T}_{i-1})^{\ltimes_i - 1}}{\Gamma(\ltimes_i)} \\
& \ \int_{\mathsf{T}_{i-1}}^{\mathsf{T}_i} (\mathsf{T}_i - t)^{\ltimes_i - 1} t^{-\delta} (K\|z_i(t) - \widetilde{\varkappa}_i(t)\| + L I^{\ltimes_i}_{\mathsf{T}^+_{i-1}} \|z_i(t) - \widetilde{\varkappa}_i(t)\|)dt \\
+ & \ \frac{1}{\Gamma(\ltimes_i)} \int_{\mathsf{T}_{i-1}}^{s} (s - t)^{\ltimes_i - 1} t^{-\delta} (K\|z_i(t) - \widetilde{\varkappa}_i(t)\| + L I^{\ltimes_i}_{\mathsf{T}^+_{i-1}} \|z_i(t) - \widetilde{\varkappa}_i(t)\|)dt \\
\leq & \ \lambda_{\vartheta(s)} \epsilon \vartheta(s) + \frac{(\mathsf{T}_i - \mathsf{T}_{i-1})^{\ltimes_i - 1}}{\Gamma(\ltimes_i)} (K\|z_i - \widetilde{\varkappa}_i\|_{E_i} + L I^{\ltimes_i}_{\mathsf{T}^+_{i-1}} \|z_i - \widetilde{\varkappa}_i\|_{E_i}) \int_{\mathsf{T}_{i-1}}^{\mathsf{T}_i} t^{-\delta} dt \\
+ & \ \frac{(\mathsf{T}_i - \mathsf{T}_{i-1})^{\ltimes_i - 1}}{\Gamma(\ltimes_i)} (K\|z_i - \widetilde{\varkappa}_i\|_{E_i} + L I^{\ltimes_i}_{\mathsf{T}^+_{i-1}} \|z_i - \widetilde{\varkappa}_i\|_{E_i}) \int_{\mathsf{T}_{i-1}}^{s} t^{-\delta} dt \\
\leq & \ \lambda_{\vartheta(s)} \epsilon \vartheta(s) + \frac{(\mathsf{T}_i - \mathsf{T}_{i-1})^{\ltimes_i - 1}({\mathsf{T}_i}^{1-\delta} - {\mathsf{T}_{i-1}}^{1-\delta})}{(1-\delta)\Gamma(\ltimes_i)} (K\|z_i - \widetilde{\varkappa}_i\|_{E_i} + L \frac{(\mathsf{T}_i - \mathsf{T}_{i-1})^{\ltimes_i}}{\Gamma(\ltimes_i + 1)} \|z_i - \widetilde{\varkappa}_i\|_{E_i}) \\
+ & \ \frac{(\mathsf{T}_i - \mathsf{T}_{i-1})^{\ltimes_i - 1}(s^{1-\delta} - {\mathsf{T}_{i-1}}^{1-\delta})}{(1-\delta)\Gamma(\ltimes_i)} (K\|z_i - \widetilde{\varkappa}_i\|_{E_i} + L \frac{(\mathsf{T}_i - \mathsf{T}_{i-1})^{\ltimes_i}}{\Gamma(\ltimes_i + 1)} \|z_i - \widetilde{\varkappa}_i\|_{E_i}) \\
\leq & \ \lambda_{\vartheta(s)} \epsilon \vartheta(s) + \frac{2(\mathsf{T}_i - \mathsf{T}_{i-1})^{\ltimes_i - 1}({\mathsf{T}_i}^{1-\delta} - {\mathsf{T}_{i-1}}^{1-\delta})}{(1-\delta)\Gamma(\ltimes_i)} (K + L \frac{(\mathsf{T}_i - \mathsf{T}_{i-1})^{\ltimes_i}}{\Gamma(\ltimes_i + 1)}) \|z_i - \widetilde{\varkappa}_i\|_{E_i} \\
\leq & \ \lambda_{\vartheta(s)} \epsilon \vartheta(s) + \mu \|z - \varkappa\|,
\end{aligned}$$

where

$$\mu = \max_{i=1,2,\dots,n} \frac{2(\mathsf{T}_i - \mathsf{T}_{i-1})^{\ltimes_i - 1}(\mathsf{T}_i{}^{1-\delta} - \mathsf{T}_{i-1}{}^{1-\delta})}{(1-\delta)\Gamma(\ltimes_i)}(K + L\frac{(\mathsf{T}_i - \mathsf{T}_{i-1})^{\ltimes_i}}{\Gamma(\ltimes_i + 1)}).$$

Then,

$$\|z - \varkappa\|(1-\mu) \leq \lambda_{\vartheta(s)}\epsilon\vartheta(s).$$

We find that, for every $s \in J$,

$$|z(s) - \varkappa(s)| \leq \|z - \varkappa\| \leq \frac{\lambda_{\vartheta(s)}}{(1-\mu)}\epsilon\vartheta(s) := c_f\epsilon\vartheta(s)$$

Consequently, the initial and terminal value problem (ITVP) (3.1) is Ulam-Hyers-$\mathfrak{R}$assias stable with respect to ϑ.

3.1.3 Numerical Application

Our focus in this case is on the fractional initial and terminal value problem.

$$\begin{cases} F_{0^+}^{\ltimes(s)}\varkappa(s) = \dfrac{s^{-\frac{1}{3}}e^{-s}}{(e^{e^{\frac{s^2}{1+s}}}+4e^{2s}+1)(1+|\varkappa(s)|+|I_0^{\ltimes(s)}\varkappa(s)|)}, & s \in J := [0,2], \\ \varkappa(0) = 0, \;\; \varkappa(2) = 0. \end{cases} \tag{3.9}$$

Let

$$f(s,y,z) = \frac{s^{-\frac{1}{3}}e^{-s}}{(e^{e^{\frac{s^2}{1+s}}} + 4e^{2s} + 1)(1 + y + z)}, \; (s,y,z) \in [0,2]\times[0,+\infty)\times[0,+\infty).$$

$$\ltimes(s) = \begin{cases} \frac{3}{2}, & s \in J_1 := [0,1], \\ \frac{9}{5}, & s \in J_2 :=]1,2]. \end{cases} \tag{3.10}$$

Therefore, we have:

$$\begin{aligned} s^{\frac{1}{3}}|f(s,y_1,z_1) - f(s,y_2,z_2)| &= \left|\frac{e^{-s}}{(e^{e^{\frac{s^2}{1+s}}}+4e^{2s}+1)}\left(\frac{1}{1+y_1+z_1} - \frac{1}{1+y_2+z_2}\right)\right| \\ &\leq \frac{e^{-s}(|y_1-y_2|+|z_1-z_2|)}{(e^{e^{\frac{s^2}{1+s}}}+4e^{2s}+1)(1+y_1+z_1)(1+y_2+z_2)} \\ &\leq \frac{e^{-s}}{(e^{e^{\frac{s^2}{1+s}}}+4e^{2s}+1)}(|y_1 - y_2| + |z_1 - z_2|) \\ &\leq \frac{1}{(e+5)}|y_1 - y_2| + \frac{1}{(e+5)}|z_1 - z_2|. \end{aligned}$$

Thus, hypothesis (H3) holds with $\delta = \frac{1}{3}$, $K = L = \frac{1}{e+5}$.

By (3.10), and according to (3.4), we consider two auxiliary ITVPs:

$$\begin{cases} F_{0+}^{\frac{3}{2}}\varkappa(s) = \dfrac{s^{-\frac{1}{3}}e^{-s}}{(e^{e^{\frac{s^2}{1+s}}}+4e^{2s}+1)(1+|\varkappa(s)|+|I_0^{\frac{3}{2}}\varkappa(s)|)}, & s \in J_1, \\ \varkappa(0) = 0, \ \ \varkappa(1) = 0. \end{cases} \tag{3.11}$$

On J_2:

$$\begin{cases} F_{1+}^{\frac{9}{5}}\varkappa(s) = \dfrac{s^{-\frac{1}{3}}e^{-s}}{(e^{e^{\frac{s^2}{1+s}}}+4e^{2s}+1)(1+|\varkappa(s)|+|I_0^{\frac{9}{5}}\varkappa(s)|)}, & s \in J_2, \\ \varkappa(1) = 0, \ \ \varkappa(2) = 0. \end{cases} \tag{3.12}$$

We verify that condition (3.6) is satisfied when $i = 1$. In fact,

$$\frac{2(\mathrm{T}_1{}^{1-\delta} - \mathrm{T}_0{}^{1-\delta})(\mathrm{T}_1 - \mathrm{T}_0)^{\ltimes-1}}{(1-\delta)\Gamma(\ltimes)}\Big(K+\frac{L(\mathrm{T}_1 - \mathrm{T}_0)^{\ltimes}}{\Gamma(\ltimes+1)}\Big) = \frac{2}{\frac{2}{3}(e+5)\Gamma(\frac{3}{2})}\Big(1+\frac{1}{\Gamma(\frac{5}{2})}\Big) \simeq 0.7685 < 1$$

Let $\vartheta(s) = s^{\frac{1}{2}}$. Then,

$$\begin{aligned} I_{0+}^{\ltimes}\vartheta(s) &= \tfrac{1}{\Gamma(\frac{3}{2})}\textstyle\int_0^s (s-t)^{\frac{1}{2}}t^{\frac{1}{2}}dt \\ &\le \tfrac{1}{\Gamma(\frac{3}{2})}\textstyle\int_0^s (s-t)^{\frac{1}{2}}dt \\ &\le \tfrac{2}{3\Gamma(\frac{3}{2})}\vartheta(s) := \lambda_{\vartheta(s)}\vartheta(s). \end{aligned}$$

Thus, (H3) is satisfied with $\lambda_{\vartheta(s)} = \frac{2}{3\Gamma(\frac{3}{2})}$.

The ITVP (3.11) has a solution $\widetilde{\varkappa}_1 \in E_1$ according to Theorem 10.

For $i = 2$, we similarly verify condition (3.6):

$$\frac{2(\mathrm{T}_2{}^{1-\delta} - \mathrm{T}_1{}^{1-\delta})(\mathrm{T}_2 - \mathrm{T}_1)^{\text{ß}-1}}{(1-\delta)\Gamma(\text{ß})}\Big(K+\frac{L(\mathrm{T}_2 - \mathrm{T}_1)^{\text{ß}}}{\Gamma(\text{ß}+1)}\Big) = \frac{2(2^{\frac{2}{3}}-1)}{\frac{2}{3}\Gamma(\frac{9}{5})}\frac{1}{e+5}\Big(1+\frac{1}{\Gamma(\frac{14}{5})}\Big) \approx 0.3913 < 1$$

Moreover,

$$\begin{aligned} I_{1+}^{\text{ß}}\vartheta(s) &= \tfrac{1}{\Gamma(\frac{9}{5})}\textstyle\int_1^s (s-t)^{\frac{4}{5}}t^{\frac{1}{2}}dt \\ &\le \tfrac{1}{\Gamma(\frac{9}{5})}\textstyle\int_1^s (s-t)^{\frac{4}{5}}dt \\ &\le \tfrac{5}{9\Gamma(\frac{9}{5})}\vartheta(s) := \lambda_{\vartheta(s)}\vartheta(s). \end{aligned}$$

Thus, (H3) is fulfilled with $\lambda_{\vartheta(s)} = \frac{5}{9\Gamma(\frac{9}{5})}$.

The ITVP (3.12) has a solution $\widetilde{\varkappa}_2 \in E_2$, as per Theorem 10.

The solution to ITVP (3.9) may be constructed using Theorem 11.

$$\varkappa(s) = \begin{cases} \varkappa_1(s), & s \in J_1, \\ \varkappa_2(s), & s \in J_2, \end{cases}$$

$$\varkappa_2(s) = \begin{cases} 0, & s \in J_1, \\ \widetilde{\varkappa}_2(s), & s \in J_2. \end{cases}$$

Finally, the ITVP (3.9) is Ulam-Hyers-$\mathfrak{R}$assias stable with respect to $\vartheta(s)$, as per Theorem 12.

3.2 Implicit Initial and Terminal Value Problems of R-Liouville Fractional Differential Equations of Variable Order

We consider the following implicit ınitial and terminal $v$alue problem (ITVP for short) involving variable-order fractional differential equations. Our aim is to investigate the existence of continuous solutions and the Ulam-Hyers-$\mathfrak{R}$assias stability of the derived solution:

$$\begin{cases} F_{0^+}^{\ltimes(s)} y(s) = f(s, y(s), D_{0^+}^{\ltimes(s)} y(s)), & s \in J \\ y(0) = 0, \ \ y(\mathsf{T}) = 0, \end{cases} \tag{3.13}$$

where $0 < \mathsf{T} < +\infty$, $1 < \ltimes(s) \leq 2$, $f : J \times \mathbb{R} \times \mathbb{R} \to \mathbb{R}$ is a continuous function and $F_{0^+}^{\ltimes(s)}$ denotes the Riemann–Liouville fractional derivative of variable order $\ltimes(s)$.

3.2.1 Existence Properties

Let us introduce the following assumptions:

(H1) Let $n \in \mathbb{N}$ be a positive integer, and let $\mathcal{P} = \{J_1 := [0, \mathsf{T}_1], J_2 := (\mathsf{T}_1, \mathsf{T}_2], J_3 := (\mathsf{T}_2, \mathsf{T}_3], ... J_n := (\mathsf{T}_{n-1}, \mathsf{T}]\}$ be a partition of the interval $J := [0,]$, where $0 < +\infty$. Let $\ltimes(s) : J \to (1, 2]$ be a piecewise constant function with respect to the partition $\mathcal{P}$, defined as

$$\ltimes(s) = \sum_{i=1}^{n} \ltimes_i I_i(s) = \begin{cases} \ltimes_1, & if \ s \in J_1, \\ \ltimes_2, & if \ s \in J_2, \\ \cdot \\ \cdot \\ \cdot \\ \ltimes_n, & if \ s \in J_n, \end{cases}$$

where $1 < \ltimes_i \leq 2$ are constants, and I_i (s) is the indicator function of the interval $J_i := (\mathsf{T}_{i-1}, \mathsf{T}_i]$, $i = 1, 2, \ldots, n$, (with $\mathsf{T}_0 = 0$ and $\mathsf{T}_n = \mathsf{T}$), such that,

$$I_i(s) = \begin{cases} 1, & for \ s \in J_i, \\ 0, & for \ elsewhere. \end{cases}$$

(H2) For each $i = 1, 2, \ldots, n$, and let $f : J_i \times \mathbb{R} \times \mathbb{R} \to \mathbb{R}$ be a continuous function. Suppose there exist constants $K, L > 0$ such that for all $s \in J_i$ and any $y_1, y_2, z_1, z_2 \in \mathbb{R}$, the following inequality holds:

$$|f(s, y_1, z_1) - f(s, y_2, z_2)| \leq K\,|y_1 - y_2| + L\,|z_1 - z_2|,$$

Let $E_i = C(J_i, \mathbb{R})$, we denote the space of continuous functions from J_i into $\mathbb{R}$ with the norm

$$\|y\|_{E_i} = \sup_{s \in J_i} |y(s)|,$$

for each $i \in \{1, 2, ..., n\}$.

We now proceed with the essential analysis of the ITVP (3.13) to establish our main results.

The left-sided Riemann–Liouville fractional derivative of variable order function $\ltimes(s)$, applied a function $\varkappa(s) \in C(J, \mathbb{R})$ can be expressed for all $s \in J_i$, where $i = 1, 2, \ldots, n$, as a sum of left-sided Riemann–Liouville fractional derivatives of fixed orders $\ltimes_k$, $k = 1, \ldots, i$.

$$\digamma_{0^+}^{\ltimes(s)} y(s) = \frac{d^2}{ds^2}\Big(\int_0^{\mathsf{T}_1} \frac{(s-t)^{1-\ltimes_1}}{\Gamma(2-\ltimes_1)} y(t)dt + \ldots + \int_{\mathsf{T}_{i-1}}^{s} \frac{(s-t)^{1-\ltimes_i}}{\Gamma(2-\ltimes_i)} y(t)dt\Big). \tag{3.14}$$

Thus, according to equation (3.14), the initial and terminal value problem (3.13) can be reformulated for any $s \in J_i$, $i = 1, 2, \ldots, n$, as follows:

$$\frac{d^2}{ds^2}\Big(\int_0^{\mathsf{T}_1} \frac{(s-t)^{1-\ltimes_1}}{\Gamma(2-\ltimes_1)} y(t)dt + \ldots + \int_{\mathsf{T}_{i-1}}^{s} \frac{(s-t)^{1-\ltimes_i}}{\Gamma(2-\ltimes_i)} y(t)dt\Big) = f(s, y(s), \digamma_{0^+}^{\ltimes_i} y(s)), \quad s \in J_i. \tag{3.15}$$

We now present the definition of a solution to the ITVP (3.13), which plays a fundamental role in our analysis.

Definition 45 *We say that the initial and terminal value problem (ITVP) (3.13) admits a solution in $C([0, \mathsf{T}_i], \mathbb{R})$, if there exist functions $y_i \in C([0, \mathsf{T}_i], \mathbb{R})$, for $i = 1, 2, ..., n$, satisfying relation 3.15 and the boundary conditions $y_i(0) = 0 = y_i(\mathsf{T}_i)$.*

According to the above analysis, the expression of the ITVP (3.13) can be written over each subinterval J_i, $i \in \{1, 2, \ldots, n\}$ in the form of equation (3.15).

For $0 \leq s \leq \mathsf{T}_{i-1}$, we take $\widetilde{y}(s) \equiv 0$, and then equation (3.15) reduces to:

$$\digamma_{\mathsf{T}_{i-1}^+}^{\ltimes_i} \widetilde{y}(s) = f(s, \widetilde{y}(s), \digamma_{\mathsf{T}_{i-1}^+}^{\ltimes_i} \widetilde{y}(s)), \quad s \in J_i.$$

We now consider the following auxiliary ITVP:

$$\begin{cases} \digamma^{\ltimes_i}_{\mathsf{T}^+_{i-1}} y(s) = f(s, y(s), \digamma^{\ltimes_i}_{\mathsf{T}^+_{i-1}} y(s)), \quad s \in J_i \\ y(\mathsf{T}_{i-1}) = 0, y(\mathsf{T}_i) = 0. \end{cases} \tag{3.16}$$

To establish the existence of solutions to problem (3.16), we require the following auxiliary lemma.

Lemma 13 *A function $y \in E_i$ is a solution to the problem (3.16) if and only it satisfies the following integral relation:*

$$y(s) = -(\mathsf{T}_i - \mathsf{T}_{i-1})^{1-\ltimes_i}(s - \mathsf{T}_{i-1})^{\ltimes_i - 1} I^{\ltimes_i}_{\mathsf{T}^+_{i-1}} \varkappa(\mathsf{T}_i)) + I^{\ltimes_i}_{\mathsf{T}^+_{i-1}} \varkappa(s), \quad s \in J_i, \tag{3.17}$$

where

$$\varkappa(s) = f\Big(s, -(\mathsf{T}_i - \mathsf{T}_{i-1})^{1-\ltimes_i}(s - \mathsf{T}_{i-1})^{\ltimes_i - 1} I^{\ltimes_i}_{\mathsf{T}^+_{i-1}} \varkappa(\mathsf{T}_i)) + I^{\ltimes_i}_{\mathsf{T}^+_{i-1}} \varkappa(s), \varkappa(s)\Big), \quad s \in J_i.$$

Proof Let $y \in E_i$ be a solution to the initial and terminal value problem (3.16). Define $\digamma^{\ltimes_i}_{\mathsf{T}^+_{i-1}} y(s) = \varkappa(s)$ and apply the operator $I^{\ltimes_i}_{\mathsf{T}^+_{i-1}}$ to both sides of the equation in (3.16). By Lemma 1, we obtain

$$y(s) = \omega_1 (s - \mathsf{T}_{i-1})^{\ltimes_i - 1} + \omega_2 (s - \mathsf{T}_{i-1})^{\ltimes_i - 2} + I^{\ltimes_i}_{\mathsf{T}^+_{i-1}} \varkappa(s), \quad s \in J_i.$$

Given that $y(\mathsf{T}_{i-1}) = 0$ and under the assumptions on the function f, it follows that $\omega_2 = 0$.

Now, imposing the condition $y(\mathsf{T}_i) = 0$, we deduce $\omega_1 = -(\mathsf{T}_i - \mathsf{T}_{i-1})^{1-\ltimes_i} I^{\ltimes_i}_{\mathsf{T}^+_{i-1}} \varkappa(\mathsf{T}_i)$. Hence, the solution takes the form

$$y(s) = -(\mathsf{T}_i - \mathsf{T}_{i-1})^{1-\ltimes_i}(s - \mathsf{T}_{i-1})^{\ltimes_i - 1} I^{\ltimes_i}_{\mathsf{T}^+_{i-1}} \varkappa(\mathsf{T}_i) + I^{\ltimes_i}_{\mathsf{T}^+_{i-1}} \varkappa(s), \quad s \in J_i,$$

where

$$\varkappa(s) = f\Big(s, -(\mathsf{T}_i - \mathsf{T}_{i-1})^{1-\ltimes_i}(s - \mathsf{T}_{i-1})^{\ltimes_i - 1} I^{\ltimes_i}_{\mathsf{T}^+_{i-1}} \varkappa(\mathsf{T}_i) + I^{\ltimes_i}_{\mathsf{T}^+_{i-1}} \varkappa(s), \varkappa(s)\Big), \quad s \in J_i.$$

Conversely, if $y \in E_i$ satisfies the integral equation (3.17), hence by Lemma 1 and the continuity of the function $s^\delta f$, it follows that y is a solution to the initial-terminal value problem (3.16).

Our existence result is established via the Krasnoselskii fixed point theorem.

Theorem 13 *Assume that conditions $(H1)$ and $(H2)$ hold. If the following inequality is satisfied:*

$$\frac{(\mathsf{T}_i - \mathsf{T}_{i-1})^{\ltimes_i - 1}(\mathsf{T}_i{}^{1-\delta} - \mathsf{T}_{i-1}{}^{1-\delta})}{(1-\delta)\Gamma(\ltimes_i)}\left(\frac{K(\mathsf{T}_i - \mathsf{T}_{i-1})^{\ltimes_i}}{\Gamma(\ltimes_i + 1)} + \frac{L}{2}\right) < \frac{1}{4}, \tag{3.18}$$

then the initial and terminal value problem (3.16) admits a solution in E_i.

Proof We convert problem (3.16) into a fixed-point formulation by examining

$$W_1, W_2 : E_i \to E_i$$

defined as follows:

$$W_1 \varkappa(s) = -(\mathsf{T}_i - \mathsf{T}_{i-1})^{1-\ltimes_i}(s - \mathsf{T}_{i-1})^{\ltimes_i - 1} I^{\ltimes_i}_{\mathsf{T}_{i-1}^+} \varkappa(\mathsf{T}_i), \quad W_2 \varkappa(s) = I^{\ltimes_i}_{\mathsf{T}_{i-1}^+} \varkappa(s), \tag{3.19}$$

where

$$\varkappa(s) \quad = \quad f(s, y(s), \varkappa(s)).$$

The continuity of the function $s^\delta f$, along with the properties of fractional integrals, ensures that the mappings $W_1, W_2 : E_i \to E_i$ defined in (3.19) are well-defined.

Let

$$R_i \quad \geq \quad \frac{\frac{2f^*(\mathsf{T}_i - \mathsf{T}_{i-1})^{\ltimes_i}}{\Gamma(\ltimes_i)}}{1 - \frac{2(\mathsf{T}_i - \mathsf{T}_{i-1})^{\ltimes_i - 1}(\mathsf{T}_i{}^{1-\delta} - \mathsf{T}_{i-1}{}^{1-\delta})}{(1-\delta)\Gamma(\ltimes_i)}\left(\frac{2K(\mathsf{T}_i - \mathsf{T}_{i-1})^{\ltimes_i}}{\Gamma(\ltimes_i + 1)} + L\right)},$$

where

$$f^* = \sup |f(s, 0, 0)|, B_{R_i} = \{y \in E_i, \ \|y\|_{E_i} \leq R_i\}.$$

It is evident that B_{R_i} is closed, convex, bounded, and nonempty.

We now proceed to demonstrate that the mappings W_1 and W_2 satisfy the assumptions of Krasnoselskii's fixed point theorem. The proof will be carried out in four steps.

Step 1: $W_1(B_{R_i}) + W_2(B_{R_i}) \subseteq (B_{R_i})$.

Let $\varkappa \in B_{R_i}$. We aim to show that $W_1(\varkappa) + W_2(\varkappa) \in B_{R_i}$. For $s \in J_i$, we estimate:

$$|(W_1\varkappa)(s) + (W_2\varkappa)(s)|$$

$$\begin{aligned}
\leq\; & \frac{(\mathsf{T}_i - \mathsf{T}_{i-1})^{1-\ltimes_i}(s - \mathsf{T}_{i-1})^{\ltimes_i - 1}}{\Gamma(\ltimes_i)} \int_{\mathsf{T}_{i-1}}^{\mathsf{T}_i} (\mathsf{T}_i - t)^{\ltimes_i - 1} \\
& \left| f\Big(t, -(\mathsf{T}_i - \mathsf{T}_{i-1})^{1-\ltimes_i}(t - \mathsf{T}_{i-1})^{\ltimes_i - 1} I^{\ltimes_i}_{\mathsf{T}^+_{i-1}} \varkappa_n(\mathsf{T}_i)) + I^{\ltimes_i}_{\mathsf{T}^+_{i-1}} \varkappa(t), \varkappa(t)\Big)\right| dt \\
& + \frac{1}{\Gamma(\ltimes_i)} \int_{\mathsf{T}_{i-1}}^{s} (s - t)^{\ltimes_i - 1} \\
& \left| f\Big(t, -(\mathsf{T}_i - \mathsf{T}_{i-1})^{1-\ltimes_i}(t - \mathsf{T}_{i-1})^{\ltimes_i - 1} I^{\ltimes_i}_{\mathsf{T}^+_{i-1}} \varkappa(\mathsf{T}_i)) + I^{\ltimes_i}_{\mathsf{T}^+_{i-1}} \varkappa(t), \varkappa(t)\Big)\right| dt \\
\leq\; & \frac{2}{\Gamma(\ltimes_i)} \int_{\mathsf{T}_{i-1}}^{\mathsf{T}_i} (\mathsf{T}_i - t)^{\ltimes_i - 1} \\
& \left| f\Big(t, -(\mathsf{T}_i - \mathsf{T}_{i-1})^{1-\ltimes_i}(t - \mathsf{T}_{i-1})^{\ltimes_i - 1} I^{\ltimes_i}_{\mathsf{T}^+_{i-1}} \varkappa(\mathsf{T}_i)) + I^{\ltimes_i}_{\mathsf{T}^+_{i-1}} \varkappa(t), \varkappa(t)\Big)\right| dt \\
\leq\; & \frac{2}{\Gamma(\ltimes_i)} \int_{\mathsf{T}_{i-1}}^{\mathsf{T}_i} (\mathsf{T}_i - t)^{\ltimes_i - 1} \\
& \left| f\Big(t, -(\mathsf{T}_i - \mathsf{T}_{i-1})^{1-\ltimes_i}(t - \mathsf{T}_{i-1})^{\ltimes_i - 1} I^{\ltimes_i}_{\mathsf{T}^+_{i-1}} \varkappa(\mathsf{T}_i)) + I^{\ltimes_i}_{\mathsf{T}^+_{i-1}} \varkappa(t), \varkappa(t)\Big)\right. \\
& \left. - f(t,0,0)\right| dt + \frac{2}{\Gamma(\ltimes_i)} \int_{\mathsf{T}_{i-1}}^{\mathsf{T}_i} (\mathsf{T}_i - t)^{\ltimes_i - 1} |f(t,0,0)| dt \\
\leq\; & \frac{2(\mathsf{T}_i - \mathsf{T}_{i-1})^{\ltimes_i - 1}}{\Gamma(\ltimes_i)} \int_{\mathsf{T}_{i-1}}^{\mathsf{T}_i} t^{-\delta}(K| - (\mathsf{T}_i - \mathsf{T}_{i-1})^{1-\ltimes_i}(t - \mathsf{T}_{i-1})^{\ltimes_i - 1} I^{\ltimes_i}_{\mathsf{T}^+_{i-1}} \varkappa(\mathsf{T}_i)) + I^{\ltimes_i}_{\mathsf{T}^+_{i-1}} \varkappa\,(t)| \\
& + L|\varkappa(t)|) dt + \frac{2f^*(\mathsf{T}_i - \mathsf{T}_{i-1})^{\ltimes_i}}{\Gamma(\ltimes_i)} \\
\leq\; & \frac{2(\mathsf{T}_i - \mathsf{T}_{i-1})^{\ltimes_i - 1}}{\Gamma(\ltimes_i)} \int_{\mathsf{T}_{i-1}}^{\mathsf{T}_i} t^{-\delta}\Big(K(|I^{\ltimes_i}_{\mathsf{T}^+_{i-1}} \varkappa(\mathsf{T}_i))| + |I^{\ltimes_i}_{\mathsf{T}^+_{i-1}} \varkappa(t)|) + L|\varkappa(t)|\Big) dt \\
& + \frac{2f^*(\mathsf{T}_i - \mathsf{T}_{i-1})^{\ltimes_i}}{\Gamma(\ltimes_i)} \\
\leq\; & \frac{2(\mathsf{T}_i - \mathsf{T}_{i-1})^{\ltimes_i - 1}({\mathsf{T}_i}^{1-\delta} - {\mathsf{T}_{i-1}}^{1-\delta})}{(1-\delta)\Gamma(\ltimes_i)} \Big(2K\|I^{\ltimes_i}_{\mathsf{T}^+_{i-1}} \varkappa\|_{E_i} + L\|\varkappa\|_{E_i}\Big) + \frac{2f^*(\mathsf{T}_i - \mathsf{T}_{i-1})^{\ltimes_i}}{\Gamma(\ltimes_i)} \\
\leq\; & \frac{2(\mathsf{T}_i - \mathsf{T}_{i-1})^{\ltimes_i - 1}({\mathsf{T}_i}^{1-\delta} - {\mathsf{T}_{i-1}}^{1-\delta})}{(1-\delta)\Gamma(\ltimes_i)} \Big(\frac{2K(\mathsf{T}_i - \mathsf{T}_{i-1})^{\ltimes_i}}{\Gamma(\ltimes_i + 1)} + L\Big) R_i + \frac{2f^*(\mathsf{T}_i - \mathsf{T}_{i-1})^{\ltimes_i}}{\Gamma(\ltimes_i)} \\
\leq\; & R_i,
\end{aligned}$$

which implies that $W_1(B_{R_i}) + W_2(B_{R_i}) \subseteq B_{R_i}$.

Step 2: W_1 is continuous.

Let $(\varkappa_n)$ be a sequence such that $\varkappa_n \to \varkappa$ in E_i. Then

$$\|(W_1\varkappa_n) - (W_1\varkappa)\|_{E_i} \to 0.$$

For $s \in J_i$, we have

$$|(W_1\varkappa_n)(s) - (W_1\varkappa)(s)|$$

$$\begin{aligned}
\leq\ & \frac{(\mathsf{T}_i - \mathsf{T}_{i-1})^{1-\ltimes_i}(s-\mathsf{T}_{i-1})^{\ltimes_i - 1}}{\Gamma(\ltimes_i)} \int_{\mathsf{T}_{i-1}}^{\mathsf{T}_i} (\mathsf{T}_i - t)^{\ltimes_i - 1} \\
& \Big| f\Big(t, -(\mathsf{T}_i - \mathsf{T}_{i-1})^{1-\ltimes_i}(t-\mathsf{T}_{i-1})^{\ltimes_i - 1} I^{\ltimes_i}_{\mathsf{T}^+_{i-1}} \varkappa_n(\mathsf{T}_i)) + I^{\ltimes_i}_{\mathsf{T}^+_{i-1}} \varkappa_n(t), \varkappa_n(t)\Big) \\
& - f\Big(t, -(\mathsf{T}_i - \mathsf{T}_{i-1})^{1-\ltimes_i}(t-\mathsf{T}_{i-1})^{\ltimes_i - 1} I^{\ltimes_i}_{\mathsf{T}^+_{i-1}} \varkappa(\mathsf{T}_i)) + I^{\ltimes_i}_{\mathsf{T}^+_{i-1}} \varkappa(t), \varkappa(t)\Big)\Big| dt \\
\leq\ & \frac{(\mathsf{T}_i - \mathsf{T}_{i-1})^{1-\ltimes_i}(s-\mathsf{T}_{i-1})^{\ltimes_i - 1}}{\Gamma(\ltimes_i)} \int_{\mathsf{T}_{i-1}}^{\mathsf{T}_i} (\mathsf{T}_i - t)^{\ltimes_i - 1} t^{-\delta} \\
& \Big(K(\mathsf{T}_i - \mathsf{T}_{i-1})^{1-\ltimes_i}(t-\mathsf{T}_{i-1})^{\ltimes_i - 1} I^{\ltimes_i}_{\mathsf{T}^+_{i-1}} |\varkappa_n(\mathsf{T}_i) - \varkappa(\mathsf{T}_i)| \\
& + K I^{\ltimes_i}_{\mathsf{T}^+_{i-1}} |\varkappa_n(t) - \varkappa(t)| + L|\varkappa_n(t) - \varkappa(t)|\Big) dt \\
\leq\ & \frac{1}{\Gamma(\ltimes_i)} \int_{\mathsf{T}_{i-1}}^{\mathsf{T}_i} (\mathsf{T}_i - t)^{\ltimes_i - 1} t^{-\delta} \Big(2K I^{\ltimes_i}_{\mathsf{T}^+_{i-1}} \|\varkappa_n - \varkappa\|_{E_i} + L\|\varkappa_n - \varkappa\|_{E_i}\Big) dt \\
\leq\ & \frac{(\mathsf{T}_i^{1-\delta} - \mathsf{T}_{i-1}^{1-\delta})(\mathsf{T}_i - \mathsf{T}_{i-1})^{\ltimes_i - 1}}{(1-\delta)\Gamma(\ltimes_i)} \Big(\frac{2K(\mathsf{T}_i - \mathsf{T}_{i-1})^{\ltimes_i}}{\Gamma(\ltimes_i + 1)} \|\varkappa_n - \varkappa\|_{E_i} + L\|\varkappa_n - \varkappa\|_{E_i}\Big) \\
\leq\ & \frac{(\mathsf{T}_i^{1-\delta} - \mathsf{T}_{i-1}^{1-\delta})(\mathsf{T}_i - \mathsf{T}_{i-1})^{\ltimes_i - 1}}{(1-\delta)\Gamma(\ltimes_i)} \Big(\frac{2K(\mathsf{T}_i - \mathsf{T}_{i-1})^{\ltimes_i}}{\Gamma(\ltimes_i + 1)} + L\Big) \|\varkappa_n - \varkappa\|_{E_i}.
\end{aligned}$$

$$\|(W_1\varkappa_n) - (W_1\varkappa)\|_{E_i} \to 0 \quad as \quad n \to \infty.$$

Hence, W_1 is a continuous mapping on E_i.

Step 3: $W_1(B_{R_i})$ is relatively closed and bounded.

Now, we will demonstrate that $W_1(B_{R_i})$ is relatively compact. Clearly, $W_1(B_{R_i})$ is equally bounded because, as shown in Step 1, we have $W_1(B_{R_i}) = \{W_1(y) : y \in B_{R_i}\} \subset W_1(B_{R_i}) + W_2(B_{R_i}) \subseteq B_{R_i}$. Thus, for each $y \in B_{R_i}$, we have $\|W_1(y)\|_{E_i} \leq R_i$, which implies that $W_1(B_{R_i})$ is equally bounded. It remains to be shown that $W_1(B_{R_i})$ is equicontinuous.

For $s_1, s_2 \in J_i$, $s_1 < s_2$ and $y \in B_{R_i}$, we have

$$|(W_1\varkappa)(s_2) - (W_1\varkappa)(s_1)|$$

$$= \Big| -\frac{(\mathsf{T}_i-\mathsf{T}_{i-1})^{1-\ltimes_i}(s_2-\mathsf{T}_{i-1})^{\ltimes_i-1}}{\Gamma(\ltimes_i)}\int_{\mathsf{T}_{i-1}}^{\mathsf{T}_i}(\mathsf{T}_i-t)^{\ltimes_i-1}$$
$$f\Big(t,-(\mathsf{T}_i-\mathsf{T}_{i-1})^{1-\ltimes_i}(t-\mathsf{T}_{i-1})^{\ltimes_i-1}I^{\ltimes_i}_{\mathsf{T}^+_{i-1}}\varkappa(\mathsf{T}_i))+I^{\ltimes_i}_{\mathsf{T}^+_{i-1}}\varkappa(t),\varkappa(t)\Big)dt$$
$$+\frac{(\mathsf{T}_i-\mathsf{T}_{i-1})^{1-\ltimes_i}(s_1-\mathsf{T}_{i-1})^{\ltimes_i-1}}{\Gamma(\ltimes_i)}\int_{\mathsf{T}_{i-1}}^{\mathsf{T}_i}(\mathsf{T}_i-t)^{\ltimes_i-1}$$
$$f\Big(t,-(\mathsf{T}_i-\mathsf{T}_{i-1})^{1-\ltimes_i}(t-\mathsf{T}_{i-1})^{\ltimes_i-1}I^{\ltimes_i}_{\mathsf{T}^+_{i-1}}\varkappa(\mathsf{T}_i))+I^{\ltimes_i}_{\mathsf{T}^+_{i-1}}\varkappa(t),\varkappa(t)\Big)dt\Big|$$
$$\leq \frac{(\mathsf{T}_i-\mathsf{T}_{i-1})^{1-\ltimes_i}}{\Gamma(\ltimes_i)}\Big((s_2-\mathsf{T}_{i-1})^{\ltimes_i-1}-(s_1-\mathsf{T}_{i-1})^{\ltimes_i-1}\Big)\int_{\mathsf{T}_{i-1}}^{\mathsf{T}_i}(\mathsf{T}_i-t)^{\ltimes_i-1}$$
$$\Big|f\Big(t,-(\mathsf{T}_i-\mathsf{T}_{i-1})^{1-\ltimes_i}(t-\mathsf{T}_{i-1})^{\ltimes_i-1}I^{\ltimes_i}_{\mathsf{T}^+_{i-1}}\varkappa(\mathsf{T}_i))+I^{\ltimes_i}_{\mathsf{T}^+_{i-1}}\varkappa(t),\varkappa(t)\Big)\Big|dt$$

$$\leq \frac{(\mathsf{T}_i-\mathsf{T}_{i-1})^{1-\ltimes_i}}{\Gamma(\ltimes_i)}\Big((s_2-\mathsf{T}_{i-1})^{\ltimes_i-1}-(s_1-\mathsf{T}_{i-1})^{\ltimes_i-1}\Big)\int_{\mathsf{T}_{i-1}}^{\mathsf{T}_i}(\mathsf{T}_i-t)^{\ltimes_i-1}$$
$$\Big|f\Big(t,-(\mathsf{T}_i-\mathsf{T}_{i-1})^{1-\ltimes_i}(t-\mathsf{T}_{i-1})^{\ltimes_i-1}I^{\ltimes_i}_{\mathsf{T}^+_{i-1}}\varkappa(\mathsf{T}_i))+I^{\ltimes_i}_{\mathsf{T}^+_{i-1}}\varkappa(t),\varkappa(t)\Big)-f(t,0,0)\Big|dt$$
$$+\frac{(\mathsf{T}_i-\mathsf{T}_{i-1})^{1-\ltimes_i}}{\Gamma(\ltimes_i)}\Big((s_2-\mathsf{T}_{i-1})^{\ltimes_i-1}-(s_1-\mathsf{T}_{i-1})^{\ltimes_i-1}\Big)\int_{\mathsf{T}_{i-1}}^{\mathsf{T}_i}(\mathsf{T}_i-t)^{\ltimes_i-1}|f(t,0,0)|dt$$
$$\leq \frac{(\mathsf{T}_i-\mathsf{T}_{i-1})^{1-\ltimes_i}}{\Gamma(\ltimes_i)}\Big((s_2-\mathsf{T}_{i-1})^{\ltimes_i-1}-(s_1-\mathsf{T}_{i-1})^{\ltimes_i-1}\Big)\int_{\mathsf{T}_{i-1}}^{\mathsf{T}_i}(\mathsf{T}_i-t)^{\ltimes_i-1}t^{-\delta}$$
$$\Big(K(\mathsf{T}_i-\mathsf{T}_{i-1})^{1-\ltimes_i}(t-\mathsf{T}_{i-1})^{\ltimes_i-1}|I^{\ltimes_i}_{\mathsf{T}^+_{i-1}}\varkappa(\mathsf{T}_i))+I^{\ltimes_i}_{\mathsf{T}^+_{i-1}}\varkappa(t)|+L|\varkappa(t)|\Big)dt$$
$$+\frac{(\mathsf{T}_i-\mathsf{T}_{i-1})^{1-\ltimes_i}f^*}{\Gamma(\ltimes_i)}\Big((s_2-\mathsf{T}_{i-1})^{\ltimes_i-1}-(s_1-\mathsf{T}_{i-1})^{\ltimes_i-1}\Big)\int_{\mathsf{T}_{i-1}}^{\mathsf{T}_i}(\mathsf{T}_i-t)^{\ltimes_i-1}t^{-\delta}dt$$
$$\leq \frac{(\mathsf{T}_i-\mathsf{T}_{i-1})^{1-\ltimes_i}}{\Gamma(\ltimes_i)}\Big((s_2-\mathsf{T}_{i-1})^{\ltimes_i-1}-(s_1-\mathsf{T}_{i-1})^{\ltimes_i-1}\Big)$$
$$\int_{\mathsf{T}_{i-1}}^{\mathsf{T}_i}(\mathsf{T}_i-t)^{\ltimes_i-1}t^{-\delta}\Big(2KI^{\ltimes_i}_{\mathsf{T}^+_{i-1}}\|\varkappa\|_{E_i}+L\|\varkappa\|_{E_i}\Big)dt$$
$$+\frac{(\mathsf{T}_i-\mathsf{T}_{i-1})^{1-\ltimes_i}f^*}{\Gamma(\ltimes_i)}\Big((s_2-\mathsf{T}_{i-1})^{\ltimes_i-1}-(s_1-\mathsf{T}_{i-1})^{\ltimes_i-1}\Big)\int_{\mathsf{T}_{i-1}}^{\mathsf{T}_i}(\mathsf{T}_i-t)^{\ltimes_i-1}t^{-\delta}dt$$
$$\leq \frac{({\mathsf{T}_i}^{1-\delta}-{\mathsf{T}_{i-1}}^{1-\delta})}{(1-\delta)\Gamma(\ltimes_i)}\Big((s_2-\mathsf{T}_{i-1})^{\ltimes_i-1}-(s_1-\mathsf{T}_{i-1})^{\ltimes_i-1}\Big)\Big(2KI^{\ltimes_i}_{\mathsf{T}^+_{i-1}}\|\varkappa\|_{E_i}+L\|\varkappa\|_{E_i}\Big)$$
$$+\frac{({\mathsf{T}_i}^{1-\delta}-{\mathsf{T}_{i-1}}^{1-\delta})f^*}{(1-\delta)\Gamma(\ltimes_i)}\Big((s_2-\mathsf{T}_{i-1})^{\ltimes_i-1}-(s_1-\mathsf{T}_{i-1})^{\ltimes_i-1}\Big)$$
$$\leq \frac{({\mathsf{T}_i}^{1-\delta}-{\mathsf{T}_{i-1}}^{1-\delta})}{(1-\delta)\Gamma(\ltimes_i)}\Big(2KI^{\ltimes_i}_{\mathsf{T}^+_{i-1}}\|\varkappa\|_{E_i}+L\|\varkappa\|_{E_i}+f^*\Big)\Big((s_2-\mathsf{T}_{i-1})^{\ltimes_i-1}-(s_1-\mathsf{T}_{i-1})^{\ltimes_i-1}\Big).$$

Hence $|(W_1\varkappa)(s_2)-(W_1\varkappa)(s_1)|\to 0$ as $|s_2-s_1|\to 0$. This implies (W_1R_i) is equicontinuous.

Step 4: W_2 is a strict contraction on B_{R_i}.

For $\varkappa_1(s),\varkappa_2(s)\in B_{R_i}$, we obtain that

$$|(W_2\varkappa_2)(s)-(W_2\varkappa_1)(s)|$$

$$\leq \frac{1}{\Gamma(\ltimes_i)}\int_{\mathsf{T}_{i-1}}^{s}(s-t)^{\ltimes_i-1}$$
$$\Big|f\Big(t,-(\mathsf{T}_i-\mathsf{T}_{i-1})^{1-\ltimes_i}(t-\mathsf{T}_{i-1})^{\ltimes_i-1}I^{\ltimes_i}_{\mathsf{T}^+_{i-1}}\varkappa_2(\mathsf{T}_i))+I^{\ltimes_i}_{\mathsf{T}^+_{i-1}}\varkappa_2(t),\varkappa_2(t)\Big)$$
$$-f\Big(t,-(\mathsf{T}_i-\mathsf{T}_{i-1})^{1-\ltimes_i}(t-\mathsf{T}_{i-1})^{\ltimes_i-1}I^{\ltimes_i}_{\mathsf{T}^+_{i-1}}\varkappa_1(\mathsf{T}_i))+I^{\ltimes_i}_{\mathsf{T}^+_{i-1}}\varkappa_1(t),\varkappa_1(t)\Big)\Big|dt$$

$$\leq \frac{1}{\Gamma(\ltimes_i)}\int_{\mathsf{T}_{i-1}}^{s}(s-t)^{\ltimes_i-1}t^{-\delta}\Big(K(\mathsf{T}_i-\mathsf{T}_{i-1})^{1-\ltimes_i}(t-\mathsf{T}_{i-1})^{\ltimes_i-1}I^{\ltimes_i}_{\mathsf{T}^+_{i-1}}|\varkappa_2(\mathsf{T}_i)-\varkappa_1(\mathsf{T}_i)|$$
$$+KI^{\ltimes_i}_{\mathsf{T}^+_{i-1}}|\varkappa_2(t)-\varkappa_1(t)|+L|\varkappa_2(t)-\varkappa_1(t)|\Big)dt$$
$$\leq \frac{1}{\Gamma(\ltimes_i)}\int_{\mathsf{T}_{i-1}}^{s}(s-t)^{\ltimes_i-1}t^{-\delta}\Big(2KI^{\ltimes_i}_{\mathsf{T}^+_{i-1}}\|\varkappa_2-\varkappa_1\|_{E_i}+L\|\varkappa_2-\varkappa_1\|_{E_i}\Big)dt$$
$$\leq \frac{(\mathsf{T}_i^{1-\delta}-\mathsf{T}_{i-1}^{1-\delta})(\mathsf{T}_i-\mathsf{T}_{i-1})^{\ltimes_i-1}}{(1-\delta)\Gamma(\ltimes_i)}\Big(\frac{2K(\mathsf{T}_i-\mathsf{T}_{i-1})^{\ltimes_i}}{\Gamma(\ltimes_i+1)}\|\varkappa_2-\varkappa_1\|_{E_i}+L\|\varkappa_2-\varkappa_1\|_{E_i}\Big)$$
$$\leq \frac{(\mathsf{T}_i^{1-\delta}-\mathsf{T}_{i-1}^{1-\delta})(\mathsf{T}_i-\mathsf{T}_{i-1})^{\ltimes_i-1}}{(1-\delta)\Gamma(\ltimes_i)}\Big(\frac{2K(\mathsf{T}_i-\mathsf{T}_{i-1})^{\ltimes_i}}{\Gamma(\ltimes_i+1)}+L\Big)\|\varkappa_2-\varkappa_1\|_{E_i}.$$

Consequently by inequality (3.18), the operator W_2 is a strictly contractive mapping.Therefore, by Krasnoselskii's fixed point theorem, there exists a function $\widetilde{y_i}\in B_{R_i}$, such that $W_1(\widetilde{y_i})+W_2(\widetilde{y_i})=\widetilde{y_i}$, which constitutes a solution to the initial and terminal value problem (3.16).

We now proceed to demonstrate the existence result for the ITVP (3.13).

Let us present the following premise:

(H3) Let $s^\delta f: J\times\mathbb{R}\times\mathbb{R}\to\mathbb{R}$ be a smooth function, where $0\leq\delta\leq 1$. Suppose there exist fixed constants $K,L>0$ such that

$$s^\delta|f(s,y_1,z_1)-f(s,y_2,z_2)|\leq K|y_1-y_2|+L|z_1-z_2|,$$

for all $y_1,y_2,z_1,z_2\in\mathbb{R}$ and $s\in J$.

Theorem 14 *Suppose that conditions (H1), (H3), and inequality (3.18) are satisfied for all $i\in\{1,2,...,n\}$. Then the initial and terminal value problem (ITVP) (3.13) admits at least one solution in $C(J,\mathbb{R})$.*

Proof For any $i \in \{1, 2, ..., n\}$, by Theorem 13, the initial and terminal value problem (ITVP) (3.16) possesses at least one solution $\widetilde{y}_i \in E_i$.

Define the function y_i for each $i \in \{1, 2, ..., n\}$ as follows:

$$y_i(s) = \begin{cases} 0, & s \in [0, \mathsf{T}_{i-1}], \\ \widetilde{y}_i, & s \in J_i. \end{cases}$$

so that $y_i \in C([0, \mathsf{T}_i], \mathbb{R})$ satisfies the integral equation (3.15) for $s \in J_i$, with boundary conditions $y_i(0) = 0$ and $y_i(\mathsf{T}_i) = \widetilde{y}_i(\mathsf{T}_i) = 0$.
Now define the function $y(s)$ by

$$y(s) = \begin{cases} y_1(s), & s \in J_1, \\ y_2(s) = \begin{cases} 0, & s \in J_1, \\ \widetilde{y}_2, & s \in J_2, \end{cases} \\ \cdot \\ \cdot \\ \cdot \\ \cdot \\ y_n(s) = \begin{cases} 0, & s \in [0, \mathsf{T}_{i-1}], \\ \widetilde{y}_i, & s \in J_i, \end{cases} \end{cases}$$

which constitutes a solution of the ITVP (3.13).

3.2.2 Study of the Solutions' Behavior

Theorem 15 *Assume that conditions (H1), (H3), inequality (3.18), and*

(H4) *hold. Let the function $\varphi \in C(J_i, \mathbb{R}_+)$ be non-decreasing, and suppose there exists a constant $\lambda_\varphi > 0$ such that, for each $s \in J_i$, we have*

$$I^{\ltimes_i}_{\mathsf{T}_{i-1}^+} \varphi(s) \leq \lambda_{\varphi(s)} \varphi(s).$$

Then the ITVP (3.13) *is Ulam-Hyers-$\mathfrak{R}$assias stable with respect to φ.*

Proof Let $\epsilon > 0$ be an arbitrary positive number, and let the function $z(s)$ from $z \in C(J, \mathbb{R})$ satisfy inequality (2.21).

For each $i \in \{1, 2, ..., n\}$, define the functions: $z_1(s) \equiv z(s)$, for $s \in [0, \mathsf{T}_1]$ and for $i = 2, 3, ..., n$,

$$z_i(s) = \begin{cases} 0, & s \in [0, \mathsf{T}_{i-1}], \\ z(s), & s \in J_i, \end{cases}$$

For each $i \in \{1, 2, ..., n\}$, based on equation (3.14) for $s \in J_i$, we get

$$F^{\ltimes(s)}_{\mathsf{T}_{i-1}+} z_i(s) = \frac{d^2}{ds^2} \int_{\mathsf{T}_{i-1}}^{s} \frac{(s-t)^{1-\ltimes_i}}{\Gamma(2-\ltimes_i)} z(t)dt.$$

By integrating inequality (2.21) and applying assumption (H4), we obtain:

$$\Big| z(s) + \frac{(\mathsf{T}_i - \mathsf{T}_{i-1})^{1-\ltimes_i}(s-\mathsf{T}_{i-1})^{\ltimes_i - 1}}{\Gamma(\ltimes_i)}$$
$$\int_{\mathsf{T}_{i-1}}^{\mathsf{T}_i} (\mathsf{T}_i - t)^{\ltimes_i - 1} f\Big(t, -(\mathsf{T}_i - \mathsf{T}_{i-1})^{1-\ltimes_i}(t-\mathsf{T}_{i-1})^{\ltimes_i - 1} I^{\ltimes_i}_{\mathsf{T}^+_{i-1}} z(\mathsf{T}_i)) + I^{\ltimes_i}_{\mathsf{T}^+_{i-1}} z(t), z(t)\Big) dt$$
$$- \frac{1}{\Gamma(\ltimes_i)} \int_{\mathsf{T}_{i-1}}^{s} (s-t)^{\ltimes_i - 1}$$
$$f\Big(t, -(\mathsf{T}_i - \mathsf{T}_{i-1})^{1-\ltimes_i}(t-\mathsf{T}_{i-1})^{\ltimes_i - 1} I^{\ltimes_i}_{\mathsf{T}^+_{i-1}} z(\mathsf{T}_i)) + I^{\ltimes_i}_{\mathsf{T}^+_{i-1}} z(t), z(t)\Big) dt \Big|$$
$$\le \epsilon \int_{\mathsf{T}_{i-1}}^{s} \frac{(s-t)^{u(i)-1}}{\Gamma(u(i))} \varphi(t) dt$$
$$\le \lambda_{\varphi(s)} \epsilon \varphi(s).$$

Based on Theorem 14, the initial-terminal value problem (ITVP) (3.13) admits a solution $y \in C(J, \mathbb{R})$ given by $y(s) = y_i(s)$ for $s \in J_i$, $i = 1, 2, ..., n$, where

$$y_i = \begin{cases} 0, & s \in [0, \mathsf{T}_{i-1}], \\ \widetilde{y}_i, & s \in J_i, \end{cases} \tag{3.20}$$

and $\widetilde{y}_i \in E_i$ is a solution of ITVP (3.16).

By using Lemma 13, we obtain:

$$\widetilde{y}_i(s) = -\frac{(\mathsf{T}_i - \mathsf{T}_{i-1})^{1-\ltimes_i}(s-\mathsf{T}_{i-1})^{\ltimes_i - 1}}{\Gamma(\ltimes_i)}$$
$$\int_{\mathsf{T}_{i-1}}^{\mathsf{T}_i} (\mathsf{T}_i - t)^{\ltimes_i - 1} f\Big(t, -(\mathsf{T}_i - \mathsf{T}_{i-1})^{1-\ltimes_i}(t-\mathsf{T}_{i-1})^{\ltimes_i - 1} I^{\ltimes_i}_{\mathsf{T}^+_{i-1}} \widetilde{y}_i(\mathsf{T}_i)) + I^{\ltimes_i}_{\mathsf{T}^+_{i-1}} \widetilde{y}_i(t), \widetilde{y}_i(t)\Big) dt$$
$$+ \frac{1}{\Gamma(\ltimes_i)} \int_{\mathsf{T}_{i-1}}^{s} (s-t)^{\ltimes_i - 1}$$
$$f\Big(t, -(\mathsf{T}_i - \mathsf{T}_{i-1})^{1-\ltimes_i}(t-\mathsf{T}_{i-1})^{\ltimes_i - 1} I^{\ltimes_i}_{\mathsf{T}^+_{i-1}} \widetilde{y}_i(\mathsf{T}_i)) + I^{\ltimes_i}_{\mathsf{T}^+_{i-1}} \widetilde{y}_i(t), \widetilde{y}_i(t)\Big) dt.$$

On the other hand, we have, for each $s \in J_i$

$$
\begin{aligned}
&|z(s) - y(s)| = |z(s) - y_i(s)| = |z_i(s) - \widetilde{y}_i(s)| \\
= \ & \Big|(z_i)(s) + \frac{(\mathsf{T}_i - \mathsf{T}_{i-1})^{1-\ltimes_i}(s - \mathsf{T}_{i-1})^{\ltimes_i - 1}}{\Gamma(\ltimes_i)} \int_{\mathsf{T}_{i-1}}^{\mathsf{T}_i} (\mathsf{T}_i - t)^{\ltimes_i - 1} \\
& f\Big(t, -(\mathsf{T}_i - \mathsf{T}_{i-1})^{1-\ltimes_i}(t - \mathsf{T}_{i-1})^{\ltimes_i - 1} I^{\ltimes_i}_{\mathsf{T}^+_{i-1}} \widetilde{y}_i(\mathsf{T}_i)) + I^{\ltimes_i}_{\mathsf{T}^+_{i-1}} \widetilde{y}_i(t), \widetilde{y}_i(t)\Big) dt \\
& - \frac{1}{\Gamma(\ltimes_i)} \int_{\mathsf{T}_{i-1}}^{s} (s - t)^{\ltimes_i - 1} f\Big(t, -(\mathsf{T}_i - \mathsf{T}_{i-1})^{1-\ltimes_i}(t - \mathsf{T}_{i-1})^{\ltimes_i - 1} I^{\ltimes_i}_{\mathsf{T}^+_{i-1}} \widetilde{y}_i(\mathsf{T}_i)) + I^{\ltimes_i}_{\mathsf{T}^+_{i-1}} \widetilde{y}_i(t), \widetilde{y}_i(t)\Big) dt\Big| \\
\leq \ & \Big|(z_i)(s) + \frac{(\mathsf{T}_i - \mathsf{T}_{i-1})^{1-\ltimes_i}(s - \mathsf{T}_{i-1})^{\ltimes_i - 1}}{\Gamma(\ltimes_i)} \int_{\mathsf{T}_{i-1}}^{\mathsf{T}_i} (\mathsf{T}_i - t)^{\ltimes_i - 1} \\
& f\Big(t, -(\mathsf{T}_i - \mathsf{T}_{i-1})^{1-\ltimes_i}(t - \mathsf{T}_{i-1})^{\ltimes_i - 1} I^{\ltimes_i}_{\mathsf{T}^+_{i-1}} z_i(\mathsf{T}_i)) + I^{\ltimes_i}_{\mathsf{T}^+_{i-1}} z_i(t), z_i(t)\Big) dt \\
& - \frac{1}{\Gamma(\ltimes_i)} \int_{\mathsf{T}_{i-1}}^{s} (s - t)^{\ltimes_i - 1} f\Big(t, -(\mathsf{T}_i - \mathsf{T}_{i-1})^{1-\ltimes_i}(t - \mathsf{T}_{i-1})^{\ltimes_i - 1} I^{\ltimes_i}_{\mathsf{T}^+_{i-1}} z_i(\mathsf{T}_i)) + I^{\ltimes_i}_{\mathsf{T}^+_{i-1}} z_i(t), z_i(t)\Big) dt\Big| \\
& + \frac{(\mathsf{T}_i - \mathsf{T}_{i-1})^{1-\ltimes_i}(s - \mathsf{T}_{i-1})^{\ltimes_i - 1}}{\Gamma(\ltimes_i) \int_{\mathsf{T}_{i-1}}^{\mathsf{T}_i} (\mathsf{T}_i - t)^{\ltimes_i - 1}} \\
& \Big|f\Big(t, -(\mathsf{T}_i - \mathsf{T}_{i-1})^{1-\ltimes_i}(t - \mathsf{T}_{i-1})^{\ltimes_i - 1} I^{\ltimes_i}_{\mathsf{T}^+_{i-1}} z_i(\mathsf{T}_i) + I^{\ltimes_i}_{\mathsf{T}^+_{i-1}} z_i(t), z_i(t)\Big) dt \\
& - f\Big(t, -(\mathsf{T}_i - \mathsf{T}_{i-1})^{1-\ltimes_i}(t - \mathsf{T}_{i-1})^{\ltimes_i - 1} I^{\ltimes_i}_{\mathsf{T}^+_{i-1}} \widetilde{y}_i(\mathsf{T}_i) + I^{\ltimes_i}_{\mathsf{T}^+_{i-1}} \widetilde{y}_i(t), \widetilde{y}_i(t)\Big) dt\Big| \\
& + \frac{1}{\Gamma(\ltimes_i)} \int_{\mathsf{T}_{i-1}}^{s} (s - t)^{\ltimes_i - 1} \Big|f\Big(t, -(\mathsf{T}_i - \mathsf{T}_{i-1})^{1-\ltimes_i}(t - \mathsf{T}_{i-1})^{\ltimes_i - 1} I^{\ltimes_i}_{\mathsf{T}^+_{i-1}} z_i(\mathsf{T}_i) + I^{\ltimes_i}_{\mathsf{T}^+_{i-1}} z_i(t), z_i(t)\Big) dt \\
& - f\Big(t, -(\mathsf{T}_i - \mathsf{T}_{i-1})^{1-\ltimes_i}(t - \mathsf{T}_{i-1})^{\ltimes_i - 1} I^{\ltimes_i}_{\mathsf{T}^+_{i-1}} \widetilde{y}_i(\mathsf{T}_i) + I^{\ltimes_i}_{\mathsf{T}^+_{i-1}} \widetilde{y}_i(t), \widetilde{y}_i(t)\Big) dt\Big| \\
\leq \ & \lambda_{\varphi(s)} \epsilon \varphi(s) + \frac{(\mathsf{T}_i - \mathsf{T}_{i-1})^{1-\ltimes_i}(s - \mathsf{T}_{i-1})^{\ltimes_i - 1}}{\Gamma(\ltimes_i)} \\
& \int_{\mathsf{T}_{i-1}}^{\mathsf{T}_i} (\mathsf{T}_i - t)^{\ltimes_i - 1} t^{-\delta} \Big(K\Big[(\mathsf{T}_i - \mathsf{T}_{i-1})^{1-\ltimes_i}(t - \mathsf{T}_{i-1})^{\ltimes_i - 1} I^{\ltimes_i}_{\mathsf{T}^+_{i-1}} |(z_i(\mathsf{T}_i) - \widetilde{y}_i(\mathsf{T}_i)| \\
& + I^{\ltimes_i}_{\mathsf{T}^+_{i-1}} |z_i(t) - \widetilde{y}_i(t)|\Big] + L|z_i(t) - \widetilde{y}_i(t)|\Big) dt \\
& + \frac{1}{\Gamma(\ltimes_i)} \int_{\mathsf{T}_{i-1}}^{s} (s - t)^{\ltimes_i - 1} t^{-\delta} \Big(K(\mathsf{T}_i - \mathsf{T}_{i-1})^{1-\ltimes_i}(t - \mathsf{T}_{i-1})^{\ltimes_i - 1} I^{\ltimes_i}_{\mathsf{T}^+_{i-1}} |z_i(\mathsf{T}_i) - \widetilde{y}_i(\mathsf{T}_i)| \\
& + I^{\ltimes_i}_{\mathsf{T}^+_{i-1}} |z_i(t) - \widetilde{y}_i(t)| + L|z_i(t) - \widetilde{y}_i(t)|\Big) dt
\end{aligned}
$$

$$
\begin{aligned}
&\leq \lambda_{\varphi(s)}\epsilon\varphi(s) + \frac{1}{\Gamma(\ltimes_i)}\int_{\mathsf{T}_{i-1}}^{\mathsf{T}_i}(\mathsf{T}_i - t)^{\ltimes_i - 1}t^{-\delta}\Big(KI_{\mathsf{T}_{i-1}^+}^{\ltimes_i}|z_i(\mathsf{T}_i) - \widetilde{y}_i(\mathsf{T}_i)| + KI_{\mathsf{T}_{i-1}^+}^{\ltimes_i}|z_i(t) - \widetilde{y}_i(t) \\
&\quad + L|z_i(t) - \widetilde{y}_i(t)|\Big)dt + \frac{1}{\Gamma(\ltimes_i)}\int_{\mathsf{T}_{i-1}}^{s}(s - t)^{\ltimes_i - 1}t^{-\delta}\Big(KI_{\mathsf{T}_{i-1}^+}^{\ltimes_i}|z_i(\mathsf{T}_i) - \widetilde{y}_i(\mathsf{T}_i)| \\
&\quad + KI_{\mathsf{T}_{i-1}^+}^{\ltimes_i}|z_i(t) - \widetilde{y}_i(t)| + L|z_i(t) - \widetilde{y}_i(t)|\Big)dt \\
&\leq \lambda_{\varphi(s)}\epsilon\varphi(s) + \frac{(\mathsf{T}_i - \mathsf{T}_{i-1})^{\ltimes_i - 1}}{\Gamma(\ltimes_i)}\Big(2KI_{\mathsf{T}_{i-1}^+}^{\ltimes_i}\|z_i - \widetilde{y}_i\|_{E_i} + L\|z_i - \widetilde{y}_i\|_{E_i}\Big)\int_{\mathsf{T}_{i-1}}^{\mathsf{T}_i} t^{-\delta}dt \\
&\quad + \frac{(s - \mathsf{T}_{i-1})^{\ltimes_i - 1}}{\Gamma(\ltimes_i)}\Big(2KI_{\mathsf{T}_{i-1}^+}^{\ltimes_i}\|z_i - \widetilde{y}_i\|_{E_i} + L\|z_i - \widetilde{y}_i\|_{E_i}\Big)\int_{\mathsf{T}_{i-1}}^{s} t^{-\delta}dt \\
&\leq \lambda_{\varphi(s)}\epsilon\varphi(s) + \frac{({\mathsf{T}_i}^{1-\delta} - {\mathsf{T}_{i-1}}^{1-\delta})(\mathsf{T}_i - \mathsf{T}_{i-1})^{\ltimes_i - 1}}{(1-\delta)\Gamma(\ltimes_i)} \\
&\quad \Big(2K\frac{(\mathsf{T}_i - \mathsf{T}_{i-1})^{\ltimes_i}}{\Gamma(\ltimes_i + 1)}\|z_i - \widetilde{y}_i\|_{E_i} + L\|z_i - \widetilde{y}_i\|_{E_i}\Big) \\
&\quad + \frac{(s - \mathsf{T}_{i-1})^{\ltimes_i - 1}}{\Gamma(\ltimes_i)}\Big(2K\frac{(s^{1-\delta} - {\mathsf{T}_{i-1}}^{1-\delta})(\mathsf{T}_i - \mathsf{T}_{i-1})^{\ltimes_i}}{(1-\delta)\Gamma(\ltimes_i + 1)}\|z_i - \widetilde{y}_i\|_{E_i} + L\|z_i - \widetilde{y}_i\|_{E_i}\Big) \\
&\leq \lambda_{\varphi(s)}\epsilon\varphi(s) + \frac{2({\mathsf{T}_i}^{1-\delta} - {\mathsf{T}_{i-1}}^{1-\delta})(\mathsf{T}_i - \mathsf{T}_{i-1})^{\ltimes_i - 1}}{(1-\delta)\Gamma(\ltimes_i)}\Big(2K\frac{(\mathsf{T}_i - \mathsf{T}_{i-1})^{\ltimes_i}}{\Gamma(\ltimes_i + 1)} + L\Big)\|z_i - \widetilde{y}_i\|_{E_i} \\
&\leq \lambda_{\varphi(s)}\epsilon\varphi(s) + \mu\|z - y\|,
\end{aligned}
$$

where

$$
\mu = \max_{i=1,2,\ldots,n}\frac{2({\mathsf{T}_i}^{1-\delta} - {\mathsf{T}_{i-1}}^{1-\delta})(\mathsf{T}_i - \mathsf{T}_{i-1})^{\ltimes_i - 1}}{(1-\delta)\Gamma(\ltimes_i)}\Big(2K\frac{(\mathsf{T}_i - \mathsf{T}_{i-1})^{\ltimes_i}}{\Gamma(\ltimes_i + 1)} + L\Big).
$$

Then,

$$
\|z - y\|(1 - \mu) \leq \lambda_{\varphi(s)}\epsilon\varphi(s).
$$

We get, for each $s \in J_i$

$$
\|z - y\| \leq \frac{\lambda_{\varphi(s)}}{(1 - \mu)}\epsilon\varphi(s) := c_f\epsilon\varphi(s).
$$

Then, the ITVP (3.13) is Ulam-Hyers-$\Re$assias stable with respect to φ.

3.2.3 Numerical Example

Let us examine the following ITVP,

$$\begin{cases} \mathit{F}_{0+}^{\ltimes(s)} y(s) = \frac{s^{\frac{1}{4}}}{\sqrt{2}}\left(\frac{1}{10}|y^{\frac{1}{2}}(s)| + \frac{2}{15}|\mathit{F}_{0+}^{\ltimes(s)} y(s)| + \frac{1}{3}\right), & s \in J := [0,2]. \\ y(0) = 0, \quad y(2) = 0. \end{cases} \tag{3.21}$$

Let

$$f(s,y,z) = \frac{s^{\frac{1}{4}}}{\sqrt{2}}\left(\frac{1}{10}y^{\frac{1}{2}} + \frac{2}{15}z + \frac{1}{3}\right), \ (s,y,z) \in [0,2] \times [1,+\infty) \times [1,+\infty).$$

$$\ltimes(s) = \begin{cases} \frac{8}{5}, & s \in J_1 := [0,1], \\ \frac{9}{5}, & s \in J_2 :=]1,2]. \end{cases} \tag{3.22}$$

Consequently, we get

$$\begin{aligned} s^{\frac{1}{4}}|f(s,y_1,z_1) - f(s,y_2,z_2)| &= \left|\frac{\sqrt{s}}{10\sqrt{2}}(y_1^{\frac{1}{2}} - y_2^{\frac{1}{2}}) + \frac{2\sqrt{s}}{15\sqrt{2}}(z_1 - z_2)\right| \\ &\leq \frac{1}{10}|y_1 - y_2| + \frac{2}{15}|z_1 - z_2|. \end{aligned}$$

Therefore, condition $(H3)$ applies using $\delta = \frac{1}{4}$, $K = \frac{1}{10}$, and $L = \frac{2}{15}$.

Based on (3.22), the expression for the ITVP (3.21) can be written in two separate forms, as detailed below

$$\begin{cases} \mathit{F}_{0+}^{\frac{8}{5}} y(s) = \frac{s^{\frac{1}{4}}}{\sqrt{2}}\left(\frac{1}{10}|y^{\frac{1}{2}}(s)| + \frac{2}{15}|\mathit{F}_{0+}^{\frac{8}{5}} y(s)| + \frac{1}{3}\right), & s \in J_1, \\ \mathit{F}_{1+}^{\frac{9}{5}} y(s) = \frac{s^{\frac{1}{4}}}{\sqrt{2}}\left(\frac{1}{10}|y^{\frac{1}{2}}(s)| + \frac{2}{15}|\mathit{F}_{1+}^{\frac{9}{5}} y(s)| + \frac{1}{3}\right), & s \in J_2. \end{cases}$$

For $s \in J_1$, the ITVP (3.21) is equivalent to the following ITVP

$$\begin{cases} \mathit{F}_{0+}^{\frac{8}{5}} y(s) = \frac{s^{\frac{1}{4}}}{\sqrt{2}}\left(\frac{1}{10}|y^{\frac{1}{2}}(s)| + \frac{2}{15}|\mathit{F}_{0+}^{\frac{8}{5}} y(s)| + \frac{1}{3}\right), & s \in J_1, \\ y(0) = 0, \quad y(1) = 0. \end{cases} \tag{3.23}$$

We proceed to verify that condition (3.18) holds.

$$\frac{(\mathsf{T}_1^{1-\delta} - \mathsf{T}_0^{1-\delta})(\mathsf{T}_1 - \mathsf{T}_0)^{\ltimes_1 - 1}}{(1-\delta)\Gamma(\ltimes_1)}\left(\frac{2K(\mathsf{T}_1 - \mathsf{T}_0)^{\ltimes_1}}{\Gamma(\ltimes_1 + 1)} + L\right) = \frac{1}{\frac{3\Gamma(\frac{8}{5})}{4}}\left(\frac{\frac{1}{5}}{\Gamma(\frac{13}{5})} + \frac{2}{15}\right) \simeq 0.4076 < 1$$

Let $\varphi(s) = s^{\frac{1}{2}}$. Then

$$\begin{aligned} I_{0+}^{\ltimes_1}\varphi(s) &= \frac{1}{\Gamma(\frac{8}{5})}\int_0^s (s-t)^{\frac{3}{5}} t^{\frac{1}{2}}\,dt \\ &\leq \frac{1}{\Gamma(\frac{8}{5})}\int_0^s (s-t)^{\frac{3}{5}}\,dt \\ &\leq \frac{5}{8\Gamma(\frac{8}{5})}\varphi(s) := \lambda_{\varphi(s)}\varphi(s). \end{aligned}$$

Thus condition (H4) is satisfied with $\varphi(s) = s^{\frac{1}{2}}$ and $\lambda_{\varphi(s)} = 0.6994$.

By Theorem 13, the ITVP (3.23) has a solution $\widetilde{y}_1 \in E_1$.

For $s \in J_2$, the ITVP (3.21) can be rewritten as follows:

$$\begin{cases} F_{1+}^{\frac{9}{5}} y(s) = \frac{s^{\frac{1}{4}}}{\sqrt{2}}\left(\frac{1}{10}|y^{\frac{1}{2}}(s)| + \frac{2}{15}|F_{1+}^{\frac{9}{5}} y(s)| + \frac{1}{3}\right), & s \in J_2, \\ y(1) = 0, \quad y(2) = 0. \end{cases} \tag{3.24}$$

We see that

$$\frac{(\mathsf{T}_2{}^{1-\delta} - \mathsf{T}_1^{1-\delta})(\mathsf{T}_2 - \mathsf{T}_1)^{\ltimes_2 - 1}}{(1-\delta)\Gamma(\ltimes_2)}\Big(\frac{2K(\mathsf{T}_2 - \mathsf{T}_1)^{\ltimes_2}}{\Gamma(\ltimes_2 + 1)} + L\Big) = \frac{2^{\frac{3}{4}} - 1}{\frac{3\Gamma(\frac{9}{5})}{4}}\Big(\frac{\frac{1}{5}}{\Gamma(2.8)} + \frac{2}{15}\Big) \simeq 0.2465 < 1.$$

Thus, the condition (3.18) is satisfied.

And

$$\begin{aligned} I_{1+}^{\ltimes_2}\varphi(s) &= \frac{1}{\Gamma(\frac{9}{5})}\int_1^s (s-t)^{\frac{4}{5}} t^{\frac{1}{2}}\,dt \\ &\leq \frac{1}{\Gamma(\frac{9}{5})}\int_1^s (s-t)^{\frac{4}{5}}\,dt \\ &\leq \frac{5}{9\Gamma(\frac{9}{5})}\varphi(s) := \lambda_{\varphi(s)}\varphi(s). \end{aligned}$$

Thus, condition (H4) is satisfied with $\varphi(s) = s^{\frac{1}{2}}$ and $\lambda_{\varphi(s)} = 1.1244$.

By Theorem 13, the ITVP (3.24) has a solution $\widetilde{y}_2 \in E_2$.

Afterward, by Theorem 14, the ITVP (3.21) has an answer

$$y(s) = \begin{cases} \widetilde{y}_1(s), & s \in J_1, \\ y_2(s), & s \in J_2, \end{cases}$$

where

$$y_2(s) = \begin{cases} 0, & s \in J_1, \\ \widetilde{y}_2(s), & s \in J_2. \end{cases}$$

By Theorem 15, the ITVP (3.21) is Ulam-Hyers-$\mathfrak{R}$assias stable with respect to φ.

3.3 Existence, Uniqueness, and Stability of Solutions to Variable Fractional Order Initial and Terminal Value Problems

We aim to study the following fractional initial and terminal value problem (ITVP) of variable order in the Riemann–Liouville sense, incorporating both variable order initial and terminal conditions:

$$\begin{cases} \digamma^{\ltimes(s)}y(s) = g(s, y(s)), & s \in J := [0, \mathsf{T}], \\ I^{2-\ltimes(s)}y(0) = aI^{2-\ltimes(s)}y(a), & \digamma^{\ltimes(s)-1}y(0) = b\digamma^{\ltimes(s)-1}y(a) \end{cases} \tag{3.25}$$

where $0 < \mathsf{T} < \infty$, $\ltimes(s) : J \to (1, 2]$ is the variable order of the fractional derivatives, $g : J \times \mathbb{R} \to \mathbb{R}$ is a given function and $\digamma^{\ltimes(s)}$ denotes the Riemann–Liouville fractional derivative of order $\ltimes(s)$. The constants a and b are real numbers, with $a = 1$ and $b = 1$.

Our primary focus will be the existence and singularity solutions to the initial and terminal value problem (3.25). Additionally, we will investigate the robustness of the solution in the sense of Ulam-Hyers-$\mathfrak{R}$assias stability.

3.3.1 Results of Existence

Hypothesis (H1) Let $n \in \mathbb{N}$ be a positive integer, and let $\mathcal{P} = \{J_1 := [0, \mathsf{T}_1], J_2 := (\mathsf{T}_1, \mathsf{T}_2], J_3 := (\mathsf{T}_2, \mathsf{T}_3], \ldots, J_n := (\mathsf{T}_{n-1}, \mathsf{T}]\}$ denote a partition of the interval J. Additionally, let $\ltimes(s) : J \to (1, 2]$ be a function that is piecewise constant with respect to $\mathcal{P}$, i.e.,

$$\ltimes(s) = \sum_{i=1}^{n} \ltimes_i I_i(s) = \begin{cases} \ltimes_1, & if\ s \in J_1, \\ \ltimes_2, & if\ s \in J_2, \\ \cdot \\ \cdot \\ \cdot \\ \ltimes_n, & if\ s \in J_n, \end{cases}$$

where $1 < \ltimes_i \leq 2$ are constants, and $I_i(s)$ denotes the indicator function for the interval $J_i := (\mathsf{T}_{i-1}, \mathsf{T}_i]$, with $\mathsf{T}_0 = 0$ and $\mathsf{T}_n = \mathsf{T}$, such that

$$I_i(s) = \begin{cases} 1, & for\ s \in J_i, \\ 0, & for\ elsewhere. \end{cases}$$

For each i in $\{1, 2, \ldots, n\}$, $E_i = C(J_i, X)$ denote the $\mathfrak{B}$anach space of continuous functions $\varkappa : J_i \to X$, equipped with the norm

$$\|\varkappa\|_{E_i} = \sup_{s \in J_i} \|\varkappa(s)\|.$$

For any $s \in J_i$, $i = 1, \ldots, n$, the Riemann–Liouville fractional variable-order derivative $\ltimes(s)$ of the function $y(s) \in C(J, \mathbb{R})$ can be expressed as the sum of Riemann–Liouville fractional derivatives of constant orders $\ltimes_k,\ k = 1, \ldots, i$

$$\digamma_{0+}^{\ltimes(s)} y(s) = \frac{1}{\Gamma(2-\ltimes(s))} \frac{d^2}{ds^2} \int_0^s (s-t)^{1-\ltimes(s)} y(t)dt$$

$$= \frac{1}{\Gamma(2-\ltimes(s))} \left(\sum_{k=1}^{i-1} \frac{d^2}{ds^2} \int_{\mathsf{T}_{k-1}}^{\mathsf{T}_k} (s-t)^{1-\ltimes_k} y(t)dt + \frac{d^2}{ds^2} \int_{\mathsf{T}_{i-1}}^{s} (s-t)^{1-\ltimes_i} y(t)dt \right) \tag{3.26}$$

As a consequence, the ITVP for ITVP (3.25) can be reformulated on each subinterval J_i as:

$$\frac{1}{\Gamma(2-\ltimes(s))} \Big(\sum_{k=1}^{i-1} \frac{d^2}{ds^2} \int_{\mathsf{T}_{k-1}}^{\mathsf{T}_k} (s-t)^{1-\ltimes_k} y(t)dt + \frac{d^2}{ds^2} \int_{\mathsf{T}_{i-1}}^{s} (s-t)^{1-\ltimes_i} y(t)dt \Big) = g(s, y(s)) \tag{3.27}$$

Let $\tilde{y} \in C(J_i, \mathbb{R})$ be a function such that $\tilde{y}(s) \equiv 0$ for $s \in [0, \mathsf{T}_{i-1}]$ and satisfies the integral equation (3.27). Then, equation (3.27) simplifies to

$$\digamma_{\mathsf{T}_{i-1}^+}^{\ltimes_i} \tilde{y}(s) = g(s, \tilde{y}(s)), \quad s \in J_i$$

We now consider the auxiliary ITVP for integer-order Riemann–Liouville fractional differential equations, corresponding to the above formulation for any $i = 1, 2, \ldots, n$.

$$\begin{cases} \digamma_{\mathsf{T}_{i-1}^+}^{\ltimes_i} y(s) = g(s, y(s)), & s \in J_i \\ \digamma_{\mathsf{T}_{i-1}^+}^{\ltimes_i - 2} y(\mathsf{T}_{i-1}) = a \digamma_{\mathsf{T}_{i-1}^+}^{\ltimes_i - 2} y(\mathsf{T}_i), & \digamma_{\mathsf{T}_{i-1}^+}^{\ltimes_i - 1} y(\mathsf{T}_{i-1}) = b \digamma_{\mathsf{T}_{i-1}^+}^{\ltimes_i - 1} y(\mathsf{T}_i) \end{cases} \tag{3.28}$$

Lemma 14 *Let $a = b = 1$, $g \in C(J_i \times \mathbb{R}, \mathbb{R})$ for $i = 1, \ldots, n$, and let $\delta \in (0, 1)$ be a number such that $s^\delta g \in C(J_i \times \mathbb{R}, \mathbb{R})$.*

Then the function $\varkappa \in E_i$ satisfies (3.28) if and only if y solves the integral equation

$$\varkappa(s) = \frac{1}{\Gamma(\ltimes_i)} \int_{\mathsf{T}_{i-1}}^{s} (s-t)^{\ltimes_i - 1} g(t, \varkappa(t))dt + \frac{b s^{\ltimes_i - 1}}{(1-b)\Gamma(\ltimes_i)} \int_{\mathsf{T}_{i-1}}^{\mathsf{T}_i} g(t, \varkappa(t))dt$$

$$\frac{s^{\ltimes_i - 2}}{(1-a)(1-b)\Gamma(\ltimes_i - 1)} \int_{\mathsf{T}_{i-1}}^{\mathsf{T}_i} [(a\,\mathsf{T}_i - b\,\mathsf{T}_{i-1} - a(1-b)t] g(t, \varkappa(t))dt \tag{3.29}$$

Proof

Let $\varkappa \in E_i$ be an answer of ITVP (3.28). Applying the operator $I_{\mathsf{T}_{i-1}^+}^{\ltimes_i}$ to both sides of (3.28), we obtain:

$$\varkappa(s) = \omega_1 s^{\ltimes_i - 1} + \omega_2 s^{\ltimes_i - 2} + I_{\mathsf{T}_{i-1}^+}^{\ltimes_i} g(s, \varkappa(s)) \tag{3.30}$$

where ω_1, ω_2 are constants.

Using equation (3.30), we find:

$$F^{\ltimes_i-1}\varkappa(s)=\omega_1\Gamma(\ltimes_i)+I^1 g(s,\varkappa(s))$$

In view of assumptions on the function g, and using initial and terminal condition

$$F^{\ltimes_i-1}_{\mathsf{T}_{i-1}^+}\varkappa(\mathsf{T}_{i-1})=bF^{\ltimes_i-1}_{\mathsf{T}_{i-1}^+}\varkappa(\mathsf{T}_i)$$

we conclude that:

$$\omega_1=\frac{b}{(1-b)\Gamma(\ltimes_i)}\int_{\mathsf{T}_{i-1}}^{\mathsf{T}_i} g(t,\varkappa(t))dt$$

Since $I^{2-\ltimes_i}(s^{\ltimes_i-1})=\Gamma(\ltimes_i)s$ and $I^{2-\ltimes_i}(s^{\ltimes_i-2})=\Gamma(\ltimes_i-1)$, from the initial and terminal condition:

$$F^{\ltimes_i-2}_{\mathsf{T}_{i-1}^+}\varkappa(\mathsf{T}_{i-1})=aF^{\ltimes_i-2}_{\mathsf{T}_{i-1}^+}\varkappa(\mathsf{T}_i)$$

we obtain:

$$\omega_2=\frac{b(a\,\mathsf{T}_i-\mathsf{T}_{i-1})}{(1-b)(1-a)\Gamma(\ltimes_i-1)}\int_{\mathsf{T}_{i-1}}^{\mathsf{T}_i} g(t,\varkappa(t))dt+\frac{a}{(1-a)\Gamma(\ltimes_i-1)}\int_{\mathsf{T}_{i-1}}^{\mathsf{T}_i}(\mathsf{T}_i-t)g(t,\varkappa(t))dt$$

Thus, we have:

$$\varkappa(s)=\int_{\mathsf{T}_{i-1}}^{\mathsf{T}_i} G_i(s,t)g(t,\varkappa(t))dt$$

where $G_i(s,t)$ is the Green's function given by:

$$G_i(s,t)=\begin{cases}\frac{bs^{\ltimes_i-1}}{(1-b)\Gamma(\ltimes_i)}+\frac{s^{\ltimes_i-2}[a\mathsf{T}_i-b\mathsf{T}_{i-1}-a(1-b)t]}{(1-a)(1-b)\Gamma(\ltimes_i-1)}+\frac{1}{\Gamma(\ltimes_i)}(s-t)^{\ltimes_i-1} & \\ \qquad \mathsf{T}_{i-1}\le t\le s\le \mathsf{T}_i & \\ \frac{bs^{\ltimes_i-1}}{(1-b)\Gamma(\ltimes_i)}+\frac{s^{\ltimes_i-2}[a\mathsf{T}_i-b\mathsf{T}_{i-1}-a(1-b)t]}{(1-a)(1-b)\Gamma(\ltimes_i-1)} & \\ \qquad \mathsf{T}_{i-1}\le s\le t\le \mathsf{T}_i & \end{cases}$$

where, $i=1,2,...,n$.

Hence, $\varkappa$ solves the integral equation (3.29).

Conversely, let $\varkappa\in E_i$ be a solution to the integral equation (3.29). Since $s^{\delta}g$ is continuous, we infer that $\varkappa$ is a solution to the ITVP (3.28).

We will now demonstrate the existence of a solution to the ITVP for the constant order Riemann–Liouville fractional differential equations (3.28). Theorem 8 will be employed to assist in the proof.

Theorem 16 *Suppose that the conditions of Lemma 14 hold, and there exists a constant $N > 0$ such that*

$$s^{\delta}|g(s,y)| \leq N, \quad \forall s \in J_i, y \in \mathbb{R}$$

with $\delta = 2 - \ltimes$. Then, the initial and terminal value problem (ITVP) for the Riemann–Liouville fractional differential equations of integer order(3.28) admits at least one solution in the space $C_\delta[\mathsf{T}_{i-1}, \mathsf{T}_i]$.

Proof

For any function $y \in C_\delta[\mathsf{T}_{i-1}, \mathsf{T}_i]$, we construct the operator

$$\mathcal{S}y(s) = \frac{1}{\Gamma(\ltimes_i)} \int_{\mathsf{T}_{i-1}}^{s} (s-t)^{\ltimes_i - 1} g(t, y(t))dt$$

$$+\frac{bs^{\ltimes_i - 1}}{(1-b)\Gamma(\ltimes_i)} \int_{\mathsf{T}_{i-1}}^{\mathsf{T}_i} g(t, y(t))dt + \frac{bs^{\ltimes_i - 1}}{(1-b)\Gamma(\ltimes_i)} \int_{\mathsf{T}_{i-1}}^{\mathsf{T}_i} g(t, y(t))dt$$

$$+\frac{s^{\ltimes_i - 2}}{(1-a)(1-b)\Gamma(\ltimes_i - 1)} \int_{\mathsf{T}_{i-1}}^{\mathsf{T}_i} [(a\,\mathsf{T}_i - b\,\mathsf{T}_{i-1} - a(1-b)t]g(t, y(t))dt \quad (3.31)$$

It follows directly from the properties of fractional integrals and the continuity of the function $s^{\delta}g$ that the operator $\mathcal{S} : C_\delta[\mathsf{T}_{i-1}, \mathsf{T}_i] \to C_\delta[\mathsf{T}_{i-1}, \mathsf{T}_i]$ defined by (3.31) is well-defined. Let

$$R_i \geq \frac{N(\mathsf{T}_i^{1-\delta} - \mathsf{T}_{i-1}^{1-\delta})}{(1-\delta)\Gamma(\ltimes_i)} \Big[\mathsf{T}_i^{\delta} (\mathsf{T}_i - \mathsf{T}_{i-1})^{\ltimes_i - 1} + \Big|\frac{b\mathsf{T}_i}{1-b}\Big| + \Big|\frac{(a\,\mathsf{T}_i - b\mathsf{T}_{i-1})(\ltimes_i - 1)}{(1-a)(1-b)}\Big|$$
$$+\Big|\frac{a(1-b)(\ltimes_i - 1)(1-\delta)(\mathsf{T}_i^{2-\delta} - \mathsf{T}_{i-1})^{2-\delta}}{(1-a)(1-b)(2-\delta)(\mathsf{T}_i^{1-\delta} - \mathsf{T}_{i-1})^{1-\delta}}\Big|\Big]$$

Consider the set

$$B_{R_i} = \{y \in C_\delta[\mathsf{T}_{i-1}, \mathsf{T}_i], \ \|y\|_\delta \leq R_i\}$$

For all $m \in \{1, 2, ..., n\}$, the ball B_{R_i} is a nonempty, closed, convex subset of $C_\delta[\mathsf{T}_{i-1}, \mathsf{T}_i]$.

We are in position to examine the assumptions of Theorem 16 for the operator $\mathcal{S}$. We shall demonstrate this in three stages.

Step 1: Let B_{R_i} be a bounded set in $C_\delta[\mathsf{T}_{i-1}, \mathsf{T}_i]$. Hence, B_{R_i} is bounded in $C[\mathsf{T}_{i-1}, \mathsf{T}_i]$, and there exists a constant N ¿ 0 such that $s^\delta|g(s, y(s))| \leq N, \forall y \in B_{R_i}, s \in [\mathsf{T}_{i-1}, \mathsf{T}_i]$. Thus,

$$s^\delta|(\mathcal{S}y)(s)|$$

$$\begin{aligned}
\leq\ & \frac{Ns^\delta}{\Gamma(\ltimes_i)} \int_{\mathsf{T}_{i-1}}^{s} t^{-\delta}(s-t)^{\ltimes_i - 1} dt + \Big|\frac{bNs}{(1-b)\Gamma(\ltimes_i)}\Big| \int_{\mathsf{T}_{i-1}}^{\mathsf{T}_i} t^{-\delta} dt \\
& + \Big|\frac{N}{(1-a)(1-b)\Gamma(\ltimes_i - 1)}\Big| \int_{\mathsf{T}_{i-1}}^{\mathsf{T}_i} \Big|a\,\mathsf{T}_i - b\,\mathsf{T}_{i-1} - a(1-b)t\Big| t^{-\delta} dt \\
\leq\ & \frac{N(\mathsf{T}_i^{1-\delta} - \mathsf{T}_{i-1}^{1-\delta})}{(1-\delta)\Gamma(\ltimes_i)} \Big[\mathsf{T}_i^\delta (\mathsf{T}_i - \mathsf{T}_{i-1})^{\ltimes_i - 1} + \Big|\frac{b\mathsf{T}_i}{1-b}\Big| + \Big|\frac{(a\,\mathsf{T}_i - b\mathsf{T}_{i-1})(\ltimes_i - 1)}{(1-a)(1-b)}\Big| \\
& + \Big|\frac{a(1-b)(\ltimes_i - 1)(1-\delta)(\mathsf{T}_i^{2-\delta} - \mathsf{T}_{i-1})^{2-\delta}}{(1-a)(1-b)(2-\delta)(\mathsf{T}_i^{1-\delta} - \mathsf{T}_{i-1})^{1-\delta}}\Big|\Big]
\end{aligned}$$

This yields the estimate:

$$\|(\mathcal{S}y)\|_\delta$$

$$\begin{aligned}
\leq\ & \frac{N(\mathsf{T}_i^{1-\delta} - \mathsf{T}_{i-1}^{1-\delta})}{(1-\delta)\Gamma(\ltimes_i)} \Big[\mathsf{T}_i^\delta (\mathsf{T}_i - \mathsf{T}_{i-1})^{\ltimes_i - 1} + \Big|\frac{b\mathsf{T}_i}{1-b}\Big| + \Big|\frac{(a\,\mathsf{T}_i - b\mathsf{T}_{i-1})(\ltimes_i - 1)}{(1-a)(1-b)}\Big| \\
& + \Big|\frac{a(1-b)(\ltimes_i - 1)(1-\delta)(\mathsf{T}_i^{2-\delta} - \mathsf{T}_{i-1})^{2-\delta}}{(1-a)(1-b)(2-\delta)(\mathsf{T}_i^{1-\delta} - \mathsf{T}_{i-1})^{1-\delta}}\Big|\Big]
\end{aligned}$$

Hence, $\mathcal{S}(B_{R_i})$ is uniformly bounded.

Step 2: Let $s_1, s_2 \in J_i$ with $s_1 < s_2$, and let $y \in B_{R_i}$. Then we have

$$\Big|s_1^\delta(\mathcal{S}y)(s_1) - s_2^\delta(\mathcal{S}y)(s_2)\Big|$$

$$\begin{aligned}
=\ & \Big|\frac{1}{\Gamma(\ltimes_i)} \int_{\mathsf{T}_{i-1}}^{s_1} \Big[s_1^\delta(s_1 - t)^{\ltimes_i - 1} - s_2^\delta(s_2 - t)^{\ltimes_i - 1}\Big] g(t, y(t)) dt \\
-\ & \frac{1}{\Gamma(\ltimes_i)} \int_{s_1}^{s_2} s_2^\delta(s_2 - t)^{\ltimes_i - 1} g(t, u(t)) dt + \frac{b(s_1 - s_2)}{(1-b)\Gamma(\ltimes_i - 1)} \int_{\mathsf{T}_{i-1}}^{\mathsf{T}_i} g(t, y(t)) dt\Big| \\
\leq\ & N\Big(\Big|\frac{1}{\Gamma(\ltimes_i)} \int_{\mathsf{T}_{i-1}}^{s_1} \Big[s_1^\delta(s_1 - t)^{\ltimes_i - 1} - s_2^\delta(s_2 - t)^{\ltimes_i - 1}\Big] dt \\
-\ & \frac{1}{\Gamma(\ltimes_i)} \int_{s_1}^{s_2} s_2^\delta(s_2 - t)^{\ltimes_i - 1} dt\Big| + \Big|\frac{b(s_1 - s_2)}{(1-b)\Gamma(\ltimes_i - 1)} \int_{\mathsf{T}_{i-1}}^{\mathsf{T}_i} dt\Big|\Big)
\end{aligned}$$

Hence, $\Big|s_1^\delta(\mathcal{S}y)(s_1) - s_2^\delta(\mathcal{S}y)(s_2)\Big| \to 0$ as $|s_1 - s_2| \to 0$. Thus, the set $s^\delta\mathcal{S}(B_{R_i})$ is equicontinuous. Consequently, the operator $\mathcal{S}$ is compact.

Step 3: Consider the set

$$\Omega = \{y \in \mathbb{R} \setminus y = \eta \mathcal{S} y, 0 < \eta < 1\}$$

and let us show that Ω is bounded. Suppose $y \in \Omega$, then $y = \eta \mathcal{S} y$ for some $0 << 1$. For any $s \in [\mathsf{T}_i - 1, \mathsf{T}_i]$, we have

$$|y(s)| \le \eta \Big[\tfrac{1}{\Gamma(\ltimes_i)} \int_{\mathsf{T}_i-1}^{s} (s-t)^{\ltimes_i -1} |g(t,y(t))| dt + \Big| \tfrac{b s^{\ltimes_i -1}}{(1-b)\Gamma(\ltimes_i)} \Big| \int_{\mathsf{T}_{i-1}}^{\mathsf{T}_i} |g(t,y(t))| dt$$
$$+ \Big| \tfrac{s^{\ltimes_i -2}}{(1-a)(1-b)\Gamma(\ltimes_i -1} \Big| \int_{\mathsf{T}_{i-1}}^{\mathsf{T}_i} \Big| a\, \mathsf{T}_i - b\, \mathsf{T}_{i-1} - a(1-b)t \Big| dt \Big]$$

From previous estimates, we know:

$$\|(\mathcal{S}y)\|_\delta \le \frac{N(\mathsf{T}_i^{1-\delta} - \mathsf{T}_{i-1}^{1-\delta})}{(1-\delta)\Gamma(\ltimes_i)} \Big[\mathsf{T}_i^{\delta} (\mathsf{T}_i - \mathsf{T}_{i-1})^{\ltimes_i -1} + \Big| \frac{b \mathsf{T}_i}{1-b} \Big| + \Big| \frac{(a\, \mathsf{T}_i - b \mathsf{T}_{i-1})(\ltimes_i - 1)}{(1-a)(1-b)} \Big|$$
$$+ \Big| \frac{a(1-b)(\ltimes_i - 1)(1-\delta)(\mathsf{T}_i^{2-\delta} - \mathsf{T}_{i-1})^{2-\delta}}{(1-a)(1-b)(2-\delta)(\mathsf{T}_i^{1-\delta} - \mathsf{T}_{i-1})^{1-\delta}} \Big| \Big]$$

This suggests that the factor $\eta \in (0,1)$ does not affect the boundedness of the set Ω. Therefore, we conclude the operator $\mathcal{S}$ has at least one fixed point by virtue of Theorem 8. This implies that the problem (3.28) admits at least one solution.

3.3.2 Study of the Solution's Uniqueness

Consider the following assumption:

(H2) Let $g \in C(J \times \mathbb{R}, \mathbb{R})$, and suppose there exists a constant $K > 0$ such that

$$s^{\delta} |g(s,u) - g(s,v)| \le K|u - v|,$$

$\forall$ for any $u,\ v \in \mathbb{R}$, $s \in J$ and $\delta = 2 - \ltimes$.

Theorem 17 *Suppose that conditions (H1) and (H2) are satisfied. Then the problem (3.28) admits a unique solution in* $C_\delta[\mathsf{T}_{i-1}, \mathsf{T}_i]$ *if*

$$K < \frac{1}{\rho} \tag{3.32}$$

where

$$\rho = \frac{(\mathsf{T}_i^{1-2\delta} - \mathsf{T}_{i-1}^{1-2\delta})}{(1-2\delta)\Gamma(\ltimes_i)} \Big[\mathsf{T}_i^{\delta} (\mathsf{T}_i - \mathsf{T}_{i-1})^{\ltimes_i -1} + \Big| \frac{b}{1-b} \Big| \mathsf{T}_i + \Big| \frac{(a\, \mathsf{T}_i - b \mathsf{T}_{i-1})(\ltimes_i - 1)}{(1-a)(1-b)} \Big|$$
$$+ \Big| \frac{a(1-b)(\ltimes_i - 1)}{(1-a)(1-b)} \Big| \Big(\frac{(1-2\delta)(\mathsf{T}_i^{2-2\delta} - \mathsf{T}_{i-1}^{2-2\delta})}{(2-2\delta)(\mathsf{T}_i^{1-2\delta} - \mathsf{T}_{i-1}^{1-2\delta})} \Big) \Big]$$

Proof

In view of (H2), for every $s \in [\mathsf{T}_{i-1}, \mathsf{T}_i]$, we have

$$s^{\delta}|(\mathcal{S}u)(s) - (\mathcal{S}v)(s)|$$

$$\leq \frac{s^{\delta}}{\Gamma(\ltimes_i)} \int_{\mathsf{T}_i-1}^{s} (s-t)^{\ltimes_i - 1} |g(t,u(t)) - g(t,\text{ß}(t))| dt$$

$$+ \left| \frac{bs}{(1-b)\Gamma(\ltimes_i)} \right| \int_{\mathsf{T}_{i-1}}^{\mathsf{T}_i} |g(t,u(t)) - g(t,\text{ß}(t))| dt$$

$$+ \left| \frac{1}{(1-a)(1-b)\Gamma(\ltimes_i - 1)} \right| \int_{\mathsf{T}_{i-1}}^{\mathsf{T}_i} \left| a\, \mathsf{T}_i - b\, \mathsf{T}_{i-1} - a(1-b)t \right|$$

$$|g(t,u(t)) - g(t,\text{ß}(t))| dt$$

$$\leq K \Bigg[\frac{s^{\delta}}{\Gamma(\ltimes_i)} \int_{\mathsf{T}_i-1}^{s} t^{-\delta} (s-t)^{\ltimes_i - 1} |u(t) - \text{ß}(t)| dt + \left| \frac{bs}{(1-b)\Gamma(\ltimes_i)} \right| \int_{\mathsf{T}_{i-1}}^{\mathsf{T}_i} t^{-\delta} |u(t) - \text{ß}(t)| dt$$

$$+ \left| \frac{1}{(1-a)(1-b)\Gamma(\ltimes_i - 1)} \right| \int_{\mathsf{T}_{i-1}}^{\mathsf{T}_i} t^{-\delta} \left| a\, \mathsf{T}_i - b\, \mathsf{T}_{i-1} - a(1-b)t \right| |u(t) - \text{ß}(t)| dt \Bigg]$$

Applying condition (H2), we obtain

$$\|(\mathcal{S}u)(t) - (\mathcal{S}v)(s)\|_{\delta}$$

$$\leq K \Bigg[\frac{s^{\delta}}{\Gamma(\ltimes_i)} \int_{\mathsf{T}_i-1}^{s} t^{-2\delta} (s-t)^{\ltimes_i - 1} dt + \left| \frac{bs}{(1-b)\Gamma(\ltimes_i)} \right| \int_{\mathsf{T}_{i-1}}^{\mathsf{T}_i} t^{-2\delta} dt$$

$$+ \left| \frac{1}{(1-a)(1-b)\Gamma(\ltimes_i - 1)} \right| \int_{\mathsf{T}_{i-1}}^{\mathsf{T}_i} t^{-2\delta} \left| a\, \mathsf{T}_i - b\, \mathsf{T}_{i-1} - a(1-b)t \right| dt \Bigg] \|u - v\|_{\delta}$$

$$\leq \frac{K(\mathsf{T}_i^{1-2\delta} - \mathsf{T}_{i-1}^{1-2\delta})}{(1-2\delta)\Gamma(\ltimes_i)} \Bigg[\mathsf{T}_i^{\delta} (\mathsf{T}_i - \mathsf{T}_{i-1})^{\ltimes_i - 1} + \left| \frac{b}{1-b} \right| \mathsf{T}_i + \left| \frac{(a\mathsf{T}_i - b\mathsf{T}_{i-1})(\ltimes_i - 1)}{(1-a)(1-b)} \right|$$

$$+ \left| \frac{a(1-b)(\ltimes_i - 1)}{(1-a)(1-b)} \right| \left(\frac{(1-2\delta)(\mathsf{T}_i^{2-2\delta} - \mathsf{T}_{i-1}^{2-2\delta})}{(2-2\delta)(\mathsf{T}_i^{1-2\delta} - \mathsf{T}_{i-1}^{1-2\delta})} \right) \Bigg] \|u - v\|_{\delta}$$

The condition (3.32) ensures that the operator $\mathcal{S}$ is a contraction based on the previous estimate. Consequently, $\mathcal{S}$ possesses a unique fixed point, which corresponds to a unique solution for the problem (3.28), in accordance with the ßanach fixed point theorem.

We shall now proceed to demonstrate the existence result for the ITVP stated in (3.25).

Theorem 18 *Suppose that conditions (H1), (H2), and inequality (3.32) are satisfied for all* $i \in \{1, 2, \ldots, n\}$. *Then the problem (3.25) admits a unique solution in* $C_{\delta}[0, a]$.

Proof

For each $i \in \{1, 2, \ldots, n\}$, by Theorem 17, the initial terminal value problem (ITVP) for the Riemann–Liouville fractional differential equation of integer order (3.28) admits a unique solution $\widetilde{y}_i \in C_{\delta}[\mathsf{T}_{i-1}, \mathsf{T}_i]$. For any $i \in \{1, 2, \ldots, n\}$, we define the function

$$y_i = \begin{cases} 0, & s \in [0, \mathsf{T}_{i-1}] \\ \widetilde{y}_i, & s \in J_i \end{cases} \tag{3.33}$$

As a result, the function $y_i \in C_\delta[\mathsf{T}_{i-1}, \mathsf{T}_i]$ satisfies the integral equation (3.27) on J_i, implying that $y_i(0) = 0,\ y_i(\mathsf{T}_i) = \widetilde{y}_i(\mathsf{T}_i) = 0$ and solves (3.27) for $s \in J_i,\ \ i \in \{1, 2, \ldots, n\}$. Consequently, the function

$$y(s) = \begin{cases} y_1(s), s \in J_1 \\ y_2(s) = \begin{cases} 0, & s \in J_1 \\ \widetilde{y}_2, & s \in J_2 \end{cases} \\ \vdots \\ y_n(s) = \begin{cases} 0, & s \in [0, \mathsf{T}_{n-1}] \\ \widetilde{y}_n, & s \in J_n \end{cases} \end{cases}$$

is a solution to the ITVP (3.25) in the space $C_\delta[0, a]$.

3.3.3 Conditions of Ulam-Hyers-$\mathfrak{R}$assias Stability

Theorem 19 *Assume that conditions (H1), (H2), and inequality (3.32) are satisfied. Additionally, suppose:*

(H3) $\vartheta \in C(J, \mathbb{R}_+)$ *is an increasing function, and there exists a constant* $\lambda_\vartheta > 0$ *such that*

$$I^{\ltimes_i}_{\mathsf{T}_{i-1}^+} \vartheta(s) \le \lambda_\vartheta\ \vartheta(s)$$

for all $s \in J_i, i = 1, 2, \ldots$. *Then, the initial terminal value problem (3.25) is Ulam-Hyers-$\mathfrak{R}$assias stable with respect to* ϑ.

Proof
Suppose $\epsilon > 0$ is an arbitrary positive constant, and the function $z(s) \in C(J, \mathbb{R})$ satisfies inequality (2.22).

For any $i \in \{1, 2, \ldots, n\}$, define the functions:

$$z_i(s) = \begin{cases} 0, & s \in [0, \mathsf{T}_{i-1}] \\ z(s), & s \in J_i \end{cases}$$

Then, for any $s \in J_i$, $i \in \{1, 2, \ldots, n\}$, we have:

$$F^{\ltimes(s)}_{0^+} z_i(s) = \frac{1}{\Gamma(2 - \ltimes(s))} \frac{d^2}{ds^2} \int_{\mathsf{T}_{i-1}}^{s} (s - t)^{1-\ltimes_i} \frac{z(t)}{t} dt \qquad (3.34)$$

Applying the Riemann–Liouville fractional integral $I^{\ltimes_i}_{\mathsf{T}_{i-1}^+}$ to both sides of inequality (2.22), and using assumption (H3), we obtain:

$$\left| z_i(s) - \frac{1}{\Gamma(\ltimes_i)} \int_{\mathsf{T}_{i-1}}^{s} (s - t)^{\ltimes_i - 1} g(t, z_i(t)) dt - \frac{b s^{\ltimes_i - 1}}{(1 - b)\Gamma(\ltimes_i)} \int_{\mathsf{T}_{i-1}}^{\mathsf{T}_i} g(t, z_i(t)) dt \right.$$
$$\left. - \frac{s^{\ltimes_i - 2}}{(1 - a)(1 - b)\Gamma(\ltimes_i - 1)} \int_{\mathsf{T}_{i-1}}^{\mathsf{T}_i} [(a\,\mathsf{T}_i - b\,\mathsf{T}_{i-1} - a(1 - b)t] g(t, z_i(t)) dt \right|$$
$$\le \epsilon\ I^{\ltimes_i}_{\mathsf{T}_{i-1}^+} \vartheta(s)$$
$$\le \epsilon\ \lambda_\vartheta\ \vartheta(s)$$

According to Theorem 18, the initial terminal value problem (3.25) admits a solution $y \in C(J, \mathbb{R})$ given by $y(s) = y_i(s)$ for $s \in J_i,\ i = 1, 2, \ldots, n$, where:

$$y_i(s) = \begin{cases} 0, & s \in [0, \mathsf{T}_{i-1}] \\ \widetilde{y}_i, & s \in J_i \end{cases} \tag{3.35}$$

and $\widetilde{y}_i \in E_i$ is a solution of (3.28). Based on Lemma 13, the integral form

$$\widetilde{y}_i(s) = \frac{1}{\Gamma(\ltimes_i)} \int_{\mathsf{T}_{i-1}}^{s} (s-t)^{\ltimes_i - 1} g(t, \widetilde{y}_i(t))dt + \frac{b s^{\ltimes_i - 1}}{(1-b)\Gamma(\ltimes_i)} \int_{\mathsf{T}_{i-1}}^{\mathsf{T}_i} g(t, \widetilde{y}_i(t))dt$$

$$+ \frac{s^{\ltimes_i - 2}}{(1-a)(1-b)\Gamma(\ltimes_i - 1)} \int_{\mathsf{T}_{i-1}}^{\mathsf{T}_i} [(a\,\mathsf{T}_i - b\,\mathsf{T}_{i-1} - a(1-b)t] g(t, \widetilde{y}_i(t))dt$$

holds.

Let $s \in J_i$ where $i \in \{1, 2, \ldots, n\}$. Then, we obtain:

$$|z(s) - y(s)| = |z(s) - y_i(s)| = |z_i(s) - \widetilde{y}_i(s)|$$

$$\leq \left| z_i(s) - \frac{1}{\Gamma(\ltimes_i)} \int_{\mathsf{T}_{i-1}}^{\mathsf{s}} (\mathsf{s} - t)^{\ltimes_i - 1} g(t, z_i(t))dt - \frac{b s^{\ltimes_i - 1}}{(1-b)\Gamma(\ltimes_i)} \int_{\mathsf{T}_{i-1}}^{\mathsf{T}_i} g(t, z_i(t))dt \right.$$

$$\left. - \frac{s^{\ltimes_i - 2}}{(1-a)(1-b)\Gamma(\ltimes_i - 1)} \int_{\mathsf{T}_{i-1}}^{\mathsf{T}_i} [(a\,\mathsf{T}_i - b\,\mathsf{T}_{i-1} - a(1-b)t] g(t, z_i(t))dt \right|$$

$$+ \frac{1}{\Gamma(\ltimes_i)} \int_{\mathsf{T}_{i-1}}^{s} (s-t)^{\ltimes_i - 1} \Big| g(t, z_i(t)) - g(t, \widetilde{y}_i(t)) \Big| dt$$

$$+ \frac{b s^{\ltimes_i - 1}}{(1-b)\Gamma(\ltimes_i)} \int_{\mathsf{T}_{i-1}}^{\mathsf{T}_i} \Big| g(t, z_i(t)) - g(t, \widetilde{y}_i(t)) \Big| dt$$

$$+ \frac{s^{\ltimes_i - 2}}{(1-a)(1-b)\Gamma(\ltimes_i - 1)} \int_{\mathsf{T}_{i-1}}^{\mathsf{T}_i} [(a\,\mathsf{T}_i - b\,\mathsf{T}_{i-1} - a(1-b)t]$$

$$\Big| g(t, z_i(t)) - g(t, \widetilde{y}_i(t)) \Big| dt$$

$$
\begin{aligned}
&\leq \lambda_\vartheta\, \epsilon\, \vartheta(s) + \frac{1}{\Gamma(\ltimes_i)} \int_{\mathsf{T}_{i-1}}^{s} t^{-\gamma}(s-t)^{\ltimes_i - 1}(K|z_i(t) - \widetilde{y}_i(t)|)dt \\
&+ \frac{b s^{\ltimes_i - 1}}{(1-b)\Gamma(\ltimes_i)} \int_{\mathsf{T}_{i-1}}^{\mathsf{T}_i} t^{-\gamma}(K|z_i(t) - \widetilde{y}_i(t)|)dt \\
&+ \frac{s^{\ltimes_i - 2}}{(1-a)(1-b)\Gamma(\ltimes_i - 1)} \int_{\mathsf{T}_{i-1}}^{\mathsf{T}_i} [(a\,\mathsf{T}_i - b\,\mathsf{T}_{i-1} - a(1-b)t] \\
&t^{-\delta}(K|z_i(t) - \widetilde{y}_i(t)|)dt \\
&\leq \lambda_\vartheta\, \epsilon\, \vartheta(s) + \frac{(s - \mathsf{T}_{i-1})^{\ltimes_i - 1}(s s^{1-\delta} - \mathsf{T}_{i-1}{}^{1-\delta})}{(1-\delta)\Gamma(\ltimes_i)}(K\|z_i - \widetilde{y}_i\|_{E_i}) \\
&+ \frac{b s^{\ltimes_i - 1}(\mathsf{T}_i^{1-\delta} - \mathsf{T}_{i-1}{}^{1-\delta})}{(1-\delta)(1-b)\Gamma(\ltimes_i)}(K\|z_i - \widetilde{y}_i\|_{E_i}) \\
&+ \frac{s^{\ltimes_i - 2}\, \mathsf{T}_{i-1}^{-\delta}\,(\mathsf{T}_i - \mathsf{T}_{i-1})}{2(1-a)(1-b)\Gamma(\ltimes_i - 1)}[2(a\,\mathsf{T}_i - b\mathsf{T}_{i-1}) - a(1-\varpi_1)(\mathsf{T}_i - \mathsf{T}_{i-1})] \\
&(K\|z_i - \widetilde{y}_i\|_{E_i}) \\
&\leq \lambda_\vartheta\, \epsilon\, \vartheta(s) + \mu\|z - y\|_J.
\end{aligned}
$$

where

$$
\begin{aligned}
\mu &= \max_{i=1,2,\ldots,n} \frac{K(s - \mathsf{T}_{i-1})^{\ltimes_i - 1}(s^{1-\delta} - \mathsf{T}_{i-1}{}^{1-\delta})}{(1-\delta)\Gamma(\ltimes_i)} + \frac{K\varpi_1 s^{\ltimes_i - 1}(\mathsf{T}_i^{1-\delta} - \mathsf{T}_{i-1}{}^{1-\delta})}{(1-\delta)(1-b)\Gamma(\ltimes_i)} \\
&+ \frac{K s^{\ltimes_i - 2}\, \mathsf{T}_{i-1}^{-\delta}\,(\mathsf{T}_i - \mathsf{T}_{i-1})}{2(1-a)(1-b)\Gamma(\ltimes_i - 1)}[2(a\,\mathsf{T}_i - b\mathsf{T}_{i-1}) - a(1-\varpi_1)(\mathsf{T}_i - \mathsf{T}_{i-1})].
\end{aligned}
$$

Thus,

$$
\|z - y\|_J\Big(1 - \mu\Big) \leq \lambda_\vartheta\, \epsilon\, \vartheta(s)
$$

or for any $s \in J$

$$
|z(s) - y(s) \leq \|z - y\|_J \leq \frac{\lambda_\vartheta}{1-\mu}\epsilon\vartheta(s)
$$

Therefore, the initial terminal value problem (3.25) is Ulam-Hyers-$\mathfrak{R}$assias stable with respect to ϑ.

3.3.4 Illustrative Application

Let $J := [0,4]$, $\mathrm{T}_0 = 0$, $\mathrm{T}_1 = 2$, $\mathrm{T}_2 = 4$. Consider the following fractional problem with variable-order fractional initial and terminal conditions:

$$\begin{cases} F_{0+}^{\ltimes(s)} y(s) = \frac{\sin y(s) + 2\cos(s)}{s^2}, & s \in J, \\ F_{0+}^{\ltimes(s)-2} y(0) = aF_{0+}^{\ltimes(s)-2} y(4), \ F_{0+}^{\ltimes(s)-1} y(0) = bF_{0+}^{\ltimes(s)-2} y(4), \end{cases} \tag{3.36}$$

where

$$\ltimes(s) = \begin{cases} \frac{3}{2}, & s \in J_1 := [0,2] \\ \frac{10}{9}, & s \in J_2 :=]2,4] \end{cases} \tag{3.37}$$

Define

$$g(s,y) = \frac{\sin y + 2\cos(s)}{s^2}, \ (s,y) \in [0,4] \times \mathbb{R}.$$

Taking into account (3.37), we can construct two auxiliary initial terminal value problems (ITVPs) using equation (3.28) for fractional differential equations of constant order in the Riemann–Liouville sense.

On J_1:

$$\begin{cases} F_{0+}^{\frac{3}{2}} y(s) = \frac{\sin y(s) + 2\cos(s)}{s^2}, & s \in J_1, \\ F_{0+}^{\frac{-1}{2}} y(0) = aF_{0+}^{\frac{-1}{2}} y(2), & F_{0+}^{\frac{1}{2}} y(0) = bF_{0+}^{\frac{1}{2}} y(2) \end{cases} \tag{3.38}$$

On J_2:

$$\begin{cases} F_{2+}^{\frac{10}{9}} y(s) = \frac{\sin y(s) + 2\cos(s)}{s^2}, & s \in J_2, \\ F_{2+}^{\frac{-8}{9}} y(2) = aF_{2+}^{\frac{-8}{9}} y(4), & F_{2+}^{\frac{1}{9}} y(2) = bF_{2+}^{\frac{1}{9}} y(4) \end{cases} \tag{3.39}$$

For $i = 1$, clearly:

$$s^{\delta}\Big|g(s,y)\Big| \ \leq \ s^{\frac{1}{2}}\left|\frac{\sin y(s) + 2\cos(s)}{s^2}\right| \leq \frac{3}{2\sqrt{2}} = N$$

Let $\vartheta(s) = s^{\frac{1}{2}}$. Then, we compute:

$$\begin{aligned} I_{0+}^{\ltimes_1}\vartheta(s) &= \tfrac{1}{\Gamma(\frac{3}{2})}\int_0^s (s-t)^{\frac{1}{2}} t^{\frac{1}{2}}\,dt \\ &\leq \tfrac{\sqrt{2}}{\Gamma(\frac{3}{2})}\int_0^s (s-t)^{\frac{1}{2}}\,dt \\ &\leq \tfrac{2\sqrt{2}}{3\Gamma(\frac{3}{2})}\vartheta(s) := \lambda_{\vartheta(s)}\vartheta(s) \end{aligned}$$

which implies $\lambda_{\vartheta} = \frac{2\sqrt{2}}{3\Gamma(\frac{3}{2})}$, and thus condition (H3) holds.

By Theorem 16, the ITVP (3.38) admits a solution $\widetilde{y}_1 \in E_1$. For $i = 2$, similarly:

$$s^{\delta}\Big|g(s,y)\Big| \ \leq \ s^{\frac{8}{9}}\left|\frac{\sin y(s) + 2\cos(s)}{s^2}\right| \leq 3.4^{\frac{-10}{9}} = N$$

Let $\vartheta(s) = s^{\frac{1}{2}}$. Then:

$$\begin{array}{rcl} I_{2^+}^{\ltimes_2}\vartheta(s) & = & \frac{1}{\Gamma(\frac{10}{9})}\int_2^s (s-t)^{\frac{1}{9}} t^{\frac{1}{2}} dt \\ & \leq & \frac{2}{\Gamma(\frac{10}{9})}\int_2^s (s-t)^{\frac{1}{9}} dt \\ & \leq & \frac{9}{5\Gamma(\frac{10}{9})}\vartheta(s) := \lambda_{\vartheta(s)}\vartheta(s) \end{array}$$

which implies $\lambda_{\vartheta} = \frac{9}{5\Gamma(\frac{10}{9})}$, and condition (H3) is satisfied.

According to Theorem 16, the ITVP (3.39) possesses a solution $\widetilde{y}_2 \in E_2$. Thus, by Theorem 18, the ITVP (3.36) has a solution

$$y(s) = \begin{cases} \widetilde{y}_1(s), & s \in J_1 \\ y_2(s), & s \in J_2 \end{cases}$$

with

$$y_2(s) = \begin{cases} 0, & s \in J_1 \\ \widetilde{y}_2(s), & s \in J_2 \end{cases}$$

In view of Theorem 19, the ITVP (3.36) is additionally Ulam-Hyers-$\mathfrak{R}$assias stable with respect to ϑ.

3.4 Initial and Terminal Value Problem for Nonlinear Fractional Differential Equations of Variable Order with Finite Delay Via Kuratowski's Measure of Noncompactness

We examine whether there exist any solutions in the style described below, for the nonlinear fractional differential equation of variable order with finite delay, subject to initial and terminal value conditions:

$$\begin{cases} F_{0+}^{\ltimes(s)}\varkappa(s) = f(s, \varkappa_s), & s \in J := [0, \mathsf{T}], \\ \varkappa(s) = \chi(s), & s \in [-\gamma, 0], \gamma > 0, \end{cases} \tag{3.40}$$

This formulation addresses initial and terminal value problems for Riemann–Liouville fractional differential equations of variable order, where $1 < \ltimes(s) \leq 2$, and $F_{0+}^{\ltimes(s)}$ denotes the Riemann–Liouville fractional derivative of variable order $\ltimes(.)$. The function $f : J \times C([-\gamma, 0], \mathbb{R}) \to \mathbb{R}$ is given, and the initial function $\chi \in C([-\gamma, 0], \mathbb{R})$ satisfies $\chi(0) = 0$. For any $\varkappa_s$ defined on $C([-\gamma, T]$ and any $s \in J$, the history function $\varkappa_s \in C([-\gamma, 0], \mathbb{R})$ is defined by

$$\varkappa_s(t) := \varkappa(s+t), \ t \in [-\gamma, 0]. \tag{3.41}$$

for any function $\varkappa$ defined on $[-\gamma, \mathsf{T}]$ and any $s \in J$.

3.4.1 Study of Existence

We begin by introducing the main hypotheses:

(H1) Let $n \in \mathbb{N}$ be a positive integer, and let $\mathcal{P} = \{J_1 := [0, \mathsf{T}_1], J_2 := (\mathsf{T}_1, \mathsf{T}_2], J_3 := (\mathsf{T}_2, \mathsf{T}_3], ...J_n := (\mathsf{T}_{n-1}, \mathsf{T}]\}$ be a partition of the interval J. Let $\ltimes(s) : J \to (1, 2]$ be a piecewise constant function with respect to $\mathcal{P}$, i.e.,

$$\ltimes(s) = \sum_{i=1}^{n} \ltimes_i I_i(s) = \begin{cases} \ltimes_1, & if\ s \in J_1, \\ \ltimes_2, & if\ s \in J_2, \\ \cdot & \\ \cdot & \\ \cdot & \\ \ltimes_n, & if\ s \in J_n, \end{cases}$$

where $1 < \ltimes_i \leq 2$ are fixed values, and I_i denotes the characteristic function of the interval $J_i := (\mathsf{T}_{i-1}, \mathsf{T}_i]$, $i = 1, 2, \ldots, n$, (with $\mathsf{T}_0 = 0$ and $\mathsf{T}_n = \mathsf{T}$), such that:

$$I_i(s) = \begin{cases} 1, & for\ s \in J_i, \\ 0, & otherwise. \end{cases}$$

(H2) Let $f : J \times C([-\gamma, 0], \mathbb{R}) \to \mathbb{R}$ be a continuous function for $0 < \delta < 1$. Then there exists a constant $K > 0$ such that $\delta\,|f(s, y_s) - f(s, z_s)| \leq K\|y_s - z_s\|_{[-\gamma,0]}$, for any functions $y, z \in C([-\gamma, \mathsf{T}], \mathbb{R})$ and for all $s \in J$.

The following definition of a solution to problem (3.40) will be essential in this work.

Definition 46 *Problem (3.40) has a solution if there exist functions $\varkappa_i, i = 1, 2, \ldots, n$, such that $\varkappa_i \in C([-\gamma, \mathsf{T}_i], \mathbb{R})$ which satisfy equation (3.43) for $s \in [0, \mathsf{T}_i]$, $\varkappa_i(s) = \chi(s)$ for $s \in [-\gamma, 0]$, and $\varkappa_i(0) = \varkappa_i(\mathsf{T}_i) = 0$.*

In order to apply Darbo's fixed point theorem and the Kuratowski measure of noncompactness, we perform an essential analysis of problem (3.40).

We represent the equation corresponding to problem (3.40) in the following form:

$$\frac{d^2}{dt^2} \int_0^s \frac{(s-t)^{1-\ltimes(t)}}{\Gamma(2-\ltimes(t))} \varkappa(t)dt = f(s, \varkappa_s), \quad s \in J. \tag{3.42}$$

Based on assumption $(H1)$, equation (3.42) can be expressed on the interval $J_i, i = 1, 2, \ldots, n$, as:

$$\frac{d^2}{dt^2}\Big(\int_0^{\mathsf{T}_1} \frac{(s-t)^{1-\ltimes_1}}{\Gamma(2-\ltimes_1)} \varkappa(t)dt + ... + \int_{\mathsf{T}_{i-1}}^{s} \frac{(s-t)^{1-\ltimes_i}}{\Gamma(2-\ltimes_i)} \varkappa(t)dt\Big) = f(s, \varkappa_s), \tag{3.43}$$

for $s \in J_i$.

For $0 \leq s \leq \mathsf{T}_{i-1}$, by taking $\varkappa(s) \equiv 0$, equation (3.43) reduces to:

$$F^{\ltimes_i}_{\mathsf{T}_{i-1}^+}\varkappa(s) = f(s, \varkappa_s), \quad s \in J_i. \tag{3.44}$$

Let us consider the following problem:

$$\begin{cases} F^{\ltimes_i}_{\mathsf{T}_{i-1}^+}\varkappa(s) = f(s, \varkappa_s), \quad s \in J_i, \\ \varkappa(\mathsf{T}_{i-1}) = 0, \qquad \varkappa(\mathsf{T}_i) = 0, \\ \varkappa(s) = \chi_i(s), \;\; s \in [\mathsf{T}_{i-1} - \gamma', \mathsf{T}_{i-1}], \end{cases} \tag{3.45}$$

where $\gamma' = \mathsf{T}_{i-1} + \gamma$, and

$$\chi_i(s) = \begin{cases} 0, \;\; if \;\; s \in [0, \mathsf{T}_{i-1}] \\ \chi(s), \;\; if \;\; s \in [-\gamma, 0]. \end{cases}$$

The following auxiliary lemma will provide existence criteria for solutions to problem (3.45).

Lemma 15 *The function $\varkappa \in C([-\gamma, \mathsf{T}_i], \mathbb{R})$ is a solution to problem (3.45) if and only if $\varkappa$ fulfills the integral equation:*

$$\varkappa(s) = \begin{cases} -\int_{\mathsf{T}_{i-1}}^{\mathsf{T}_i} G_i(s,t) f(t, \varkappa_t) dt, \quad s \in J_i, \\ \chi_i(s), \quad s \in [-\gamma, \mathsf{T}_{i-1}], \end{cases} \tag{3.46}$$

where $G_i(s,t)$ is the Green's function defined by:

$$G_i(s,t) = \begin{cases} \frac{1}{\Gamma(\ltimes_i)}\Big[(\mathsf{T}_i - \mathsf{T}_{i-1})^{1-\ltimes_i}(s - \mathsf{T}_{i-1})^{\ltimes_i - 1}(\mathsf{T}_i - t)^{\ltimes_i - 1} - (s-t)^{\ltimes_i - 1}\Big], \;\; \mathsf{T}_{i-1} \leq t \leq s \leq \mathsf{T}_i, \\ \frac{1}{\Gamma(\ltimes_i)}(\mathsf{T}_i - \mathsf{T}_{i-1})^{1-\ltimes_i}(s - \mathsf{T}_{i-1})^{\ltimes_i - 1}(\mathsf{T}_i - t)^{\ltimes_i - 1}, \;\; \mathsf{T}_{i-1} \leq s \leq t \leq \mathsf{T}_i, \end{cases} \tag{3.47}$$

$i = 1, 2, \ldots, n.$

Proof
Let $\varkappa \in C([-\gamma, \mathsf{T}_i], \mathbb{R})$ be a solution to problem (3.45). From Lemma 1, we have

$$\varkappa(s) = \omega_1(s - \mathsf{T}_{i-1})^{\ltimes_i - 1} + \omega_2(s - \mathsf{T}_{i-1})^{\ltimes_i - 2} + I^{\ltimes_i}_{\mathsf{T}_{i-1}^+} f(s, \varkappa_s), \quad s \in J_i, \; i \in \{1, 2, \ldots, n\}. \tag{3.48}$$

Using the conditions $\varkappa(\mathsf{T}_{i-1}) = \varkappa(\mathsf{T}_i) = 0$ we find that $\omega_2 = 0$ and

$$\omega_1 = -(\mathsf{T}_i - \mathsf{T}_{i-1})^{1-\ltimes_i} \, I^{\ltimes_i}_{\mathsf{T}_{i-1}^+} f(\mathsf{T}_i, \varkappa_{\mathsf{T}_i}).$$

Substituting the values of ω_1 and ω_2 into equation (3.48), we obtain:

$$\varkappa(s) = -(\mathsf{T}_i - \mathsf{T}_{i-1})^{1-\ltimes_i}(s - \mathsf{T}_{i-1})^{\ltimes_i - 1} \, I^{\ltimes_i}_{\mathsf{T}_{i-1}^+} f(\mathsf{T}_i, \varkappa_{\mathsf{T}_i}) + I^{\ltimes_i}_{\mathsf{T}_{i-1}^+} f(s, \varkappa_s), \quad s \in J_i. \tag{3.49}$$

Next, Problem (3.45) is solved as follows:

$$\begin{aligned}\varkappa(s) &= -(\mathsf{T}_i-\mathsf{T}_{i-1})^{1-\ltimes_i}(s-\mathsf{T}_{i-1})^{\ltimes_i-1}\frac{1}{\Gamma(\ltimes_i)}\int_{\mathsf{T}_{i-1}}^{\mathsf{T}_i}(\mathsf{T}_i-t)^{\ltimes_i-1}f(t,\varkappa_t)dt\\
&+ \frac{1}{\Gamma(\ltimes_i)}\int_{\mathsf{T}_{i-1}}^{s}(s-t)^{\ltimes_i-1}f(t,\varkappa_t)dt\\
&= -\frac{1}{\Gamma(\ltimes_i)}\Big[\int_{\mathsf{T}_{i-1}}^{s}\Big[(\mathsf{T}_i-\mathsf{T}_{i-1})^{1-\ltimes_i}(s-\mathsf{T}_{i-1})^{\ltimes_i-1}(\mathsf{T}_i-t)^{\ltimes_i-1}-(s-t)^{\ltimes_i-1}\Big]f(t,\varkappa_t)dt\\
&+ \int_{s}^{\mathsf{T}_i}(\mathsf{T}_i-\mathsf{T}_{i-1})^{1-\ltimes_i}(s-\mathsf{T}_{i-1})^{\ltimes_i-1}(\mathsf{T}_i-t)^{\ltimes_i-1}f(t,\varkappa_t)dt\Big]\\
&= -\Big[\int_{\mathsf{T}_{i-1}}^{s}G_i(s,t)f(t,\varkappa_t)dt+\int_{s}^{\mathsf{T}_i}G_i(s,t)f(t,\varkappa_t)dt\Big]\end{aligned}$$

By the continuity of the Green's function, this simplifies to:

$$\varkappa(s) = -\int_{\mathsf{T}_{i-1}}^{\mathsf{T}_i}G_i(s,t)f(t,\varkappa_t)dt,\ s\in J_i.$$

On the other hand, if $\varkappa \in C([-\gamma,\mathsf{T}_i],\mathbb{R})$ is a solution to the integral equation (3.46), then by the continuity of δf and Lemma 1, we can readily conclude that κ is a solution to Problem (3.45).

Proposition 4 *[123]. Let* $0<\delta<1$*, and suppose that* $s^{\delta}f : J_i\times C([-\gamma,0],\mathbb{R})\to\mathbb{R}$ *is continuous, and that* $\ltimes : J_i\to(1,2]$ *satisfies condition* $(H1)$*. Then, the Green's function for the problem* (3.45) *exhibits the following properties:*

(1) $G_i(s,t)\geq 0$ *for all* $\mathsf{T}_{i-1}\leq s,t\leq\mathsf{T}_i$,

(2) $\max_{s\in J_i}G_i(s,t)=G_i(t,t), t\in J_i$,

(3) $G_i(s,s)$ *has a singular maximum given by*

$$\max_{t\in J_i}G_i(t,t)=\frac{1}{\Gamma(\ltimes_i)}\left(\frac{\mathsf{T}_i-\mathsf{T}_{i-1}}{4}\right)^{\ltimes_i-1}, \tag{3.50}$$

where $i=1,2,\ldots,n$.

We now proceed to establish existence results for the Riemann–Liouville constant-order fractional problem (3.45). The first result is based on Darbo's fixed point theorem.

Theorem 20 *Suppose that both conditions $(H1)$ and $(H2)$ hold, and that*

$$\frac{K\Big(\mathsf{T}_i^{1-\delta}-\mathsf{T}_{i-1}^{1-\delta}\Big)\Big(\mathsf{T}_i-\mathsf{T}_{i-1}\Big)^{\ltimes_i-1}}{4^{\ltimes_i-1}(1-\delta)\Gamma(\ltimes_i)}<1. \tag{3.51}$$

Then, at least one solution to the R-Liouville constant-order fractional problem (3.45) exists on $C([-\gamma,\mathsf{T}_i],\mathbb{R})$.

Proof

Consider the operator

$$W: C([-\gamma,\mathsf{T}_i],\mathbb{R})\to C([-\gamma,\mathsf{T}_i],\mathbb{R}),$$

defined by

$$(W\varkappa)(s)=\begin{cases}\chi_i(s), & s\in[-\gamma,\mathsf{T}_{i-1}],\\ -\int_{\mathsf{T}_{i-1}}^{\mathsf{T}_i}G_i(s,t)f(t,\varkappa_t)dt, & s\in J_i.\end{cases}$$

Let $v(.):\ [-\gamma,\mathsf{T}_i]\to\mathbb{R}$ be a function defined by

$$v(s)=\begin{cases}0, & if\ s\in J_i,\\ \chi_i(s), & if\ s\in[-\gamma,\mathsf{T}_{i-1}].\end{cases}$$

For each $z\in C([\mathsf{T}_{i-1},\mathsf{T}_i],\mathbb{R})$, with $z(\mathsf{T}_{i-1})=0$, we define the extended function

$$\overline{z}(s)=\begin{cases}z(s), & if\ s\in J_i,\\ 0, & if\ s\in[-\gamma,\mathsf{T}_{i-1}].\end{cases}$$

If $\varkappa(\cdot)$ satisfies the integral equation

$$\varkappa(s)=-\int_{\mathsf{T}_{i-1}}^{\mathsf{T}_i}G_i(s,t)f(t,\varkappa_t)dt,$$

then we can decompose $\varkappa(s)=z(s)+v(s)$, which implies $\varkappa_s=\overline{z}_s+v_s$ for every $\mathsf{T}_{i-1}\le s\le\mathsf{T}_i$. The function $z(.)$ then satisfies

$$z(s)=-\int_{\mathsf{T}_{i-1}}^{\mathsf{T}_i}G_i(s,t)f(t,z_t+v_t)dt.$$

Let

$$C_{\mathsf{T}_{i-1}}=\{z\in C([\mathsf{T}_{i-1},\mathsf{T}_i],\mathbb{R}) \colon z(\mathsf{T}_{i-1})=0\}$$

and define the norm

$$\|z\|_{\mathsf{T}_i}=\sup_{s\in J_i}|z(s)|,\ z\in C_{\mathsf{T}_{i-1}}.$$

Thus, $C_{\mathsf{T}_{i-1}}$ is a ßanach space with the norm $\|.\|_{\mathsf{T}_i}$. Define the operator $\mathcal{P}: C_{\mathsf{T}_{i-1}} \to C_{\mathsf{T}_{i-1}}$,

$$(\mathcal{P}z)(s) = -\int_{\mathsf{T}_{i-1}}^{\mathsf{T}_i} G_i(s,t) f(t, \overline{z}_t + v_t)dt, \; s \in J_i. \tag{3.52}$$

The continuity of function $s^\delta f$ and the properties of fractional integrals imply that the operator $\mathcal{P}$ is well-defined.

It is therefore sufficient to show that the operator $\mathcal{P}$ has a fixed point z, which guarantees that the operator W has a fixed point $\varkappa = \overline{z} + v$. Consequently, this fixed point corresponds to a solution of problem (3.45). Indeed:

$$\begin{aligned}
\varkappa(s) &= \overline{z}(s) + v(s) \\
&= \begin{cases} z(s), & if\ s \in J_i, \\ \chi_i(s), & if\ s \in [-\gamma, \mathsf{T}_{i-1}] \end{cases} \\
&= \begin{cases} -\int_{\mathsf{T}_{i-1}}^{\mathsf{T}_i} G_i(s,t) f(t, \overline{z}_t + v_t)dt, & if\ mathfrakt \in J_i, \\ \chi_i(s), & if\ mathfrakt \in [-\gamma, \mathsf{T}_{i-1}] \end{cases} \\
&= \begin{cases} -\int_{\mathsf{T}_{i-1}}^{\mathsf{T}_i} G_i(s,t) f(t, \varkappa_t)dt, & if\ s \in J_i, \\ \chi_i(s), & if\ s \in [-\gamma, \mathsf{T}_{i-1}] \end{cases} \\
&= (W\varkappa)(s).
\end{aligned}$$

Let

$$R_i \geq \frac{\dfrac{(K\|\chi\|_{[-\gamma,0]} + f^\star)(\mathsf{T}_i - \mathsf{T}_{i-1})^{\Bbbk_i - 1}\left(\mathsf{T}_i^{1-\delta} - \mathsf{T}_{i-1}^{1-\delta}\right)}{4^{\Bbbk_i - 1}\Gamma(\Bbbk_i)(1-\delta)}}{1 - \dfrac{K\left(\mathsf{T}_i^{1-\delta} - \mathsf{T}_{i-1}^{1-\delta}\right)\left(\mathsf{T}_i - \mathsf{T}_{i-1}\right)^{\Bbbk_i - 1}}{4^{\Bbbk_i - 1}(1-\delta)\Gamma(\Bbbk_i)}} \tag{3.53}$$

where $f^\star = \sup_{t \in J} s^\delta |f(s,0)|$, and consider the set

$$B_{R_i} = \{z \in C_{\mathsf{T}_{i-1}}, \; \|z\|_{\mathsf{T}_i} \leq R_i\}.$$

Clearly, B_{R_i} is nonempty, convex, bounded, and closed.

For $z \in B_{R_i}$ and $s \in J_i$, we have:

$$\begin{aligned}
\|\overline{z}_s\|_{[-\gamma',0]} &= \sup_{-\mathsf{T}_{i-1}-\gamma \leq \theta \leq 0} |\overline{z}_s(\theta)| \\
&= \sup_{-\mathsf{T}_{i-1}-\gamma \leq \theta \leq 0} |\overline{z}(s+\theta)| \\
&\leq \sup_{-\gamma \leq t \leq \mathsf{T}_i} |\overline{z}(t)| \\
&= \sup_{t \in J_i} |z(t)| = \|z\|_{\mathsf{T}_i}
\end{aligned}$$

Similarly,

$$\begin{aligned}
\|v_s\|_{[-\gamma',0]} &= \sup_{-\mathsf{T}_{i-1}-\gamma\leq\theta\leq 0} |v_s(\theta)| \\
&= \sup_{-\mathsf{T}_{i-1}-\gamma\leq\theta\leq 0} |v(s+\theta)| \\
&\leq \sup_{-\gamma\leq t\leq \mathsf{T}_i} |v(t)| \\
&= \sup_{-\gamma\leq t\leq 0} |v(t)| = \sup_{-\gamma\leq t\leq 0} |\chi(t)| = \|\chi\|_{[-\gamma,0]}.
\end{aligned}$$

We shall show that the operator $\mathcal{P}$ satisfies the conditions of Darbo's fixed point theorem in five steps.

Step 1: We aim to show that $P(B_{R_i}) \subset B_{R_i}$.

Let $z \in B_{R_i}$. Based on Proposition 4 and assumption $(H2)$, we obtain:

$$\begin{aligned}
|\mathcal{P}z(s)| &= \Big| \int_{\mathsf{T}_{i-1}}^{\mathsf{T}_i} G_i(s,t) f(t, \overline{z}_t + v_t) dt \Big| \\
&\leq \int_{\mathsf{T}_{i-1}}^{\mathsf{T}_i} G_i(s,t) |f(t, \overline{z}_t + v_t)| dt \\
&\leq \frac{1}{\Gamma(\ltimes_i)} \Big(\frac{\mathsf{T}_i - \mathsf{T}_{i-1}}{4} \Big)^{\ltimes_i - 1} \int_{\mathsf{T}_{i-1}}^{\mathsf{T}_i} |f(t, \overline{z}_t + v_t)| dt \\
&\leq \frac{1}{\Gamma(\ltimes_i)} \Big(\frac{\mathsf{T}_i - \mathsf{T}_{i-1}}{4} \Big)^{\ltimes_i - 1} \int_{\mathsf{T}_{i-1}}^{\mathsf{T}_i} t^{-\delta} t^{\delta} \Big| f(t, \overline{z}_t + v_t) - f(t,0) \Big| dt \\
&+ \frac{1}{\Gamma(\ltimes_i)} \Big(\frac{\mathsf{T}_i - \mathsf{T}_{i-1}}{4} \Big)^{\ltimes_i - 1} \int_{\mathsf{T}_{i-1}}^{\mathsf{T}_i} t^{-\delta} t^{\delta} |f(t,0)| dt \\
&\leq \frac{1}{\Gamma(\ltimes_i)} \Big(\frac{\mathsf{T}_i - \mathsf{T}_{i-1}}{4} \Big)^{\ltimes_i - 1} \int_{\mathsf{T}_{i-1}}^{\mathsf{T}_i} t^{-\delta} (K \|\overline{z}_t + v_t\|_{[-\gamma',0]}) dt \\
&+ \frac{f^\star (\mathsf{T}_i - \mathsf{T}_{i-1})^{\ltimes_i - 1}}{\Gamma(\ltimes_i) 4^{\ltimes_i - 1}} \int_{\mathsf{T}_{i-1}}^{\mathsf{T}_i} t^{-\delta} dt
\end{aligned}$$

$$
\begin{aligned}
&\leq \frac{K}{\Gamma(\ltimes_i)}\Big(\frac{\mathsf{T}_i-\mathsf{T}_{i-1}}{4}\Big)^{\ltimes_i-1}\int_{\mathsf{T}_{i-1}}^{\mathsf{T}_i}(\|\overline{z}_t\|_{[-\gamma',0]}+\|v_t\|_{[-\gamma',0]})t^{-\delta}dt\\
&+ \frac{f^\star\Big(\mathsf{T}_i-\mathsf{T}_{i-1}\Big)^{\ltimes_i-1}\Big(\mathsf{T}_i^{1-\delta}-\mathsf{T}_{i-1}^{1-\delta}\Big)}{4^{\ltimes_i-1}\Gamma(\ltimes_i)(1-\delta)}\\
&\leq \frac{K}{\Gamma(\ltimes_i)}\Big(\frac{\mathsf{T}_i-\mathsf{T}_{i-1}}{4}\Big)^{\ltimes_i-1}(\|z\|_{\mathsf{T}_i}+\|\chi\|_{[-\gamma,0]})\int_{\mathsf{T}_{i-1}}^{\mathsf{T}_i}t^{-\delta}dt\\
&+ \frac{f^\star\Big(\mathsf{T}_i-\mathsf{T}_{i-1}\Big)^{\ltimes_i-1}\Big(\mathsf{T}_i^{1-\delta}-\mathsf{T}_{i-1}^{1-\delta}\Big)}{4^{\ltimes_i-1}\Gamma(\ltimes_i)(1-\delta)}\\
&\leq \frac{K}{\Gamma(\ltimes_i)}\Big(\frac{\mathsf{T}_i-\mathsf{T}_{i-1}}{4}\Big)^{\ltimes_i-1}R_i\Big(\frac{\mathsf{T}_i^{1-\delta}-\mathsf{T}_{i-1}^{1-\delta}}{1-\delta}\Big)\\
&+ \frac{K}{\Gamma(\ltimes_i)}\Big(\frac{\mathsf{T}_i-\mathsf{T}_{i-1}}{4}\Big)^{\ltimes_i-1}\|\chi\|_{[-\gamma,0]}\Big(\frac{\mathsf{T}_i^{1-\delta}-\mathsf{T}_{i-1}^{1-\delta}}{1-\delta}\Big)\\
&+ \frac{f^\star\Big(\mathsf{T}_i-\mathsf{T}_{i-1}\Big)^{\ltimes_i-1}\Big(\mathsf{T}_i^{1-\delta}-\mathsf{T}_{i-1}^{1-\delta}\Big)}{4^{\ltimes_i-1}\Gamma(\ltimes_i)(1-\delta)}\\
&\leq \frac{K\Big(\mathsf{T}_i^{1-\delta}-\mathsf{T}_{i-1}^{1-\delta}\Big)\Big(\mathsf{T}_i-\mathsf{T}_{i-1}\Big)^{\ltimes_i-1}}{4^{\ltimes_i-1}(1-\delta)\Gamma(\ltimes_i)}R_i\\
&+ \frac{\Big(\mathsf{T}_i-\mathsf{T}_{i-1}\Big)^{\ltimes_i-1}\Big(\mathsf{T}_i^{1-\delta}-\mathsf{T}_{i-1}^{1-\delta}\Big)}{4^{\ltimes_i-1}\Gamma(\ltimes_i)(1-\delta)}\Big(K\|\chi\|_{[-\gamma,0]}+f^\star\Big)\\
&\leq R_i,
\end{aligned}
$$

which implies $\mathcal{P}(B_{R_i})\subseteq B_{R_i}$.

Step 2: $\mathcal{P}$ is continuous. Let $(z_n)\subset C_{\mathsf{T}_{i-1}}$ be a sequence converging to $z\in C_{\mathsf{T}_{i-1}}$, and let $s\in J_i$. Then,

$$
\begin{aligned}
|\mathcal{P}(z_n)(s)-(\mathcal{P}z)(s)| &\leq \int_{\mathsf{T}_{i-1}}^{\mathsf{T}_i}G_i(s,t)\Big|f(t,\overline{z_n}_t+v_t)-f(t,\overline{z}_t+v_t)\Big|dt\\
&\leq \frac{1}{\Gamma(\ltimes_i)}\Big(\frac{\mathsf{T}_i-\mathsf{T}_{i-1}}{4}\Big)^{\ltimes_i-1}\int_{\mathsf{T}_{i-1}}^{\mathsf{T}_i}\Big|f(t,\overline{z_n}_t+v_t)-f(t,\overline{z}_t+v_t)\Big|dt\\
&\leq \frac{1}{\Gamma(\ltimes_i)}\Big(\frac{\mathsf{T}_i-\mathsf{T}_{i-1}}{4}\Big)^{\ltimes_i-1}\int_{\mathsf{T}_{i-1}}^{\mathsf{T}_i}t^{-\delta}K\|\overline{z_n}_t-\overline{z}_t\|_{[-\gamma',0]}dt
\end{aligned}
$$

$$\le \frac{1}{\Gamma(\ltimes_i)}\Big(\frac{\mathsf{T}_i-\mathsf{T}_{i-1}}{4}\Big)^{\ltimes_i-1}(K\|z_n-z\|_{\mathsf{T}_i})\int_{\mathsf{T}_{i-1}}^{\mathsf{T}_i} t^{-\delta}dt$$

$$\le \frac{K\Big(\mathsf{T}_i^{1-\delta}-\mathsf{T}_{i-1}^{1-\delta}\Big)\Big(\mathsf{T}_i-\mathsf{T}_{i-1}\Big)^{\ltimes_i-1}}{4^{\ltimes_i-1}(1-\delta)\Gamma(\ltimes_i)}\|z_n-z\|_{\mathsf{T}_i}.$$

Hence, we get

$$\|(\mathcal{P}z_n)-(\mathcal{P}z)\|_{\mathsf{T}_i}\to 0 \text{ as } n\to\infty. \tag{3.54}$$

This proves that the operator $\mathcal{P}$ is a relatively compact.

Step 3: $\mathcal{P}(B_{R_i})$ is restricted set in $C_{\mathsf{T}_{i-1}}$.
As in Step 1, we have $\mathcal{P}(B_{R_i})\subset B_{R_i}$. It implies that $\mathcal{P}(B_{R_i})$ is bounded set in $C_{\mathsf{T}_{i-1}}$.

Step 4: $\mathcal{P}(B_{R_i})$ is an Equicontinuous Set in $C_{\mathsf{T}_{i-1}}$.
Let $z\in B_{R_i}$, and let $s_1,s_2\in J_i$ with $s_1<s_2$. We estimate

$$|\mathcal{P}(z)(s_2)-(\mathcal{P}z)(s_1)|$$

$$\begin{aligned}
&= \Big|\int_{\mathsf{T}_{i-1}}^{\mathsf{T}_i} G_i(s_2,t)f(t,\overline{z}_t+v_t)dt-\int_{\mathsf{T}_{i-1}}^{\mathsf{T}_i} G_i(s_1,t)f(t,\overline{z}_t+v_t)dt\Big|\\
&\le \int_{\mathsf{T}_{i-1}}^{\mathsf{T}_i}\Big|\Big(G_i(s_2,t)-G_i(s_1,t)\Big)f(t,\overline{z}_t+v_t)\Big|dt\\
&\le \int_{\mathsf{T}_{i-1}}^{\mathsf{T}_i}\Big|G_i(s_2,t)-G_i(s_1,t)\Big||f(t,\overline{z}_t+v_t)|dt\\
&\le \int_{\mathsf{T}_{i-1}}^{\mathsf{T}_i}\Big|G_i(s_2,t)-G_i(s_1,t)\Big|t^{-\delta}\Big(t^{\delta}\Big|f(t,\overline{z}_t+v_t)-f(t,0)\Big|+t^{\delta}|f(t,0)|\Big)dt\\
&\le \int_{\mathsf{T}_{i-1}}^{\mathsf{T}_i}\Big|G_i(s_2,t)-G_i(s_1,t)\Big|\Big[t^{-\delta}(K\|\overline{z}_t+v_t\|_{[-\gamma',0]})+t^{-\delta}f^\star\Big]dt\\
&\le \int_{\mathsf{T}_{i-1}}^{\mathsf{T}_i}\Big|G_i(s_2,t)-G_i(s_1,t)\Big|\Big[t^{-\delta}K(\|\overline{z}_t\|_{[-\gamma',0]}+\|v_t\|_{[-\gamma',0]})+t^{-\delta}f^\star\Big]dt\\
&\le \int_{\mathsf{T}_{i-1}}^{\mathsf{T}_i}\Big|G_i(s_2,t)-G_i(s_1,t)\Big|\Big[t^{-\delta}K(\|z\|_{\mathsf{T}_i}+\|\chi\|_{[-\gamma,0]})+f^\star\Big]dt\\
&\le K\,\mathsf{T}_{i-1}^{-\delta}\,(R+\|\chi\|_{[-\gamma,0]})\int_{\mathsf{T}_{i-1}}^{\mathsf{T}_i}\Big|G_i(s_2,t)-G_i(s_1,t)\Big|dt\\
&+ f^\star\,\mathsf{T}_{i-1}^{-\delta}\int_{\mathsf{T}_{i-1}}^{\mathsf{T}_i}\Big|G_i(s_2,t)-G_i(s_1,t)\Big|dt.
\end{aligned}$$

Therefore, $|\mathcal{P}(z)(s_2) - (\mathcal{P}z)(s_1)| \to 0$ as $|s_2 - s_1| \to 0$, uniformly for all $z \in B_{R_i}$. This implies that $\mathcal{P}(B_{R_i})$ is equicontinuous in $C_{\mathsf{T}_{i-1}}$.

Remark 10 *As noted in [23] the inequality*

$$\varsigma\Big(s^{\delta} f(s, B_1)\Big) \leq K\varsigma_{[-\gamma,0]}(B_1)$$

is equivalent to condition (H2) for each bounded subset $B_1 \subset C([-\gamma, 0], \mathbb{R})$ and for all $s \in J$, where B_1 is limited.

Step 5: $\mathcal{P}$ is an L-contraction mapping.
For $U \subset B_{R_i}$, $s \in J_i$, we obtain:

$$\begin{aligned}
\varsigma(\mathcal{P}(U)(s)) &= \varsigma(\Big\{(\mathcal{P}z)(s), z \in U\Big\}) \\
&= \varsigma(\Big\{ - \textstyle\int_{\mathsf{T}_{i-1}}^{\mathsf{T}_i} G_i(s,t) f(t, \overline{z}_t + v_t)dt, z \in U\Big\}) \\
&\leq \int_{\mathsf{T}_{i-1}}^{\mathsf{T}_i} G_i(s,t)\varsigma(\Big\{f(t, \overline{z}_t + v_t), z \in U\Big\}) \\
&\leq \int_{\mathsf{T}_{i-1}}^{\mathsf{T}_i} G_i(s,t)t^{-\delta}\varsigma(\Big\{t^{\delta} f(t, \overline{z}_t + v_t), z \in U\Big\}).
\end{aligned}$$

By Remark 10, it follows that:

$$\begin{aligned}
\varsigma(\mathcal{P}(U)(s)) &\leq \int_{\mathsf{T}_{i-1}}^{\mathsf{T}_i} G_i(s,t)t^{-\delta}[K(\varsigma_{[-\gamma',0]}\{\overline{z}_t + v_t, z \in U\})]dt \\
&\leq \int_{\mathsf{T}_{i-1}}^{\mathsf{T}_i} G_i(s,t)t^{-\delta}[K\varsigma_{[-\gamma',0]}(\{\overline{z}_t, z \in U\} + v_t)]dt \\
&\leq \int_{\mathsf{T}_{i-1}}^{\mathsf{T}_i} G_i(s,t)t^{-\delta}K[\varsigma_{[-\gamma',0]}(\{\overline{z}_t, z \in U\})]dt \\
&\leq \int_{\mathsf{T}_{i-1}}^{\mathsf{T}_i} G_i(s,t)t^{-\delta}K \sup_{-\gamma' \leq \theta \leq 0} \varsigma(\{\overline{z}_t(\theta), z \in U\}dt \\
&\leq \int_{\mathsf{T}_{i-1}}^{\mathsf{T}_i} G_i(s,t)t^{-\delta}K \sup_{-\gamma' \leq \theta \leq 0} \varsigma(\{\overline{z}(t+\theta), z \in U\})dt \\
&\leq \int_{s_{i-1}}^{s_i} G_i(s,t)t^{-\delta}K \sup_{-r \leq s \leq \mathsf{T}_i} \varsigma(\{\overline{z}(s), z \in U\})dt \\
&= \int_{\mathsf{T}_{i-1}}^{\mathsf{T}_i} G_i(s,t)t^{-\delta}K \sup_{\mathsf{T}_{i-1} \leq s \leq \mathsf{T}_i} \varsigma(\{\overline{z}(s), z \in U\} \cup \{0\})dt \\
&\leq \int_{\mathsf{T}_{i-1}}^{\mathsf{T}_i} G_i(s,t)t^{-\delta}K \sup_{\mathsf{T}_{i-1} \leq s \leq \mathsf{T}_i} \varsigma(\{\overline{z}(s), z \in U\})dt \\
&\leq \int_{\mathsf{T}_{i-1}}^{\mathsf{T}_i} G_i(s,t)t^{-\delta}K \sup_{\mathsf{T}_{i-1} \leq s \leq \mathsf{T}_i} \varsigma(\{z(s), z \in U\})dt \\
&\leq \int_{\mathsf{T}_{i-1}}^{\mathsf{T}_i} G_i(s,t)t^{-\delta}K \sup_{s \in J_i} \varsigma(U(s))dt
\end{aligned}$$

$$\leq \frac{1}{\Gamma(\ltimes_i)}\left(\frac{\mathsf{T}_i - \mathsf{T}_{i-1}}{4}\right)^{\ltimes_i - 1}\left[K\varsigma_{J_i}(U)\int_{\mathsf{T}_{i-1}}^{\mathsf{T}_i} t^{-\delta}dt\right],$$

$$\leq \frac{K\left(\mathsf{T}_i^{1-\delta} - \mathsf{T}_{i-1}^{1-\delta}\right)\left(\mathsf{T}_i - \mathsf{T}_{i-1}\right)^{\ltimes_i - 1}}{4^{\ltimes_i - 1}(1-\delta)\Gamma(\ltimes_i)}\varsigma_{J_i}(U).$$

Hence,

$$\varsigma_{J_i}(\mathcal{P}U) \leq \frac{K\left(\mathsf{T}_i^{1-\delta} - \mathsf{T}_{i-1}^{1-\delta}\right)\left(\mathsf{T}_i - \mathsf{T}_{i-1}\right)^{\ltimes_i - 1}}{4^{\ltimes_i - 1}(1-\delta)\Gamma(\ltimes_i)}\varsigma_{J_i}(U).$$

Thus, based on (3.51), it follows that $\mathcal{P}$ is an L-contractive mapping, where

$$L := \frac{K\left(\mathsf{T}_i^{1-\delta} - \mathsf{T}_{i-1}^{1-\delta}\right)\left(\mathsf{T}_i - \mathsf{T}_{i-1}\right)^{\ltimes_i - 1}}{4^{\ltimes_i - 1}(1-\delta)\Gamma(\ltimes_i)}. \tag{3.55}$$

Therefore, since all conditions of Darbo's fixed point theorem are satisfied, we deduce that $\mathcal{P}$ has a fixed point $z_i \in B_{R_i}$.

Hence, there exists at least one solution to the Riemann–Liouville constant-order fractional initial and terminal value problem (3.45), given by the formula $\varkappa_i = \overline{z_i} + v \in C([-\gamma, \mathsf{T}_i], \mathbb{R})$, because W has a fixed point.

We now proceed to demonstrate the existence result for the variable order Riemann–Liouville fractional problem (3.40).

Theorem 21 *Assume that the conditions (H1), (H2), and inequality (3.51) hold for all $i \in \{1, 2, \dots, n\}$. Then, the Riemann–Liouville fractional problem of variable order (3.40) admits at least one solution in $C([-\gamma, \mathsf{T}], \mathbb{R})$.*

Proof

For all $i \in \{1, 2, \dots, n\}$, as stated in Theorem 20, the Riemann–Liouville constant-order fractional initial and terminal value problem (3.45) admits at least one solution $\varkappa_i \in C([-\gamma, \mathsf{T}_i], \mathbb{R})$. For any $i \in \{1, 2, \dots, n\}$, we have

$$\varkappa_1(s) = \overline{z_1}(s) + v(s) = \begin{cases} \chi(s), & s \in [-\gamma, 0], \\ z_1(s), & s \in J_1 \end{cases}$$

For any $i \in \{2, \dots, n\}$, we have

$$\varkappa_i(s) = \overline{z_i}(s) + v(s) = \begin{cases} \chi(s), & s \in [-\gamma, 0], \\ 0, & s \in [0, \mathsf{T}_{i-1}], \\ z_i(s), & s \in J_i. \end{cases}$$

Therefore, the function $\varkappa_i \in C([-\gamma, \mathsf{T}_i], \mathbb{R})$ satisfies the integral equation (3.43) for $s \in J_i$, with initial and terminal conditions $\varkappa_i(0) = 0$, $\varkappa_i(\mathsf{T}_i) = z_i(\mathsf{T}_i) = 0$, and the initial condition $\varkappa_i(s) = \chi(s)$ for $s \in [-\gamma, 0]$.

As a result, the function $\kappa(s)$ defined by

$$\varkappa_1(s) = \begin{cases} \chi(s), & s \in [-\gamma, 0], \\ z_1(s), & s \in J_1, \end{cases}$$
$$\varkappa_2(s) = \begin{cases} \chi(s), & s \in [-\gamma, 0], \\ 0, & s \in J_1, \\ z_2(s), & s \in J_2, \end{cases} \quad .$$
$$\vdots$$
$$\varkappa_n(s) = \begin{cases} \chi(s), & s \in [-\gamma, 0], \\ 0, & s \in [0, \mathsf{T}_{n-1}], \\ z_n(s), & s \in J_n, \end{cases}$$

yields the solution to the Riemann–Liouville fractional problem of variable order (3.40).

3.4.2 Ulam-Hyers Stability

To illustrate the relevance of our solutions, we establish Ulam–Hyers stability criteria—essential for understanding their qualitative behavior.

Theorem 22 *Assume that conditions* $(H1)$, $(H2)$, *and* (3.51) *hold. Then the equation* (3.40) *is Ulam-Hyers stable.*

Proof
Let $\epsilon > 0$ be an arbitrary positive value, and assume that the function $y \in C([-\gamma, \mathsf{T}], \mathbb{R})$ satisfies the inequality (2.23).

We define the functions

$$y_1(s) = \begin{cases} y(s), & s \in [0, \mathsf{T}_1], \\ \chi(s), & s \in [-\gamma, 0] \end{cases} \tag{3.56}$$

and for $i = 2, 3, \ldots, n$:

$$y_i(s) = \begin{cases} \chi(s), & s \in [-\gamma, 0], \\ 0, & s \in [0, \mathsf{T}_{i-1}], \\ y(s), & s \in J_i. \end{cases} \tag{3.57}$$

For all $i \in \{1, 2, \ldots, n\}$, from equation (3.43) for $s \in J_i$, we conclude that for $s \in J_i$:

$$F_{0+}^{\ltimes(s)} y_i(s) = \frac{1}{\Gamma(2-\ltimes_i)}\left(\frac{d}{dt}\right)^2 \int_{\mathsf{T}_{i-1}}^{t} (s-t)^{1-\ltimes_i} y(t)dt. \tag{3.58}$$

Taking the operator $I_{\mathsf{T}_{i-1}^+}^{\ltimes_i}$ on both sides of (2.23), we obtain

$$\begin{aligned}
&\left| y(s) + \int_{\mathsf{T}_{i-1}}^{\mathsf{T}_i} G_i(s,t) f(t, y_t)dt \right| \\
&\qquad \leq \frac{\epsilon}{\Gamma(\ltimes_i)} \int_{\mathsf{T}_{i-1}}^{s} (s-t)^{\ltimes_i - 1} dt \\
&\qquad \leq \epsilon \frac{\left(\mathsf{T}_i - \mathsf{T}_{i-1}\right)^{\ltimes_i}}{\Gamma(\ltimes_i + 1)}.
\end{aligned}$$

As stated in Theorem 21, the Riemann–Liouville fractional problem (3.40) of variable order admits a solution $\varkappa \in C([-\gamma, \mathsf{T}], \mathbb{R})$, defined piecewise by $\varkappa(s) = \varkappa_i(s)$ for $s \in [0, \mathsf{T}_i]$, where $i = 1, 2, \ldots, n$, and

$$\varkappa_1(s) = \begin{cases} \chi(s), & s \in [-\gamma, 0] \\ z_1(s), & s \in J_1, \end{cases} \tag{3.59}$$

and for any $i \in \{2, \ldots, n\}$

$$\varkappa_i(s) = \begin{cases} \chi(s), & s \in [-\gamma, 0], \\ 0, & s \in [0, \mathsf{T}_{i-1}], \\ z_i(s), & s \in J_i \end{cases} \tag{3.60}$$

Moreover, each $\varkappa_i \in C([-\gamma, \mathsf{T}_i], \mathbb{R})$ satisfies the constant-order Riemann–Liouville fractional problem (3.45). By Lemma 15, it follows that

$$\varkappa_i(s) = -\int_{\mathsf{T}_{i-1}}^{\mathsf{T}_i} G_i(s,t) f(t, (\varkappa_i)_t)dt. \tag{3.61}$$

Let $s \in J_i$, and $k \in \{1, 2, \ldots, n\}$. Then, by equations (3.57), (3.59), (3.60) and (3.61), we obtain:

$$\begin{aligned}
|y(s) - \varkappa(s)| &= |y(s) - \varkappa_i(s)| = |y_i(s) - \varkappa_i(s)| \\
&= |y_i(s) + \int_{\mathsf{T}_{i-1}}^{\mathsf{T}_i} G_i(s,t) f(t, (\varkappa_i)_t) dt| \\
&\leq \Big| y_i(s) + \int_{\mathsf{T}_{i-1}}^{\mathsf{T}_i} G_i(s,t) f(t, (y_i)_t) dt \Big| + \int_{\mathsf{T}_{i-1}}^{\mathsf{T}_i} G_i(s,t) \Big| f(t, (y_i)_t) - f(t, (\varkappa_i)_t) \Big| dt \\
&\leq \epsilon \frac{\Big(\mathsf{T}_i - \mathsf{T}_{i-1} \Big)^{\ltimes_i}}{\Gamma(\ltimes_i + 1)} \\
&+ K \frac{1}{\Gamma(\ltimes_i)} \Big(\frac{\mathsf{T}_i - \mathsf{T}_{i-1}}{4} \Big)^{\ltimes_i - 1} \int_{\mathsf{T}_{i-1}}^{\mathsf{T}_i} t^{-\delta} \|(y_i)_t - (\varkappa_i)_t\|_{[-\gamma', 0]} dt \\
&\leq \epsilon \frac{\Big(\mathsf{T}_i - \mathsf{T}_{i-1} \Big)^{\ltimes_i}}{\Gamma(\ltimes_i + 1)} \\
&+ K \frac{1}{\Gamma(\ltimes_i)} \Big(\frac{\mathsf{T}_i - \mathsf{T}_{i-1}}{4} \Big)^{\ltimes_i - 1} \int_{\mathsf{T}_{i-1}}^{\mathsf{T}_i} t^{-\delta} \sup_{-\mathsf{T}_{i-1} - \gamma \leq \theta \leq 0} |(y_i)_t(\theta) - (\varkappa_i)_t(\theta)| dt \\
&\leq \epsilon \frac{\Big(\mathsf{T}_i - \mathsf{T}_{i-1} \Big)^{\ltimes_i}}{\Gamma(\ltimes_i + 1)} \\
&+ K \frac{1}{\Gamma(\ltimes_i)} \Big(\frac{\mathsf{T}_i - \mathsf{T}_{i-1}}{4} \Big)^{\ltimes_i - 1} \int_{\mathsf{T}_{i-1}}^{\mathsf{T}_i} t^{-\delta} \sup_{-\mathsf{T}_{i-1} - \gamma \leq \theta \leq 0} |y_i(t + \theta) - \varkappa_i(t + \theta)| dt \\
&\leq \epsilon \frac{\Big(\mathsf{T}_i - \mathsf{T}_{i-1} \Big)^{\ltimes_i}}{\Gamma(\ltimes_i + 1)} \\
&+ K \frac{1}{\Gamma(\ltimes_i)} \Big(\frac{\mathsf{T}_i - \mathsf{T}_{i-1}}{4} \Big)^{\ltimes_i - 1} \int_{\mathsf{T}_{i-1}}^{\mathsf{T}_i} t^{-\delta} \sup_{-\gamma \leq s \leq \mathsf{T}_i} |y_i(s) - \varkappa_i(s)| dt
\end{aligned}$$

$$\begin{aligned}
&\leq \epsilon \frac{\Big(\mathsf{T}_i - \mathsf{T}_{i-1} \Big)^{\ltimes_i}}{\Gamma(\ltimes_i + 1)} \\
&+ K \frac{1}{\Gamma(\ltimes_i)} \Big(\frac{\mathsf{T}_i - \mathsf{T}_{i-1}}{4} \Big)^{\ltimes_i - 1} \int_{\mathsf{T}_{i-1}}^{\mathsf{T}_i} t^{-\delta} \|y_i - \varkappa_i\|_{[-\gamma, \mathsf{T}_i]} dt \\
&\leq \epsilon \frac{\Big(\mathsf{T}_i - \mathsf{T}_{i-1} \Big)^{\ltimes_i}}{\Gamma(\ltimes_i + 1)} \\
&+ K \frac{1}{\Gamma(\ltimes_i)} \Big(\frac{\mathsf{T}_i - \mathsf{T}_{i-1}}{4} \Big)^{\ltimes_i - 1} \|y_i - \varkappa_i\|_{[-\gamma, \mathsf{T}_i]} \int_{\mathsf{T}_{i-1}}^{\mathsf{T}_i} t^{-\delta} dt \\
&\leq \epsilon \frac{\Big(\mathsf{T}_i - \mathsf{T}_{i-1} \Big)^{\ltimes_i}}{\Gamma(\ltimes_i + 1)} + \frac{K \Big(\mathsf{T}_i^{1-\delta} - \mathsf{T}_{i-1}^{1-\delta} \Big) \Big(\mathsf{T}_i - \mathsf{T}_{i-1} \Big)^{\ltimes_i - 1}}{4^{\ltimes_i - 1} (1 - \delta) \Gamma(\ltimes_i)} \|y_i - \varkappa_i\|_{[-\gamma, \mathsf{T}_i]} \\
&\leq \epsilon \frac{\Big(\mathsf{T}_i - \mathsf{T}_{i-1} \Big)^{\ltimes_i}}{\Gamma(\ltimes_i + 1)} + \mu \|y_i - \varkappa_i\|_{[-\gamma, \mathsf{T}_i]},
\end{aligned}$$

where

$$\mu = \max_{i=1,2,\dots,n} \frac{K\left(\mathsf{T}_i^{1-\delta} - \mathsf{T}_{i-1}^{1-\delta}\right)\left(\mathsf{T}_i - \mathsf{T}_{i-1}\right)^{\ltimes_i - 1}}{4^{\ltimes_i - 1}(1-\delta)\Gamma(\ltimes_i)}. \tag{3.62}$$

and setting

$$\|y - \varkappa\|_{[-\gamma,\mathsf{T}_i]}(1-\mu) \leq \epsilon \frac{\left(\mathsf{T}_i - \mathsf{T}_{i-1}\right)^{\ltimes_i}}{\Gamma(\ltimes_i + 1)}, \tag{3.63}$$

we obtain $c_f := \frac{\left(\mathsf{T}_i - \mathsf{T}_{i-1}\right)^{\ltimes_i}}{(1-\mu)\Gamma(\ltimes_i+1)}$,

$$\|y - \varkappa\|_{[-\gamma,\mathsf{T}_i]} \leq c_f \epsilon, \tag{3.64}$$

i.e.,

$$|y(s) - \varkappa(s)| \leq c_f \epsilon, \quad s \in [-\gamma, \mathsf{T}_i]. \tag{3.65}$$

Accordingly, the variable-order Riemann–Liouville fractional problem (3.40) is Ulam–Hyers stable.

3.4.3 Numerical Approaches

Numerical approach 1 Let $\gamma > 0$, and define the variable-order function:

$$\ltimes(s) = \begin{cases} \frac{7}{5}, & s \in J_1 := [0,1], \\ \frac{3}{2}, & s \in J_2 :=]1,2] \end{cases} \tag{3.66}$$

Consider the following Riemann–Liouville fractional variable-order initial and terminal value problem:

$$\begin{cases} F_{0^+}^{\ltimes(s)} \varkappa(s) = \frac{s^{-\frac{1}{2}}}{4e^s(1+\|\varkappa_s\|_{[-\gamma,0]})}, & s \in J :=]0,2], \\ \varkappa(s) = \chi(s), & s \in [-\gamma, 0]. \end{cases} \tag{3.67}$$

The choice of $\ltimes(s)$ ensures that condition $(H1)$ holds. Let

$$f(s, y_s) = \frac{s^{-\frac{1}{2}}}{4e^s(1 + \|y_s\|_{[-\gamma,0]})}, \quad (s, y_s) \in [0,2] \times C([-\gamma,0],\mathbb{R}).$$

For $y, z \in C([-\gamma, 2], \mathbb{R})$ and $s \in J$, we have

$$\begin{aligned} s^{\frac{1}{2}}|f(s,y_s) - f(s,z_s)| &= \left|\frac{1}{4e^s}\left(\frac{1}{1+\|y_s\|_{[-\gamma,0]}} - \frac{1}{1+\|z_s\|_{[-\gamma,0]}}\right)\right| \\ &\leq \frac{\left|\|y_s\|_{[-\gamma,0]} - \|z_s\|_{[-\gamma,0]}\right|}{4e^s(1+\|y_s\|_{[-\gamma,0]})(1+\|y_s\|_{[-\gamma,0]})} \\ &\leq \frac{1}{4e^s}(\|y_s - z_s\|_{[-\gamma,0]}) \\ &\leq \frac{1}{4}\|y_s - z_s\|_{[-\gamma,0]}. \end{aligned}$$

Hence, condition $(H2)$ holds for $\delta = \frac{1}{2}$ and $K = \frac{1}{4}$.

We now analyze two auxiliary problems for constant-order Riemann–Liouville fractional differential equations consistent with (3.45):

Problem 1:

$$\begin{cases} F_{0+}^{\frac{7}{5}}\varkappa(s) = \frac{s^{-\frac{1}{2}}}{4e^{s}(1+\|\varkappa_s\|_{[-\gamma,0]})}, & s \in J_1, \\ \varkappa(0) = 0, \ \varkappa(1) = 0, \\ \varkappa(s) = \chi_1(s), & s \in [-\gamma, 0] \end{cases} \tag{3.68}$$

Problem 2:

$$\begin{cases} F_{0+}^{\frac{3}{2}}\varkappa(s) = \frac{s^{-\frac{1}{2}}}{4e^{s}(1+\|\varkappa_s\|_{[-\gamma,0]})}, & s \in J_2, \\ \varkappa(1) = 0, \ \varkappa(2) = 0, \\ \varkappa(s) = \chi_2(s), & s \in [-\gamma, 1], \end{cases} \tag{3.69}$$

where $\chi_1 = \chi$,

$$\chi_2(s) = \begin{cases} 0, \ , \ if \ s \in [0,1], \\ \chi(s), \ \ if \ s \in [-\gamma, 0]. \end{cases}$$

We verify that condition (3.51) is satisfied for $i = 1$:

$$\frac{K\Big(\mathsf{T}_1^{1-\delta} - \mathsf{T}_0^{1-\delta}\Big)\Big(\mathsf{T}_1 - \mathsf{T}_0\Big)^{\ltimes_1 - 1}}{4^{\ltimes_1 - 1}(1-\delta)\Gamma(\ltimes_1)} = \frac{\frac{1}{4}\Big(1^{1-\frac{1}{2}} - 0^{1-\frac{1}{2}}\Big)\Big(1-0\Big)^{\frac{7}{5}-1}}{4^{\frac{7}{5}-1}(1-\frac{1}{2})\Gamma(\frac{7}{5})} \simeq 0.323663 < 1.$$

By Theorem 20, the problem (3.68) admits a solution $\varkappa_1 \in C([-\gamma, 1], \mathbb{R})$, where

$$\varkappa_1(s) = \begin{cases} \chi(s), & s \in [-\gamma, 0], \\ z_1(s), & s \in J_1. \end{cases}$$

Similarly, for $i = 2$:

$$\frac{K\Big(\mathsf{T}_2^{1-\delta} - \mathsf{T}_1^{1-\delta}\Big)\Big(\mathsf{T}_2 - \mathsf{T}_1\Big)^{\ltimes_2 - 1}}{4^{\ltimes_2 - 1}(1-\delta)\Gamma(\ltimes_2)} = \frac{\frac{1}{4}\Big(2^{1-\frac{1}{2}} - 1^{1-\frac{1}{2}}\Big)\Big(2-1\Big)^{\frac{3}{2}-1}}{4^{\frac{3}{2}-1}(1-\frac{1}{2})\Gamma(\frac{3}{2})}$$

$$\simeq 0.11684748 < 1.$$

Thus, by Theorem 20, problem (3.69) admits a solution $\varkappa_2 \in C([-\gamma, 2], \mathbb{R})$, where

$$\varkappa_2(s) = \begin{cases} \chi(s), & s \in [-\gamma, 0], \\ 0, & s \in [0,1], \\ z_2(s), & s \in J_2. \end{cases}$$

Hence, by Theorem 21, the original problem (3.67) admits a solution:

$$\varkappa(s) = \begin{cases} \varkappa_1(s) = \begin{cases} \chi(s), & s \in [-\gamma, 0], \\ z_1(s), & s \in J_1, \end{cases} \\ \varkappa_2(s) = \begin{cases} \chi(s), & s \in [-\gamma, 0], \\ 0, & s \in J_1, \\ z_2(s), & s \in J_2, \end{cases} \end{cases}$$

Moreover, according to Theorem 22, problem (3.67) is Ulam-Hyers stable.

Numerical approach 2 Let $\gamma > 0$, and define the piecewise function $\varrho(s)$ as follows:

$$\ltimes(s) = \begin{cases} \frac{7}{5}, & s \in J_1 := [0,1], \\ \frac{6}{5}, & s \in J_2 :=]1, \frac{3}{2}], \\ \frac{3}{2}, & s \in J_3 :=]\frac{3}{2}, 2]. \end{cases} \tag{3.70}$$

Consider the following Riemann–Liouville fractional variable-order initial and terminal value problem:

$$\begin{cases} F_{0^+}^{\ltimes(s)} \varkappa(s) = \frac{s^{-\frac{1}{3}}}{(e^{e^{\frac{s^3}{1+s}}}+6)(1+\|\varkappa_s\|_{[-\gamma,0]})}, & s \in J :=]0,2], \\ \varkappa(s) = \chi(s), & s \in [-\gamma, 0], \end{cases} \tag{3.71}$$

The choice of $\ltimes(s)$ ensures that condition $(H1)$ holds. Define

$$f(s, y_s) = \frac{s^{-\frac{1}{3}}}{(e^{e^{\frac{s^3}{1+s}}} + 6)(1 + \|y_s\|_{[-\gamma,0]})}, \quad (s, y_s) \in [0,2] \times C([-\gamma, 0], \mathbb{R}).$$

For $y, z \in C([-\gamma, 2], \mathbb{R})$ and $s \in J$, we have

$$\begin{aligned} s^{\frac{1}{3}}|f(s, y_s) - f(s, z_s)| &= \left| \frac{1}{(e^{e^{\frac{s^3}{1+s}}} + 6)} \left(\frac{1}{1 + \|y_s\|_{[-\gamma,0]}} - \frac{1}{1 + \|z_s\|_{[-\gamma,0]}} \right) \right| \\ &\leq \frac{|\|y_s\|_{[-\gamma,0]} - \|z_s\|_{[-\gamma,0]}|}{(e^{e^{\frac{s^3}{1+s}}} + 6)(1 + \|y_s\|_{[-\gamma,0]})(1 + \|y_s\|_{[-\gamma,0]})} \\ &\leq \frac{1}{(e^{e^{\frac{s^3}{1+s}}} + 6)} \|y_s - z_s\|_{[-\gamma,0]} \\ &\leq \frac{1}{e + 6} \|y_s - z_s\|_{[-\gamma,0]}. \end{aligned}$$

Hence, condition $(H2)$ holds for $\delta = \frac{1}{3}$ and $K = \frac{1}{e+6}$.

We analyze three auxiliary problems for Riemann–Liouville fractional differential equations of constant order as specified by (3.70), in accordance with (3.45):

$$\begin{cases} F_{0^+}^{\frac{7}{5}} \varkappa(s) = \frac{s^{-\frac{1}{3}}}{(e^{e^{\frac{s^3}{1+s}}}+6)(1+\|\varkappa_s\|_{[-\gamma,0]})}, & s \in J_1, \\ \varkappa(0) = 0, \ \varkappa(1) = 0, & \\ \varkappa(s) = \chi_1(s), & s \in [-\gamma, 0], \end{cases} \tag{3.72}$$

Problem on J_2:

$$\begin{cases} F_{0^+}^{\frac{6}{5}} \varkappa(s) = \frac{s^{-\frac{1}{3}}}{(e^{e^{\frac{s^3}{1+s}}}+6)(1+\|\varkappa_s\|_{[-\gamma,0]})}, & s \in J_2, \\ \varkappa(1) = 0, \ \varkappa(\frac{3}{2}) = 0, & \\ \varkappa(s) = \chi_2(s), & s \in [-\gamma, 1], \end{cases} \tag{3.73}$$

Problem on J_3:

$$\begin{cases} F_{0+}^{\frac{3}{2}} = \frac{s^{-\frac{1}{3}}}{(e^{e^{\frac{s^3}{1+s}}}+6)(1+\|\varkappa_s\|_{[-\gamma,0]})}, & s \in J_3, \\ \varkappa(\frac{3}{2}) = 0, \ \varkappa(2) = 0, \\ \varkappa(s) = \chi_2(s), & s \in [-\gamma, \frac{3}{2}], \end{cases} \tag{3.74}$$

Define: $\chi_1 = \chi$,

$$\chi_2(s) = \begin{cases} 0, \ , \ if \ s \in [0,1], \\ \chi(s), \ \ if \ s \in [-\gamma, 0] \end{cases}$$

and

$$\chi_3(s) = \begin{cases} 0, \ , \ if \ s \in [0, \frac{3}{2}], \\ \chi(s), \ \ if \ s \in [-\gamma, 0]. \end{cases}$$

We verify condition (3.51) is satisfied for $i = 1$. Clearly,

$$\frac{K\Big(\mathsf{T}_1^{1-\delta} - \mathsf{T}_0^{1-\delta}\Big)\Big(\mathsf{T}_1 - \mathsf{T}_0\Big)^{\ltimes_1 - 1}}{4^{\ltimes_1 - 1}(1-\delta)\Gamma(\ltimes_1)} = \frac{\frac{1}{e+6}\Big(1^{1-\frac{1}{3}} - 0^{1-\frac{1}{3}}\Big)\Big(1-0\Big)^{\frac{7}{5}-1}}{4^{\frac{7}{5}-1}(1-\frac{1}{3})\Gamma(\frac{7}{5})} \simeq 0.11137 < 1.$$

By Theorem 20, the problem (3.72) admits a solution $\varkappa_1 \in C([-\gamma, 1], \mathbb{R})$, where

$$\varkappa_1(s) \quad = \quad \begin{cases} \chi(s), & s \in [-\gamma, 0] \\ z_1(s), & s \in J_1. \end{cases}$$

We further verify that condition (3.51) is satisfied for $i = 2$. Specifically,

$$\frac{K\Big(\mathsf{T}_2^{1-\delta} - \mathsf{T}_1^{1-\delta}\Big)\Big(\mathsf{T}_2 - \mathsf{T}_1\Big)^{\ltimes_2 - 1}}{4^{\ltimes_2 - 1}(1-\delta)\Gamma(\ltimes_2)} = \frac{\frac{1}{e+6}\Big(\frac{3}{2}^{1-\frac{1}{3}} - 1^{1-\frac{1}{3}}\Big)\Big(\frac{3}{2}-1\Big)^{\frac{6}{5}-1}}{4^{\frac{6}{5}-1}(1-\frac{1}{3})\Gamma(\frac{6}{5})}$$

$$\simeq 0.03837 < 1.$$

Thus, the initial-terminal value problem (3.73) has a solution $\varkappa_2 \in C([-\gamma, \frac{3}{2}], \mathbb{R})$, according to Theorem 20. where

$$\varkappa_2(s) = \begin{cases} \chi(s), & s \in [-\gamma, 0], \\ 0, & s \in [0, 1], \\ z_2(s), & s \in J_2. \end{cases}$$

We further verify that condition (3.51) is satisfied for $i = 3$. Specifically,

$$\frac{K\Big(\mathsf{T}_3^{1-\delta} - \mathsf{T}_2^{1-\delta}\Big)\Big(\mathsf{T}_3 - \mathsf{T}_2\Big)^{\ltimes_3 - 1}}{4^{\ltimes_3 - 1}(1-\delta)\Gamma(\ltimes_3)} = \frac{\frac{1}{e+6}\Big(2^{1-\frac{1}{3}} - \frac{3}{2}^{1-\frac{1}{3}}\Big)\Big(2-\frac{3}{2}\Big)^{\frac{3}{2}-1}}{4^{\frac{3}{2}-1}(1-\frac{1}{3})\Gamma(\frac{3}{2})}$$

$$\simeq 0.01901 < 1.$$

For $i = 3 : K \cdot expression \approx 0.01901 < 1.$

Hence, by Theorem 20, each auxiliary problem has a solution $\varkappa_i \in C([-\gamma, b_i], \mathbb{R})$, where b_i is the endpoint of J_i. By Theorem 21, the full problem (3.71) has a solution:

$$\varkappa(s) = \begin{cases} \varkappa_1(s) = \begin{cases} \chi(s), & s \in [-\gamma, 0], \\ z_1(s), & s \in J_1, \end{cases} \\ \varkappa_2(s) = \begin{cases} \chi(s), & s \in [-\gamma, 0], \\ 0, & s \in]0, \mathsf{T}_1], \\ z_2(s), & s \in J_2, \end{cases} \\ \varkappa_3(s) = \begin{cases} \chi(s), & s \in [-\gamma, 0], \\ 0, & s \in]0, \mathsf{T}_2], \\ z_3(s), & s \in J_3. \end{cases} \end{cases}$$

Moreover, by Theorem 22, the problem (3.71) is Ulam-Hyers stable.

Remark 11 *The examples demonstrate the effectiveness of the proposed existence and stability results.*

3.5 Variable-Order Implicit Fractional Differential Equations based on the Kuratowski MNC Technique

We address the following initial terminal value problem (ITVP):

$$\begin{cases} \digamma_{0+}^{\ltimes(s)} \varkappa(s) + f(s, \varkappa(s), \digamma_{0+}^{\ltimes(s)} \varkappa(s)) = 0, & s \in J := [0, \mathsf{T}], \\ \varkappa(0) = 0, \quad \varkappa(\mathsf{T}) = 0. \end{cases} \tag{3.75}$$

Here, $1 < \ltimes(s) \leq 2$, the mapping $f : J \times X \times X \to X$ is continuous, X denotes a real or complex ßanach space, and $\digamma_{0+}^{\ltimes(s)}$ represents the Riemann–Liouville fractional derivative of variable order $\ltimes(s)$.

3.5.1 Existence of Solutions

Let's introduce these presumptions:

(H1) Let $n \in \mathbb{N}$, and define the partition $\mathcal{P} = \{J_1 := [0, \mathsf{T}_1], J_2 := (\mathsf{T}_1, \mathsf{T}_2], J_3 := (\mathsf{T}_2, \mathsf{T}_3], \ldots, J_n := (\mathsf{T}_{n-1}, \mathsf{T}]\}$ of the interval $J = [0, \tau]$. Let $\ltimes(s) : J \to (1, 2]$ be a piecewise constant function with respect to $\mathcal{P}$, defined by

$$\ltimes(s) = \sum_{i=1}^{n} \ltimes_i I_i(s) = \begin{cases} \ltimes_1, & if\ s \in J_1, \\ \ltimes_2, & if\ s \in J_2, \\ \cdot \\ \cdot \\ \cdot \\ \ltimes_n, & if\ s \in J_n, \end{cases}$$

with constants $1 < \ltimes_i \leq 2$ and $I_i(s)$ denotes the indicator function of the interval $J_i := (\tau_i - 1\tau_i]$, defined by

$$I_i(s) = \begin{cases} 1, & for\ s \in J_i, \\ 0, & otherwise. \end{cases}$$

(H2) Assume that $s^\delta f : J \times X \times X \to X$ is a continuous function, where $0 \leq \delta \leq \min_{s \in J} | \ltimes(s)|$. There exist constants $K > 0$ and $L > 0$ such that

$$s^\delta \|f(s, y_1, z_1) - f(s, y_2, z_2)\| \leq K\|y_1 - y_2\| + L\|z_1 - z_2\|,$$

for all $y_1, y_2, z_1, z_2 \in X$.

Remark 12 *According to the previous observation in [23], it is straightforward to demonstrate that the following inequality and the condition (H2)*

$$\vartheta(\mathsf{T}^\delta \|f(\mathsf{T}, F_1, F_2)\|) \leq K\vartheta(F_1) + L\vartheta(F_2),$$

are equivalent for any bounded sets $F_1, F_2 \subset X$ *and every* $s \in J$.

Furthermore, let us define $\ltimes : J \to X$ for a given set of functions χ, denoted by

$$\chi(s) = \{\varkappa(s), \varkappa \in \chi\}, \ s \in J,$$

and

$$\chi(J) = \{\chi(s) : \varkappa \in \chi, \ s \in J\}.$$

Using the concepts of the Kuratowski measure of noncompactness and Darbo's fixed point theorem, we can demonstrate the existence a solution to the initial terminal value problem (ITVP) defined in equation (3.75).

Let $E_i = C(J_i, X)$ denote the ßanach space of a continuous functions $\varkappa : J_i \to X$ for each $i \in \{1, 2, \ldots, n\}$, equipped with the norm

$$\|\varkappa\|_{E_i} = \sup_{s \in J_i} \|\varkappa(s)\|.$$

We begin by analyzing the ITVP as defined in equation (3.75), which can be expressed as

$$\frac{d^2}{ds^2} \int_0^s \frac{(s-t)^{1-\ltimes(s)}}{\Gamma(2 - \ltimes(s))} \varkappa(t) dt + f(s, \varkappa(s), F_{0+}^{\ltimes(s)} \varkappa(s)) = 0, \quad s \in J. \tag{3.76}$$

In the subinterval J_i, where $i = 1, 2, ..., n$, and under assumption $(H1)$, equation (3.76) may be reformulated as

$$\frac{d^2}{ds^2}\Big(\int_0^{\mathsf{T}_1} \frac{(s-t)^{1-\ltimes_1}}{\Gamma(2-\ltimes_1)} \varkappa(t) dt + ... + \int_{\mathsf{T}_{i-1}}^s \frac{(s-t)^{1-\ltimes_i}}{\Gamma(2-\ltimes_i)} \varkappa(t) dt \Big) + f(s, \varkappa(s), F_{0+}^{\ltimes_i} \varkappa(s)) = 0, \ \ s \in J_i. \tag{3.77}$$

Definition 47 *The initial terminal value problem (ITVP) (3.75) admits a solution if there exist functions* $\kappa_i \in C([0, \mathsf{T}_i], X)$ $i = 1, 2, ..., n$, *such that each* κ_i *satisfies equation (3.77) and the boundary conditions* $\varkappa_i(0) = 0 = \varkappa_i(\mathsf{T}_i)$.

If we consider $\varkappa(s) \equiv 0$ for $0 \leq s \leq \mathsf{T}_{i-1}$, then equation (3.77) takes the form

$$\digamma^{\ltimes_i}_{\mathsf{T}_{i-1}^+} \varkappa(s) + f(s, \varkappa(s), \digamma^{\ltimes_i}_{\mathsf{T}_{i-1}^+} \varkappa(s)) = 0, \quad s \in J_i.$$

We now address the following ITVPs:

$$\begin{cases} \digamma^{\ltimes_i}_{\mathsf{T}_{i-1}^+} \varkappa(s) + f(s, \varkappa(s), \digamma^{\ltimes_i}_{\mathsf{T}_{i-1}^+} \varkappa(s)) = 0, \quad s \in J_i, \\ \varkappa(\mathsf{T}_{i-1}) = 0, \varkappa(\mathsf{T}_i) = 0. \end{cases} \tag{3.78}$$

Lemma 16 *The solution* $\varkappa$ *(3.78) can be expressed in the form of the following integral equation:*

$$\varkappa(s) = \int_{\mathsf{T}_{i-1}}^{\mathsf{T}_i} G_i(s,t) f\Big(t, \int_{\mathsf{T}_{i-1}}^{\mathsf{T}_i} G_i(s,\sigma)\varkappa(\sigma)d\sigma, \varkappa(t)\Big)dt, \tag{3.79}$$

where $\digamma^{\ltimes_i}_{\mathsf{T}_{i-1}^+} \varkappa(s) = \varkappa(s)$, $G_i(s,t)$ *is the Green's function defined as follows:*

$$G_i(s,t) = \begin{cases} \frac{1}{\Gamma(\ltimes_i)}\Big[(\mathsf{T}_i - \mathsf{T}_{i-1})^{1-\ltimes_i}(s - \mathsf{T}_{i-1})^{\ltimes_i - 1}(\mathsf{T}_i - t)^{\ltimes_i - 1} - (s-t)^{\ltimes_i - 1}\Big], \\ \qquad\qquad \mathsf{T}_{i-1} \leq t \leq s \leq \mathsf{T}_i, \\ \frac{1}{\Gamma(\ltimes_i)}(\mathsf{T}_i - \mathsf{T}_{i-1})^{1-\ltimes_i}(s - \mathsf{T}_{i-1})^{\ltimes_i - 1}(\mathsf{T}_i - t)^{\ltimes_i - 1}, \\ \qquad\qquad \mathsf{T}_{i-1} \leq s \leq t \leq \mathsf{T}_i, \end{cases}$$

where $i = 1, 2, ..., n$.

Proof
Assume that $\varkappa \in E_i$ is a solution to problem (3.78), and let $\digamma^{\ltimes_i}_{\mathsf{T}_{i-1}^+} \varkappa(s) = \varkappa(s)$. Applying the operator $I^{\ltimes_i}_{\mathsf{T}_{i-1}^+}$ to both sides of equation (3.78), we obtain (see Lemma 1):

$$\varkappa(s) = \omega_1(s - \mathsf{T}_{i-1})^{\ltimes_i - 1} + \omega_2(s - \mathsf{T}_{i-1})^{\ltimes_i - 2} - I^{\ltimes_i}_{\mathsf{T}_{i-1}^+} \varkappa(s), s \in J_i.$$

Given the assumption $\varkappa(\mathsf{T}_{i-1}) = 0$ assumptions, we deduce that $\omega_2 = 0$.

Now, let $\varkappa(s)$ satisfy the terminal condition $\varkappa(\mathsf{T}_i) = 0$. Then we find:

$$\omega_1 = (\mathsf{T}_i - \mathsf{T}_{i-1})^{1-\ltimes_i} I^{\ltimes_i}_{\mathsf{T}_{i-1}^+} \varkappa(\mathsf{T}_i).$$

Substituting back, we observe:

$$\varkappa(s) \;=\; (\mathsf{T}_i - \mathsf{T}_{i-1})^{1-\ltimes_i}(s - \mathsf{T}_{i-1})^{\ltimes_i - 1} I^{\ltimes_i}_{\mathsf{T}_{i-1}^+} \varkappa(\mathsf{T}_i) - I^{\ltimes_i}_{\mathsf{T}_{i-1}^+} \varkappa(s), s \in J_i.$$

Hence, the solution to problem (3.78) is given by:

$$\begin{aligned}
\varkappa(s) &= (\mathsf{T}_i - \mathsf{T}_{i-1})^{1-\ltimes_i}(s-\mathsf{T}_{i-1})^{\ltimes_i-1}\frac{1}{\Gamma(\ltimes_i)}\int_{\mathsf{T}_{i-1}}^{\mathsf{T}_i}(\mathsf{T}_i-t)^{\ltimes_i-1}\varkappa(t)dt \\
&- \frac{1}{\Gamma(\ltimes_i)}\int_{\mathsf{T}_{i-1}}^{s}(s-t)^{\ltimes_i-1}\varkappa(t)dt, \\
&= \frac{1}{\Gamma(\ltimes_i)}\Big[\int_{\mathsf{T}_{i-1}}^{s}\Big((\mathsf{T}_i-\mathsf{T}_{i-1})^{1-\ltimes_i}(s-\mathsf{T}_{i-1})^{\ltimes_i-1}(\mathsf{T}_i-t)^{\ltimes_i-1}-(s-t)^{\ltimes_i-1}\Big)\varkappa(t)dt \\
&+ \int_{s}^{\mathsf{T}_i}(\mathsf{T}_i-\mathsf{T}_{i-1})^{1-\ltimes_i}(s-\mathsf{T}_{i-1})^{\ltimes_i-1}(\mathsf{T}_i-t)^{\ltimes_i-1}\varkappa(t)dt\Big],
\end{aligned}$$

The continuation of the Green's function formulation suggests that

$$\varkappa(s) = \int_{\mathsf{T}_{i-1}}^{\mathsf{T}_i} G_i(s,t) f\Big(t, \int_{\mathsf{T}_{i-1}}^{\mathsf{T}_i} G_i(s,\sigma)\varkappa(\sigma)d\sigma, \varkappa(t)\Big)dt.$$

Thus, the solution to the integral equation (3.79) is $\varkappa \in E_i$. We conclude that $\varkappa$ solves problem (3.78), based on the continuity of the function f, its behavior δf, and Lemma 1.

The following claim will be required.

Proposition 5 *[123] Assume that $s^{\delta} f : J \times X \times X \to X$, $(0 \leq \delta \leq \min_{s\in J} |(\ltimes(s))|)$, is a continuous function, and that $\ltimes(s) : J \to (1,2]$ satisfies hypothesis $(H1)$. Then, the Green's functions associated with the initial and terminal value problem (6) satisfy the following properties:*

(1) $G_i(s,t) \geq 0 \ \forall \mathsf{T}_{i-1} \leq s,t \leq \mathsf{T}_i$,

(2) $\max\limits_{s\in J_i} G_i(s,t) = G_i(t,t), t \in J_i$,

(3) *$G_i(t,t)$ attains a unique maximum, given by*

$$\max_{t\in J_i} G_i(t,t) = \frac{1}{\Gamma(\ltimes_i)}\Big(\frac{\mathsf{T}_i - \mathsf{T}_{i-1}}{4}\Big)^{\ltimes_i-1},$$

for $i = 1, 2, ..., n$.

The first existence result we derive is based on Darbo's fixed point theorem.

Theorem 23 *Let conditions (H1)–(H2) be satisfied. If*

$$\frac{(\mathsf{T}_i - \mathsf{T}_{i-1})^{\ltimes_i-1}\Big(\mathsf{T}_i^{1-\delta} - \mathsf{T}_{i-1}^{1-\delta}\Big)}{(1-\delta)4^{\ltimes_i-1}\Gamma(\ltimes_i)}\Big(\frac{K(\mathsf{T}_i-\mathsf{T}_{i-1})^{\ltimes_i}}{4^{\ltimes_i-1}\Gamma(\ltimes_i)} + L\Big) < 1, \tag{3.80}$$

then there exists at least one solution to the problem (3.78) within the interval J.

Proof
We define the mapping

$$\mathcal{S} : E_i \to E_i$$

as follows:

$$\mathcal{S}\varkappa(s) = \int_{\mathsf{T}_{i-1}}^{\mathsf{T}_i} G_i(s,t)\varkappa(t)dt, \quad t \in J_i, \tag{3.81}$$

where

$$\varkappa(t) \quad = \quad f\Big(t, \int_{\mathsf{T}_{i-1}}^{\mathsf{T}_i} G_i(s,\sigma)\varkappa(\sigma)d\sigma, \varkappa(t)\Big).$$

The operator $\mathcal{S}$ defined in (3.81) is well-defined due to the continuity of the function $s^{\delta} f$ and the properties of fractional integrals.

Let

$$R_i \quad \geq \quad \frac{\dfrac{f^{\star}(\mathsf{T}_i - \mathsf{T}_{i-1})^{\ltimes_i}}{4^{\ltimes_i - 1}\Gamma(\ltimes_i)}}{1 - \dfrac{(\mathsf{T}_i - \mathsf{T}_{i-1})^{\ltimes_i - 1}\Big(\mathsf{T}_i^{1-\delta} - \mathsf{T}_{i-1}^{1-\delta}\Big)}{(1-\delta)4^{\ltimes_i - 1}\Gamma(\ltimes_i)}\Big(\dfrac{K(\mathsf{T}_i - \mathsf{T}_{i-1})^{\ltimes_i}}{4^{\ltimes_i - 1}\Gamma(\ltimes_i)} + \gamma_2\Big)},$$

where

$$f^{\star} = \sup_{s\in J_i} \|f(s,0,0)\|.$$

We consider the set

$$B_{R_i} = \{\varkappa \in E_i, \ \|\varkappa\|_{E_i} \leq R_i\}.$$

The set B_{R_i} contains all its limit points possesses the convex-hull property, and is bounded within a finite region.

We now proceed to demonstrate the applicability of Darbo's fixed point theorem, which will be presented in four stages.

STEP 1: $S(B_{R_i}) \subseteq (B_{R_i})$.
By Proposition (5) and (H2), we have:

$$\begin{aligned}
&\|S\varkappa(s)\| = \Big\| \int_{\mathsf{T}_{i-1}}^{\mathsf{T}_i} G_i(s,t)\varkappa(t)dt \Big\| \\
\leq\ & \int_{\mathsf{T}_{i-1}}^{\mathsf{T}_i} G_i(s,t)\|\varkappa(t)\|dt, \\
\leq\ & \frac{1}{\Gamma(\ltimes_i)}\Big(\frac{\mathsf{T}_i - K_{i-1}}{4}\Big)^{\ltimes_i - 1} \int_{\mathsf{T}_{i-1}}^{\mathsf{T}_i} \|\varkappa(t)\|dt, \\
\leq\ & \frac{1}{\Gamma(\ltimes_i)}\Big(\frac{\mathsf{T}_i - K_{i-1}}{4}\Big)^{\ltimes_i - 1} \int_{\mathsf{T}_{i-1}}^{\mathsf{T}_i} \Big\|f\Big(t, \int_{\mathsf{T}_{i-1}}^{\mathsf{T}_i} G_i(s,\sigma)\varkappa(\sigma)d\sigma, \varkappa(t)\Big)\Big\|dt,
\end{aligned}$$

$$\begin{aligned}
&\leq \frac{1}{\Gamma(\ltimes_i)}\Big(\frac{K_i-\mathsf{T}_{i-1}}{4}\Big)^{\ltimes_i-1}\int_{\mathsf{T}_{i-1}}^{\mathsf{T}_i}\Big\|f\Big(t,\int_{\mathsf{T}_{i-1}}^{\mathsf{T}_i}G_i(s,\sigma)\varkappa(\sigma)d\sigma,y(t)\Big)-f(t,0,0)\Big\|dt\\
&+ \frac{1}{\Gamma(\ltimes_i)}\Big(\frac{K_i-\mathsf{T}_{i-1}}{4}\Big)^{\ltimes_i-1}\int_{\mathsf{T}_{i-1}}^{\mathsf{T}_i}\|f(t,0,0)\|dt,\\
&\leq \frac{1}{\Gamma(\ltimes_i)}\Big(\frac{\mathsf{T}_i-\mathsf{T}_{i-1}}{4}\Big)^{\ltimes_i-1}\int_{\mathsf{T}_{i-1}}^{\mathsf{T}_i}t^{-\delta}\Big(K\Big\|\int_{\mathsf{T}_{i-1}}^{\mathsf{T}_i}G_i(s,\sigma)\varkappa(\sigma)d\sigma\Big\|+L\|\varkappa(t)\|\Big)dt\\
&+ \frac{f^\star(\mathsf{T}_i-\mathsf{T}_{i-1})^{\ltimes_i}}{4^{\ltimes_i-1}\Gamma(\ltimes_i)},\\
&\leq \frac{1}{\Gamma(\ltimes_i)}\Big(\frac{\mathsf{T}_i-\mathsf{T}_{i-1}}{4}\Big)^{\ltimes_i-1}\int_{\mathsf{T}_{i-1}}^{\mathsf{T}_i}t^{-\delta}\Big(\frac{K(\mathsf{T}_i-\mathsf{T}_{i-1})^{\ltimes_i}}{4^{\ltimes_i-1}\Gamma(\ltimes_i)}\|\varkappa\|_{E_i}+\gamma_2\|\varkappa\|_{E_i}\Big)dt\\
&+ \frac{f^\star(\mathsf{T}_i-\mathsf{T}_{i-1})^{\ltimes_i}}{4^{\ltimes_i-1}\Gamma(\ltimes_i)},\\
&\leq \frac{(\mathsf{T}_i-\mathsf{T}_{i-1})^{\ltimes_i-1}\Big(\mathsf{T}_i^{1-\delta}-\mathsf{T}_{i-1}^{1-\delta}\Big)}{(1-\delta)4^{\ltimes_i-1}\Gamma(\ltimes_i)}\Big(\frac{K(\mathsf{T}_i-\mathsf{T}_{i-1})^{\ltimes_i}}{4^{\ltimes_i-1}\Gamma(\ltimes_i)}+\gamma_2\Big)R_i+\frac{f^\star(\mathsf{T}_i-\mathsf{T}_{i-1})^{\ltimes_i}}{4^{\ltimes_i-1}\Gamma(\ltimes_i)},\\
&\leq R_i,
\end{aligned}$$

Hence, $\mathcal{S}(B_{R_i}) \subseteq B_{R_i}$.

STEP 2: $\mathcal{S}$ is continuous.
Let the sequence $(\varkappa_n)$ is converge to $\varkappa$ in E_i and $s \in J_i$. Then,

$$\begin{aligned}
&\|(s\varkappa_n)(s)-(s\varkappa)(s)\| \leq \int_{\mathsf{T}_{i-1}}^{\mathsf{T}_i}G_i(s,t)\Big\|\varkappa_n(t)-\varkappa(t)\Big\|dt,\\
&\leq \frac{1}{\Gamma(\ltimes_i)}\Big(\frac{\mathsf{T}_i-\mathsf{T}_{i-1}}{4}\Big)^{\ltimes_i-1}\int_{\mathsf{T}_{i-1}}^{\mathsf{T}_i}\Big\|\varkappa_n(t)-\varkappa(t)\Big\|dt,\\
&\leq \frac{1}{\Gamma(\ltimes_i)}\Big(\frac{\mathsf{T}_i-\mathsf{T}_{i-1}}{4}\Big)^{\ltimes_i-1}\int_{\mathsf{T}_{i-1}}^{\mathsf{T}_i}\Big\|f\Big(t,\int_{\mathsf{T}_{i-1}}^{\mathsf{T}_i}G_i(s,\sigma)\varkappa_n(\sigma)d\sigma,\varkappa_n(t)\Big)\\
&- f\Big(t,\int_{\mathsf{T}_{i-1}}^{\mathsf{T}_i}G_i(s,\sigma)y(\sigma)d\sigma,\varkappa(t)\Big)\Big\|dt,\\
&\leq \frac{1}{\Gamma(\ltimes_i)}\Big(\frac{K_i-\mathsf{T}_{i-1}}{4}\Big)^{\ltimes_i-1}\int_{\mathsf{T}_{i-1}}^{\mathsf{T}_i}t^{-\delta}\Big(K\Big\|\int_{\mathsf{T}_{i-1}}^{\mathsf{T}_i}G_i(s,\sigma)\varkappa_n(\sigma)d\sigma-\int_{\mathsf{T}_{i-1}}^{\mathsf{T}_i}G_i(s,\sigma)\varkappa(\sigma)d\sigma\Big\|\\
&+ L\|\varkappa_n(t)-\varkappa(t)\|\Big)dt,\\
&\leq \frac{1}{\Gamma(\ltimes_i)}\Big(\frac{\mathsf{T}_i-K_{i-1}}{4}\Big)^{\ltimes_i-1}\int_{\mathsf{T}_{i-1}}^{\mathsf{T}_i}t^{-\delta}\Big(K\int_{\mathsf{T}_{i-1}}^{\mathsf{T}_i}G_i(s,\sigma)\Big\|\varkappa_n(\sigma)-\varkappa(\sigma)\Big\|d\sigma+L\|\varkappa_n(t)-\varkappa(t)\|\Big)dt,\\
&\leq \frac{1}{\Gamma(\ltimes_i)}\Big(\frac{\mathsf{T}_i-\mathsf{T}_{i-1}}{4}\Big)^{\ltimes_i-1}\int_{\mathsf{T}_{i-1}}^{\mathsf{T}_i}t^{-\delta}\Big(\frac{K(\mathsf{T}_i-\mathsf{T}_{i-1})^{\ltimes_i}}{4^{\ltimes_i-1}\Gamma(\ltimes_i)}\|\varkappa_n-\varkappa\|_{E_i}+L\|\varkappa_n-\varkappa\|_{E_i}\Big)dt,\\
&\leq \frac{(\mathsf{T}_i-\mathsf{T}_{i-1})^{\ltimes_i-1}\Big(\mathsf{T}_i^{1-\delta}-\mathsf{T}_{i-1}^{1-\delta}\Big)}{(1-\delta)4^{\ltimes_i-1}\Gamma(\ltimes_i)}\Big(\frac{K(\mathsf{T}_i-\mathsf{T}_{i-1})^{\ltimes_i}}{4^{\ltimes_i-1}\Gamma(\ltimes_i)}+L\Big)\|\varkappa_n-\varkappa\|_{E_i},
\end{aligned}$$

i.e., we get

$$\|(s\varkappa_n)-(s\varkappa)\|_{E_i} \to 0 \quad as \quad n\to\infty.$$

Therefore, $\mathcal{S}$ is a continuous operator.

STEP 3: The operator $\mathcal{S}$ is bounded and uniformly continuous.
From Step 2, we have $\mathcal{S}(B_{R_i}) = \{\mathcal{S}(\varkappa) : \varkappa \in B_{R_i}\} \subset B_{R_i}$. Consequently, for all $\varkappa \in B_{R_i}$, $\|\mathcal{S}(\varkappa)\|_{E_i} \leq R_i$, which shows that $\mathcal{S}(B_{R_i})$ is bounded. It remains to prove that $\mathcal{S}(B_{R_i})$ is equicontinuous.

Let $s_1, s_2 \in J_i$, with $s_1 < s_2$, and let $\varkappa \in B_{R_i}$. Then,

$$\begin{aligned}
&\|(s\varkappa)(s_2) - (s\varkappa)(s_1)\| = \Big\| \int_{\mathsf{T}_{i-1}}^{\mathsf{T}_i} G_i(s_2,t)\varkappa(t)dt - \int_{\mathsf{T}_{i-1}}^{\mathsf{T}_i} G_i(s_1,t)\varkappa(t)dt \Big\|,\\
\leq\ & \int_{\mathsf{T}_{i-1}}^{\mathsf{T}_i} \Big\|\Big(G_i(s_2,t) - G_i(s_1,t)\Big)\varkappa(t)\Big\|dt,\\
\leq\ & \int_{\mathsf{T}_{i-1}}^{\mathsf{T}_i} \|G_i(s_2,t) - G_i(s_1,t)\| \Big\| f\Big(t, \int_{\mathsf{T}_{i-1}}^{\mathsf{T}_i} G_i(s,\sigma)\varkappa(\sigma)d\sigma, \varkappa(t)\Big)\Big\|dt,\\
\leq\ & \int_{\mathsf{T}_{i-1}}^{\mathsf{T}_i} \Big\|G_i(s_2,t) - G_i(s_1,t)\Big\| \Big(\Big\| f\Big(t, \int_{\mathsf{T}_{i-1}}^{\mathsf{T}_i} G_i(s,\sigma)\varkappa(\sigma)d\sigma, \varkappa(t)\Big) - f(t,0,0)\Big\| + \|f(t,0,0)\|\Big)dt,\\
\leq\ & \int_{\mathsf{T}_{i-1}}^{\mathsf{T}_i} \Big\|G_i(s_2,t) - G_i(s_1,t)\Big\| \Big[t^{-\delta}\Big(K\Big\| \int_{\mathsf{T}_{i-1}}^{\mathsf{T}_i} G_i(s,\sigma)\varkappa(\sigma)d\sigma\Big\| + \gamma_2\|\varkappa(t)\|\Big) + f^\star\Big]dt,\\
\leq\ & \int_{\mathsf{T}_{i-1}}^{\mathsf{T}_i} \Big\|G_i(s_2,t) - G_i(s_1,t)\Big\| \Big[t^{-\delta}\Big(\frac{K(\mathsf{T}_i - \mathsf{T}_{i-1})^{\ltimes_i}}{4^{\ltimes_i - 1}\Gamma(\ltimes_i)}\|\varkappa\|_{E_i} + \gamma_2\|\varkappa\|_{E_i}\Big) + f^\star\Big]dt,\\
\leq\ & \Big(\frac{K(\mathsf{T}_i - \mathsf{T}_{i-1})^{\ltimes_i}(\mathsf{T}_{i-1})^{-\delta}}{4^{\ltimes_i - 1}\Gamma(\ltimes_i)} + L(\mathsf{T}_{i-1})^{-\delta}\Big)\|\varkappa\|_{E_i} \int_{\mathsf{T}_{i-1}}^{\mathsf{T}_i} \Big\|G_i(s_2,t) - G_i(s_1,t)\Big\|dt\\
+\ & f^\star \int_{\mathsf{T}_{i-1}}^{\mathsf{T}_i} \Big\|G_i(s_2,t) - G_i(s_1,t)\Big\|dt.
\end{aligned}$$

By the continuity of the Green's function, we have: $\|(s\varkappa)(s_2) - (s\varkappa)(s_1)\|_{E_i} \to 0$ as $|s_2 - s_1| \to 0$. This implies that $\mathcal{S}(B_{R_i})$ is equicontinuous.

STEP 4: $\mathcal{S}$ is a *k-set contractions.* Let $\chi \in B_{R_i}$ and $s \in J_i$. Then we have:

$$\begin{aligned}
\vartheta(s(\chi)(s)) &= \vartheta((s\varkappa)(s), \varkappa \in \chi)\\
&\leq \Big\{\int_{\mathsf{T}_{i-1}}^{\mathsf{T}_i} G_i(s,t)\vartheta f\Big(t, \int_{\mathsf{T}_{i-1}}^{\mathsf{T}_i} G_i(s,\sigma)\varkappa(\sigma)d\sigma, \varkappa(t)\Big)dt, \varkappa \in \chi\Big\}.
\end{aligned}$$

According to Remark 12, we obtain:

$$\begin{aligned}
\vartheta(s(\chi)(s)) &\leq \Big\{\int_{\mathsf{T}_{i-1}}^{\mathsf{T}_i} G_i(s,t)t^{-\delta}\Big[K\vartheta(\int_{\mathsf{T}_{i-1}}^{\mathsf{T}_i} G_i(s,\sigma)\varkappa(\sigma)d\sigma) + L\vartheta(\varkappa(t))\Big]dt, y \in \chi\Big\},\\
&\leq \Big\{\frac{1}{\Gamma(\ltimes_i)}\Big(\frac{\mathsf{T}_i - K_{i-1}}{4}\Big)^{\ltimes_i - 1}\Big[\frac{K(\mathsf{T}_i - \mathsf{T}_{i-1})^{\ltimes_i}}{4^{\ltimes_i - 1}\Gamma(\ltimes_i)}\widehat{\vartheta}(\chi)\int_{\mathsf{T}_{i-1}}^{\mathsf{T}_i} t^{-\delta}dt\\
&+ L\widehat{\vartheta}(\chi)\int_{\mathsf{T}_{i-1}}^{\mathsf{T}_i} t^{-\delta}dt\Big], \varkappa \in \chi\Big\},\\
&\leq \frac{\Big({\mathsf{T}_i}^{1-\delta} - {\mathsf{T}_{i-1}}^{1-\delta}\Big)(K_i - \mathsf{T}_{i-1})^{\ltimes_i - 1}}{4^{\ltimes_i - 1}(1-\delta)\Gamma(\ltimes_i)}\Big(\frac{K(\mathsf{T}_i - K_{i-1})^{\ltimes_i}}{4^{\ltimes_i - 1}\Gamma(\ltimes_i)} + L\Big)\widehat{\vartheta}(\chi).
\end{aligned}$$

Evaluating the integral:

$$\widehat{\vartheta}(s\chi) \leq \frac{\left(\mathsf{T}_i^{1-\delta} - \mathsf{T}_{i-1}^{1-\delta}\right)(\mathsf{T}_i - \mathsf{T}_{i-1})^{\ltimes_i - 1}}{4^{\ltimes_i - 1}(1-\delta)\Gamma(\ltimes_i)} \left(\frac{K(\mathsf{T}_i - \mathsf{T}_{i-1})^{\ltimes_i}}{4^{\ltimes_i - 1}\Gamma(\ltimes_i)} + L\right)\widehat{\vartheta}(\chi).$$

By condition (3.80), the coefficient is strictly less than 1. Therefore, $\mathcal{S}$ is a k-set contraction. According to Darbo's fixed point theorem, problem (3.78) admits at least one solution $\kappa_i \in B_{R_i}$.

Define:

$$\kappa_i(s) = \begin{cases} 0, & s \in [0, \mathsf{T}_{i-1}], \\ \widetilde{\varkappa}_i, & s \in J_i. \end{cases} \tag{3.82}$$

with $\kappa_i \in C([0, \mathsf{T}_i], X)$ defined by (3.82). Then κ_i satisfies the equation:

$$\frac{d^2}{ds^2}\left(\int_0^{\mathsf{T}_1} \frac{(s-t)^{1-\ltimes_1}}{\Gamma(2-\ltimes_1)}\varkappa_i(t)dt + \ldots + \int_{\mathsf{T}_{i-1}}^{s} \frac{(s-t)^{1-\ltimes_i}}{\Gamma(2-\ltimes_i)}\varkappa_i(t)dt\right) + f(t, \varkappa_i(t), F_{0+}^{\ltimes_i}\varkappa_i(t)) = 0,$$

for $s \in J_i$, with boundary conditions $\varkappa_i(0) = 0$, $\varkappa_i(\mathsf{T}_i) = \widetilde{\varkappa}_i(\mathsf{T}_i) = 0$. This confirms that $\varkappa_i$ is a solution to equation (3.77).

Finally, define the global solution:

$$\varkappa(s) = \begin{cases} \varkappa_1(s), & s \in J_1, \\ \varkappa_2(s) = \begin{cases} 0, & s \in J_1, \\ \widetilde{\varkappa}_2, & s \in J_2 \end{cases} \\ \vdots \\ \varkappa_n(s) = \begin{cases} 0, & s \in [0, \mathsf{T}_{i-1}], \\ \widetilde{\varkappa}_i, & s \in J_i \end{cases} \end{cases}.$$

This function $\kappa(s)$ constitutes a solution to the initial terminal value problem (ITVP) (3.75).

3.5.2 Solutions Stability

Theorem 24 *Consider assumptions (H1), (H2), (3.80) and*

(H3). *Let $\psi \in C(J_i, X)$ be an increasing function, and suppose there exists a constant $\lambda_\psi > 0$ exists such that:*

$$I_{\mathsf{T}_{i-1}+}^{\ltimes_i}\psi(s) \leq \lambda_{\psi(s)}\psi(s), \ for\ any\ s \in J_i.$$

Then, the equation of (3.75) is Ulam-Hyers $\Re$assias stable with respect to ψ.

Proof Let $z \in C(J_i, X)$ be a function satisfying the inequality:

$$\|\digamma^{\ltimes_i}_{\mathsf{T}_{i-1}^+} z(s) + f(s, z(s), \digamma^{\ltimes_i}_{\mathsf{T}_{i-1}^+} z(s))\| \leq \epsilon\psi(s), s \in J_i. \tag{3.83}$$

Let $\kappa \in C(J_i, X)$ be the exact solution to the problem

$$\begin{cases} \digamma^{\ltimes_i}_{\mathsf{T}_{i-1}^+} \varkappa(s) + f(s, \varkappa(s), \digamma^{\ltimes_i}_{\mathsf{T}_{i-1}^+} \varkappa(s)) = 0, & s \in J_i, \\ \varkappa(\mathsf{T}_{i-1}) = 0, \quad \varkappa(\mathsf{T}_i) = 0. \end{cases}$$

with boundary conditions $\varkappa(\mathsf{T}_{i-1}) = 0, \varkappa(\mathsf{T}_i) = 0$. By Lemma 16, the solution $\varkappa(s)$ admits the representation:

$$\varkappa(s) = \int_{\mathsf{T}_{i-1}}^{\mathsf{T}_i} G_i(s,t) f\Big(t, \int_{\mathsf{T}_{i-1}}^{\mathsf{T}_i} G_i(s,\sigma)\varkappa(\sigma)d\sigma, \varkappa(t)\Big)dt.$$

From inequality (3.83) and assumption (H3), we obtain:

$$\begin{aligned} \|z(s) + \int_{\mathsf{T}_{i-1}}^{\mathsf{T}_i} G_i(s,t) f(t, \int_{\mathsf{T}_{i-1}}^{\mathsf{T}_i} G_i(s,\sigma)z(\sigma)d\sigma, z(t))dt\| &\leq \epsilon \int_{\mathsf{T}_{i-1}}^{s} \frac{(s-t)^{\ltimes(i)-1}}{\Gamma(\ltimes(i))}\psi(t)dt \\ &\leq \epsilon\lambda_{\psi(s)}\psi(s). \end{aligned}$$

By Proposition 5, for each $s \in J_i$, we have:

$$\begin{aligned} \|z(s) - \varkappa(s)\| &\leq \Big\|z(s) - \int_{\mathsf{T}_{i-1}}^{\mathsf{T}_i} G_i(s,t) f\Big(t, \int_{\mathsf{T}_{i-1}}^{\mathsf{T}_i} G_i(s,\sigma)z(\sigma)d\sigma, z(t)\Big)dt\Big\| \\ &+ \int_{\mathsf{T}_{i-1}}^{\mathsf{T}_i} G_i(s,t)\Big\|f\Big(t, \int_{\mathsf{T}_{i-1}}^{\mathsf{T}_i} G_i(s,\sigma)z(\sigma)d\sigma, z(t)\Big) - f\Big(t, \int_{\mathsf{T}_{i-1}}^{\mathsf{T}_i} G_i(s,\sigma)y(\sigma)d\sigma, \varkappa(t)\Big)\Big\|dt, \\ &\leq \Big\|z(s) + \int_{\mathsf{T}_{i-1}}^{\mathsf{T}_i} G_i(s,t) f\Big(t, \int_{\mathsf{T}_{i-1}}^{\mathsf{T}_i} G_i(s,\sigma)z(\sigma)d\sigma, z(t)\Big)dt\Big\| \\ &+ \int_{\mathsf{T}_{i-1}}^{\mathsf{T}_i} G_i(s,t)\Big\|f\Big(t, \int_{\mathsf{T}_{i-1}}^{\mathsf{T}_i} G_i(s,\sigma)z(\sigma)d\sigma, z(t)\Big) - f\Big(t, \int_{\mathsf{T}_{i-1}}^{\mathsf{T}_i} G_i(s,\sigma)y(\sigma)d\sigma, \varkappa(t)\Big)\Big\|dt, \\ &\leq \lambda_{\psi(s)}\epsilon\psi(s) \\ &+ \frac{1}{\Gamma(\ltimes_i)}\Big(\frac{\mathsf{T}_i - \mathsf{T}_{i-1}}{4}\Big)^{\ltimes_i - 1} \int_{\mathsf{T}_{i-1}}^{\mathsf{T}_i} t^{-\delta}\Big(K\int_{\mathsf{T}_{i-1}}^{\mathsf{T}_i} G_i(s,\sigma)\|z(\sigma) - \varkappa(\sigma)\|d\sigma + L\|z(t) - \varkappa(t)\|\Big)dt, \\ &\leq \lambda_{\psi(s)}\epsilon\psi(s) + \frac{1}{\Gamma(\ltimes_i)}\Big(\frac{\mathsf{T}_i - \mathsf{T}_{i-1}}{4}\Big)^{\ltimes_i - 1} \int_{\mathsf{T}_{i-1}}^{\mathsf{T}_i} t^{-\delta}\Big(K\frac{(\mathsf{T}_i - \mathsf{T}_{i-1})^{\ltimes_i}}{4^{\ltimes_i - 1}\Gamma(\ltimes_i)}\|z - \varkappa\|_{E_i} + L\|z - \varkappa\|_{E_i}\Big)dt, \\ &\leq \lambda_{\psi(s)}\epsilon\psi(s) + \frac{(\mathsf{T}_i - \mathsf{T}_{i-1})^{\ltimes_i - 1}\Big(\mathsf{T}_i^{1-\delta} - \mathsf{T}_{i-1}^{1-\delta}\Big)}{(1-\delta)4^{\ltimes_i - 1}\Gamma(\ltimes_i)}\Big(\frac{K(\mathsf{T}_i - \mathsf{T}_{i-1})^{\ltimes_i}}{4^{\ltimes_i - 1}\Gamma(\ltimes_i)} + L\Big)\|z - \varkappa\|_{E_i}. \end{aligned}$$

Hence,

$$\|z-\varkappa\|_{E_i}\Big[1-\frac{(\mathsf{T}_i-\mathsf{T}_{i-1})^{\ltimes_i-1}\Big(\mathsf{T}_i^{1-\delta}-\mathsf{T}_{i-1}^{1-\delta}\Big)}{(1-\delta)4^{\ltimes_i-1}\Gamma(\ltimes_i)}\Big(\frac{K(\mathsf{T}_i-\mathsf{T}_{i-1})^{\ltimes_i}}{4^{\ltimes_i-1}\Gamma(\ltimes_i)}+L\Big)\Big]\leq\lambda_{\psi(s)}\epsilon\psi(s).$$

This yields:

$$\begin{aligned}\|z-\varkappa\|_{E_i} &\leq \Big[1-\frac{(\mathsf{T}_i-\mathsf{T}_{i-1})^{\ltimes_i-1}\Big(\mathsf{T}_i^{1-\delta}-\mathsf{T}_{i-1}^{1-\delta}\Big)}{(1-\delta)4^{\ltimes_i-1}\Gamma(\ltimes_i)}\Big(\frac{K(\mathsf{T}_i-\mathsf{T}_{i-1})^{\ltimes_i}}{4^{\ltimes_i-1}\Gamma(\ltimes_i)}+L\Big)\Big]^{-1}\lambda_{\psi(s)}\epsilon\psi(s)\\ &:= c_f\epsilon\psi(s).\end{aligned}$$

Therefore, the initial terminal value problem (ITVP) (3.75) is Ulam-Hyers ℜassias stable with respect to ψ.

3.5.3 Approximate Numerical Examples

3.5.3.1 Example 1

Consider the following initial terminal value problem (ITVP):

$$\begin{cases}\digamma_{0+}^{\ltimes(s)}\varkappa(s)+\dfrac{|\varkappa(s)|+|\digamma_{0+}^{\ltimes(s)}\varkappa(s)|}{(s+1)^{\frac{1}{2}}(9+e^s)(1+\varkappa^2(s))}=0, \quad s\in J:=[0,2],\\ \\ \varkappa(0)=0, \;\; \varkappa(2)=0.\end{cases} \tag{3.84}$$

Let the function

$$f(s,y,z)=\frac{y+z}{(s+1)^{\frac{1}{2}}(9+e^s)(1+y^2)}, \quad (s,y,z)\in[0,2]\times[0,+\infty)\times[0,+\infty),$$

and define the variable-order function:

$$\ltimes(s)=\begin{cases}\frac{9}{8}, & s\in J_1:=[0,1],\\ \frac{7}{4}, & s\in J_2:=]1,2].\end{cases} \tag{3.85}$$

We verify the Lipschitz condition (H2). For $s\in J$, we have:

$$\begin{aligned}s^{\frac{1}{2}}\Big|f(s,y_1,z_1)-f(s,y_2,z_2)\Big| &\leq \frac{1}{9+e^s}\left|\frac{y_1+z_1}{1+y_1^2}-\frac{y_2+z_2}{1+y_2^2}\right|\\ &\leq \frac{1}{10}\Big|(y_1+z_1)-(y_2+z_2)\Big|\\ &\leq \frac{1}{10}|y_1-y_2|+\frac{1}{10}|z_1-z_2|.\end{aligned}$$

Thus, (H2) holds with $\delta=\frac{1}{2}$, $K=L=\frac{1}{10}$.

The equation (3.84) is decomposed into two subproblems according to (3.85):

$$\begin{cases} F^{\frac{9}{8}}_{0^+}\varkappa(s) + \dfrac{|\varkappa(s)| + |D^{\frac{9}{8}}_{0^+}\varkappa(s)|}{(s+1)^{\frac{1}{2}}(9+e^s)(1+\varkappa^2(s))} = 0, \quad s \in J_1, \\[2ex] F^{\frac{7}{4}}_{1^+}\varkappa(s) + \dfrac{|\varkappa(s)| + |F^{\frac{7}{4}}_{1^+}\varkappa(s)|}{(s+1)^{\frac{1}{2}}(9+e^s)(1+\varkappa^2(s))} = 0, \quad s \in J_2. \end{cases}$$

For $s \in J_1$, equation (3.84) reduces to:

$$\begin{cases} F^{\frac{9}{8}}_{0^+}\varkappa(s) + \dfrac{|\varkappa(s)| + |F^{\frac{9}{8}}_{0^+}\varkappa(s)|}{(s+1)^{\frac{1}{2}}(9+e^s)(1+\varkappa^2(s))} = 0, \quad s \in J_1, \\[2ex] \varkappa(0) = 0, \quad \varkappa(1) = 0. \end{cases} \tag{3.86}$$

We verify condition (3.80):

$$\frac{(\mathrm{T}_1 - \mathrm{T}_0)^{\ltimes_1 - 1}\left(\mathrm{T}_1^{1-\delta} - \mathrm{T}_0^{1-\delta}\right)}{(1-\delta)4^{\ltimes_1 - 1}\Gamma(\ltimes_1)}\left(\frac{K(\mathrm{T}_1 - \mathrm{T}_0)^{\ltimes_1}}{4^{\ltimes_1 - 1}\Gamma(\ltimes_1)} + L\right) = \frac{(1)^{\frac{1}{8}}(1^{\frac{1}{2}})}{\frac{1}{2}4^{\frac{1}{8}}\Gamma(\frac{9}{8})}\left(\frac{\frac{1}{10}(1)^{\frac{9}{8}}}{4^{\frac{1}{8}}\Gamma(\frac{9}{8})} + \frac{1}{10}\right) \simeq 0.3380 < 1.$$

Let $\psi(s) = s^{\frac{1}{2}}$. Then:

$$\begin{aligned} I^{\ltimes_1}_{0^+}\psi(s) &= \frac{1}{\Gamma(\frac{9}{8})}\int_0^s (s-t)^{\frac{1}{8}} t^{\frac{1}{2}}\, dt, \\ &\le \frac{1}{\Gamma(\frac{9}{8})}\int_0^s (s-t)^{\frac{1}{8}}\, dt, \\ &\le \frac{8}{9\Gamma(\frac{9}{8})}\psi(s). \end{aligned}$$

So (H3) is satisfied with $\lambda_{\psi(s)} = \dfrac{8}{9\Gamma(\frac{9}{8})}$.

Thus, by Theorem 23, problem (3.86) admits a solution $\varkappa_1 \in E_1$.

For $s \in J_2$, equation (3.84) becomes:

$$\begin{cases} F^{\frac{7}{4}}_{1^+}\varkappa(s) + \dfrac{|\varkappa(s)| + |F^{\frac{7}{4}}_{1^+}\varkappa(s)|}{(s+1)^{\frac{1}{2}}(9+e^s)(1+\varkappa^2(s))} = 0, \quad s \in J_2, \\[2ex] \varkappa(1) = 0, \quad \varkappa(2) = 0. \end{cases} \tag{3.87}$$

We verify condition (3.80) again:

$$\frac{(\mathsf{T}_2-\mathsf{T}_1)^{\ltimes_2-1}\left(\mathsf{T}_2^{1-\delta}-\mathsf{T}_1^{1-\delta}\right)}{(1-\delta)4^{\ltimes_2-1}\Gamma(\ltimes_2)}\left(\frac{K(\mathsf{T}_2-\mathsf{T}_1)^{\ltimes_2}}{4^{\ltimes_2-1}\Gamma(\ltimes_2)}+L\right)=\frac{(1)^{\frac{3}{4}}(2^{\frac{1}{2}}-1^{\frac{1}{2}})}{\frac{1}{2}4^{\frac{3}{4}}\Gamma(\frac{7}{4})}\left(\frac{\frac{1}{10}(1)^{\frac{7}{4}}}{4^{\frac{3}{4}}\Gamma(\frac{7}{4})}+\frac{1}{10}\right)\simeq 0.0441<1.$$

Let $\psi(s)=s^{\frac{1}{2}}$. Then:

$$\begin{aligned} I_{1^+}^{\ltimes_2}\psi(s) &= \frac{1}{\Gamma(\frac{7}{4})}\int_1^s (s-t)^{\frac{3}{4}}t^{\frac{1}{2}}\,dt, \\ &\le \frac{\sqrt{2}}{\Gamma(\frac{7}{4})}\int_1^s (s-t)^{\frac{3}{4}}\,dt, \\ &\le \frac{4\sqrt{2}}{7\Gamma(\frac{7}{4})}\psi(s) := \lambda_{\psi(s)}\psi(s). \end{aligned}$$

So (H3) is satisfied with $\lambda_{\psi(s)}=\frac{4\sqrt{2}}{7\Gamma(\frac{7}{4})}$.

Thus, by Theorem 23, problem (3.87) admits a solution $\varkappa_2 \in E_2$.

It's well known that:

$$\varkappa_2(s)=\begin{cases} 0, & s\in J_1, \\ \widetilde{\varkappa}_2(s), & s\in J_2. \end{cases}$$

Therefore, by Definition 47, the initial and terminal value problem (3.84) admits a solution

$$\varkappa(s)=\begin{cases} \varkappa_1(s), & s\in J_1, \\ \varkappa_2(s)=\begin{cases} 0, & s\in J_1, \\ \widetilde{\varkappa}_2(s), & s\in J_2. \end{cases} \end{cases}$$

Furthermore, by Theorem 24 and the function ψ, the equation (3.84) is Ulam-Hyers $\mathfrak{R}$assias stable.

Figures below illustrate the plot of the function $f(s,y,z)$ over the intervals $[0,1]$ and $[1,2]$.

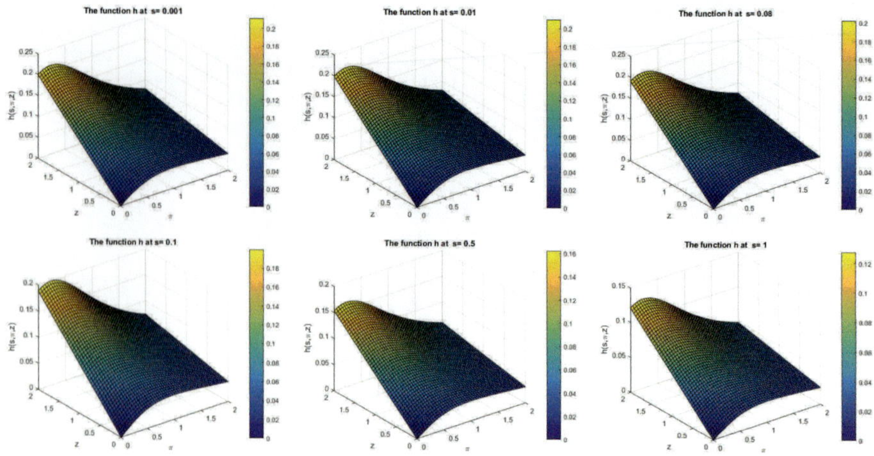

FIGURE 3.1: A plot of $f(s, y, z)$ for different values of $s \in [0, 1]$.

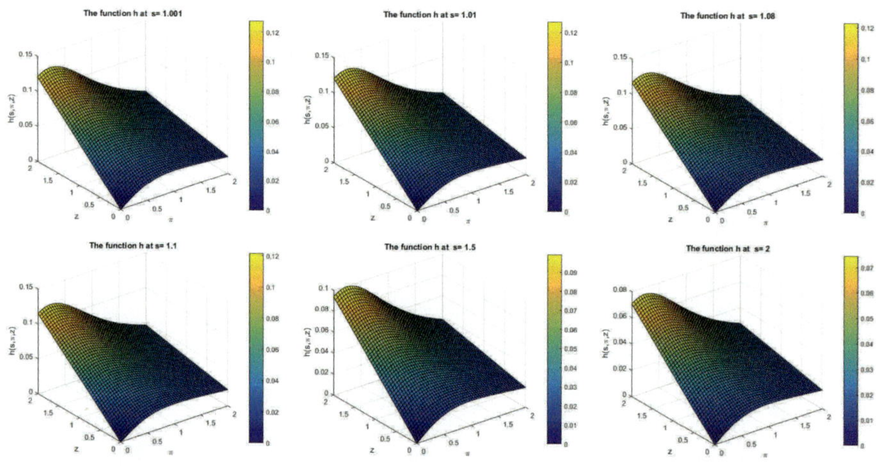

FIGURE 3.2: A plot of $f(s, y, z)$ for different values of $s \in]1, 2]$.

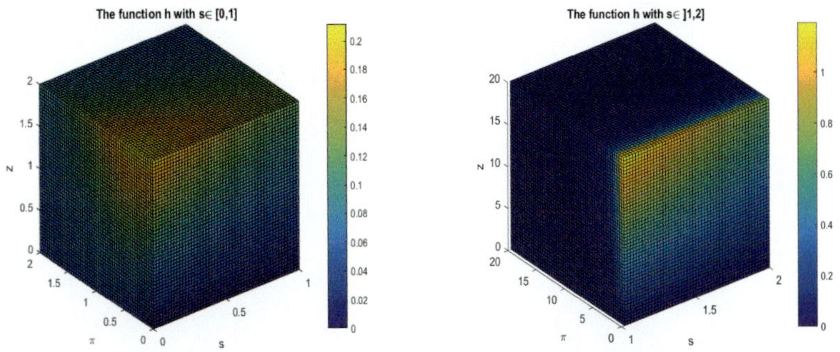

FIGURE 3.3: The 3D plot of $f(s, y, z)$ for $s \in [0, 1]$ and $]1, 2]$.

In this table, we present the *value* of $\max_{y,z} f(s, y, z)$ for different *values* of $s \in [0, 1]$ and $]1, 2]$.

$s \in [0, 1]$	0.0010	0.0100	0.0200	0.0800	0.1000	0.5000	1.0000
$\max_{y,z} f(s, y, z)$	0.2117	0.2105	0.2093	0.2021	0.1998	0.1624	0.1278
$s \in]1, 2]$	1.0010	1.0100	1.0200	1.0800	1.1000	1.5000	2.0000
$\max_{y,z} f(s, y, z)$	0.1277	0.1272	0.1266	0.1229	0.1217	0.0994	0.0746

We observe that the *value* of $\max_{y,z} f(s, y, z)$ is decreasing with respect to the *value* of s in the two intervals.

3.5.3.2 Example 2

Consider the following initial terminal value problem (ITVP):

$$\begin{cases} F_{0^+}^{\ltimes(s)}\varkappa(s) + \dfrac{s^{\frac{-1}{4}}e^{-s}}{(4e^{\frac{s}{1+s^2}} + 5e^s + 2)(1 + |\varkappa(s)| + |F_{0^+}^{\ltimes(s)}\varkappa(s)|)} = 0, \quad s \in J := [0, 3], \\ \\ \varkappa(0) = 0, \ \ \varkappa(3) = 0. \end{cases} \tag{3.88}$$

Let the function

$$f(s, y, z) = \frac{s^{\frac{-1}{4}}e^{-s}}{(4e^{\frac{s}{1+s^2}} + 5e^s + 2)(1 + y + z)}, \ (s, y, z) \in [0, 3]\times[0, +\infty)\times[0, +\infty).$$

Define the variable-order function:

$$\ltimes(s) = \begin{cases} \frac{3}{2}, & s \in J_1 := [0,1], \\ \frac{8}{5}, & s \in J_2 :=]1,2], \\ \frac{9}{5}, & s \in J_3 :=]2,3]. \end{cases} \tag{3.89}$$

$s \in J$ we have

$$\begin{aligned} s^{\frac{1}{4}}\Big|f(s,y_1,z_1) - f(s,y_2,z_2)\Big| &\le \frac{1}{e^s(4e^{\frac{s}{1+s^2}} + 5e^s + 2)}\Big|\frac{1}{1+y_1+z_1} - \frac{1}{1+y_2+z_2}\Big| \\ &\le \frac{1}{11}\Big|(y_2+z_2) - (y_1+z_1)\Big| \\ &\le \frac{1}{11}|y_1 - y_2| + \frac{1}{11}|z_1 - z_2|. \end{aligned}$$

Thus, (H2) holds with $\delta = \frac{1}{4}$, $K = L = \frac{1}{11}$.

The equation (3.88) is decomposed into three subproblems according to (3.89)

$$\begin{cases} F_{0^+}^{\frac{3}{2}} \varkappa(s) + \dfrac{s^{\frac{-1}{4}} e^{-s}}{(4e^{\frac{s}{1+s^2}} + 5e^s + 2)(1 + |\varkappa(s)| + |F_{0^+}^{\ltimes(s)} \varkappa(s)|)} = 0, \quad s \in J_1, \\[2ex] F_{1^+}^{\frac{8}{5}} \varkappa(s) + \dfrac{s^{\frac{-1}{4}} e^{-s}}{(4e^{\frac{s}{1+s^2}} + 5e^s + 2)(1 + |\varkappa(s)| + |F_{1^+}^{\ltimes(s)} \varkappa(s)|)} = 0, \quad s \in J_2, \\[2ex] F_{2^+}^{\frac{9}{5}} \varkappa(s) + \dfrac{s^{\frac{-1}{4}} e^{-s}}{(4e^{\frac{s}{1+s^2}} + 5e^s + 2)(1 + |\varkappa(s)| + |F_{2^+}^{\ltimes(s)} \varkappa(s)|)} = 0, \quad s \in J_3. \end{cases}$$

For $s \in J_1$, equation (3.88) reduces to:

$$\begin{cases} F_{0^+}^{\frac{3}{2}} \varkappa(s) + \dfrac{s^{\frac{-1}{4}} e^{-s}}{(4e^{\frac{s}{1+s^2}} + 5e^s + 2)(1 + |\varkappa(s)| + |F_{0^+}^{\ltimes(s)} \varkappa(s)|)} = 0, \quad s \in J_1, \\[2ex] \varkappa(0) = 0, \quad \varkappa(1) = 0. \end{cases} \tag{3.90}$$

We verify condition (3.80):

$$\frac{(\mathsf{T}_1 - \mathsf{T}_0)^{\ltimes_1 - 1}\Big(\mathsf{T}_1^{1-\delta} - \mathsf{T}_0^{1-\delta}\Big)}{(1-\delta)4^{\ltimes_1 - 1}\Gamma(\ltimes_1)}\Big(\frac{K(\mathsf{T}_1 - \mathsf{T}_0)^{\ltimes_1}}{4^{\ltimes_1 - 1}\Gamma(\ltimes_1)} + L\Big) = \frac{(1)^{\frac{1}{2}}(1^{\frac{3}{4}})}{\frac{3}{4}4^{\frac{1}{2}}\Gamma(\frac{3}{2})}\Big(\frac{\frac{1}{11}(1)^{\frac{3}{2}}}{4^{\frac{1}{2}}\Gamma(\frac{3}{2})} + \frac{1}{11}\Big) \simeq 0.1070 < 1.$$

Thus, by Theorem 23, problem (3.90) admits a solution $\varkappa_1 \in E_1$.

The problem (3.88) may be formulated as follows for $s \in J_2$:

$$\begin{cases} \mathsf{F}_{1^+}^{\frac{8}{5}} \varkappa(s) + \dfrac{s^{\frac{-1}{4}} e^{-s}}{(4e^{\frac{s}{1+s^2}} + 5e^s + 2)(1 + |\varkappa(s)| + |\mathsf{F}_{1^+}^{\ltimes(s)} \varkappa(s)|)} = 0, \quad s \in J_2, \\ \\ \varkappa(1) = 0, \ \ \varkappa(2) = 0. \end{cases} \tag{3.91}$$

We observe that:

$$\frac{(\mathsf{T}_2 - \mathsf{T}_1)^{\ltimes_2 - 1}\left(\mathsf{T}_2^{1-\delta} - \mathsf{T}_1^{1-\delta}\right)}{(1-\delta)4^{\ltimes_2 - 1}\Gamma(\ltimes_2)}\left(\frac{K(\mathsf{T}_2 - \mathsf{T}_1)^{\ltimes_2}}{4^{\ltimes_2 - 1}\Gamma(\ltimes_2)} + L\right) = \frac{(1)^{\frac{3}{5}}(2^{\frac{3}{4}} - 1^{\frac{3}{4}})}{\frac{3}{4}4^{\frac{3}{5}}\Gamma(\frac{8}{5})}\left(\frac{\frac{1}{11}(1)^{\frac{8}{5}}}{4^{\frac{3}{5}}\Gamma(\frac{8}{5})} + \frac{1}{11}\right) \simeq 0.0599 < 1.$$

Therefore, condition (3.80) is satisfied.

The problem (3.91) has to have a solution $\widetilde{\varkappa}_2 \in E_2$ according to Theorem (23).

Similarly, the problem (3.88) may be formulated as follows for $s \in J_3$ as:

$$\begin{cases} \mathsf{F}_{2^+}^{\frac{9}{5}} \varkappa(s) + \dfrac{s^{\frac{-1}{4}} e^{-s}}{(4e^{\frac{s}{1+s^2}} + 5e^s + 2)(1 + |\varkappa(s)| + |\mathsf{F}_{2^+}^{\ltimes(s)} \varkappa(s)|)} = 0, \quad s \in J_3, \\ \\ \varkappa(2) = 0, \ \ \varkappa(3) = 0. \end{cases} \tag{3.92}$$

We observe that:

$$\frac{(\mathsf{T}_3 - \mathsf{T}_2)^{\ltimes_3 - 1}\left(\mathsf{T}_3^{1-\delta} - \mathsf{T}_2^{1-\delta}\right)}{(1-\delta)4^{\ltimes_3 - 1}\Gamma(\ltimes_3)}\left(\frac{K(\mathsf{T}_3 - \mathsf{T}_2)^{\ltimes_3}}{4^{\ltimes_3 - 1}\Gamma(\ltimes_3)} + L\right) = \frac{(1)^{\frac{4}{5}}(3^{\frac{3}{4}} - 2^{\frac{3}{4}})}{\frac{3}{4}4^{\frac{4}{5}}\Gamma(\frac{9}{5})}\left(\frac{\frac{1}{11}(1)^{\frac{9}{5}}}{4^{\frac{4}{5}}\Gamma(\frac{9}{5})} + \frac{1}{11}\right) \simeq 0.0347 < 1.$$

Hence, condition (3.80) is satisfied.

According to Theorem 23, the problem (3.92) admits a solution $\widetilde{\varkappa}_3 \in E_3$.

Therefore, the initial and terminal value problem (3.88) admits a solution by Definition 47, given by:

$$\varkappa(s) = \begin{cases} \varkappa_1(s), \quad s \in J_1, \\ \varkappa_2(s) = \begin{cases} 0, \quad s \in J_1, \\ \widetilde{\varkappa}_2(s), \quad s \in J_2, \end{cases} \\ \varkappa_3(s) = \begin{cases} 0, \quad s \in J_1 \cup J_2, \\ \widetilde{\varkappa}_3(s), \quad s \in J_3. \end{cases} \end{cases}$$

In the accompanying figures, we present the plots of the function $f(s, y, z)$ over the three subintervals: $[0, 1]$, $]1, 2]$ and $]2, 3]$.

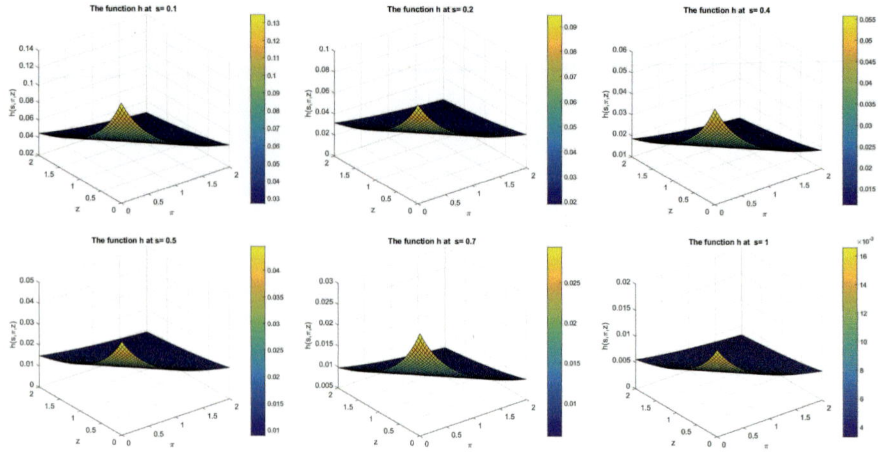

FIGURE 3.4: A plot of $f(s, y, z)$ for different values of $s \in [0, 1]$.

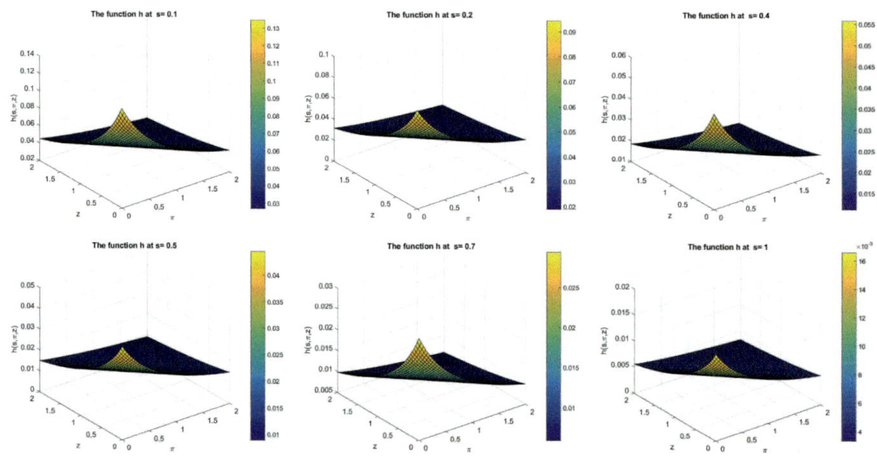

FIGURE 3.5: A plot of $f(s, y, z)$ for different values of $s \in]1, 2]$.

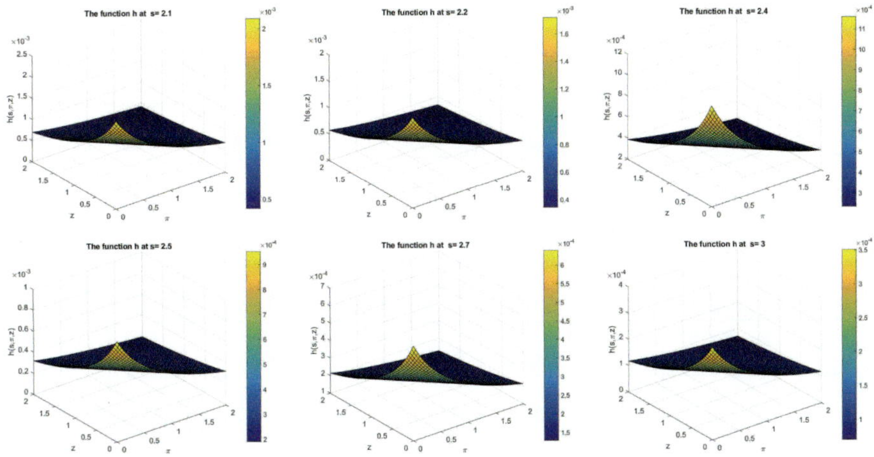

FIGURE 3.6: A plot of $f(s, y, z)$ for different values of $s \in]2, 3]$.

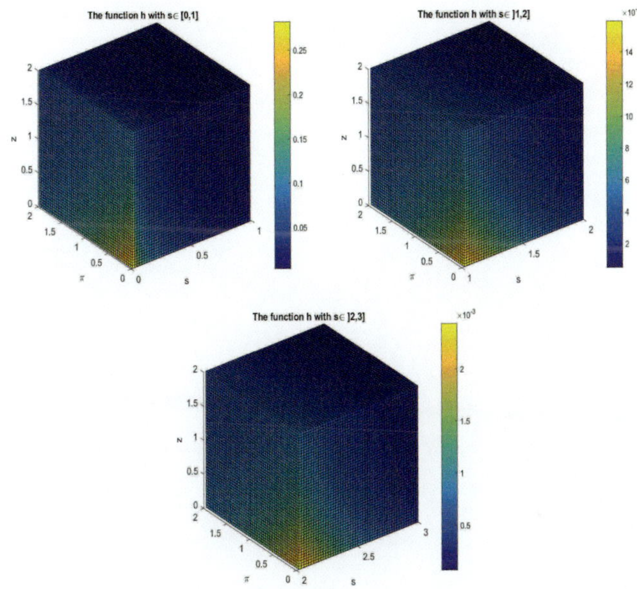

FIGURE 3.7: The 3D plot of $f(s, y, z)$ for $s \in [0, 1]$, $]1, 2]$ and $]2, 3]$.

In this table, we present the *value* of $\max_{y,z} f(s,y,z)$ for different *values* of $s \in [0,1]$, $]1,2]$ and $]2,3]$.

$s \in [0,1]$	0.1000	0.2000	0.3000	0.4000	0.5000	0.7000	1.0000
$\max_{y,z} f(s,y,z)$	0.1347	0.0945	0.0714	0.0558	0.0445	0.0294	0.0166
$s \in]1,2]$	1.1000	1.2000	1.3000	1.4000	1.5000	1.7000	2.0000
$\max_{y,z} f(s,y,z)$	0.0138	0.0114	0.0095	0.0079	0.0066	0.0045	0.0025
$s \in]2,3]$	2.1000	2.2000	2.3000	2.4000	2.5000	2.7000	3.0000
$\max_{y,z} f(s,y,z)$	0.0021	0.0017	0.0014	0.0012	0.0010	0.0006	0.0004

We observe a similar pattern as in Example 1.

4

Problems for Caputo's Fractional Differential Equations with Variable Order

This chapter focuses on the qualitative analysis of solutions to Caputo variable-order fractional differential equations—particularly impulsive and thermostat-type models—under initial and terminal value conditions. The study employs fixed point theorems as the core analytical tool.

4.1 Initial and Terminal Value Problems for Variable-Order Caputo Fractional Differential Equations via Kuratowski's Measure of Noncompactness

The initial and terminal value problem (ITVP) under consideration is given by

$$\begin{cases} {}^{c}\digamma_{0+}^{\ltimes(s)}\varkappa(s) = f(s,\varkappa(s)), & s \in J := [0, \mathsf{T}] \\ \varkappa(0) = 0, \quad \varkappa(\mathsf{T}) = 0, \end{cases} \tag{4.1}$$

where $1 < \ltimes(s) \le 2$, $f : J \times \mathbb{R} \to \mathbb{R}$ is a continuous function, and ${}^{c}\digamma_{0+}^{\ltimes(s)}$ denotes the Caputo fractional derivative of variable-order $\ltimes(s)$, and we shall look for a solution of (4.1). Moreover, we examine the Ulam-Hyers stability of the solution obtained for (4.1).

DOI: 10.1201/9781003687900-4

4.1.1 Existence Criteria

Let us state the premise as follows:

(H1) Let $n \in \mathbb{N}$ be a positive integer, and let $\mathcal{P} = \{J_1 := [0, \mathsf{T}_1], J_2 := (\mathsf{T}_1, \mathsf{T}_2], J_3 := (\mathsf{T}_2, \mathsf{T}_3], ...J_n := (\mathsf{T}_{n-1}, \mathsf{T}]\}$ be a partition of the interval $J := [0, \tau]\}$. Define a piecewise constant function $\ltimes(s) : J \to (1, 2]$ with respect to the partition $\mathcal{P}$, such that

$$\ltimes(s) = \sum_{i=1}^{n} \ltimes_i I_i(s) = \begin{cases} \ltimes_1, & if\ s \in J_1, \\ \ltimes_2, & if\ s \in J_2, \\ \cdot & \\ \cdot & \\ \cdot & \\ \ltimes_n, & if\ s \in J_n, \end{cases}$$

where $1 < \ltimes_i \leq 2$ are constants, and I_i denotes the interval $J_i := (\mathsf{T}_{i-1}, \mathsf{T}_i]$, $i = 1, 2, \ldots, n$, with $\mathsf{T}_0 = 0$ and $\mathsf{T}_n = \mathsf{T}$.

$$I_i(s) = \begin{cases} 1, & for\ s \in J_i, \\ 0, & for\ elsewhere. \end{cases}$$

For each $i \in \{1, 2, \ldots, n\}$, let $E_i = C(J_i, \mathbb{R})$ denote the ßanach space of continuous functions $\varkappa : J_i \to \mathbb{R}$, equipped with the norm

$$\|\varkappa\|_{E_i} = \sup_{s \in J_i} |\varkappa(s)|,$$

Then, for $s \in J_i$ $i = 1, 2, \ldots, n$, the Caputo fractional derivative (CFD) of variable order $\varkappa(s)$, for a function $\varkappa(s) \in C(J, \mathbb{R})$, as defined in equation (2.10), can be expressed as a sum of left Caputo fractional derivatives with constant orders $\ltimes_i$, namely:

$$^{c}\digamma_{0+}^{\ltimes(s)} \varkappa(s) = \int_0^{\mathsf{T}_1} \frac{(s-t)^{1-\ltimes_1}}{\Gamma(2-\ltimes_1)} \varkappa^{(2)}(t)dt + ... + \int_{\mathsf{T}_{i-1}}^{s} \frac{(s-t)^{1-\ltimes_i}}{\Gamma(2-\ltimes_i)} \varkappa^{(2)}(t)dt. \tag{4.2}$$

Thus, in accordance with equation (4.2), the ınitial and terminal value problem (ITVP) described in (4.1) can be formulated for any $s \in J_i$, where $i = 1, 2, \ldots, n$, as follows:

$$\int_0^{\mathsf{T}_1} \frac{(s-t)^{1-\ltimes_1}}{\Gamma(2-\ltimes_1)} \varkappa^{(2)}(t)dt + ... + \int_{\mathsf{T}_{i-1}}^{s} \frac{(s-t)^{1-\ltimes_i}}{\Gamma(2-\ltimes_i)} \varkappa^{(2)}(t)dt = f(s, \varkappa(s)), \quad s \in J_i. \tag{4.3}$$

The solution to the ITVP will be presented in the sections that follow equation (4.1).

Definition 48 *A solution to the initial and terminal value problem (ITVP) given in equation (4.1) exists if there exist functions $\varkappa_i$, $i = 1, 2, \ldots, n$, such that*

- $\varkappa_i \in C([0, \mathsf{T}_i], \mathbb{R})$,
- $\varkappa_i$ *satisfies equation (4.3), and*
- *the initial and terminal conditions $\varkappa_i(0) = 0$ and $\varkappa_i(\mathsf{T}_i) = 0$ are fulfilled.*

If $\varkappa \in C(J, \mathbb{R})$ is such that $\varkappa(s) \equiv 0$ on $[0, \mathsf{T}_{i-1}]$, and $\varkappa$ satisfies the integral equation (4.3), then equation (4.3) simplifies to

$$ {}^{c}\digamma^{\ltimes_i}_{\mathsf{T}_{i-1}^{+}} \varkappa(s) = f(s, \varkappa(s)), \quad s \in J_i. $$

We now consider the corresponding initial and terminal value problem ITVP

$$ \begin{cases} {}^{c}\digamma^{\ltimes_i}_{\mathsf{T}_{i-1}^{+}} \varkappa(s) = f(s, \varkappa(s)), \quad s \in J_i \\ \varkappa(\mathsf{T}_{i-1}) = 0, \varkappa(\mathsf{T}_i) = 0. \end{cases} \tag{4.4} $$

The forthcoming lemma will provide the foundation for our approach to solving the ITVP given in equation (4.4).

Lemma 17 *Suppose $i \in \{1, 2, \ldots, n\}$ is a natural number, $f \in C(J_i \times \mathbb{R}, \mathbb{R})$, and there exists a number $\delta \in (0, 1)$ such that $s^{\delta} f \in C(J_i \times \mathbb{R}, \mathbb{R})$. Then, $\varkappa \in E_i$ is a solution to the initial and terminal value problem (ITVP) given in equation (4.4) if and only if $\varkappa$ satisfies the integral equation*

$$ \varkappa(s) = -(\mathsf{T}_i - \mathsf{T}_{i-1})^{-1}(s - \mathsf{T}_{i-1}) I^{\ltimes_i}_{\mathsf{T}_{i-1}^{+}} f(\mathsf{T}_i, \varkappa(\mathsf{T}_i)) + I^{\ltimes_i}_{\mathsf{T}_{i-1}^{+}} f(s, \varkappa(s)). \tag{4.5} $$

Proof: Assume that, $\varkappa \in E_i$ is a solution to the initial and terminal value problem (ITVP) given in equation (4.4). By applying Lemma 2 and the fractional integral operator $I^{\ltimes_i}_{\mathsf{T}_{i-1}^{+}}$ to both sides of equation (4.4), we obtain:

$$ \varkappa(s) = \omega_1 + \omega_2(s - \mathsf{T}_{i-1}) + \frac{1}{\Gamma(\ltimes_i)} \int_{\mathsf{T}_{i-1}}^{s} (s - t)^{\ltimes_i - 1} f(t, \varkappa(t)) dt, \quad s \in J_i. $$

Given the initial condition $\varkappa(\mathsf{T}_{i-1}) = 0$, it follows that $\omega_1 = 0$.

Now, let $\varkappa(s)$ also satisfy the terminal condition $\varkappa(\mathsf{T}_i) = 0$. Substituting $s = \mathsf{T}_i$ into the expression above, we find:

$$ \omega_2 = -(\mathsf{T}_i - \mathsf{T}_{i-1})^{-1} I^{\ltimes_i}_{\mathsf{T}_{i-1}^{+}} f(\mathsf{T}_i, \varkappa(\mathsf{T}_i)). $$

Then, we obtain the nonlinear fractional differential equation of variable order:

$$\varkappa(s) = -(\mathsf{T}_i - \mathsf{T}_{i-1})^{-1}(s - \mathsf{T}_{i-1})I^{\ltimes_i}_{\mathsf{T}^+_{i-1}} f(\mathsf{T}_i, \varkappa(\mathsf{T}_i)) + I^{\ltimes_i}_{\mathsf{T}^+_{i-1}} f(s, \varkappa(s)), \;\; s \in J_i.$$

On the other hand, let $\varkappa \in E_i$ denote the solution to the integral equation (4.5). By Lemma 2 and the continuity of function $s^\delta f$, we conclude that $\varkappa$ is also a solution to the (4.4).

We now proceed to demonstrate the existence of a solution to the ITVP (4.4) using Darbo's fixed point theorem and the Kuratowski measure of noncompactness.

Theorem 25 *Assume the conditions of Lemma 17 are satisfied, and there exists a constant $K > 0$ such that*

$$s^\delta |f(s, y_1) - f(s, y_2)| \leq K|y_1 - y_2|, \tag{4.6}$$

for all $y_1, \; y_2 \in \mathbb{R}$, $s \in J_i$, and the inequality

$$\frac{2K(\mathsf{T}_i - \mathsf{T}_{i-1})^{\ltimes_i - 1}(\mathsf{T}_i^{1-\delta} - \mathsf{T}_{i-1}^{1-\delta})}{(1-\delta)\Gamma(\ltimes_i)} < 1, \tag{4.7}$$

holds. Then, the initial and terminal value problem (ITVP) (4.4) admits at least one solution in E_i.

Proof: We construct the mapping as follows:

$$W : E_i \to E_i,$$

$$W\varkappa(s) = -(\mathsf{T}_i - \mathsf{T}_{i-1})^{-1}(s - \mathsf{T}_{i-1})I^{\ltimes_i}_{\mathsf{T}^+_{i-1}} f(\mathsf{T}_i, \varkappa(\mathsf{T}_i))$$
$$+\frac{1}{\Gamma(\ltimes_i)}\int_{\mathsf{T}_{i-1}}^{s} (s-t)^{\ltimes_i - 1} f(t, \varkappa(t))dt, \;\; s \in J_i. \tag{4.8}$$

Given the properties of fractional integrals and the continuity of $s^\delta f$, the operator $W : E_i \to E_i$ defined in (4.8) is therefore well-defined.

Let

$$R_i \geq \frac{\frac{2f^\star(\mathsf{T}_i - \mathsf{T}_{i-1})^{\ltimes_i}}{\Gamma(\ltimes_i)}}{1 - \frac{2K(\mathsf{T}_i - \mathsf{T}_{i-1})^{\ltimes_i - 1}(\mathsf{T}_i^{1-\delta} - \mathsf{T}_{i-1}^{1-\delta})}{(1-\delta)\Gamma(\ltimes_i)}},$$

where we define

$$f^\star = \sup_{s \in J_i} |f(s, 0)|.$$

We consider the set

$$B_{R_i} = \{\varkappa \in E_i, \ \|\varkappa\|_{E_i} \leq R_i\}.$$

It is evident that B_{R_i} is bounded, closed, convex, and nonempty.

We now show that W satisfies the assumptions of Darbo's fixed point theorem. This will be demonstrated in four stages.

STEP 1: Claim: $W(B_{R_i}) \subseteq (B_{R_i})$.

Let $\varkappa \in B_{R_i}$. By assumption (H2), we obtain:

$$\begin{aligned}
|W\varkappa(s)| &\leq \tfrac{(\mathsf{T}_i-\mathsf{T}_{i-1})^{-1}(s-\mathsf{T}_{i-1})}{\Gamma(\kappa_i)}\int_{\mathsf{T}_{i-1}}^{\mathsf{T}_i}(\mathsf{T}_i-t)^{\kappa_i-1}|f(t,\varkappa(t))|dt + \tfrac{1}{\Gamma(\kappa_i)}\int_{\mathsf{T}_{i-1}}^{s}(s-t)^{\kappa_i-1}|f(t,\varkappa(t))|dt\\
&\leq \tfrac{2}{\Gamma(\kappa_i)}\int_{\mathsf{T}_{i-1}}^{\mathsf{T}_i}(\mathsf{T}_i-t)^{\kappa_i-1}|f(t,\varkappa(t))|dt\\
&\leq \tfrac{2}{\Gamma(\kappa_i)}\int_{\mathsf{T}_{i-1}}^{\mathsf{T}_i}(\mathsf{T}_i-t)^{\kappa_i-1}|f(t,\varkappa(t))-f(t,0)|dt + \tfrac{2}{\Gamma(\kappa_i)}\int_{\mathsf{T}_{i-1}}^{\mathsf{T}_i}(\mathsf{T}_i-t)^{\kappa_i-1}|f(t,0)|dt\\
&\leq \tfrac{2}{\Gamma(\kappa_i)}\int_{\mathsf{T}_{i-1}}^{\mathsf{T}_i}(\mathsf{T}_i-t)^{\kappa_i-1}t^{-\delta}(K|\varkappa(t)|)dt + \tfrac{2f^{*}(\mathsf{T}_i-\mathsf{T}_{i-1})^{\kappa_i}}{\Gamma(\kappa_i)}\\
&\leq \tfrac{2(\mathsf{T}_i-\mathsf{T}_{i-1})^{\kappa_i-1}}{\Gamma(\kappa_i)}\int_{\mathsf{T}_{i-1}}^{\mathsf{T}_i}t^{-\delta}(K|\varkappa(t)|)dt + \tfrac{2f^{*}(\mathsf{T}_i-\mathsf{T}_{i-1})^{\kappa_i}}{\Gamma(\kappa_i)}\\
&\leq \tfrac{2K(\mathsf{T}_i-\mathsf{T}_{i-1})^{\kappa_i-1}(\mathsf{T}_i^{1-\delta}-\mathsf{T}_{i-1}^{1-\delta})}{(1-\delta)\Gamma(\kappa_i)}R_i + \tfrac{2f^{*}(\mathsf{T}_i-\mathsf{T}_{i-1})^{\kappa_i}}{\Gamma(\kappa_i)}\\
&\leq R_i,
\end{aligned}$$

which means: $W(B_{R_i}) \subseteq B_{R_i}$.

STEP 2: Claim: The operator W is continuous.

Assume that $(\varkappa_n)$ is a sequence in E_i converging to $\varkappa$, with $s \in J_i$. Then,

$$\begin{aligned}
|(W\varkappa_n)(s)-(W\varkappa)(s)| &\leq \tfrac{(\mathsf{T}_i-\mathsf{T}_{i-1})^{-1}(s-\mathsf{T}_{i-1})}{\Gamma(\kappa_i)}\int_{\mathsf{T}_{i-1}}^{\mathsf{T}_i}(\mathsf{T}_i-t)^{\kappa_i-1}|f(t,\varkappa_n(t))-f(t,\varkappa(t))|dt\\
&\quad+\tfrac{1}{\Gamma(\kappa_i)}\int_{\mathsf{T}_{i-1}}^{s}(s-t)^{\kappa_i-1}|f(t,\varkappa_n(t))-f(t,\varkappa(t))|dt\\
&\leq \tfrac{2}{\Gamma(\kappa_i)}\int_{\mathsf{T}_{i-1}}^{\mathsf{T}_i}(\mathsf{T}_i-t)^{\kappa_i-1}|f(t,\varkappa_n(t))-f(t,\varkappa(t))|dt\\
&\leq \tfrac{2}{\Gamma(\kappa_i)}\int_{\mathsf{T}_{i-1}}^{\mathsf{T}_i}t^{-\delta}(\mathsf{T}_i-t)^{\kappa_i-1}(K|\varkappa_n(t)-\varkappa(t)|)dt\\
&\leq \tfrac{2K(\mathsf{T}_i-\mathsf{T}_{i-1})^{\kappa_i-1}}{\Gamma(\kappa_i)}\|\varkappa_n-\varkappa\|_{E_i}\int_{\mathsf{T}_{i-1}}^{\mathsf{T}_i}t^{-\delta}dt\\
&\leq \tfrac{2K(\mathsf{T}_i-\mathsf{T}_{i-1})^{\kappa_i-1}({\mathsf{T}_i}^{1-\delta}-{\mathsf{T}_{i-1}}^{1-\delta})}{(1-\delta)\Gamma(\kappa_i)}\|\varkappa_n-\varkappa\|_{E_i},
\end{aligned}$$

Hence,

$$\|(W\varkappa_n)-(W\varkappa)\|_{E_i} \to 0 \ \ as \ \ n \to \infty.$$

Consequently, W is a continuous operator on E_i.

STEP 3: Claim: The operator W possesses the properties of being bounded and equicontinuous.

From Step 1, we have $W(B_{R_i}) = \{W(\varkappa) : \varkappa \in B_{R_i}\} \subset B_{R_i}$. Thus, for each $\varkappa \in B_{R_i}$, it follows that $\|W(\varkappa)\|_{E_i} \leq R_i$, which implies that $W(B_{R_i})$ is bounded. It remains to show that $W(B_{R_i})$ is equicontinuous.

For $s_1, s_2 \in J_i$ with $s_1 < s_2$ and $\varkappa \in B_{R_i}$, we have:

$$
\begin{aligned}
&|(W\varkappa)(s_2) - (W\varkappa)(s_1)| \\
&= |-\frac{(\mathsf{T}_i - \mathsf{T}_{i-1})^{-1}(s_2 - \mathsf{T}_{i-1})}{\Gamma(\ltimes_i)} \int_{\mathsf{T}_{i-1}}^{\mathsf{T}_i} (\mathsf{T}_i - t)^{\ltimes_i - 1} f(t, \varkappa(t))dt \\
&\quad + \frac{1}{\Gamma(\ltimes_i)} \int_{\mathsf{T}_{i-1}}^{s_2} (s_2 - t)^{\ltimes_i - 1} f(t, \varkappa(t))dt + \frac{(\mathsf{T}_i - \mathsf{T}_{i-1})^{-1}(s_1 - \mathsf{T}_{i-1})}{\Gamma(\ltimes_i)} \\
&\quad \int_{\mathsf{T}_{i-1}}^{\mathsf{T}_i} (\mathsf{T}_i - t)^{\ltimes_i - 1} f(t, \varkappa(t))dt - \frac{1}{\Gamma(\ltimes_i)} \int_{\mathsf{T}_{i-1}}^{s_1} (s_1 - t)^{\ltimes_i - 1} f(t, \varkappa(t))dt| \\
&\leq \frac{(\mathsf{T}_i - \mathsf{T}_{i-1})^{-1}}{\Gamma(\ltimes_i)} \Big((s_2 - \mathsf{T}_{i-1}) - (s_1 - \mathsf{T}_{i-1})\Big) \int_{\mathsf{T}_{i-1}}^{\mathsf{T}_i} (\mathsf{T}_i - t)^{\ltimes_i - 1} |f(t, \varkappa(t))|dt \\
&\quad + \frac{1}{\Gamma(\ltimes_i)} \int_{\mathsf{T}_{i-1}}^{s_1} \Big((s_2 - t)^{\ltimes_i - 1} - (s_1 - t)^{\ltimes_i - 1}\Big) |f(t, \varkappa(t))|dt + \frac{1}{\Gamma(\ltimes_i)} \int_{s_1}^{s_2} (s_2 - t)^{\ltimes_i - 1} |f(t, \varkappa(t))|dt \\
&\leq \frac{(\mathsf{T}_i - \mathsf{T}_{i-1})^{-1}}{\Gamma(\ltimes_i)} \Big((s_2 - \mathsf{T}_{i-1}) - (s_1 - \mathsf{T}_{i-1})\Big) \int_{\mathsf{T}_{i-1}}^{\mathsf{T}_i} (\mathsf{T}_i - t)^{\ltimes_i - 1} |f(t, \varkappa(t)) - f(t, 0)|dt \\
&\quad + \frac{(\mathsf{T}_i - \mathsf{T}_{i-1})^{-1}}{\Gamma(\ltimes_i)} \Big((s_2 - \mathsf{T}_{i-1}) - (s_1 - \mathsf{T}_{i-1})\Big) \int_{\mathsf{T}_{i-1}}^{\mathsf{T}_i} (\mathsf{T}_i - t)^{\ltimes_i - 1} |f(t, 0)|dt \\
&\quad + \frac{1}{\Gamma(\ltimes_i)} \int_{\mathsf{T}_{i-1}}^{s_1} \Big((s_2 - t)^{\ltimes_i - 1} - (s_1 - t)^{\ltimes_i - 1}\Big) |f(t, \varkappa(t)) - f(t, 0)|dt \\
&\quad + \frac{1}{\Gamma(\ltimes_i)} \int_{\mathsf{T}_{i-1}}^{s_1} \Big((s_2 - t)^{\ltimes_i - 1} - (s_1 - t)^{\ltimes_i - 1}\Big) |f(t, 0)|dt \\
&\quad + \frac{1}{\Gamma(\ltimes_i)} \int_{s_1}^{s_2} (s_2 - t)^{\ltimes_i - 1} |f(t, \varkappa(t)) - f(t, 0)|dt + \frac{1}{\Gamma(\ltimes_i)} \int_{s_1}^{s_2} (s_2 - t)^{\ltimes_i - 1} |f(t, 0)|dt.
\end{aligned}
$$

$$
\begin{aligned}
\leq\;& \frac{(\mathsf{T}_i-\mathsf{T}_{i-1})^{-1}}{\Gamma(\ltimes_i)}\Big((s_2-\mathsf{T}_{i-1})-(s_1-\mathsf{T}_{i-1})\Big)\int_{\mathsf{T}_{i-1}}^{\mathsf{T}_i}(\mathsf{T}_i-t)^{\ltimes_i-1}t^{-\delta}(K|\varkappa(t)|)dt\\
&+\frac{f^\star(\mathsf{T}_i-\mathsf{T}_{i-1})^{-1}}{\Gamma(\ltimes_i)}\Big((s_2-\mathsf{T}_{i-1})-(s_1-\mathsf{T}_{i-1})\Big)\int_{\mathsf{T}_{i-1}}^{\mathsf{T}_i}(\mathsf{T}_i-t)^{\ltimes_i-1}dt\\
&+\frac{1}{\Gamma(\ltimes_i)}\int_{\mathsf{T}_{i-1}}^{s_1}t^{-\delta}\Big((s_2-t)^{\ltimes_i-1}-(s_1-t)^{\ltimes_i-1}\Big)(K|\varkappa(t)|)dt\\
&+\frac{f^\star}{\Gamma(\ltimes_i)}\int_{\mathsf{T}_{i-1}}^{s_1}\Big((s_2-t)^{\ltimes_i-1}-(s_1-t)^{\ltimes_i-1}\Big)dt\\
&+\frac{1}{\Gamma(\ltimes_i)}\int_{s_1}^{s_2}t^{-\delta}(s_2-t)^{\ltimes_i-1}(K|\varkappa(t)|)dt+\frac{f^\star}{\Gamma(\ltimes_i)}\int_{s_1}^{s_2}(s_2-t)^{\ltimes_i-1}dt\\
\leq\;& \frac{K(\mathsf{T}_i-\mathsf{T}_{i-1})^{\ltimes_i-2}}{\Gamma(\ltimes_i)}\Big((s_2-\mathsf{T}_{i-1})-(s_1-\mathsf{T}_{i-1})\Big)\|\varkappa\|_{E_i}\int_{\mathsf{T}_{i-1}}^{\mathsf{T}_i}t^{-\delta}dt\\
&+\frac{f^\star(\mathsf{T}_i-\mathsf{T}_{i-1})^{\ltimes_i-1}}{\Gamma(\ltimes_i+1)}\Big((s_2-\mathsf{T}_{i-1})-(s_1-\mathsf{T}_{i-1})\Big)+\frac{K}{\Gamma(\ltimes_i)}\|\varkappa\|_{E_i}\int_{\mathsf{T}_{i-1}}^{s_1}t^{-\delta}((s_2-s_1)^{\ltimes_i-1})dt\\
&+\frac{f^\star}{\Gamma(\ltimes_i)}\Big(\frac{(s_2-\mathsf{T}_{i-1})^{\ltimes_i}}{\ltimes_i}-\frac{(s_2-s_1)^{\ltimes_i}}{\ltimes_i}-\frac{(s_1-\mathsf{T}_{i-1})^{\ltimes_i}}{\ltimes_i}\Big)\\
&+\frac{K(s_2-s_1)^{\ltimes_i-1}}{\Gamma(\ltimes_i)}\|\varkappa\|_{E_i}\int_{s_1}^{s_2}t^{-\delta}dt+\frac{f^\star}{\Gamma(\ltimes_i)}\frac{(s_2-s_1)^{\ltimes_i}}{\ltimes_i}\\
\leq\;& \frac{K(\mathsf{T}_i-\mathsf{T}_{i-1})^{\ltimes_i-2}({\mathsf{T}_i}^{1-\delta}-{\mathsf{T}_{i-1}}^{1-\delta})}{(1-\delta)\Gamma(\ltimes_i)}\Big((s_2-\mathsf{T}_{i-1})-(s_1-\mathsf{T}_{i-1})\Big)\|\varkappa\|_{E_i}\\
&+\frac{f^\star(\mathsf{T}_i-\mathsf{T}_{i-1})^{\ltimes_i-1}}{\Gamma(\ltimes_i+1)}\Big((s_2-\mathsf{T}_{i-1})-(s_1-\mathsf{T}_{i-1})\Big)+\Big(\frac{K({s_1}^{1-\delta}-{\mathsf{T}_{i-1}}^{1-\delta})(s_2-s_1)^{\ltimes_i-1}}{(1-\delta)\Gamma(\ltimes_i)}\Big)\|\varkappa\|_{E_i}\\
&+\frac{f^\star}{\Gamma(\ltimes_i+1)}\Big((s_2-\mathsf{T}_{i-1})^{\ltimes_i}-(s_2-s_1)^{\ltimes_i}-(s_1-\mathsf{T}_{i-1})^{\ltimes_i}\Big)+\frac{K({s_2}^{1-\delta}-{s_1}^{1-\delta})(s_2-s_1)^{\ltimes_i-1}}{(1-\delta)\Gamma(\ltimes_i)}\|\varkappa\|_{E_i}\\
&+\frac{f^\star(s_2-s_1)^{\ltimes_i}}{\Gamma(\ltimes_i+1)}\\
\leq\;& \Big(\frac{K(\mathsf{T}_i-\mathsf{T}_{i-1})^{\ltimes_i-2}({\mathsf{T}_i}^{1-\delta}-{\mathsf{T}_{i-1}}^{1-\delta})}{(1-\delta)\Gamma(\ltimes_i)}\|\varkappa\|_{E_i}+\frac{f^\star(\mathsf{T}_i-\mathsf{T}_{i-1})^{\ltimes_i-1}}{\Gamma(\ltimes_i+1)}\Big)\Big((s_2-\mathsf{T}_{i-1})-(s_1-\mathsf{T}_{i-1})\Big)\\
&+\Big(\frac{K({s_2}^{1-\delta}-{\mathsf{T}_{i-1}}^{1-\delta})}{(1-\delta)\Gamma(\ltimes_i)}\|\varkappa\|_{E_i}\Big)(s_2-s_1)^{\ltimes_i-1}+\frac{f^\star}{\Gamma(\ltimes_i+1)}\Big((s_2-\mathsf{T}_{i-1})^{\ltimes_i}-(s_1-\mathsf{T}_{i-1})^{\ltimes_i}\Big).
\end{aligned}
$$

Therefore $\|(W\varkappa)(s_2)-(W\varkappa)(s_1)\|_{E_i}\to 0$ as $|s_2-s_1|\to 0$. It implies that $\mathsf{T}(B_{R_i})$ is equicontinuous.

Remark 13 *According to the remark on page 20 of [23], it is straightforward to show that inequality (4.6) is equivalent to the following inequality:*

$$\zeta(s^\delta|f(s,B)|)\leq K\zeta(B),$$

for any bounded set $B\subset X$ and for each $s\in J_i$.

STEP 4: Claim: W is a k-*set contractions.*

For any $U \in B_{R_i}$, $s \in J_i$, we obtain:

$$\begin{aligned}
\zeta(W(U)(s)) &= \zeta((W\varkappa)(s), \varkappa \in U)\\
&\leq \Big\{\tfrac{(\mathsf{T}_i-\mathsf{T}_{i-1})^{-1}(s-\mathsf{T}_{i-1})}{\Gamma(\Bbbk_i)}\int_{\mathsf{T}_{i-1}}^{\mathsf{T}_i}(\mathsf{T}_i-t)^{\Bbbk_i-1}\zeta f(t,\varkappa(t))dt\\
&+ \tfrac{1}{\Gamma(\Bbbk_i)}\int_{\mathsf{T}_{i-1}}^{s}(s-t)^{\Bbbk_i-1}\zeta f(t,\varkappa(t))dt, \varkappa \in U\Big\}.
\end{aligned}$$

Then, by Remark 13 implies for each $t \in J_i$ we have:

$$\begin{aligned}
\zeta(W(U)(s)) &\leq \Big\{\tfrac{(\mathsf{T}_i-\mathsf{T}_{i-1})^{-1}(s-\mathsf{T}_{i-1})}{\Gamma(\Bbbk_i)}\int_{\mathsf{T}_{i-1}}^{\mathsf{T}_i}(\mathsf{T}_i-t)^{\Bbbk_i-1}\Big[K\widehat{\zeta}(U)\int_{\mathsf{T}_{i-1}}^{\mathsf{T}_i}t^{-\delta}dt\Big]\\
&\quad+\tfrac{1}{\Gamma(\Bbbk_i)}\int_{\mathsf{T}_{i-1}}^{s}(s-t)^{\Bbbk_i-1}\Big[K\widehat{\zeta}(U)\int_{\mathsf{T}_{i-1}}^{s}t^{-\delta}dt\Big], \varkappa \in U\Big\}.\\
&\leq \Big\{\tfrac{(\mathsf{T}_i-\mathsf{T}_{i-1})^{\Bbbk_i-2}(s-\mathsf{T}_{i-1})}{\Gamma(\Bbbk_i)}\int_{\mathsf{T}_{i-1}}^{\mathsf{T}_i}\Big[K\widehat{\zeta}(U)\int_{\mathsf{T}_{i-1}}^{\mathsf{T}_i}t^{-\delta}dt\Big]\\
&\quad+\tfrac{(s-\mathsf{T}_{i-1})^{\Bbbk_i-1}}{\Gamma(\Bbbk_i)}\int_{\mathsf{T}_{i-1}}^{s}\Big[K\widehat{\zeta}(U)\int_{\mathsf{T}_{i-1}}^{s}t^{-\delta}dt\Big], \varkappa \in U\Big\}.\\
&\leq \tfrac{K({\mathsf{T}_i}^{1-\delta}-{\mathsf{T}_{i-1}}^{1-\delta})(\mathsf{T}_i-\mathsf{T}_{i-1})^{\Bbbk_i-2}(s-\mathsf{T}_{i-1})}{(1-\delta)\Gamma(\Bbbk_i)}\widehat{\zeta}(U)+\tfrac{K(s^{1-\delta}-{\mathsf{T}_{i-1}}^{1-\delta})(s-\mathsf{T}_{i-1})^{\Bbbk_i-1}}{(1-\delta)\Gamma(\Bbbk_i)}\widehat{\zeta}(U)\\
&\leq \tfrac{2K({\mathsf{T}_i}^{1-\delta}-{\mathsf{T}_{i-1}}^{1-\delta})(\mathsf{T}_i-\mathsf{T}_{i-1})^{\Bbbk_i-1}}{(1-\delta)\Gamma(\Bbbk_i)}\widehat{\zeta}(U).
\end{aligned}$$

Therefore,

$$\widehat{\zeta}(WU) \leq \tfrac{2K({\mathsf{T}_i}^{1-\delta}-{\mathsf{T}_{i-1}}^{1-\delta})(\mathsf{T}_i-\mathsf{T}_{i-1})^{\Bbbk_i-1}}{(1-\delta)\Gamma(\Bbbk_i)}\widehat{\zeta}(U).$$

Darbo's fixed point theorem requirements are thus all satisfied, and as a result, the initial and terminal value problem (4.4) admits at least one solution, $\widetilde{\varkappa}_i \in B_{R_i}$. Since $B_{R_i} \subset E_i$, the assertion of Theorem 25 is thereby supported.

We will now demonstrate the existence result for the initial and terminal value problem (ITVP) (4.1).

We begin by presenting the following premise:

(H2) Let $f \in C(J \times \mathbb{R}, \mathbb{R})$ and suppose there exists a number $\delta \in (0,1)$ such that $s^{\delta} f \in C(J \times \mathbb{R}, \mathbb{R})$. Additionally, assume there exists a constant $K > 0$ such that

$$s^{\delta}|f(s,y_1) - f(s,y_2)| \leq K|y_1 - y_2|,$$

for any $y_1, y_2 \in \mathbb{R}$ and $s \in J$.

Theorem 26 *Let the conditions (H1), (H2), and inequality (4.7) be satisfied for all $i \in \{1, 2, \ldots, n\}$. Then, the problem (4.1) admits at least one solution in $C(J, \mathbb{R})$.*

Proof: According to Theorem 25, for each $i \in \{1, 2, \ldots, n\}$, the initial and terminal value problem (ITVP) (4.4) possesses at least one solution $\widetilde{\varkappa}_i \in E_i$.

Given any $i \in \{1, 2, \ldots, n\}$, we define the function κ_i as follows:

$$\kappa_i(s) = \begin{cases} 0, & s \in [0, \mathsf{T}_{i-1}], \\ \widetilde{\varkappa}_i, & s \in J_i, \end{cases}$$

Therefore, $\varkappa_i \in C([0, \mathsf{T}_i], \mathbb{R})$ is a solution to the integral equation (4.3) on $s \in J_i$, satisfying the initial and terminal conditions $\varkappa_i(0) = 0$ and $\varkappa_i(\mathsf{T}_i) = \widetilde{\varkappa}_i(\mathsf{T}_i) = 0$.

Next, define the function $\varkappa(s)$ by

$$\varkappa(s) = \begin{cases} \varkappa_1(s), \quad s \in J_1, \\ \varkappa_2(s) = \begin{cases} 0, & s \in J_1, \\ \widetilde{\varkappa}_2, & s \in J_2 \end{cases} \\ \cdot \\ \cdot \\ \cdot \\ \varkappa_n(s) = \begin{cases} 0, & s \in [0, \mathsf{T}_{i-1}], \\ \widetilde{\varkappa}_i, & s \in J_i \end{cases} \end{cases} \tag{4.9}$$

which constitutes a solution to the initial and terminal value problem (4.1) in $C(J, \mathbb{R})$.

4.1.2 Stability Study

Theorem 27 *Under the fulfillment of conditions (H1), (H2), and inequality (4.7), the initial and terminal value problem (ITVP) (4.1) exhibits Ulam-Hyers stability.*

Proof: Let $\epsilon > 0$ to be any positive value, and assume that the function $z(s) \in C(J_i, \mathbb{R})$ satisfies the inequality (2.25).
For any $i \in \{1, 2, ..., n\}$ we define the functions as follows:

- $z_1(s) \equiv z(s), s \in [0, \mathsf{T}_1]$
- For $i = 2, 3, ..., n$:

$$z_i(s) = \begin{cases} 0, & s \in [0, \mathsf{T}_{i-1}], \\ z(s), & s \in J_i, \end{cases}$$

For each $i \in \{1, 2, \ldots, n\}$, based on equality (4.2), we derive:

$$ {}^{c}F^{\ltimes(s)}_{\mathsf{T}_{i-1}+} z_i(s) = \int_{\mathsf{T}_{i-1}}^{s} \frac{(s-t)^{1-\ltimes_i}}{\Gamma(2-\ltimes_i)} z^{(2)}(t)\, dt. $$

By applying the Riemann–Liouville fractional integral $I^{\ltimes_i}_{\mathsf{T}_{i-1}^+}$ to both sides of inequality (2.25), we obtain:

$$ \begin{aligned} \Big| z_i(s) \quad &+ \quad \frac{(\mathsf{T}_i - \mathsf{T}_{i-1})^{-1}(s-\mathsf{T}_{i-1})}{\Gamma(\ltimes_i)} \int_{\mathsf{T}_{i-1}}^{\mathsf{T}_i} (\mathsf{T}_i - t)^{\ltimes_i - 1} f(t, z_i(t))dt \\ &- \frac{1}{\Gamma(\ltimes_i)} \int_{\mathsf{T}_{i-1}}^{s} (s-t)^{\ltimes_i-1} f(t, z_i(t))dt \Big| \\ &\leq \epsilon \int_{\mathsf{T}_{i-1}}^{s} \frac{(s-t)^{\ltimes_i - 1}}{\Gamma(\ltimes_i)} dt \\ &\leq \epsilon \frac{(\mathsf{T}_i - \mathsf{T}_{i-1})^{\ltimes_i}}{\Gamma(\ltimes_i + 1)}. \end{aligned} $$

As established by Theorem 26, the initial and terminal value problem (ITVP) (4.1) admits a solution $\varkappa \in C(J, \mathbb{R})$ described by $\varkappa(s) = \varkappa_i(s)$, $s \in J_i$, $i = 1, 2, ..., n$, where

$$ \varkappa_i = \begin{cases} 0, & s \in [0, \mathsf{T}_{i-1}], \\ \widetilde{\varkappa}_i, & s \in J_i, \end{cases} \tag{4.10} $$

and $\widetilde{\varkappa}_i \in E_i$ is a solution of ITVP (4.4). According to Lemma 17 the integral equation

$$ \begin{aligned} \widetilde{\varkappa}_i(s) \quad &= \quad -\frac{(\mathsf{T}_i - \mathsf{T}_{i-1})^{-1}(s-\mathsf{T}_{i-1})}{\Gamma(\ltimes_i)} \int_{\mathsf{T}_{i-1}}^{\mathsf{T}_i} (\mathsf{T}_i - t)^{\ltimes_i - 1} f(t, \widetilde{\varkappa}_i(t))dt \\ &+ \frac{1}{\Gamma(\ltimes_i)} \int_{\mathsf{T}_{i-1}}^{s} (s-t)^{\ltimes_i-1} f(t, \widetilde{\varkappa}_i(t))dt \end{aligned} \tag{4.11} $$

holds.

Let $s \in J_i$, $i = 1, 2, ..., n$. Then by equations (4.10) and (4.11) we get

$$ |z(s) - \varkappa(s)| = |z(s) - \varkappa_i(s)| = |z_i(s) - \widetilde{\varkappa}_i(s)| $$

$$
\begin{aligned}
= \; & \Big| z_i(s) + \frac{(\mathsf{T}_i - \mathsf{T}_{i-1})^{-1}(s - \mathsf{T}_{i-1})}{\Gamma(\ltimes_i)} \int_{\mathsf{T}_{i-1}}^{\mathsf{T}_i} (\mathsf{T}_i - t)^{\ltimes_i - 1} f(t, \widetilde{\varkappa}_i(t))dt \\
& - \frac{1}{\Gamma(\ltimes_i)} \int_{\mathsf{T}_{i-1}}^{s} (s-t)^{\ltimes_i - 1} f(t, \widetilde{\varkappa}_i(t))dt \Big| \\
= \; & \Big| z_i(s) + \frac{(\mathsf{T}_i - \mathsf{T}_{i-1})^{-1}(s - \mathsf{T}_{i-1})}{\Gamma(\ltimes_i)} \int_{\mathsf{T}_{i-1}}^{\mathsf{T}_i} (\mathsf{T}_i - t)^{\ltimes_i - 1} f(t, z_i(t))dt \\
& - \frac{1}{\Gamma(\ltimes_i)} \int_{\mathsf{T}_{i-1}}^{s} (s-t)^{\ltimes_i - 1} f(t, z_i(t))dt \Big| \\
& + \frac{(\mathsf{T}_i - \mathsf{T}_{i-1})^{-1}(s - \mathsf{T}_{i-1})}{\Gamma(\ltimes_i)} \int_{\mathsf{T}_{i-1}}^{\mathsf{T}_i} (\mathsf{T}_i - t)^{\ltimes_i - 1} |f(t, z_i(t)) - f(t, \widetilde{\varkappa}_i(t))|dt \\
& + \frac{1}{\Gamma(\ltimes_i)} \int_{\mathsf{T}_{i-1}}^{s} (s-t)^{\ltimes_i - 1} |f(t, z_i(t)) - f(t, \widetilde{\varkappa}_i(t))|dt \\
\leq \; & \epsilon \frac{(\mathsf{T}_i - \mathsf{T}_{i-1})^{\ltimes_i}}{\Gamma(\ltimes_i + 1)} + \frac{(\mathsf{T}_i - \mathsf{T}_{i-1})^{-1}(s - \mathsf{T}_{i-1})}{\Gamma(\ltimes_i)} \int_{\mathsf{T}_{i-1}}^{\mathsf{T}_i} (\mathsf{T}_i - t)^{\ltimes_i - 1} t^{-\delta} (K|z_i(t) - \widetilde{\varkappa}_i(t)|)dt \\
& + \frac{1}{\Gamma(\ltimes_i)} \int_{\mathsf{T}_{i-1}}^{s} (s-t)^{\ltimes_i - 1} t^{-\delta} (K|z_i(t) - \widetilde{\varkappa}_i(t)|)dt \\
\leq \; & \epsilon \frac{(\mathsf{T}_i - \mathsf{T}_{i-1})^{\ltimes_i}}{\Gamma(\ltimes_i + 1)} + \frac{(\mathsf{T}_i - \mathsf{T}_{i-1})^{\ltimes_i - 1}}{\Gamma(\ltimes_i)} (K\|z_i - \widetilde{\varkappa}_i\|_{E_i}) \int_{\mathsf{T}_{i-1}}^{\mathsf{T}_i} t^{-\delta} dt \\
& + \frac{(\mathsf{T}_i - \mathsf{T}_{i-1})^{\ltimes_i - 1}}{\Gamma(\ltimes_i)} (K\|z_i - \widetilde{\varkappa}_i\|_{E_i}) \int_{\mathsf{T}_{i-1}}^{s} t^{-\delta} dt
\end{aligned}
$$

$$
\begin{aligned}
\leq \; & \epsilon \frac{(\mathsf{T}_i - \mathsf{T}_{i-1})^{\ltimes_i}}{\Gamma(\ltimes_i + 1)} + \frac{(\mathsf{T}_i - \mathsf{T}_{i-1})^{\ltimes_i - 1}({\mathsf{T}_i}^{1-\delta} - {\mathsf{T}_{i-1}}^{1-\delta})}{(1-\delta)\Gamma(\ltimes_i)} (K\|z_i - \widetilde{\varkappa}_i\|_{E_i}) \\
& + \frac{(\mathsf{T}_i - \mathsf{T}_{i-1})^{\ltimes_i - 1}(s^{1-\delta} - {\mathsf{T}_{i-1}}^{1-\delta})}{(1-\delta)\Gamma(\ltimes_i)} (K\|z_i - \widetilde{\varkappa}_i\|_{E_i}) \\
\leq \; & \epsilon \frac{(\mathsf{T}_i - \mathsf{T}_{i-1})^{\ltimes_i}}{\Gamma(\ltimes_i + 1)} + \frac{2K(\mathsf{T}_i - \mathsf{T}_{i-1})^{\ltimes_i - 1}({\mathsf{T}_i}^{1-\delta} - {\mathsf{T}_{i-1}}^{1-\delta})}{(1-\delta)\Gamma(\ltimes_i)} \|z_i - \widetilde{\varkappa}_i\|_{E_i} \\
\leq \; & \epsilon \frac{(\mathsf{T}_i - \mathsf{T}_{i-1})^{\ltimes_i}}{\Gamma(\ltimes_i + 1)} + \mu \|z - \varkappa\|,
\end{aligned}
$$

where

$$
\mu = \max_{i=1,2,\ldots,n} \frac{2K(\mathsf{T}_i - \mathsf{T}_{i-1})^{\ltimes_i - 1}({\mathsf{T}_i}^{1-\delta} - {\mathsf{T}_{i-1}}^{1-\delta})}{(1-\delta)\Gamma(\ltimes_i)}.
$$

Then,

$$
\|z - \varkappa\|(1-\mu) \leq \frac{(\mathsf{T}_i - \mathsf{T}_{i-1})^{\ltimes_i}}{\Gamma(\ltimes_i + 1)} \epsilon.
$$

It follows that, for each $s \in J_i$

$$|z(s) - \varkappa(s)| \leq \|z - \varkappa\| \leq \frac{(\mathsf{T}_i - \mathsf{T}_{i-1})^{\ltimes_i}}{(1-\mu)\Gamma(\ltimes_i + 1)}\epsilon := c_f \epsilon.$$

Consequently, the initial and terminal value problem (ITVP) (4.1) exhibits Ulam–Hyers stability.

4.1.3 Illustrative Examples

4.1.3.1 Illustrative Example 1

The following fractional initial and terminal value issue warrants our attention:

Let

$$\begin{cases} {}^{c}\digamma_{0+}^{\ltimes(s)} \varkappa(s) = \dfrac{s^{-\frac{1}{3}} e^{-s}}{(e^{e^{\frac{s^2}{1+s}}} + 4e^{2s} + 1)(1 + |\varkappa(s)|)}, & s \in J := [0,2], \\ \varkappa(0) = 0, \ \ \varkappa(2) = 0. \end{cases} \tag{4.12}$$

Define the function

$$f(s,y) = \frac{s^{-\frac{1}{3}} e^{-s}}{(e^{e^{\frac{s^2}{1+s}}} + 4e^{2s} + 1)(1+y)}, \quad (s,y) \in [0,2] \times [0,+\infty).$$

$$\text{Let } \ltimes(s) = \begin{cases} \frac{3}{2}, & s \in J_1 := [0,1], \\ \frac{9}{5}, & s \in J_2 :=]1,2]. \end{cases} \tag{4.13}$$

Therefore, it follows that

$$\begin{aligned} s^{\frac{1}{3}} |f(s,y_1) - f(s,y_2)| &\leq \left| \frac{e^{-s}}{(e^{e^{\frac{s^2}{1+s}}} + 4e^{2s} + 1)} \left(\frac{1}{1+y_1} - \frac{1}{1+y_2} \right) \right| \\ &\leq \frac{e^{-s}|y_1 - y_2|}{(e^{e^{\frac{s^2}{1+s}}} + 4e^{2s} + 1)(1+y_1)(1+y_2)} \\ &\leq \frac{e^{-s}}{(e^{e^{\frac{s^2}{1+s}}} + 4e^{2s} + 1)} |y_1 - y_2| \\ &\leq \frac{1}{(e+5)} |y_1 - y_2|. \end{aligned}$$

Hence, condition (H2) holds with $\delta = \frac{1}{3}$ and $K = \frac{1}{e+5}$.

We address two auxiliary initial and terminal value problems (ITVPs) for Caputo fractional differential equations of constant order, as specified in (4.13), in accordance with the initial and terminal value problem (4.4).

$$\begin{cases} {}^{c}F_{0+}^{\frac{3}{2}}\varkappa(s) = \dfrac{s^{-\frac{1}{3}}e^{-s}}{(e^{e^{\frac{s^2}{1+s}}}+4e^{2s}+1)(1+|\varkappa(s)|)}, & s \in J_1, \\ \varkappa(0) = 0, \ \ \varkappa(1) = 0, \end{cases} \tag{4.14}$$

and

$$\begin{cases} {}^{c}F_{1+}^{\frac{9}{5}}\varkappa(s) = \dfrac{s^{-\frac{1}{3}}e^{-s}}{(e^{e^{\frac{s^2}{1+s}}}+4e^{2s}+1)(1+|\varkappa(s)|)}, & s \in J_2, \\ \varkappa(1) = 0, \ \ \varkappa(2) = 0. \end{cases} \tag{4.15}$$

We proceed by proving that condition (4.7) is satisfied when $i = 1$. To this end,

$$\frac{2K(\mathsf{T}_1{}^{1-\delta} - \mathsf{T}_0{}^{1-\delta})(\mathsf{T}_1 - \mathsf{T}_0)^{\ltimes_1 - 1}}{(1-\delta)\Gamma(\ltimes_1)} = \frac{2}{\frac{2}{3}(e+5)\Gamma(\frac{3}{2})} \simeq 0.4385 < 1.$$

In light of this, condition (4.7) is satisfied. The initial and terminal value problem (ITVP) (4.14) admits a solution $\widetilde{\varkappa}_1 \in E_1$, according to Theorem 25. For $i = 2$, we demonstrate that condition (4.7) is also satisfied. Indeed,

$$\frac{2K(\mathsf{T}_2{}^{1-\delta} - \mathsf{T}_1{}^{1-\delta})(\mathsf{T}_2 - \mathsf{T}_1)^{\ltimes_2 - 1}}{(1-\delta)\Gamma(\ltimes_2)} = \frac{2}{e+5}\,\frac{2^{\frac{2}{3}}-1}{\frac{2}{3}\Gamma(\frac{9}{5})} \simeq 0.2451 < 1.$$

Thus, condition (4.7) is satisfied.

The initial and terminal value problem (ITVP) (4.15) admits a solution $\widetilde{\varkappa}_2 \in E_2$, as per Theorem 25.

Then, by Theorem 26, the ITVP (4.12) has a solution given by Theorem 26

$$\varkappa(s) = \begin{cases} \widetilde{\varkappa}_1(s), & s \in J_1, \\ \varkappa_2(s), & s \in J_2, \end{cases}$$

where

$$\varkappa_2(s) = \begin{cases} 0, & s \in J_1, \\ \widetilde{\varkappa}_2(s), & s \in J_2. \end{cases}$$

A Ulam-Hyers stable solution to ITVP (4.12) is established by Theorem 27.

4.1.3.2 Illustrative Example 2

The following fractional initial and terminal value problem warrants our attention:

$$\begin{cases} {}^{c}F_{0+}^{\ltimes(s)}\varkappa(s) = \frac{s^{-\frac{1}{2}}}{5e^{s}(1+|\varkappa(s)|)}, & s \in J := [0,3], \\ \varkappa(0) = 0, \ \ \varkappa(3) = 0. \end{cases} \tag{4.16}$$

Let

$$f(s,y) = \frac{s^{-\frac{1}{2}}}{5e^s(1+y)}, \quad (s,y) \in [0,3] \times [0,+\infty).$$

Define the variable-order function

$$\ltimes(s) = \begin{cases} \frac{9}{6}, & s \in J_1 := [0,1], \\ \frac{6}{5}, & s \in J_2 :=]1,3]. \end{cases} \tag{4.17}$$

Therefore, we obtain

$$\begin{aligned} s^{\frac{1}{2}}|f(s,y_1) - f(s,y_2)| &= \left| \frac{1}{5e^s} \left(\frac{1}{1+y_1} - \frac{1}{1+y_2} \right) \right| \\ &\leq \frac{|y_2 - y_1|}{5e^s(1+y_1)(1+y_2)} \\ &\leq \frac{1}{5}|y_1 - y_2|. \end{aligned}$$

As a result, the condition (H2) is satisfied with $\delta = \frac{1}{2}$ and $K = \frac{1}{5}$. Based on the formulation in (4.17) and following (4.4), we introduce two auxiliary initial and terminal value problems for Caputo fractional differential equations of constant order.

$$\begin{cases} {}^{c}F_{0^+}^{\frac{9}{6}} \varkappa(s) = \frac{s^{-\frac{1}{2}}}{5e^s(1+|\varkappa(s)|)}, & s \in J_1, \\ \varkappa(0) = 0, \ \ \varkappa(1) = 0. \end{cases} \tag{4.18}$$

and

$$\begin{cases} {}^{c}F_{1^+}^{\frac{6}{5}} \varkappa(s) = \frac{s^{-\frac{1}{2}}}{5e^s(1+|\varkappa(s)|)}, & s \in J_2, \\ \varkappa(1) = 0, \ \ \varkappa(3) = 0. \end{cases} \tag{4.19}$$

We proceed to prove that condition (4.7) is valid for $i = 1$. To this end,

$$\frac{2K(\mathsf{T}_1{}^{1-\delta} - \mathsf{T}_0{}^{1-\delta})(\mathsf{T}_1 - \mathsf{T}_0)^{\ltimes_1 - 1}}{(1-\delta)\Gamma(\ltimes_1)} = \frac{2\frac{1}{5}}{\frac{1}{2}\Gamma(\frac{9}{6})} \simeq 0.9027 < 1.$$

Therefore, condition (4.7) is met. According to Theorem 25, the initial and terminal value problem (ITVP) (4.18) has a solution $\widetilde{\varkappa}_1 \in E_1$. We will now establish that condition (4.7) is valid for $i = 2$. To this end,

$$\frac{2K(\mathsf{T}_2{}^{1-\delta} - \mathsf{T}_1{}^{1-\delta})(\mathsf{T}_2 - \mathsf{T}_1)^{\ltimes_2 - 1}}{(1-\delta)\Gamma(\ltimes_2)} = \frac{2\frac{1}{5}(\sqrt{3}-1)(2)^{0.2}}{\frac{1}{2}\Gamma(\frac{6}{5})} \simeq 0.7326 < 1.$$

Thus, the condition (4.7) is satisfied.

ITVP (4.19) has a solution $\widetilde{\varkappa}_2 \in E_2$, as per Theorem 25.

It follows that, ITVP (4.16) also has a solution according to Theorem 26.

$$\varkappa(s) = \begin{cases} \widetilde{\varkappa}_1(s), & s \in J_1, \\ \varkappa_2(s), & s \in J_2. \end{cases}$$

where

$$\varkappa_2(s) = \begin{cases} 0, & s \in J_1, \\ \widetilde{\varkappa}_2(s), & s \in J_2. \end{cases}$$

A Ulam-Hyers stable solution to ITVP (4.16) is established by Theorem 27.

4.2 Fractional Variable Order Differential Equations with Impulses

The following impulsive initial value problems (IVPs) are addressed in this section:

$$^{c}\digamma_{0+}^{\ltimes(s)}\varkappa(s) = f(s,\varkappa), \text{ for } s \in J := [0,\mathsf{T}],\ \mathsf{T} > 1,\ s \neq \mathsf{T}_i,\ i = 1,\dots,n, \tag{4.20}$$

$$\Delta\varkappa|_{s=\mathsf{T}_i} = \Phi_i(\varkappa(\mathsf{T}_i^-)),\ i = 1,\dots,n, \tag{4.21}$$

$$\varkappa(0) = \varkappa_0, \tag{4.22}$$

Here, $0 < \ltimes(s) \leq 1$, $f : J\times\mathbb{R} \to \mathbb{R}$, $\Phi_i : \mathbb{R} \to \mathbb{R}$, $i = 1,\dots,n$ are continuous functions. The operator $^{c}\digamma_{0+}^{\ltimes(s)}$ denotes the Caputo fractional derivative with of variable order $\ltimes(s)$. The impulse points satisfy $0 = \mathsf{T}_0 < \mathsf{T}_1 < \cdots < \mathsf{T}_n < \mathsf{T}_{n+1} = \mathsf{T}$. The jump condition is defined by $\Delta\varkappa|_{s=\mathsf{T}_i} = \varkappa(\mathsf{T}_i^+) - \varkappa(\mathsf{T}_i^-)$, where $\varkappa(\mathsf{T}_i^+) = \lim_{h\to 0^+}\varkappa(\mathsf{T}_i + h)$, $\varkappa(\mathsf{T}_i^-) = \lim_{h\to 0^-}\varkappa(\mathsf{T}_i + h)$ represent the right and left limits of $\varkappa(s)$ at $s = \mathsf{T}_i$ for $i = 1,\dots,n$.

4.2.1 Main Existence Results

Let us state the premise as follows:

(H1) Consider the partition $\mathcal{P} = \{[\mathsf{T}_0,\mathsf{T}_1],(\mathsf{T}_1,\mathsf{T}_2],(\mathsf{T}_2,\mathsf{T}_3],\cdots,(\mathsf{T}_n,\mathsf{T}_{n+1}]\}$ of the interval J, where $\mathsf{T}_0 = 0$ and $\mathsf{T}_{n+1} = \mathsf{T}$. Let $\ltimes(s) : J \to (0,1]$ be a piecewise constant function relative to $\mathcal{P}$. Define $\ltimes^* = \sup_{s\in J}\ltimes(s)$, where

$$\ltimes(s) = \sum_{i=0}^{n} \ltimes_i I_i(s) = \begin{cases} \ltimes_0, & \text{if } s \in [\mathsf{T}_0,\mathsf{T}_1], \\ \ltimes_1, & \text{if } s \in (\mathsf{T}_1,\mathsf{T}_2], \\ \cdot & \\ \cdot & \\ \cdot & \\ \ltimes_n, & \text{if } s \in (\mathsf{T}_n,\mathsf{T}_{n+1}], \end{cases}$$

where $0 < \ltimes_i \leq \ltimes^* \leq 1$ are constants, and

$$I_i(s) = \begin{cases} 1, & \text{for } s \in (\mathsf{T}_i,\mathsf{T}_{i+1}], \\ 0, & \text{otherwise.} \end{cases} \quad i = 0,1,\dots,n.$$

Now, we proceed to define the solution to the initial value problem (IVP) given by equations (4.20 – 4.22).

Definition 49 *A function $\varkappa \in PC(J,\mathbb{R})$ is said to solve initial value problem (IVP) given by equations (4.20 – –4.22) if it satisfies the following differential equation ${}^{c}\digamma_{0+}^{\ltimes(s)}\varkappa(s) = f(s,\varkappa)$, for each , $s \in J/\{\mathsf{T}_1, ..., \mathsf{T}_n\}$ and the jump conditions at the discontinuity points*

$$\Delta\varkappa|_{s=\mathsf{T}_i} = \Phi_i(\varkappa(\mathsf{T}_i^-)),\ i = 1,\ldots,n,$$

and the initial condition

$$\varkappa(0) = \varkappa_0.$$

First, we analyze equation (4.20) of the IVP.

Then, for any $s \in (\mathsf{T}_i, \mathsf{T}_{i+1}]$, where $i = 0, 1, ..., n$, it becomes the Caputo fractional derivative (CFD) of variable order $\ltimes(s)$ for $\varkappa(s) \in C(J,\mathbb{R})$, as given by (2.10). This is expressed as the sum of the CFDs of the constant-orders $\ltimes_0, \ltimes_1, ..., \ltimes_i$:

$$ {}^{c}\digamma_{0+}^{\ltimes(s)}\varkappa(s) = \int_0^{\mathsf{T}_1} \frac{(s-t)^{-\ltimes_0}}{\Gamma(1-\ltimes_0)}\varkappa'(t)dt + ... + \int_{\mathsf{T}_i}^{s} \frac{(s-t)^{-\ltimes_i}}{\Gamma(1-\ltimes_i)}\varkappa'(t)dt. \quad (4.23)$$

Thus, according to (4.23), equation (4.20) of the IVP can be written, for any $t \in (\mathsf{T}_i, \mathsf{T}_{i+1}]$, $i = 0, 1, ..., n$ in the form:

$$\int_0^{\mathsf{T}_1} \frac{(s-t)^{-\ltimes_0}}{\Gamma(1-\ltimes_0)}\varkappa'(t)dt + ... + \int_{\mathsf{T}_i}^{s} \frac{(s-t)^{-\ltimes_i}}{\Gamma(1-\ltimes_i)}\varkappa'(t)dt = f(s,\varkappa(s)). \quad (4.24)$$

We assume that $\varkappa(s) \equiv 0$ for $s \in [0, \mathsf{T}_i]/\{\mathsf{T}_1, ..., \mathsf{T}_{i-1}\}$. Then, the equation (4.20) of IVP reduces to:

$$ {}^{c}\digamma_{0+}^{\ltimes_i}\varkappa(s) = f(s,\varkappa(s)),\ s \in [0, \mathsf{T}_{i+1}]/\{\mathsf{T}_1, ..., \mathsf{T}_i\}.$$

Proposition 6 *Let $h : J \to \mathbb{R}$ be continuous. The solution of the following impulsive initial value problem (IVP):*

$$ {}^{c}\digamma_{0+}^{\ltimes_i}\varkappa(s) = h(s),\ \textit{for each},\ s \in [0, \mathsf{T}_{i+1}]/\{\mathsf{T}_1, ..., \mathsf{T}_i\},\quad i = 0, 1, \ldots, n, \quad (4.25)$$

$$\Delta\varkappa|_{s=\mathsf{T}_i} = \Phi_i(\varkappa(\mathsf{T}_i^-)),\quad i = 1,\ldots,n, \quad (4.26)$$

$$\varkappa(0) = \varkappa_0, \quad (4.27)$$

is given by:

$$\varkappa(s) = \begin{cases} \varkappa_0 + \int_0^s \frac{(s-t)^{\ltimes_0 - 1}}{\Gamma(\ltimes_0)} h(t)dt, & s \in [\mathsf{T}_0, \mathsf{T}_1], \\ \varkappa_0 + \sum_{j=1}^{i} \int_{\mathsf{T}_{j-1}}^{\mathsf{T}_j} \frac{(\mathsf{T}_j - t)^{\ltimes_{j-1} - 1}}{\Gamma(\ltimes_{j-1})} h(t)dt + \int_{\mathsf{T}_i}^{s} \frac{(s-t)^{\ltimes_i - 1}}{\Gamma(\ltimes_i)} h(t)dt & \\ + \sum_{j=1}^{i} \Phi_j(\varkappa(\mathsf{T}_j^-)), & s \in (\mathsf{T}_i, \mathsf{T}_{i+1}],\ i = 1, ..., n \end{cases} \quad (4.28)$$

Proof: Let $\varkappa$ be a solution of the impulsive initial value problem (IVP) (4.25–4.27). If $s \in [\mathsf{T}_0, \mathsf{T}_1]$, then by Lemma 2, we obtain:

$$\varkappa(s) = \varkappa_0 + \int_0^s \frac{(s-t)^{\ltimes_0 - 1}}{\Gamma(\ltimes_0)} h(t)dt.$$

If $s \in (\mathsf{T}_1, \mathsf{T}_2]$, then Lemma 2 implies:

$$\begin{aligned}
\varkappa(s) &= \varkappa(\mathsf{T}_1^+) + \int_{\mathsf{T}_1}^s \frac{(s-t)^{\ltimes_1 - 1}}{\Gamma(\ltimes_1)} h(t)dt \\
&= \Delta\varkappa|_{s=\mathsf{T}_1} + \varkappa(\mathsf{T}_1^-) + \int_{\mathsf{T}_1}^s \frac{(s-t)^{\ltimes_1 - 1}}{\Gamma(\ltimes_1)} h(t)dt \\
&= \Phi_1(\varkappa(\mathsf{T}_1^-)) + \varkappa_0 + \int_0^{\mathsf{T}_1} \frac{(\mathsf{T}_1-t)^{\ltimes_0 - 1}}{\Gamma(\ltimes_0)} h(t)dt + \int_{\mathsf{T}_1}^s \frac{(s-t)^{\ltimes_1 - 1}}{\Gamma(\ltimes_1)} h(t)dt.
\end{aligned}$$

If $s \in (\mathsf{T}_2, \mathsf{T}_3]$, then by Lemma 2, we obtain:

$$\begin{aligned}
\varkappa(s) &= \varkappa(\mathsf{T}_2^+) + \int_{\mathsf{T}_2}^s \frac{(s-t)^{\ltimes_2 - 1}}{\Gamma(\ltimes_2)} h(t)dt \\
&= \Delta\varkappa|_{s=\mathsf{T}_2} + \varkappa(\mathsf{T}_2^-) + \int_{\mathsf{T}_2}^s \frac{(s-t)^{\ltimes_2 - 1}}{\Gamma(\ltimes_2)} h(t)dt \\
&= \Phi_2(\varkappa(\mathsf{T}_2^-)) + \Phi_1(\varkappa(\mathsf{T}_1^-)) + \varkappa_0 + \int_0^{\mathsf{T}_1} \frac{(\mathsf{T}_1-t)^{\ltimes_0 - 1}}{\Gamma(\ltimes_0)} h(t)dt \\
&+ \int_{\mathsf{T}_1}^{\mathsf{T}_2} \frac{(\mathsf{T}_2-t)^{\ltimes_1 - 1}}{\Gamma(\ltimes_1)} h(t)dt + \int_{\mathsf{T}_2}^s \frac{(s-t)^{\ltimes_2 - 1}}{\Gamma(\ltimes_2)} h(t)dt.
\end{aligned}$$

Then, if $s \in (\mathsf{T}_i, \mathsf{T}_{i+1}]$, Lemma 2 implies the general representation given in equation (4.28).

Conversely, assume that $\varkappa$ satisfies equation (4.28). If $s \in [\mathsf{T}_0, \mathsf{T}_1]$, then clearly $\varkappa(0) = \varkappa_0$. Applying the operator ${}^{c}F_{0+}^{\ltimes_0}$, we obtain:

$${}^{c}F_{0+}^{\ltimes_0}\varkappa(s) = h(s).$$

If $s \in (\mathsf{T}_i, \mathsf{T}_{i+1}]$, $i = 0, 1, ..., n$, then using Lemma 2 and the fact that ${}^{c}F_{0+}^{\ltimes_i} C = 0$ for any constant C, we deduce:

$${}^{c}F_{0+}^{\ltimes_i}\varkappa(s) = h(s).$$

Moreover, it is straightforward to verify

$$\Delta\varkappa|_{s=\mathsf{T}_i} = \Phi_i(\varkappa(\mathsf{T}_i^-)),\ i = 1, ..., n.$$

Remark 14 *By the continuity of the function* $\ltimes(t)$, *let* $\ltimes_* = \min_{t\in J} |\ltimes(t)|$. *Then:*

- *If* $1 < \mathsf{T} < \infty$, *we have* $\mathsf{T}^{\ltimes(t)-1} < 1$.
- *If* $0 \leq \mathsf{T} \leq 1$, *it follows that* $\mathsf{T}^{\ltimes(t)-1} \leq \mathsf{T}^{\ltimes_*-1}$.

Hence, we conclude: $\mathsf{T}^{\ltimes(t)-1} \leq \max(1, \mathsf{T}^{\ltimes_*-1}) = \mathsf{T}^*$.

Now, we present our first result, assuming that the following assumptions are satisfied:

(H2) Let $s^\delta f : J\times\mathbb{R} \to \mathbb{R}$ be a continuous function, where $0 \leq \delta \leq \min_{s\in J} |\ltimes(s)|$. If there exists a constant $F_1 > 0$ such that

$$s^\delta |f(s,\varkappa_1) - f(s,\varkappa_2)| \leq F_1|\varkappa_1 - \varkappa_2|,$$

for every $\varkappa_1, \varkappa_2 \in \mathbb{R}$ and $s \in J$, then the condition holds.

(H3) Let $i = 1, ..., n$. For $\varkappa \in \mathbb{R}$ and $s \in J$, there exists a constant $F_2 > 0$ such that

$$|\Phi_i(\varkappa(s))| \leq F_2|\varkappa(s)|.$$

Theorem 28 *Suppose that conditions (H1)–(H3) are satisfied. If*

$$\left[\frac{(n+1)F_1\,\mathsf{T}^*\,\mathsf{T}^{1-\delta}}{\Gamma(\ltimes^*)}\frac{\Gamma(\ltimes_*)\Gamma(1-\delta)}{\Gamma(\ltimes_*+1-\delta)} + nF_2\right] < 1, \tag{4.29}$$

then the initial value problem $(4.20 - 4.22)$ *admits a solution in* $PC(J,\mathbb{R})$.

Proof: We construct the operator

$$\mathcal{S} : PC(J,\mathbb{R}) \to PC(J,\mathbb{R}),$$

as follows:

$$\mathcal{S}\varkappa(s) = \varkappa_0 + \sum_{0<\mathsf{T}_j<s}\int_{\mathsf{T}_{j-1}}^{\mathsf{T}_j}\frac{(\mathsf{T}_j - t)^{\ltimes_{j-1}-1}}{\Gamma(\ltimes_{j-1})}f(t,\varkappa(t))dt$$

$$+\int_{\mathsf{T}_i}^{s}\frac{(s-t)^{\ltimes_i-1}}{\Gamma(\ltimes_i)}f(t,\varkappa(t))dt + \sum_{0<\mathsf{T}_j<s}\Phi_j(\varkappa(\mathsf{T}_j^-)) \tag{4.30}$$

The operator $\mathcal{S}$, defined in (4.30), is well defined due to the continuity of function $s^\delta f$ and the properties of fractional integrals.

Let the set

$$B_R = \{\varkappa \in PC(J,\mathbb{R}),\ \|\varkappa\| \leq R\},$$

where

$$R \geq \frac{|\varkappa_0| + f^\star\left(\frac{(n+1)\mathsf{T}^{\ltimes^*}}{2(\sqrt{2}-1)}\right)}{1 - \left[\frac{(n+1)F_1\mathsf{T}^*\mathsf{T}^{1-\delta}}{\Gamma(\ltimes^*)}\frac{\Gamma(\ltimes_*)\Gamma(1-\delta)}{\Gamma(\ltimes_*+1-\delta)} + nF_2\right]},$$

and

$$f^\star = \sup_{s\in J}|f(s,0)|.$$

It is evident that B_R is bounded, closed, convex, and nonempty.

STEP 1: Claim: $\mathcal{S}(B_R) \subseteq (B_R)$.
For $\varkappa \in B_R$, we get

$$|\mathcal{S}\varkappa(s)|$$

$$\begin{aligned}
\leq\ & |\varkappa_0| + \sum_{0<\mathsf{T}_j<s}\int_{\mathsf{T}_{j-1}}^{\mathsf{T}_j}\frac{(\mathsf{T}_j-t)^{\ltimes_{j-1}-1}}{\Gamma(\ltimes_{j-1})}|f(t,\varkappa(t))-f(t,0)|dt \\
& + \sum_{0<\mathsf{T}_j<s}\int_{\mathsf{T}_{j-1}}^{\mathsf{T}_j}\frac{(\mathsf{T}_j-t)^{\ltimes_{j-1}-1}}{\Gamma(\ltimes_{j-1})}|f(t,0)|dt + \int_{\mathsf{T}_i}^{s}\frac{(s-t)^{\ltimes_i-1}}{\Gamma(\ltimes_i)}|f(t,\varkappa(t))-f(t,0)|dt \\
& + \int_{\mathsf{T}_i}^{s}\frac{(s-t)^{\ltimes_i-1}}{\Gamma(\ltimes_i)}|f(t,0)|dt + \sum_{0<\mathsf{T}_j<s}|\Phi_j(\varkappa(\mathsf{T}_j^-))| \\
\leq\ & |\varkappa_0| + \sum_{0<\mathsf{T}_j<s}\int_{\mathsf{T}_{j-1}}^{\mathsf{T}_j}\frac{(\mathsf{T}_j-t)^{\ltimes_{j-1}-1}}{\Gamma(\ltimes_{j-1})}t^{-\delta}(F_1|\varkappa(t)|)dt + f^\star\sum_{0<\mathsf{T}_j<s}\int_{\mathsf{T}_{j-1}}^{\mathsf{T}_j}\frac{(\mathsf{T}_j-t)^{\ltimes_{j-1}-1}}{\Gamma(\ltimes_{j-1})}dt \\
& + \int_{\mathsf{T}_i}^{s}\frac{(s-t)^{\ltimes_i-1}}{\Gamma(\ltimes_i)}t^{-\delta}(F_1|\varkappa(t)|) + f^\star\int_{\mathsf{T}_i}^{s}\frac{(s-t)^{\ltimes_i-1}}{\Gamma(\ltimes_i)}dt + \sum_{0<\mathsf{T}_j<s}F_2|\varkappa(\mathsf{T}_j^-)|
\end{aligned}$$

By remark 14, we have

$$\begin{aligned}
\leq\ & |\varkappa_0| + \sum_{0<\mathsf{T}_j<s}\frac{\mathsf{T}^*}{\Gamma(\ltimes_{j-1})}\int_{\mathsf{T}_{j-1}}^{\mathsf{T}_j}(\frac{\mathsf{T}_j-t}{\mathsf{T}})^{\ltimes_*-1}t^{-\delta}(F_1|\varkappa(t)|)dt + f^\star\sum_{0<\mathsf{T}_j<s}\int_{\mathsf{T}_{j-1}}^{\mathsf{T}_j}\frac{(\mathsf{T}_j-t)^{\ltimes_{j-1}-1}}{\Gamma(\ltimes_{j-1})}dt \\
& + \frac{\mathsf{T}^*}{\Gamma(\ltimes_i)}\int_{\mathsf{T}_i}^{t}(\frac{s-t}{\mathsf{T}})^{\ltimes_*-1}t^{-\delta}(F_1|\varkappa(t)|)dt + f^\star\int_{\mathsf{T}_i}^{t}\frac{(s-t)^{\ltimes_i-1}}{\Gamma(\ltimes_i)}dt + \sum_{0<\mathsf{T}_j<s}F_2|\varkappa(\mathsf{T}_j^-)| \\
\leq\ & |\varkappa_0| + \sum_{0<\mathsf{T}_j<s}\frac{\mathsf{T}^*}{\mathsf{T}^{\ltimes_*-1}\Gamma(\ltimes_{j-1})}\int_{\mathsf{T}_{j-1}}^{\mathsf{T}_j}(\mathsf{T}_j-t)^{\ltimes_*-1}t^{-\delta}(F_1|\varkappa(t)|)dt + f^\star\sum_{0<\mathsf{T}_j<s}\int_{\mathsf{T}_{j-1}}^{\mathsf{T}_j}\frac{(\mathsf{T}_j-t)^{\ltimes_{j-1}-1}}{\Gamma(\ltimes_{j-1})}dt \\
& + \frac{\mathsf{T}^*}{\mathsf{T}^{\ltimes_*-1}\Gamma(\ltimes_i)}\int_{\mathsf{T}_i}^{t}(s-t)^{\ltimes_*-1}t^{-\delta}(F_1|\varkappa(t)|)dt + f^\star\int_{\mathsf{T}_i}^{t}\frac{(s-t)^{\ltimes_i-1}}{\Gamma(\ltimes_i)}dt + \sum_{0<\mathsf{T}_j<s}F_2|\varkappa(\mathsf{T}_j^-)| \\
\leq\ & |\varkappa_0| + \frac{(n+1)F_1\mathsf{T}^*\mathsf{T}^{\ltimes_*-\delta}}{\mathsf{T}^{\ltimes_*-1}\Gamma(\ltimes^*)}\frac{\Gamma(\ltimes_*)\Gamma(1-\delta)}{\Gamma(\ltimes_*+1-\delta)}\|\varkappa\| + f^\star\Big(\frac{(n+1)\mathsf{T}^{\ltimes^*}}{2(\sqrt{2}-1)}\Big) + nF_2\|\varkappa\| \\
\leq\ & |\varkappa_0| + f^\star\Big(\frac{(n+1)\mathsf{T}^{\ltimes^*}}{2(\sqrt{2}-1)}\Big) + \Big[\frac{(n+1)F_1\mathsf{T}^*\mathsf{T}^{1-\delta}}{\Gamma(\ltimes^*)}\frac{\Gamma(\ltimes_*)\Gamma(1-\delta)}{\Gamma(\ltimes_*+1-\delta)} + nF_2\Big]\|\varkappa\| \\
\leq\ & R
\end{aligned}$$

STEP 2: Assertion: $\mathcal{S}$ is continuous.
Suppose the sequence $(\varkappa_n)$ converges to $\varkappa$ in $PC(J,\mathbb{R})$. Then,

$$\|(\mathcal{S}\varkappa_n) - (\mathcal{S}\varkappa)\| \to 0.$$

For $s \in J$, we have

$$
\begin{aligned}
&|(\mathcal{S}\varkappa_n)(s) - (\mathcal{S}\varkappa)(s)| \\
\leq\ & \sum_{0<\mathsf{T}_j<s} \int_{\mathsf{T}_{j-1}}^{\mathsf{T}_j} \frac{(\mathsf{T}_j - t)^{\ltimes_{j-1}-1}}{\Gamma(\ltimes_{j-1})} |f(t,\varkappa_n(t)) - f(t,\varkappa(t))| dt \\
&+ \int_{\mathsf{T}_i}^{s} \frac{(s-t)^{\ltimes_i - 1}}{\Gamma(\ltimes_i)} |f(t,\varkappa_n(t)) - f(t,\varkappa(t))| dt + \sum_{0<\mathsf{T}_j<s} |\Phi_j(\varkappa_n(\mathsf{T}_j^-)) - \Phi_j(\varkappa(\mathsf{T}_j^-))| \\
\leq\ & \sum_{0<\mathsf{T}_j<s} \int_{\mathsf{T}_{j-1}}^{\mathsf{T}_j} \frac{(\mathsf{T}_j - t)^{\ltimes_{j-1}-1}}{\Gamma(\ltimes_{j-1})} t^{-\delta}(F_1|\varkappa_n(t) - \varkappa(t)|) dt + \int_{\mathsf{T}_i}^{s} \frac{(s-t)^{\ltimes_i-1}}{\Gamma(\ltimes_i)} t^{-\delta}(F_1|\varkappa_n(t) - \varkappa(t)|) dt \\
&+ \sum_{0<\mathsf{T}_j<s} |\Phi_j(\varkappa_n(\mathsf{T}_j^-)) - \Phi_j(\varkappa(\mathsf{T}_j^-))| \\
\leq\ & \sum_{0<\mathsf{T}_j<s} \frac{\mathsf{T}^*}{\Gamma(\ltimes_{j-1})} \int_{\mathsf{T}_{j-1}}^{\mathsf{T}_j} (\frac{\mathsf{T}_j - t}{\mathsf{T}})^{\ltimes_*-1} t^{-\delta}(F_1|\varkappa_n(t) - \varkappa(t)|) dt \\
&+ \frac{\mathsf{T}^*}{\Gamma(\ltimes_i)} \int_{\mathsf{T}_i}^{t} (\frac{s-t}{\mathsf{T}})^{\ltimes_*-1} t^{-\delta}(F_1|\varkappa_n(t) - \varkappa(t)|) dt + \sum_{0<\mathsf{T}_j<s} |\Phi_j(\varkappa_n(\mathsf{T}_j^-)) - \Phi_j(\varkappa(\mathsf{T}_j^-))| \\
\leq\ & \sum_{0<\mathsf{T}_j<s} \frac{F_1 \mathsf{T}^*}{\mathsf{T}^{\ltimes_*-1}\Gamma(\ltimes_{j-1})} \|\varkappa_n - \varkappa\| \int_{\mathsf{T}_{j-1}}^{\mathsf{T}_j} (\mathsf{T}_j - t)^{\ltimes_*-1} t^{-\delta} dt \\
&+ \frac{F_1 \mathsf{T}^*}{\mathsf{T}^{\ltimes_*-1}\Gamma(\ltimes_i)} \|\varkappa_n - \varkappa\| \int_{\mathsf{T}_i}^{t} (s-t)^{\ltimes_*-1} t^{-\delta} dt + \sum_{0<\mathsf{T}_j<s} |\Phi_j(\varkappa_n(\mathsf{T}_j^-)) - \Phi_j(\varkappa(\mathsf{T}_j^-))| \\
\leq\ & \frac{(n+1)F_1\, \mathsf{T}^*\, \mathsf{T}^{\ltimes_*-\delta}}{\mathsf{T}^{\ltimes_*-1}\Gamma(\ltimes^*)} \frac{\Gamma(\ltimes_*)\Gamma(1-\delta)}{\Gamma(\ltimes_* + 1 - \delta)} \|\varkappa_n - \varkappa\| + \sum_{0<\mathsf{T}_j<s} |\Phi_j(\varkappa_n(\mathsf{T}_j^-)) - \Phi_j(\varkappa(\mathsf{T}_j^-))| \\
\leq\ & \frac{(n+1)F_1\, \mathsf{T}^*\, \mathsf{T}^{1-\delta}}{\Gamma(\ltimes^*)} \frac{\Gamma(\ltimes_*)\Gamma(1-\delta)}{\Gamma(\ltimes_* + 1 - \delta)} \|\varkappa_n - \varkappa\| + \sum_{0<\mathsf{T}_j<s} |\Phi_j(\varkappa_n(\mathsf{T}_j^-)) - \Phi_j(\varkappa(\mathsf{T}_j^-))|
\end{aligned}
$$

Since Φ_j is continuous,

$$\|(\mathcal{S}\varkappa_n) - (\mathcal{S}\varkappa)\| \to 0 \quad as \quad n \to \infty.$$

Hence, $\mathcal{S}$ is continuous.

STEP 3: Claim: $\mathcal{S}$ is compact.
By Step 1, we have $\|\mathcal{S}(\varkappa)\| \leq R$ for each $\varkappa \in B_R$, which means $\mathcal{S}(B_R)$ is bounded. Now we show that $\mathcal{S}(B_R)$ is equicontinuous.

For $s_1, s_2 \in J,\ s_1 < s_2$ and $\varkappa \in B_R$, we have

$$
\begin{aligned}
&|(\mathcal{S}\varkappa)(s_2) - (\mathcal{S}\varkappa)(s_1)| \\
\leq\ & \frac{1}{\Gamma(\ltimes_i)} \int_0^{s_1} \Big((s_2-t)^{\ltimes_i-1} - (s_1-t)^{\ltimes_i-1}\Big)|f(t,\varkappa(t))|dt + \frac{1}{\Gamma(\ltimes_i)} \int_{s_1}^{s_2} (s_2-t)^{\ltimes_i-1}|f(t,\varkappa(t))|dt \\
&+ \sum_{0<\mathsf{T}_j<s_2-s_1} |\Phi_j(\varkappa(\mathsf{T}_j^-))| \\
\leq\ & \frac{1}{\Gamma(\ltimes_i)} \int_0^{s_1} \Big((s_2-t)^{\ltimes_i-1} - (s_1-t)^{\ltimes_i-1}\Big)|f(t,\varkappa(t)) - f(t,0)|dt \\
+\ & \frac{1}{\Gamma(\ltimes_i)} \int_0^{s_1} \Big((s_2-t)^{\ltimes_i-1} - (s_1-t)^{\ltimes_i-1}\Big)|f(t,0)|dt \\
+\ & \frac{1}{\Gamma(\ltimes_i)} \int_{s_1}^{s_2} (s_2-t)^{\ltimes_i-1}|f(t,\varkappa(t)) - f(t,0)|dt + \frac{1}{\Gamma(\ltimes_i)} \int_{s_1}^{s_2} (s_2-t)^{\ltimes_i-1}|f(t,0)|dt \\
&+ \sum_{0<\mathsf{T}_j<s_2-s_1} |\mathcal{F}_2(\varkappa(\mathsf{T}_j^-))| \\
\leq\ & \frac{1}{\Gamma(\ltimes_i)} \int_0^{s_1} t^{-\delta}\Big((s_2-t)^{\ltimes_i-1} - (s_1-t)^{\ltimes_i-1}\Big)(\mathcal{F}_1|\varkappa(t)|)dt \\
+\ & \frac{f^\star}{\Gamma(\ltimes_i)} \int_0^{s_1} \Big((s_2-t)^{\ltimes_i-1} - (s_1-t)^{\ltimes_i-1}\Big)dt \\
+\ & \frac{1}{\Gamma(\ltimes_i)} \int_{s_1}^{s_2} t^{-\delta}(s_2-t)^{\ltimes_i-1}(\mathcal{F}_1|\varkappa(t)|)dt + \frac{f^\star}{\Gamma(\ltimes_i)} \int_{s_1}^{s_2} (s_2-t)^{\ltimes_i-1}dt + \sum_{0<\mathsf{T}_j<s_2-s_1} |\mathcal{F}_2(\varkappa(\mathsf{T}_j^-))| \\
\leq\ & \frac{1}{\Gamma(\ltimes_i)} \int_0^{s_1} t^{-\delta}\Big((s_2-s_1)^{\ltimes_i-1}\Big)(\mathcal{F}_1|\varkappa(t)|)dt + \frac{f^\star}{\Gamma(\ltimes_i)} \int_0^{s_1} \Big((s_2-t)^{\ltimes_i-1} - (s_1-t)^{\ltimes_i-1}\Big)dt \\
+\ & \frac{1}{\Gamma(\ltimes_i)} \int_{s_1}^{s_2} t^{-\delta}(s_2-t)^{\ltimes_i-1}(\mathcal{F}_1|\varkappa(t)|)dt + \frac{f^\star}{\Gamma(\ltimes_i)} \int_{s_1}^{s_2} (s_2-t)^{\ltimes_i-1}dt + \sum_{0<\mathsf{T}_j<s_2-s_1} |\mathcal{F}_2(\varkappa(\mathsf{T}_j^-))| \\
\leq\ & \frac{1}{\Gamma(\ltimes_i)}(\mathcal{F}_1\|\varkappa\|)\Big(\frac{({s_1}^{1-\delta})(s_2-s_1)^{\ltimes_i-1}}{(1-\delta)}\Big) + \frac{f^\star}{\Gamma(\ltimes_i)}\Big(\frac{s_2^{\ltimes_i}}{\ltimes_i} - \frac{(s_2-s_1)^{\ltimes_i}}{\ltimes_i} - \frac{s_1^{\ltimes_i}}{\ltimes_i}\Big) \\
+\ & \frac{1}{\Gamma(\ltimes_i)}(\mathcal{F}_1\|\varkappa\|)\frac{({s_2}^{1-\delta} - {s_1}^{1-\delta})(s_2-s_1)^{\ltimes_i-1}}{(1-\delta)} + \frac{f^\star}{\Gamma(\ltimes_i)}\frac{(s_2-s_1)^{\ltimes_i}}{\ltimes_i} + \sum_{0<\mathsf{T}_j<s_2-s_1} |\mathcal{F}_2(\varkappa(\mathsf{T}_j^-))| \\
\leq\ & \frac{1}{\Gamma(\ltimes_i)}(\mathcal{F}_1)\|\varkappa\|\frac{{s_1}^{1-\delta}}{(1-\delta)}\Big(s_2^{\ltimes_i-1} - s_1^{\ltimes_i-1}\Big) + \frac{f^\star}{\Gamma(\ltimes_i+1)}\Big(s_2^{\ltimes_i} - s_1^{\ltimes_i}\Big) \\
+\ & \frac{1}{\Gamma(\ltimes_i)}(\mathcal{F}_1)\|\varkappa\|\frac{({s_2}^{1-\delta} - {s_1}^{1-\delta})}{(1-\delta)}(s_2-s_1)^{\ltimes_i-1} + \sum_{0<\mathsf{T}_j<s_2-s_1} |\mathcal{F}_2(\varkappa(\mathsf{T}_j^-))|
\end{aligned}
$$

As $s_1 \to s_2$, the inequality above tends to zero on the right side. Hence $|(\mathcal{S}\varkappa)(s_2) - (\mathcal{S}\varkappa)(s_1)| \to 0$. It implies that $\mathcal{S}(B_R)$ is equicontinuous.

Thus, by the Schauder fixed point theorem, the initial value problem (4.20 – 4.22) possesses a solution in B_R. Since $B_R \subset PC(J,\mathbb{R})$, the assertion of Theorem 28's is validated.

4.2.2 Solution Uniqueness

Now, we will invoke the ßanach contraction principle to confirm that the solution to the initial value problem (4.20 – 4.22) is unique.

Introduce the following assumption:

(H4) For each $i = 1, \ldots, n$, there exists a constant $F_3 > 0$ such that, for any $\varkappa, y \in \mathbb{R}$ and for every $s \in J$, the following inequality holds:

$$|\Phi_i(\varkappa(s)) - \Phi_i(y(s))| \leq F_3|\varkappa(s) - y(s)|.$$

Theorem 29 *Assume that the conditions (H1), (H2), and (H4) are satisfied. If*

$$\left[\frac{(n+1)F_1\,T^*\,T^{1-\delta}}{\Gamma(\ltimes^*)}\frac{\Gamma(\ltimes_*)\Gamma(1-\delta)}{\Gamma(\ltimes_*+1-\delta)} + nF_3\right] < 1. \tag{4.31}$$

then the initial value problem (4.20 – 4.22) admits a unique solution in the space $PC(J, \mathbb{R})$.

Moreover, for $s \in J$ and $\varkappa, y \in PC(J, \mathbb{R})$, we have

$$|(S\varkappa)(s) - (Sy)(s)|$$

$$\begin{aligned}
\leq\ & \sum_{0<T_j<s} \int_{T_{j-1}}^{T_j} \frac{(T_j - t)^{\ltimes_{j-1}-1}}{\Gamma(\ltimes_{j-1})}|f(t,\varkappa(t)) - f(t,y(t))|dt \\
& + \int_{T_i}^{s} \frac{(s-t)^{\ltimes_i - 1}}{\Gamma(\ltimes_i)}|f(t,\varkappa(t)) - f(t,y(t))|dt + \sum_{0<T_j<s} |\Phi_j(\varkappa(T_j^-)) - \Phi_j(y(T_j^-))| \\
\leq\ & \sum_{0<T_j<s} \int_{T_{j-1}}^{T_j} \frac{(T_j - t)^{\ltimes_{j-1}-1}}{\Gamma(\ltimes_{j-1})} t^{-\delta}(F_1|\varkappa(t) - y(t)|)dt + \int_{T_i}^{s} \frac{(s-t)^{\ltimes_i-1}}{\Gamma(\ltimes_i)} t^{-\delta}(F_1|\varkappa(t) - y(t)|) \\
& + \sum_{0<T_j<s} F_3|\varkappa(t) - y(t)| \\
\leq\ & \sum_{0<T_j<s} \frac{T^*}{\Gamma(\ltimes_{j-1})} \int_{T_{j-1}}^{T_j} (\frac{T_j - t}{T})^{\ltimes_*-1} t^{-\delta}(F_1|\varkappa(t) - y(t)|)dt \\
& + \frac{T^*}{\Gamma(\ltimes_i)} \int_{T_i}^{t} (\frac{s-t}{T})^{\ltimes_*-1} t^{-\delta}(F_1|\varkappa(t) - y(t)|)dt + nF_3\|\varkappa - y\| \\
\leq\ & \sum_{0<T_j<s} \frac{F_1 T^*}{T^{\ltimes_*-1}\Gamma(\ltimes_{j-1})}\|\varkappa - y\| \int_{T_{j-1}}^{T_j} (T_j - t)^{\ltimes_*-1} t^{-\delta} dt \\
& + \frac{F_1 T^*}{T^{\ltimes_*-1}\Gamma(\ltimes_i)}\|\varkappa - y\| \int_{T_i}^{t} (s-t)^{\ltimes_*-1} t^{-\delta} dt + nF_3\|\varkappa - y\| \\
\leq\ & \frac{(n+1)F_1\,T^*\,T^{\ltimes_*-\delta}}{T^{\ltimes_*-1}\Gamma(\ltimes^*)} \frac{\Gamma(\ltimes_*)\Gamma(1-\delta)}{\Gamma(\ltimes_*+1-\delta)}\|\varkappa - y\| + nF_3\|\varkappa - y\| \\
\leq\ & \left[\frac{(n+1)F_1\,T^*\,T^{1-\delta}}{\Gamma(\ltimes^*)} \frac{\Gamma(\ltimes_*)\Gamma(1-\delta)}{\Gamma(\ltimes_*+1-\delta)} + nF_3\right]\|\varkappa - y\|
\end{aligned}$$

Ergo, by condition (4.31), the operator $\mathcal{S}$ constitutes a contraction. Consequently, $\mathcal{S}$ admits a unique fixed point, which corresponds to the sole solution of the initial value problem (4.20 – 4.22).

4.2.3 Sufficient Conditions for Stability

We shall establish the Ulam-Hyers stability of the proposed solutions to the ınitial υalue problem IVP (4.20 – 4.22).

Theorem 30 *Assume the hypotheses of Theorem 29 hold. Then, the initial value problem (IVP)* (4.20 – 4.22) *is Ulam-Hyers stable.*

Proof: Assume that $z(s)$ satisfies inequality (2.26). Then, the following integral inequality holds:

$$z(s) - z_0 + \sum_{0<\mathsf{T}_j<s} \int_{\mathsf{T}_{j-1}}^{\mathsf{T}_j} \frac{(\mathsf{T}_j - t)^{\ltimes_{j-1}-1}}{\Gamma(\ltimes_{j-1})} f(t, z(t))dt - \int_{\mathsf{T}_i}^{s} \frac{(s-t)^{\ltimes_i - 1}}{\Gamma(\ltimes_i)} f(t, z(t))dt$$

$$- \sum_{0<\mathsf{T}_j<s} \Phi_j(z(\mathsf{T}_j^-)) \leq \epsilon \frac{\mathsf{T}^{\ltimes^*}}{2(\sqrt{2}-1)}.$$

Let $\varkappa$ be the unique solution of the initial value problem (IVP) (4.20–4.22). According to Proposition 6, $\varkappa$ is given by

$$\varkappa = \varkappa_0 + \sum_{j=1}^{i} \int_{\mathsf{T}_{j-1}}^{\mathsf{T}_j} \frac{(\mathsf{T}_j - t)^{\ltimes_{j-1}-1}}{\Gamma(\ltimes_{j-1})} f(t)dt + \int_{\mathsf{T}_i}^{s} \frac{(s-t)^{\ltimes_i - 1}}{\Gamma(\ltimes_i)} f(t)dt$$

$$+ \sum_{j=1}^{i} \Phi_j(\varkappa(\mathsf{T}_j^-)), \quad s \in (\mathsf{T}_i, \mathsf{T}_{i+1}],\ i = 1, ..., n \tag{4.32}$$

Let $s \in (\mathsf{T}_i, \mathsf{T}_{i+1}]$, for $i = 1, ..., n$. Then, we obtain

$$|z(s) - \varkappa(s)|$$

$$= \left| z(s) - \varkappa_0 - \sum_{0<\mathsf{T}_j<s} \int_{\mathsf{T}_{j-1}}^{\mathsf{T}_j} \frac{(\mathsf{T}_j - t)^{\ltimes_{j-1}-1}}{\Gamma(\ltimes_{j-1})} f(t, \varkappa(t))dt - \int_{\mathsf{T}_i}^{s} \frac{(s-t)^{\ltimes_i - 1}}{\Gamma(\ltimes_i)} f(t, \varkappa(t))dt \right.$$

$$\left. - \sum_{0<\mathsf{T}_j<s} \Phi_j(\varkappa(\mathsf{T}_j^-)) \right|$$

$$\leq \left| z(s) - \varkappa_0 + \sum_{0<\mathsf{T}_j<s} \int_{\mathsf{T}_{j-1}}^{\mathsf{T}_j} \frac{(\mathsf{T}_j - t)^{\ltimes_{j-1}-1}}{\Gamma(\ltimes_{j-1})} f(t, z(t))dt - \int_{\mathsf{T}_i}^{s} \frac{(s-t)^{\ltimes_i - 1}}{\Gamma(\ltimes_i)} f(t, z(t))dt \right.$$

$$\left. - \sum_{0<\mathsf{T}_j<s} \Phi_j(z(\mathsf{T}_j^-)) \right| + \sum_{0<\mathsf{T}_j<s} \int_{\mathsf{T}_{j-1}}^{\mathsf{T}_j} \frac{(\mathsf{T}_j - t)^{\ltimes_{j-1}-1}}{\Gamma(\ltimes_{j-1})} \Big| f(t, z(t)) - f(t, \varkappa(t)) \Big| dt$$

$$+ \int_{\mathsf{T}_i}^{s} \frac{(s-t)^{\ltimes_i - 1}}{\Gamma(\ltimes_i)} \Big| f(t, z(t)) - f(t, \varkappa(t)) \Big| dt + \sum_{0<\mathsf{T}_j<s} \Big| \Phi_j(z(\mathsf{T}_j^-)) - \Phi_j(\varkappa(\mathsf{T}_j^-)) \Big|$$

$$\leq \epsilon \frac{\mathsf{T}^{\ltimes^*}}{2(\sqrt{2}-1)} + \sum_{0<\mathsf{T}_j<s} \int_{\mathsf{T}_{j-1}}^{\mathsf{T}_j} \frac{(\mathsf{T}_j - t)^{\ltimes_{j-1}-1}}{\Gamma(\ltimes_{j-1})} t^{-\delta}(F_1|z(t) - \varkappa(t)|)dt$$
$$+ \int_{\mathsf{T}_i}^{s} \frac{(s-t)^{\ltimes_i - 1}}{\Gamma(\ltimes_i)} t^{-\delta}(F_1|z(t) - \varkappa(t)|)dt + \sum_{0<\mathsf{T}_j<s} F_3|z(s) - \varkappa(s)|$$
$$\leq \epsilon \frac{\mathsf{T}^{\ltimes^*}}{2(\sqrt{2}-1)} + \sum_{0<\mathsf{T}_j<s} \frac{\mathsf{T}^*}{\Gamma(\ltimes_{j-1})} \int_{\mathsf{T}_{j-1}}^{\mathsf{T}_j} (\frac{\mathsf{T}_j - t}{\mathsf{T}})^{\ltimes_* - 1} t^{-\delta}(F_1|z(t) - \varkappa(t)|)dt$$
$$+ \frac{\mathsf{T}^*}{\Gamma(\ltimes_i)} \int_{\mathsf{T}_i}^{t} (\frac{s-t}{\mathsf{T}})^{\ltimes_* - 1} t^{-\delta}(F_1|z(t) - \varkappa(t)|)dt + nF_3\|z - \varkappa\|$$
$$\leq \epsilon \frac{\mathsf{T}^{\ltimes^*}}{2(\sqrt{2}-1)} + \sum_{0<\mathsf{T}_j<s} \frac{F_1 \mathsf{T}^*}{\mathsf{T}^{\ltimes_* - 1}\Gamma(\ltimes_{j-1})} \|z - \varkappa\| \int_{\mathsf{T}_{j-1}}^{\mathsf{T}_j} (\mathsf{T}_j - t)^{\ltimes_* - 1} t^{-\delta} dt$$
$$+ \frac{F_1 \mathsf{T}^*}{\mathsf{T}^{\ltimes_* - 1}\Gamma(\ltimes_i)} \|z - \varkappa\| \int_{\mathsf{T}_i}^{t} (s-t)^{\ltimes_* - 1} t^{-\delta} dt + nF_3\|z - \varkappa\|$$
$$\leq \epsilon \frac{\mathsf{T}^{\ltimes^*}}{2(\sqrt{2}-1)} + \frac{(n+1)F_1\, \mathsf{T}^*\, \mathsf{T}^{\ltimes_* - \delta}}{\mathsf{T}^{\ltimes_* - 1}\Gamma(\ltimes^*)} \frac{\Gamma(\ltimes_*)\Gamma(1-\delta)}{\Gamma(\ltimes_* + 1 - \delta)} \|z - \varkappa\| + nF_3\|z - \varkappa\|$$
$$\leq \epsilon \frac{\mathsf{T}^{\ltimes^*}}{2(\sqrt{2}-1)} + \Big[\frac{(n+1)F_1\, \mathsf{T}^*\, \mathsf{T}^{1-\delta}}{\Gamma(\ltimes^*)} \frac{\Gamma(\ltimes_*)\Gamma(1-\delta)}{\Gamma(\ltimes_* + 1 - \delta)} + nF_3\Big] \|z - \varkappa\|$$

Then, we have

$$\|z - \varkappa\| \Big[1 - \Big(\frac{(n+1)F_1\, \mathsf{T}^*\, \mathsf{T}^{1-\delta}}{\Gamma(\ltimes^*)} \frac{\Gamma(\ltimes_*)\Gamma(1-\delta)}{\Gamma(\ltimes_* + 1 - \delta)} + nF_3\Big)\Big] \leq \epsilon \frac{\mathsf{T}^{\ltimes^*}}{2(\sqrt{2}-1)}.$$

Thus, we obtain,

$$|z(t) - \varkappa(t)| \leq \|z - \varkappa\| \leq \frac{\mathsf{T}^{\ltimes^*}}{2(\sqrt{2}-1)\Big[1 - \Big(\frac{(n+1)F_1 \mathsf{T}^* \mathsf{T}^{1-\delta}}{\Gamma(\ltimes^*)} \frac{\Gamma(\ltimes_*)\Gamma(1-\delta)}{\Gamma(\ltimes_*+1-\delta)} + nF_3\Big)\Big]} \epsilon := c_{f,n}\epsilon.$$

Consequently, the initial value problem (IVP) (4.20 – 4.22) is Ulam-Hyers stable.

4.2.4 Practical Example

Consider the following impulsive initial value problem:

$$^{c}F_{\mathsf{T}_i^+}^{\ltimes(s)} \varkappa(s) = \frac{e^{-3s}}{s^{\frac{1}{3}}(e^{e^{\frac{3s^2}{1+s}}} + 5)(1 + |\varkappa(s)|)}, \quad s \in J := J_0 \cup J_1. \tag{4.33}$$

With impulsive condition at $s = \frac{1}{2}$:

$$\Delta\varkappa|_{s=\frac{1}{2}} = \frac{|(\varkappa(\frac{1}{2}^-))|}{10 + |(\varkappa(\frac{1}{2}^-))|}, \tag{4.34}$$

Initial condition:

$$\varkappa(0) = \varkappa_0, \tag{4.35}$$

Where:

- $\mathsf{T}_0 = 0, \quad \mathsf{T}_1 = \frac{1}{2}, \quad \mathsf{T}_2 = \mathsf{T} = 1,$
- $n = 1,$
- $J := [0,1], \quad J_0 = [0, \frac{1}{2}], \quad J_1 = [\frac{1}{2}, 1].$

The order function $\phi(s)$ is defined by:

$$\ltimes(s) = \begin{cases} \frac{1}{2}, & s \in J_0, \\ \frac{3}{4}, & s \in J_1. \end{cases} \tag{4.36}$$

Let the function $f : J \times R \to R$ be defined by:

$$f(s,\varkappa) = \frac{e^{-3s}}{s^{\frac{1}{3}}(e^{e^{\frac{3s^2}{1+s}}} + 5)(1 + |\varkappa(s)|)}, \quad (s,\varkappa) \in J \times \mathbb{R}.$$

For each $s \in J$ and $\varkappa, y \in \mathbb{R}$, we have:

$$s^{\frac{1}{3}}|f(s,\varkappa) - f(s,y)| \quad \leq \quad \frac{1}{e+5}|\varkappa - y|.$$

Thus, assumption (H2) is satisfied with $\digamma_1 = \frac{1}{e+5}$ and $\delta = \frac{1}{3}$.

Let

$$\Phi_1(\varkappa) = \frac{|\varkappa|}{7 + |\varkappa|}, \quad \varkappa \in \mathbb{R}.$$

For $\varkappa, y \in [0, \infty)$, we have:

$$|\Phi_1(\varkappa) - \Phi_1(y) \leq \frac{1}{7}|\varkappa - y|.$$

Then, assumption (H4) holds with $\digamma_3 = \frac{1}{7}$.

We shall verify that assumption (4.31) is fulfilled $-\ltimes^* = \frac{1}{2}$, and $\ltimes^* = \frac{3}{4}$. Indeed,

$$\begin{aligned}\left[\frac{(n+1)\digamma_1 \mathsf{T}^* \mathsf{T}^{1-\delta}}{\Gamma(\ltimes^*)} \frac{\Gamma(\ltimes_*)\Gamma(1-\delta)}{\Gamma(\ltimes_* + 1 - \delta)} + n\digamma_3\right] &= \left[\frac{\frac{2}{e+5}}{\Gamma\left(\frac{3}{4}\right)} \cdot \frac{\Gamma\left(\frac{3}{4}\right) \cdot \Gamma\left(1 - \frac{1}{3}\right)}{\Gamma\left(\frac{3}{4} + 1 - \frac{1}{3}\right)} + \frac{1}{7}\right] \\ &= \left[\frac{\frac{2}{e+5}}{\Gamma\left(\frac{3}{4}\right)} \cdot \frac{\Gamma\left(\frac{3}{4}\right) \cdot \Gamma\left(\frac{2}{3}\right)}{\Gamma\left(\frac{17}{12}\right)} + \frac{1}{7}\right] \\ &\simeq 0.4189 < 1.\end{aligned}$$

Hence, assumption (4.31) is satisfied.

In accordance with Theorem 29, initial value problem (IVP) (4.33)–(4.35) admits a unique solution in the space stable.

4.3 Existence and Stability of Solutions for a Class of Nonlinear Fractional Integro-Differential Equations of Variable Order

We consider the following initial and terminal value problem:

$$\begin{cases} {}^{c}\digamma_{0^+}^{\ltimes(s)}\varkappa(s) = f(s, \varkappa(s), I_{0^+}^{\ltimes(s)}\varkappa(s)), \quad s \in J := [0, \mathsf{T}] \\ \varkappa(0) = 0, \quad \varkappa(\mathsf{T}) = 0, \end{cases} \tag{4.37}$$

where $1 < \ltimes(s) \leq 2$, f is a continuous function defined on $J \times \mathbb{R} \times \mathbb{R}$ with values in $\mathbb{R}$, and ${}^{c}\digamma_{0^+}^{\ltimes(s)}$ and $I_{0^+}^{\ltimes(s)}$ denote the Caputo fractional derivative and integral of order $\ltimes(s)$, respectively.

Our objective is to investigate the existence and the uniqueness of solutions to problem (4.37). Furthermore, we analyze the stability of the obtained solution in the sense of Ulam-Hyers-$\mathfrak{R}$assias.

4.3.1 Properties of Existence

Let us now articulate the foundational presumptions:

(H1) Let $n \in N$ be a positive integer, and consider the partition

$$P = \{J_1 := [0, \mathsf{T}_1],\ J_2 := (\mathsf{T}_1, \mathsf{T}_2],\ J_3 := (\mathsf{T}_2, \mathsf{T}_3],\ \ldots,\ J_n := (\mathsf{T}_{n-1}, \mathsf{T}]\}$$

represent a partition of the interval $J := [0, \mathsf{T}]$. Define the $\ltimes(s) : J \to (1, 2]$ as a function to be piecewise constant on each subinterval $J_i \in P$, i.e.,

$$\ltimes(s) = \sum_{i=1}^{n} \ltimes_i I_i(s) = \begin{cases} \ltimes_1, & if\ s \in J_1, \\ \ltimes_2, & if\ s \in J_2, \\ \cdot \\ \cdot \\ \cdot \\ \ltimes_n, & if\ s \in J_n, \end{cases}$$

where $\ltimes(s) \in (1, 2]$ are constant parameters, and $I_i(s)$ denotes the characteristic function of the interval $(\mathsf{T}_{i-1}, \mathsf{T}_i]$, defined by

$$I_i(s) = \begin{cases} 1, & for\ s \in J_i, \\ 0, & otherwise, \end{cases}$$

for $i = 1, 2, \ldots, n$, with $\mathsf{T}_0 = 0$ and $\mathsf{T}_n = \mathsf{T}$).

(H2) Let $s^{\delta} f : J \times \mathbb{R} \times \mathbb{R} \to \mathbb{R}$ be a continuous mapping, where $0 \leq \delta \leq 1$. Suppose there exist positive constants K and L such that

$$s^{\delta}\, |f(s, y_1, z_1) - f(s, y_2, z_2)| \leq K \cdot |y_1 - y_2| + L \cdot |z_1 - z_2|,$$

for all $y_1, y_2, z_1, z_2 \in \mathbb{R}$ and $s \in J$.

For each $i \in \{1, 2, \ldots, n\}$, let E_i denote the Banach space $C(J_i, \mathbb{R})$, of continuous functions from J_i to $\mathbb{R}$, equipped with the norm

$$\|\varkappa\|_{E_i} = \sup_{s \in J_i} |\varkappa(s)|,$$

where $\varkappa \in E_i$ and $i \in \{1, 2, ..., n\}$.

Initially, we examine the initial and terminal value problem (ITVP) specified in (4.37).

Taking assumption $(H1)$ into account, for any $s \in J_i$, where $i = 1, 2, \ldots, n$, the left Caputo fractional derivative of variable order $\ltimes(s)$ for a function $\varkappa(s) \in C(J, \mathbb{R})$, as defined by (2.10), can be expressed as a sum of left Caputo fractional derivatives of constant orders $\ltimes_j$, where $j = 1, 2, \ldots, i$.

Specifically,

$$\int_0^s \frac{(s-t)^{1-\ltimes(s)}}{\Gamma(2-\ltimes(s))} \varkappa''(t)dt = \int_0^{\mathsf{T}_1} \frac{(s-t)^{1-\ltimes_1}}{\Gamma(2-\ltimes_1)} \varkappa''(t)dt + \ldots + \int_{\mathsf{T}_{i-1}}^s \frac{(s-t)^{1-\ltimes_i}}{\Gamma(2-\ltimes_i)} \varkappa''(t)dt. \tag{4.38}$$

Consequently, in light of (4.38), the initial and terminal value problem (ITVP) given in (4.37) can be reformulated for every $s \in J_i$, with $i = 1, 2, \ldots, n$, as

$$\int_0^{\mathsf{T}_1} \frac{(s-t)^{1-\ltimes_1}}{\Gamma(2-\ltimes_1)} \varkappa''(t)dt + \ldots + \int_{\mathsf{T}_{i-1}}^s \frac{(s-t)^{1-\ltimes_i}}{\Gamma(2-\ltimes_i)} \varkappa''(t)dt = f(s, \varkappa(s), I_{0^+}^{\ltimes_i} \varkappa(s)), \quad s \in J_i. \tag{4.39}$$

We now proceed to describe the solution to the initial and terminal value problem (ITVP) stated in (4.37).

Definition 50 *The initial and terminal value problem (ITVP) given in (4.37) admits a solution if there exist functions $\varkappa_i$, $i = 1, 2, \ldots, n$, such that $\varkappa_i \in C([0, \mathsf{T}_i], X)$, and each $\varkappa_i$ satisfies equation (4.39) along with the initial and terminal conditions $\varkappa_i(0) = 0$ and $\varkappa_i(\mathsf{T}_i) = 0$.*

Let the function $\varkappa(\mathsf{T}) \in C(J, \mathbb{R})$ be such that $\varkappa(s) \equiv 0$ for all $s \in [0, \mathsf{T}_{i-1}]$, and suppose it satisfies the integral equation (4.39). Then (4.39) simplifies, for $s \in J_i$, to

$${}^{c}F_{\mathsf{T}_{i-1}^+}^{\ltimes_i} \varkappa(s) = f(s), \varkappa(s), I_{\mathsf{T}_{i-1}^+}^{\ltimes_i} \varkappa(s)), \quad s \in J_i,$$

where $CD_{\mathsf{T}_{i-1}^+}^{\ltimes_i}$ and $I_{\mathsf{T}_{i-1}^+}^{\ltimes_i}$ denote the left Caputo fractional derivative and fractional integral of order $\varkappa_i$, respectively, starting from T_{i-1}.

We now investigate the following initial and terminal value problem:

$$\begin{cases} {}^{c}F_{\mathsf{T}_{i-1}^+}^{\ltimes_i} \varkappa(s) = f(s, \varkappa(s), I_{\mathsf{T}_{i-1}^+}^{\ltimes_i} \varkappa(s)), & s \in J_i \\ \varkappa(\mathsf{T}_{i-1}) = 0, \varkappa(\mathsf{T}_i) = 0. \end{cases} \tag{4.40}$$

The next lemma will serve as the foundation for our analysis of equation (4.40).

Lemma 18 *A function $\varkappa \in E_i$ is a solution to the boundary value problem (4.40) if and only if $\varkappa$ satisfies the integral equation:*

$$\varkappa(s) = -\frac{s-\mathsf{T}_{i-1}}{\mathsf{T}_i-\mathsf{T}_{i-1}}\left[I^{\ltimes_i}_{\mathsf{T}_{i-1}^+} f(s,\varkappa(s), I^{\ltimes_i}_{\mathsf{T}_{i-1}^+}\varkappa(s))\right]_{s=\mathsf{T}_i} + I^{\ltimes_i}_{\mathsf{T}_{i-1}^+} f(s,\varkappa(s), I^{\ltimes_i}_{\mathsf{T}_{i-1}^+}\varkappa(s)). \tag{4.41}$$

Proof: Assume $\varkappa \in E_i$ is a solution to the problem (4.40). Applying the operator $I^{\ltimes_i}_{\mathsf{T}_{i-1}^+}$ to both sides of equation (4.40), and invoking Lemma 2, we obtain:

$$\varkappa(s) = \omega_1 + \omega_2(s-\mathsf{T}_{i-1}) + \frac{1}{\Gamma(\ltimes_i)}\int_{\mathsf{T}_{i-1}}^{s}(s-t)^{\ltimes_i-1} f(t,\varkappa(t), I^{\ltimes_i}_{\mathsf{T}_{i-1}^+}\varkappa(t))dt, \quad s\in J_i.$$

By $\varkappa(\mathsf{T}_{i-1}) = 0$, we obtain $\omega_1 = 0$.

Taking into account the additional initial and terminal condition $\varkappa(\mathsf{T}_i) = 0$, it follows that

$$0 = \omega_2(\mathsf{T}_i-\mathsf{T}_{i-1}) + \frac{1}{\Gamma(\ltimes_i)}\int_{\mathsf{T}_{i-1}}^{\mathsf{T}_i}(\mathsf{T}_i-t)^{\ltimes_i-1} f(t,\varkappa(t), I^{\ltimes_i}_{\mathsf{T}_{i-1}^+}\varkappa(t))dt.$$

$$\textit{Hence,}\quad \omega_2 = -(\mathsf{T}_i-\mathsf{T}_{i-1})^{-1} I^{\ltimes_i}_{\mathsf{T}_{i-1}^+} f(s,\varkappa(s), I^{\ltimes_i}_{\mathsf{T}_{i-1}^+}\varkappa(s))\Big|_{s=\mathsf{T}_i}$$

Then, we observe that

$$\begin{aligned}\varkappa(s) &= -(\mathsf{T}_i-\mathsf{T}_{i-1})^{-1}(s-\mathsf{T}_{i-1})\left[I^{\ltimes_i}_{\mathsf{T}_{i-1}^+} f(s,\varkappa(s), I^{\ltimes_i}_{\mathsf{T}_{i-1}^+}\varkappa(s))\right]\Big|_{s=\mathsf{T}_i}\\ &+ I^{\ltimes_i}_{\mathsf{T}_{i-1}^+} f(s,\varkappa(s), I^{\ltimes_i}_{\mathsf{T}_{i-1}^+}\varkappa(s)), \quad s\in J_i.\end{aligned}$$

In contrast, take $\varkappa \in E_i$ be a solution to the integral relation (4.41). Given the continuity of the function $s^\delta f$ and invoking Lemma 2, we determine that $\varkappa$ solves the problem (4.40).

Based on the Schauder fixed point theorem, we are now able to establish our first existence result.

Theorem 31 *Assume that conditions (H1), (H2) hold, and suppose*

$$\frac{2(\mathsf{T}_i-\mathsf{T}_{i-1})^{\ltimes_i-1}(\mathsf{T}_i^{1-\delta}-\mathsf{T}_{i-1}^{1-\delta})}{(1-\delta)\Gamma(\ltimes_i)}(K + L\frac{(\mathsf{T}_i-\mathsf{T}_{i-1})^{\ltimes_i}}{\Gamma(\ltimes_i+1)}) < 1. \tag{4.42}$$

Then the problem (4.40) admits at least one solution over E_i.

Proof: We define the operator

$$W: E_i \to E_i$$

as follows:

$$W\varkappa(s) = -(\mathsf{T}_i - \mathsf{T}_{i-1})^{-1}(s - \mathsf{T}_{i-1})\left[I^{\Bbbk_i}_{\mathsf{T}^+_{i-1}} f(s, \varkappa(s), I^{\Bbbk_i}_{\mathsf{T}^+_{i-1}} \varkappa(s))\right]\Big|_{s=\mathsf{T}_i}$$

$$+\frac{1}{\Gamma(\Bbbk_i)}\int_{\mathsf{T}_{i-1}}^{s}(s-t)^{\Bbbk_i-1} f(t, \varkappa(t), I^{\Bbbk_i}_{\mathsf{T}^+_{i-1}} \varkappa(t))dt, \quad s \in J_i. \tag{4.43}$$

It follows from the continuity of the function f and the properties of fractional integrals $s^{\delta} f$ that the operator $W : E_i \to E_i$ defined in (4.43) is properly is well-defined. Consider the quantity

$$R_i \;\geq\; \frac{\frac{2f^{\star}(\mathsf{T}_i-\mathsf{T}_{i-1})^{\Bbbk_i}}{\Gamma(\Bbbk_i)}}{1-\frac{2(\mathsf{T}_i-\mathsf{T}_{i-1})^{\Bbbk_i-1}(\mathsf{T}_i^{1-\delta}-\mathsf{T}_{i-1}^{1-\delta})}{(1-\delta)\Gamma(\Bbbk_i)}(K+L\frac{(\mathsf{T}_i-\mathsf{T}_{i-1})^{\Bbbk_i}}{\Gamma(\Bbbk_i+1)})},$$

where

$$f^{\star} = \sup_{s\in J_i} |f(s, 0, 0)|.$$

We now consider the set

$$B_{R_i} = \{\varkappa \in E_i,\ \|\varkappa\|_{E_i} \leq R_i\}.$$

It is evident that B_{R_i} is bounded, closed, convex, and nonempty.

We proceed to verify that the operator W satisfies the hypotheses of the Schauder fixed point theorem. This will be demonstrated in three stages.

STEP 1: Claim: $W(B_{R_i}) \subseteq (B_{R_i})$.
Let $\varkappa \in B_{R_i}$. By assumption (H2), we estimate:

$$\begin{aligned}
|W\varkappa(s)| &\leq \frac{(\mathsf{T}_i-\mathsf{T}_{i-1})^{-1}(s-\mathsf{T}_{i-1})}{\Gamma(\Bbbk_i)}\int_{\mathsf{T}_{i-1}}^{\mathsf{T}_i}(\mathsf{T}_i-t)^{\Bbbk_i-1}|f(t,\varkappa(t),I^{\Bbbk_i}_{\mathsf{T}^+_{i-1}}\varkappa(t))|dt\\
&+ \frac{1}{\Gamma(\Bbbk_i)}\int_{\mathsf{T}_{i-1}}^{s}(s-t)^{\Bbbk_i-1}|f(t,\varkappa(t),I^{\Bbbk_i}_{\mathsf{T}^+_{i-1}}\varkappa(t))|dt\\
&\leq \frac{2}{\Gamma(\Bbbk_i)}\int_{\mathsf{T}_{i-1}}^{\mathsf{T}_i}(\mathsf{T}_i-t)^{\Bbbk_i-1}|f(t,\varkappa(t),I^{\Bbbk_i}_{\mathsf{T}^+_{i-1}}\varkappa(t))|dt\\
&\leq \frac{2}{\Gamma(\Bbbk_i)}\int_{\mathsf{T}_{i-1}}^{\mathsf{T}_i}(\mathsf{T}_i-t)^{\Bbbk_i-1}|f(t,\varkappa(t),I^{\Bbbk_i}_{\mathsf{T}^+_{i-1}}\varkappa(t))-f(t,0,0)|dt\\
&+ \frac{2}{\Gamma(\Bbbk_i)}\int_{\mathsf{T}_{i-1}}^{\mathsf{T}_i}(\mathsf{T}_i-t)^{\Bbbk_i-1}|f(t,0,0)|dt\\
&\leq \frac{2}{\Gamma(\Bbbk_i)}\int_{\mathsf{T}_{i-1}}^{\mathsf{T}_i}(\mathsf{T}_i-t)^{\Bbbk_i-1}t^{-\delta}(K|\varkappa(t)|+L|I^{\Bbbk_i}_{\mathsf{T}^+_{i-1}}\varkappa(t)|)dt+\frac{2f^{\star}(\mathsf{T}_i-\mathsf{T}_{i-1})^{\Bbbk_i}}{\Gamma(\Bbbk_i)}\\
&\leq \frac{2(\mathsf{T}_i-\mathsf{T}_{i-1})^{\Bbbk_i-1}}{\Gamma(\Bbbk_i)}\int_{\mathsf{T}_{i-1}}^{\mathsf{T}_i}t^{-\delta}(K+L\frac{(\mathsf{T}_i-\mathsf{T}_{i-1})^{\Bbbk_i}}{\Gamma(\Bbbk_i+1)})|\varkappa(t)|dt+\frac{2f^{\star}(\mathsf{T}_i-\mathsf{T}_{i-1})^{\Bbbk_i}}{\Gamma(\Bbbk_i)}\\
&\leq \frac{2(\mathsf{T}_i-\mathsf{T}_{i-1})^{\Bbbk_i-1}(\mathsf{T}_i^{1-\delta}-\mathsf{T}_{i-1}^{1-\delta})}{(1-\delta)\Gamma(\Bbbk_i)}(K+L\frac{(\mathsf{T}_i-\mathsf{T}_{i-1})^{\Bbbk_i}}{\Gamma(\Bbbk_i+1)})R_i+\frac{2f^{\star}(\mathsf{T}_i-\mathsf{T}_{i-1})^{\Bbbk_i}}{\Gamma(\Bbbk_i)}\\
&\leq R_i,
\end{aligned}$$

which proves that $W(B_{R_i}) \subseteq B_{R_i}$.

STEP 2: Assertion: The operator W is continuous.

Suppose the sequence $(\varkappa_n)$ converges to $\varkappa$ in E_i, let $s \in J_i$. Then,

$$
\begin{aligned}
|(W\varkappa_n)(s) - (W\varkappa)(s)| &\leq \tfrac{(\mathsf{T}_i - \mathsf{T}_{i-1})^{-1}(s-\mathsf{T}_{i-1})}{\Gamma(\Bbbk_i)} \int_{\mathsf{T}_{i-1}}^{\mathsf{T}_i} (\mathsf{T}_i - t)^{\Bbbk_i - 1} |f(t, \varkappa_n(t), I_{\mathsf{T}_{i-1}^+}^{\Bbbk_i} \varkappa_n(t)) \\
&- f(t, \varkappa(t), I_{\mathsf{T}_{i-1}^+}^{\Bbbk_i} \varkappa(t))| dt \\
&+ \tfrac{1}{\Gamma(\Bbbk_i)} \int_{\mathsf{T}_{i-1}}^{s} (s-t)^{\Bbbk_i - 1} |f(t, \varkappa_n(t), I_{\mathsf{T}_{i-1}^+}^{\Bbbk_i} \varkappa_n(t)) - f(t, \varkappa(t), I_{\mathsf{T}_{i-1}^+}^{\Bbbk_i} \varkappa(t))| dt \\
&\leq \tfrac{2}{\Gamma(\Bbbk_i)} \int_{\mathsf{T}_{i-1}}^{\mathsf{T}_i} (\mathsf{T}_i - t)^{\Bbbk_i - 1} |f(t, \varkappa_n(t), I_{\mathsf{T}_{i-1}^+}^{\Bbbk_i} \varkappa_n(t)) - f(t, \varkappa(t), I_{\mathsf{T}_{i-1}^+}^{\Bbbk_i} \varkappa(t))| dt \\
&\leq \tfrac{2}{\Gamma(\Bbbk_i)} \int_{\mathsf{T}_{i-1}}^{\mathsf{T}_i} t^{-\delta} (\mathsf{T}_i - t)^{\Bbbk_i - 1} (K|\varkappa_n(t) - \varkappa(t)| + L I_{\mathsf{T}_{i-1}^+}^{\Bbbk_i} |\varkappa_n(t) - \varkappa(t)|) dt \\
&\leq \tfrac{2K}{\Gamma(\Bbbk_i)} \|\varkappa_n - \varkappa\|_{E_i} \int_{\mathsf{T}_{i-1}}^{\mathsf{T}_i} t^{-\delta} (\mathsf{T}_i - t)^{\Bbbk_i - 1} dt \\
&+ \tfrac{2L}{\Gamma(\Bbbk_i)} \|I_{\mathsf{T}_{i-1}^+}^{\Bbbk_i} (\varkappa_n - \varkappa)\|_{E_i} \int_{\mathsf{T}_{i-1}}^{\mathsf{T}_i} t^{-\delta} (\mathsf{T}_i - t)^{\Bbbk_i - 1} dt \\
&\leq \tfrac{2K}{\Gamma(\Bbbk_i)} \|\varkappa_n - \varkappa\|_{E_i} \int_{\mathsf{T}_{i-1}}^{\mathsf{T}_i} t^{-\delta} (\mathsf{T}_i - t)^{\Bbbk_i - 1} dt \\
&+ \tfrac{2L(\mathsf{T}_i - \mathsf{T}_{i-1})^{\Bbbk_i}}{\Gamma(\Bbbk_i)\Gamma(\Bbbk_i + 1)} \|\varkappa_n - \varkappa\|_{E_i} \int_{\mathsf{T}_{i-1}}^{\mathsf{T}_i} t^{-\delta} (\mathsf{T}_i - t)^{\Bbbk_i - 1} dt \\
&\leq \left(\tfrac{2K}{\Gamma(\Bbbk_i)} + \tfrac{2L(\mathsf{T}_i - \mathsf{T}_{i-1})^{\Bbbk_i}}{\Gamma(\Bbbk_i)\Gamma(\Bbbk_i + 1)} \right) \|\varkappa_n - \varkappa\|_{E_i} \int_{\mathsf{T}_{i-1}}^{\mathsf{T}_i} t^{-\delta} (\mathsf{T}_i - t)^{\Bbbk_i - 1} dt \\
&\leq \tfrac{2(\mathsf{T}_i - \mathsf{T}_{i-1})^{\Bbbk_i - 1} (\mathsf{T}_i{}^{1-\delta} - \mathsf{T}_{i-1}{}^{1-\delta})}{(1-\delta)\Gamma(\Bbbk_i)} \left(K + L \tfrac{(\mathsf{T}_i - \mathsf{T}_{i-1})^{\Bbbk_i}}{\Gamma(\Bbbk_i + 1)} \right) \|\varkappa_n - \varkappa\|_{E_i}
\end{aligned}
$$

we conclude:

$$\|(W\varkappa_n) - (W\varkappa)\|_{E_i} \to 0 \quad as \quad n \to \infty.$$

Hence, the operator W is continuous on E_i.

Step 3: The operator W is compact.

We now demonstrate the relative compactness of $W(B_{R_i})$, thereby establishing the compactness of W. From Step 1, we have $W(B_{R_i}) = \{W(\varkappa) : \varkappa \in B_{R_i}\} \subset B_{R_i}$, which implies that for every $\varkappa \in B_{R_i}$, $\|W(\varkappa)\|_{E_i} \leq R_i$. Thus, $W(B_{R_i})$ is bounded. Next, we must show that $W(B_{R_i})$ is equicontinuous.

Let $s_1, s_2 \in J_i$ with $s_1 < s_2$, and let $\varkappa \in B_{R_i}$. It follows that

$$
\begin{aligned}
&|(W\varkappa)(s_2) - (W\varkappa)(s_1)| = |-\frac{(\mathsf{T}_i - \mathsf{T}_{i-1})^{-1}(s_2 - \mathsf{T}_{i-1})}{\Gamma(\ltimes_i)} \int_{\mathsf{T}_{i-1}}^{\mathsf{T}_i} (\mathsf{T}_i - t)^{\ltimes_i - 1} f(t, \varkappa(t), I^{\ltimes_i}_{\mathsf{T}^+_{i-1}} \varkappa(t))dt \\
+ \quad & \frac{1}{\Gamma(\ltimes_i)} \int_{\mathsf{T}_{i-1}}^{s_2} (s_2 - t)^{\ltimes_i - 1} f(t, \varkappa(t), I^{\ltimes_i}_{\mathsf{T}^+_{i-1}} \varkappa(t))dt + \frac{(\mathsf{T}_i - \mathsf{T}_{i-1})^{-1}(s_1 - \mathsf{T}_{i-1})}{\Gamma(\ltimes_i)} \\
& \int_{\mathsf{T}_{i-1}}^{\mathsf{T}_i} (\mathsf{T}_i - t)^{\ltimes_i - 1} f(t, \varkappa(t), I^{\ltimes_i}_{\mathsf{T}^+_{i-1}} \varkappa(t))dt - \frac{1}{\Gamma(\ltimes_i)} \int_{\mathsf{T}_{i-1}}^{s_1} (s_1 - t)^{\ltimes_i - 1} f(t, \varkappa(t), I^{\ltimes_i}_{\mathsf{T}^+_{i-1}} \varkappa(t))dt| \\
\leq \quad & \frac{(\mathsf{T}_i - \mathsf{T}_{i-1})^{-1}}{\Gamma(\ltimes_i)} \Big((s_2 - \mathsf{T}_{i-1}) - (s_1 - \mathsf{T}_{i-1})\Big) \int_{\mathsf{T}_{i-1}}^{\mathsf{T}_i} (\mathsf{T}_i - t)^{\ltimes_i - 1} |f(t, \varkappa(t), I^{\ltimes_i}_{\mathsf{T}^+_{i-1}} \varkappa(t))|dt \\
+ \quad & \frac{1}{\Gamma(\ltimes_i)} \int_{\mathsf{T}_{i-1}}^{s_1} \Big((s_2 - t)^{\ltimes_i - 1} - (s_1 - t)^{\ltimes_i - 1}\Big) |f(t, \varkappa(t), I^{\ltimes_i}_{\mathsf{T}^+_{i-1}} \varkappa(t))|dt \\
+ \quad & \frac{1}{\Gamma(\ltimes_i)} \int_{s_1}^{s_2} (s_2 - t)^{\ltimes_i - 1} |f(t, \varkappa(t), I^{\ltimes_i}_{\mathsf{T}^+_{i-1}} \varkappa(t))|dt \\
\leq \quad & \frac{(\mathsf{T}_i - \mathsf{T}_{i-1})^{-1}}{\Gamma(\ltimes_i)} \Big((s_2 - \mathsf{T}_{i-1}) - (s_1 - \mathsf{T}_{i-1})\Big) \int_{\mathsf{T}_{i-1}}^{\mathsf{T}_i} (\mathsf{T}_i - t)^{\ltimes_i - 1} |f(t, \varkappa(t), I^{\ltimes_i}_{\mathsf{T}^+_{i-1}} \varkappa(t)) - f(t, 0, 0)|dt \\
+ \quad & \frac{(\mathsf{T}_i - \mathsf{T}_{i-1})^{-1}}{\Gamma(\ltimes_i)} \Big((s_2 - \mathsf{T}_{i-1}) - (s_1 - \mathsf{T}_{i-1})\Big) \int_{\mathsf{T}_{i-1}}^{\mathsf{T}_i} (\mathsf{T}_i - t)^{\ltimes_i - 1} |f(t, 0, 0)|dt \\
+ \quad & \frac{1}{\Gamma(\ltimes_i)} \int_{\mathsf{T}_{i-1}}^{s_1} \Big((s_2 - t)^{\ltimes_i - 1} - (s_1 - t)^{\ltimes_i - 1}\Big) |f(t, \varkappa(t), I^{\ltimes_i}_{\mathsf{T}^+_{i-1}} \varkappa(t)) - f(t, 0, 0)|dt \\
+ \quad & \frac{1}{\Gamma(\ltimes_i)} \int_{\mathsf{T}_{i-1}}^{s_1} \Big((s_2 - t)^{\ltimes_i - 1} - (s_1 - t)^{\ltimes_i - 1}\Big) |f(t, 0, 0)|dt \\
+ \quad & \frac{1}{\Gamma(\ltimes_i)} \int_{s_1}^{s_2} (s_2 - t)^{\ltimes_i - 1} |f(t, \varkappa(t), I^{\ltimes_i}_{\mathsf{T}^+_{i-1}} \varkappa(t)) - f(t, 0, 0)|dt + \frac{1}{\Gamma(\ltimes_i)} \int_{s_1}^{s_2} (s_2 - t)^{\ltimes_i - 1} |f(t, 0, 0)|dt. \\
\leq \quad & \frac{(\mathsf{T}_i - \mathsf{T}_{i-1})^{-1}}{\Gamma(\ltimes_i)} \Big((s_2 - \mathsf{T}_{i-1}) - (s_1 - \mathsf{T}_{i-1})\Big) \int_{\mathsf{T}_{i-1}}^{\mathsf{T}_i} (\mathsf{T}_i - t)^{\ltimes_i - 1} t^{-\delta} (K|\varkappa(t)| + L|I^{\ltimes_i}_{\mathsf{T}^+_{i-1}} \varkappa(t)|)dt \\
+ \quad & \frac{f^\star(\mathsf{T}_i - \mathsf{T}_{i-1})^{-1}}{\Gamma(\ltimes_i)} \Big((s_2 - \mathsf{T}_{i-1}) - (s_1 - \mathsf{T}_{i-1})\Big) \int_{\mathsf{T}_{i-1}}^{\mathsf{T}_i} (\mathsf{T}_i - t)^{\ltimes_i - 1} dt \\
+ \quad & \frac{1}{\Gamma(\ltimes_i)} \int_{\mathsf{T}_{i-1}}^{s_1} t^{-\delta} \Big((s_2 - t)^{\ltimes_i - 1} - (s_1 - t)^{\ltimes_i - 1}\Big) (K|\varkappa(t)| + L|I^{\ltimes_i}_{\mathsf{T}^+_{i-1}} \varkappa(t)|)dt \\
+ \quad & \frac{f^\star}{\Gamma(\ltimes_i)} \int_{\mathsf{T}_{i-1}}^{s_1} \Big((s_2 - t)^{\ltimes_i - 1} - (s_1 - t)^{\ltimes_i - 1}\Big) dt \\
+ \quad & \frac{1}{\Gamma(\ltimes_i)} \int_{s_1}^{s_2} t^{-\delta} (s_2 - t)^{\ltimes_i - 1} (K|\varkappa(t)| + L|I^{\ltimes_i}_{\mathsf{T}^+_{i-1}} \varkappa(t)|)dt + \frac{f^\star}{\Gamma(\ltimes_i)} \int_{s_1}^{s_2} (s_2 - t)^{\ltimes_i - 1} dt \\
\leq \quad & \frac{(\mathsf{T}_i - \mathsf{T}_{i-1})^{\ltimes_i - 2}}{\Gamma(\ltimes_i)} \Big((s_2 - \mathsf{T}_{i-1}) - (s_1 - \mathsf{T}_{i-1})\Big) (K\|\varkappa\|_{E_i} + L\|I^{\ltimes_i}_{\mathsf{T}^+_{i-1}} \varkappa\|_{E_i}) \int_{\mathsf{T}_{i-1}}^{\mathsf{T}_i} t^{-\delta} dt \\
+ \quad & \frac{f^\star(\mathsf{T}_i - \mathsf{T}_{i-1})^{\ltimes_i - 1}}{\Gamma(\ltimes_i + 1)} \Big((s_2 - \mathsf{T}_{i-1}) - (s_1 - \mathsf{T}_{i-1})\Big) \\
+ \quad & \frac{1}{\Gamma(\ltimes_i)} (K\|\varkappa\|_{E_i} + L\|I^{\ltimes_i}_{\mathsf{T}^+_{i-1}} \varkappa\|_{E_i}) \int_{\mathsf{T}_{i-1}}^{s_1} t^{-\delta} ((s_2 - s_1)^{\ltimes_i - 1}) dt
\end{aligned}
$$

$$
\begin{aligned}
&+ \frac{f^\star}{\Gamma(\Bbbk_i)}\Big(\frac{(s_2-\mathsf{T}_{i-1})^{\Bbbk_i}}{\Bbbk_i} - \frac{(s_2-s_1)^{\Bbbk_i}}{\Bbbk_i} - \frac{(s_1-\mathsf{T}_{i-1})^{\Bbbk_i}}{\Bbbk_i}\Big)\\
&+ \frac{(s_2-s_1)^{\Bbbk_i-1}}{\Gamma(\Bbbk_i)}(K\|\varkappa\|_{E_i} + L\|I^{\Bbbk_i}_{\mathsf{T}^+_{i-1}}\varkappa\|_{E_i})\int_{s_1}^{s_2} t^{-\delta}dt + \frac{f^\star}{\Gamma(\Bbbk_i)}\frac{(s_2-s_1)^{\Bbbk_i}}{\Bbbk_i}\\
&\leq \frac{(\mathsf{T}_i-\mathsf{T}_{i-1})^{\Bbbk_i-2}(\mathsf{T}_i{}^{1-\delta}-\mathsf{T}_{i-1}{}^{1-\delta})}{(1-\delta)\Gamma(\Bbbk_i)}\Big((s_2-\mathsf{T}_{i-1})-(s_1-\mathsf{T}_{i-1})\Big)\\
&\quad (K\|\varkappa\|_{E_i} + L\frac{(\mathsf{T}_i-\mathsf{T}_{i-1})^{\Bbbk_i}}{\Gamma(\Bbbk_i+1)}\|\varkappa\|_{E_i}) + \frac{f^\star(\mathsf{T}_i-\mathsf{T}_{i-1})^{\Bbbk_i-1}}{\Gamma(\Bbbk_i+1)}\Big((s_2-\mathsf{T}_{i-1})-(s_1-\mathsf{T}_{i-1})\Big)\\
&+ \Big(\frac{(s_1{}^{1-\delta}-\mathsf{T}_{i-1}{}^{1-\delta})(s_2-s_1)^{\Bbbk_i-1}}{(1-\delta)\Gamma(\Bbbk_i)}\Big)(K\|\varkappa\|_{E_i} + L\frac{(\mathsf{T}_i-\mathsf{T}_{i-1})^{\Bbbk_i}}{\Gamma(\Bbbk_i+1)}\|\varkappa\|_{E_i})\\
&+ \frac{f^\star}{\Gamma(\Bbbk_i+1)}\Big((s_2-\mathsf{T}_{i-1})^{\Bbbk_i} - (s_2-s_1)^{\Bbbk_i} - (s_1-\mathsf{T}_{i-1})^{\Bbbk_i}\Big)\\
&+ \frac{(s_2{}^{1-\delta}-s_1{}^{1-\delta})(s_2-s_1)^{\Bbbk_i-1}}{(1-\delta)\Gamma(\Bbbk_i)}(K\|\varkappa\|_{E_i} + L\frac{(\mathsf{T}_i-\mathsf{T}_{i-1})^{\Bbbk_i}}{\Gamma(\Bbbk_i+1)}\|\varkappa\|_{E_i}) + \frac{f^\star(s_2-s_1)^{\Bbbk_i}}{\Gamma(\Bbbk_i+1)}\\
&\leq \Big(\frac{(\mathsf{T}_i-\mathsf{T}_{i-1})^{\Bbbk_i-2}(\mathsf{T}_i{}^{1-\delta}-\mathsf{T}_{i-1}{}^{1-\delta})}{(1-\delta)\Gamma(\Bbbk_i)}(K + L\frac{(\mathsf{T}_i-\mathsf{T}_{i-1})^{\Bbbk_i}}{\Gamma(\Bbbk_i+1)})\|\varkappa\|_{E_i} + \frac{f^\star(\mathsf{T}_i-\mathsf{T}_{i-1})^{\Bbbk_i-1}}{\Gamma(\Bbbk_i+1)}\Big)\\
&\quad \Big((s_2-\mathsf{T}_{i-1})-(s_1-\mathsf{T}_{i-1})\Big) + \Big(\frac{s_2{}^{1-\delta}-\mathsf{T}_{i-1}{}^{1-\delta}}{(1-\delta)\Gamma(\Bbbk_i)}(K + L\frac{(\mathsf{T}_i-\mathsf{T}_{i-1})^{\Bbbk_i}}{\Gamma(\Bbbk_i+1)})\|\varkappa\|_{E_i}\Big)(s_2-s_1)^{\Bbbk_i-1}\\
&+ \frac{f^\star}{\Gamma(\Bbbk_i+1)}\Big((s_2-\mathsf{T}_{i-1})^{\Bbbk_i} - (s_1-\mathsf{T}_{i-1})^{\Bbbk_i}\Big)
\end{aligned}
$$

Consequently, $\|(W\varkappa)(s_2)-(W\varkappa)(s_1)\|E_i \to 0$ as $|s_2-s_1|\to 0$. This shows that $\mathsf{T}(B_{R_i})$ is equicontinuous.

As a consequence of the Schauder fixed point theorem, the integral–terminal value problem (ITVP) (4.40) admits at least one solution $\varkappa_i \in B_{R_i}$. Since $B_{R_i} \subset E_i$, the proof of Theorem 31 is complete.

4.3.2 Uniqueness

The following result is based on the Banach contraction principle.

Theorem 32 *Assume that conditions* $(H1)$ *and* $(H2)$ *hold. If*

$$
\frac{2(\mathsf{T}_i{}^{1-\delta}-\mathsf{T}_{i-1}{}^{1-\delta})(\mathsf{T}_i-\mathsf{T}_{i-1})^{\Bbbk_i-1}}{(1-\delta)\Gamma(\Bbbk_i)}\Big(K + \frac{L(\mathsf{T}_i-\mathsf{T}_{i-1})^{\Bbbk_i}}{\Gamma(\Bbbk_i+1)}\Big) < 1, \qquad (4.44)
$$

then the problem (4.40) has a unique solution in E_i.

Proof: We will establish that the operator W, as defined in (4.43), possesses a unique fixed point through the application of the ßanach contraction principle.

Given $\varkappa(s)$ and $y(s)$ in E_i,we obtain

$$|(W\varkappa)(s) - (Wy)(s)|$$

$$\begin{aligned}
\leq\ & \frac{(\mathsf{T}_i - \mathsf{T}_{i-1})^{-1}(s - \mathsf{T}_{i-1})}{\Gamma(\ltimes_i)} \int_{\mathsf{T}_{i-1}}^{\mathsf{T}_i} (\mathsf{T}_i - t)^{\ltimes_i - 1} |f(t, \varkappa(t), I^{\ltimes_i}_{\mathsf{T}^+_{i-1}} \varkappa(t)) - f(t, y(t), I^{\ltimes_i}_{\mathsf{T}^+_{i-1}} y(t))| dt \\
+\ & \frac{1}{\Gamma(\ltimes_i)} \int_{\mathsf{T}_{i-1}}^{s} (s - t)^{\ltimes_i - 1} |f(t, \varkappa(t), I^{\ltimes_i}_{\mathsf{T}^+_{i-1}} \varkappa(t)) - f(t, y(t), I^{\ltimes_i}_{\mathsf{T}^+_{i-1}} y(t))| dt \\
\leq\ & \frac{2}{\Gamma(\ltimes_i)} \int_{\mathsf{T}_{i-1}}^{\mathsf{T}_i} (\mathsf{T}_i - t)^{\ltimes_i - 1} |f(t, \varkappa(t), I^{\ltimes_i}_{\mathsf{T}^+_{i-1}} \varkappa(t)) - f(t, y(t), I^{\ltimes_i}_{\mathsf{T}^+_{i-1}} y(t))| dt \\
\leq\ & \frac{2}{\Gamma(\ltimes_i)} \int_{\mathsf{T}_{i-1}}^{\mathsf{T}_i} (\mathsf{T}_i - t)^{\ltimes_i - 1} t^{-\delta} (K|\varkappa(t) - y(t)| + L I^{\ltimes_i}_{\mathsf{T}^+_{i-1}} |\varkappa(t) - y(t))|) dt \\
\leq\ & \frac{2K}{\Gamma(\ltimes_i)} \|\varkappa - y\|_{E_i} \int_{\mathsf{T}_{i-1}}^{\mathsf{T}_i} t^{-\delta} (\mathsf{T}_i - t)^{\ltimes_i - 1} dt + \frac{2L}{\Gamma(\ltimes_i)} \|I^{\ltimes_i}_{\mathsf{T}^+_{i-1}} (\varkappa - y)\|_{E_i} \int_{\mathsf{T}_{i-1}}^{\mathsf{T}_i} t^{-\delta} (\mathsf{T}_i - t)^{\ltimes_i - 1} dt \\
\leq\ & \frac{2K}{\Gamma(\ltimes_i)} \|\varkappa - y\|_{E_i} \int_{\mathsf{T}_{i-1}}^{\mathsf{T}_i} t^{-\delta} (\mathsf{T}_i - t)^{\ltimes_i - 1} dt + \frac{2L(\mathsf{T}_i - \mathsf{T}_{i-1})^{\ltimes_i}}{\Gamma(\ltimes_i)\Gamma(\ltimes_i + 1)} \|\varkappa - y\|_{E_i} \int_{\mathsf{T}_{i-1}}^{\mathsf{T}_i} t^{-\delta} (\mathsf{T}_i - t)^{\ltimes_i - 1} dt \\
\leq\ & \Big(\frac{2K}{\Gamma(\ltimes_i)} + \frac{2L(\mathsf{T}_i - \mathsf{T}_{i-1})^{\ltimes_i}}{\Gamma(\ltimes_i)\Gamma(\ltimes_i + 1)}\Big) \|\varkappa - y\|_{E_i} \int_{\mathsf{T}_{i-1}}^{\mathsf{T}_i} t^{-\delta} (\mathsf{T}_i - t)^{\ltimes_i - 1} dt \\
\leq\ & \frac{2({\mathsf{T}_i}^{1-\delta} - {\mathsf{T}_{i-1}}^{1-\delta})(\mathsf{T}_i - \mathsf{T}_{i-1})^{\ltimes_i - 1}}{(1-\delta)\Gamma(\ltimes_i)} \Big(K + \frac{L(\mathsf{T}_i - \mathsf{T}_{i-1})^{\ltimes_i}}{\Gamma(\ltimes_i + 1)}\Big) \|\varkappa - y\|_{E_i}
\end{aligned}$$

Hence, the operator W is a contraction, as demonstrated in (4.44). Consequently, by the ßanach fixed-point theorem, W admits a unique fixed point $\widetilde{\varkappa}_i \in E_i$, which uniquely solves the problem (4.40). Since $B_{R_i} \subset E_i$, this completes the proof of Theorem 32.

Our current objective is to demonstrate the existence of a solution to the initial terminal value problem (ITVP) given in (4.37).

(H2) Let $f : J \times \mathbb{R} \times \mathbb{R} \to \mathbb{R}$ be a continuous function, and suppose there exists $\delta \in (0, 1)$ such that $s^\delta f \in C(J \times \mathbb{R} \times \mathbb{R}, \mathbb{R})$. Additionally, assume there exist constants $K, L > 0$ such that for all $y_1, y_2, z_1, z_2 \in \mathbb{R}$ and $s \in J$, the following inequality holds:

$$s^\delta |f(s, y_1, z_1) - f(s, y_2, z_2)| \leq K|y_1 - y_2| + L|z_1 - z_2|.$$

Theorem 33 *Assuming conditions (H1) and (H2) are satisfied, and that the inequality (4.42) holds for every $i \in \{1, 2, \ldots, n\}$, it follows that the ınitial and terminal value problem (ITVP) (4.37) admits a unique solution in $C(J, \mathbb{R})$.*

Proof: By Theorem 32, the ITVP (4.40) possesses a unique solution $\widetilde{\varkappa}_i \in E_i$, for each $i \in \{1, 2, ..., n\}$.

Let us now define the mapping

$$\varkappa_i = \begin{cases} 0, & s \in [0, \mathsf{T}_{i-1}], \\ \widetilde{\varkappa}_i, & s \in J_i, \end{cases}, \quad i \in \{1, 2, ..., n\}$$

Hence, for all $s \in J_i$, the integral equation (4.39) admits a unique solution $\varkappa_i \in C([0, \mathsf{T}_i], \mathbb{R})$ such that $\varkappa_i(0) = 0$ and $\varkappa_i(\mathsf{T}_i) = \widetilde{\varkappa}_i(\mathsf{T}_i) = 0$.

Finally, define the mapping $\varkappa(s)$ by

$$\varkappa(s) = \begin{cases} \varkappa_1(s), \quad s \in J_1, \\ \varkappa_2(s) = \begin{cases} 0, & s \in J_1, \\ \widetilde{\varkappa}_2, & s \in J_2 \end{cases}. \\ \cdot \\ \cdot \\ \cdot \\ \varkappa_n(s) = \begin{cases} 0, & s \in [0, \mathsf{T}_{i-1}], \\ \widetilde{\varkappa}_i, & s \in J_i \end{cases} \end{cases} \tag{4.45}$$

which constitutes the unique solution to the initial and terminal value problem (ITVP) (4.37) in $C(J, \mathbb{R})$.

4.3.3 Stability

Theorem 34 *Suppose that assumptions (H1), (H2), and condition (4.44) hold. Assume further:*

(H3) *The function $\vartheta \in C(J_i, \mathbb{R}_+)$ is non-decreasing, and there exists a positive constant λ_ϑ such that*

$$I^{\ltimes_i}_{\mathsf{T}_{i-1}^+}\vartheta(s) \leq \lambda_\vartheta \vartheta(s), forall s \in J_i.$$

As a result, the problem (4.37) is stable in the Ulam-Hyers-$\mathfrak{R}$assias sense relative to ϑ.

Proof: Suppose $z(s) \in C(J_i, \mathbb{R})$ satisfies the condition given by inequality (2.27).

We now define the following functions:

$$z_i(s) = \begin{cases} 0, & s \in [0, \mathsf{T}_{i-1}], \\ z(s), & s \in J_i, \end{cases}, \quad i \in \{1, 2, ..., n\}$$

By equality (4.38), for $i \in \{1, 2, ..., n\}$ and $s \in J_i$ we obtain:

$$ {}^{c}F^{\ltimes(s)}_{\mathsf{T}_{i-1}^{+}} z_i(s) = \int_{\mathsf{T}_{i-1}}^{s} \frac{(s-t)^{1-\ltimes_i}}{\Gamma(2-\ltimes_i)} z^{(2)}(t)dt. $$

Applying the Riemann–Liouville fractional integral operator $I^{\ltimes_i}_{\mathsf{T}_{i-1}^{+}}$ to both sides of inequality (2.27), we obtain:

$$ \begin{aligned} \Big| z_i(s) &+ \frac{(\mathsf{T}_i - \mathsf{T}_{i-1})^{-1}(s-\mathsf{T}_{i-1})}{\Gamma(\ltimes_i)} \int_{\mathsf{T}_{i-1}}^{\mathsf{T}_i} (\mathsf{T}_i - t)^{\ltimes_{i-1}} f(t, z_i(t), I^{\ltimes_i}_{\mathsf{T}_{i-1}^{+}} z_i(t))dt \\ &- \frac{1}{\Gamma(\ltimes_i)} \int_{\mathsf{T}_{i-1}}^{s} (s-t)^{\ltimes_{i-1}} f(t, z_i(t), I^{\ltimes_i}_{\mathsf{T}_{i-1}^{+}} z_i(t))dt \Big| \\ &\leq \epsilon \int_{\mathsf{T}_{i-1}}^{s} \frac{(s-t)^{\ltimes_i - 1}}{\Gamma(\ltimes_i)} \vartheta(t)dt \\ &\leq \epsilon \lambda_{\vartheta} \vartheta(s) \end{aligned} $$

As stated in Theorem 33, the initial and terminal value problem (4.37) admits a singular solution $\varkappa \in C(J, \mathbb{R})$, defined for every $s \in J_i$, $i = 1, 2, \ldots, n$, by $\varkappa(s) = \varkappa_i(s)$, where

$$ \varkappa_i = \begin{cases} 0, & s \in [0, \mathsf{T}_{i-1}], \\ \widetilde{\varkappa}_i, & s \in J_i, \end{cases} \tag{4.46} $$

and $\widetilde{\varkappa}_i$ is a singular solution of (4.40), given according to Lemma 18 by:

$$ \begin{aligned} \widetilde{\varkappa}_i(s) &= -\frac{(\mathsf{T}_i - \mathsf{T}_{i-1})^{-1}(s-\mathsf{T}_{i-1})}{\Gamma(\ltimes_i)} \int_{\mathsf{T}_{i-1}}^{\mathsf{T}_i} (\mathsf{T}_i - t)^{\ltimes_{i-1}} f(t, \widetilde{\varkappa}_i(t), I^{\ltimes_i}_{\mathsf{T}_{i-1}^{+}} \widetilde{\varkappa}_i(t))dt \\ &+ \frac{1}{\Gamma(\ltimes_i)} \int_{\mathsf{T}_{i-1}}^{s} (s-t)^{\ltimes_{i-1}} f(t, \widetilde{\varkappa}_i(t), I^{\ltimes_i}_{\mathsf{T}_{i-1}^{+}} \widetilde{\varkappa}_i(t))dt \end{aligned} $$

On the other hand, for each $s \in J_i$, we have

$$ |z(s) - \varkappa(s)| = |z(s) - \varkappa_i(s)| = |z_i(s) - \widetilde{\varkappa}_i(s)| $$

$$
\begin{aligned}
= & \ \Big|z_i(s) + \frac{(\mathsf{T}_i - \mathsf{T}_{i-1})^{-1}(s - \mathsf{T}_{i-1})}{\Gamma(\ltimes_i)} \int_{\mathsf{T}_{i-1}}^{\mathsf{T}_i} (\mathsf{T}_i - t)^{\ltimes_i - 1} f(t, \widetilde{\varkappa}_i(t), I_{\mathsf{T}_{i-1}^+}^{\ltimes_i} \widetilde{\varkappa}_i(t))dt \\
- & \ \frac{1}{\Gamma(\ltimes_i)} \int_{\mathsf{T}_{i-1}}^{s} (s - t)^{\ltimes_i - 1} f(t, \widetilde{\varkappa}_i(t), I_{\mathsf{T}_{i-1}^+}^{\ltimes_i} \widetilde{\varkappa}_i(t))dt\Big| \\
\leq & \ \Big|z_i(s) + \frac{(\mathsf{T}_i - \mathsf{T}_{i-1})^{-1}(s - \mathsf{T}_{i-1})}{\Gamma(\ltimes_i)} \int_{\mathsf{T}_{i-1}}^{\mathsf{T}_i} (\mathsf{T}_i - t)^{\ltimes_i - 1} f(t, z_i(t), I_{\mathsf{T}_{i-1}^+}^{\ltimes_i} z_i(t))dt \\
- & \ \frac{1}{\Gamma(\ltimes_i)} \int_{\mathsf{T}_{i-1}}^{s} (s - t)^{\ltimes_i - 1} f(t, z_i(t), I_{\mathsf{T}_{i-1}^+}^{\ltimes_i} z_i(t))dt\Big| \\
+ & \ \frac{(\mathsf{T}_i - \mathsf{T}_{i-1})^{-1}(s - \mathsf{T}_{i-1})}{\Gamma(\ltimes_i)} \int_{\mathsf{T}_{i-1}}^{\mathsf{T}_i} (\mathsf{T}_i - t)^{\ltimes_i - 1} |f(t, z_i(t), I_{\mathsf{T}_{i-1}^+}^{\ltimes_i} z_i) - f(t, \widetilde{\varkappa}_i(t), I_{\mathsf{T}_{i-1}^+}^{\ltimes_i} \widetilde{\varkappa}_i)|dt \\
+ & \ \frac{1}{\Gamma(\ltimes_i)} \int_{\mathsf{T}_{i-1}}^{s} (s - t)^{\ltimes_i - 1} |f(t, z_i(t), I_{\mathsf{T}_{i-1}^+}^{\ltimes_i} z_i) - f(t, \widetilde{\varkappa}_i(t), I_{\mathsf{T}_{i-1}^+}^{\ltimes_i} \widetilde{\varkappa}_i)|dt \\
\leq & \ \lambda_\vartheta \epsilon \vartheta(s) + \frac{(\mathsf{T}_i - \mathsf{T}_{i-1})^{-1}(s - \mathsf{T}_{i-1})}{\Gamma(\ltimes_i)} \\
& \ \int_{\mathsf{T}_{i-1}}^{\mathsf{T}_i} (\mathsf{T}_i - t)^{\ltimes_i - 1} t^{-\delta} (K|z_i(t) - \widetilde{\varkappa}_i(t)| + L I_{\mathsf{T}_{i-1}^+}^{\ltimes_i} |z_i(t) - \widetilde{\varkappa}_i(t)|)dt \\
+ & \ \frac{1}{\Gamma(\ltimes_i)} \int_{\mathsf{T}_{i-1}}^{s} (s - t)^{\ltimes_i - 1} t^{-\delta} (K|z_i(t) - \widetilde{\varkappa}_i(t)| + L I_{\mathsf{T}_{i-1}^+}^{\ltimes_i} |z_i(t) - \widetilde{\varkappa}_i(t)|)dt \\
\leq & \ \lambda_\vartheta \epsilon \vartheta(s) + \frac{(\mathsf{T}_i - \mathsf{T}_{i-1})^{\ltimes_i - 1}}{\Gamma(\ltimes_i)} (K\|z_i - \widetilde{\varkappa}_i\|_{E_i} + L\|I_{\mathsf{T}_{i-1}^+}^{\ltimes_i} (z_i - \widetilde{\varkappa}_i)\|_{E_i}) \int_{\mathsf{T}_{i-1}}^{\mathsf{T}_i} t^{-\delta} dt \\
+ & \ \frac{(\mathsf{T}_i - \mathsf{T}_{i-1})^{\ltimes_i - 1}}{\Gamma(\ltimes_i)} (K\|z_i - \widetilde{\varkappa}_i\|_{E_i} + L\|I_{\mathsf{T}_{i-1}^+}^{\ltimes_i} (z_i - \widetilde{\varkappa}_i)\|_{E_i}) \int_{\mathsf{T}_{i-1}}^{s} t^{-\delta} dt \\
\leq & \ \lambda_\vartheta \epsilon \vartheta(s) + \frac{(\mathsf{T}_i - \mathsf{T}_{i-1})^{\ltimes_i - 1} ({\mathsf{T}_i}^{1-\delta} - {\mathsf{T}_{i-1}}^{1-\delta})}{(1-\delta)\Gamma(\ltimes_i)} (K\|z_i - \widetilde{\varkappa}_i\|_{E_i} + L\frac{(\mathsf{T}_i - \mathsf{T}_{i-1})^{\ltimes_i}}{\Gamma(\ltimes_i + 1)} \|z_i - \widetilde{\varkappa}_i\|_{E_i}) \\
+ & \ \frac{(\mathsf{T}_i - \mathsf{T}_{i-1})^{\ltimes_i - 1} (s^{1-\delta} - {\mathsf{T}_{i-1}}^{1-\delta})}{(1-\delta)\Gamma(\ltimes_i)} (K\|z_i - \widetilde{\varkappa}_i\|_{E_i} + L\frac{(\mathsf{T}_i - \mathsf{T}_{i-1})^{\ltimes_i}}{\Gamma(\ltimes_i + 1)} \|z_i - \widetilde{\varkappa}_i\|_{E_i}) \\
\leq & \ \lambda_\vartheta \epsilon \vartheta(s) + \frac{2(\mathsf{T}_i - \mathsf{T}_{i-1})^{\ltimes_i - 1} ({\mathsf{T}_i}^{1-\delta} - {\mathsf{T}_{i-1}}^{1-\delta})}{(1-\delta)\Gamma(\ltimes_i)} (K + L\frac{(\mathsf{T}_i - \mathsf{T}_{i-1})^{\ltimes_i}}{\Gamma(\ltimes_i + 1)}) \|z_i - \widetilde{\varkappa}_i\|_{E_i} \\
\leq & \ \lambda_\vartheta \epsilon \vartheta(s) + \mu \|z - \varkappa\|,
\end{aligned}
$$

where

$$
\mu = \max_{i=1,2,\ldots,n} \frac{2(\mathsf{T}_i - \mathsf{T}_{i-1})^{\ltimes_i - 1} ({\mathsf{T}_i}^{1-\delta} - {\mathsf{T}_{i-1}}^{1-\delta})}{(1-\delta)\Gamma(\ltimes_i)} (K + L\frac{(\mathsf{T}_i - \mathsf{T}_{i-1})^{\ltimes_i}}{\Gamma(\ltimes_i + 1)}) \|z_i - \widetilde{\varkappa}_i\|_{E_i}.
$$

Then,

$$\|z - \varkappa\|(1-\mu) \leq \lambda_\vartheta \vartheta(s)\epsilon.$$

Thus, for $s \in J_i$ we have

$$|z(s) - y(s)| \leq \|z - y\| \leq \frac{\lambda_\vartheta \vartheta(s)}{1-\mu}\epsilon := c_f \epsilon.$$

As a result, the initial terminal value problem (4.37) is stable in the sense of Ulam-Hyers-$\mathfrak{R}$assias with respect to the function ϑ.

4.3.4 Applied Representation

Consider the following initial terminal value problem (ITVP):

$$\begin{cases} {}^{c}F_{0^+}^{\ltimes(s)}\varkappa(s) = \dfrac{s^{-\frac{1}{3}}e^{-s}}{(e^{e^{\frac{s^2}{1+s}}}+4e^{2s}+1)(1+|\varkappa(s)|+|I_0^{\ltimes(s)}\varkappa(s)|)}, & s \in J := [0,2], \\ \varkappa(0) = 0, \quad \varkappa(2) = 0. \end{cases} \tag{4.47}$$

Let

$$f(s,y,z) = \frac{s^{-\frac{1}{3}}e^{-s}}{(e^{e^{\frac{s^2}{1+s}}} + 4e^{2s} + 1)(1+y+z)}, \ (s,y,z) \in [0,2]\times[0,+\infty)\times[0,+\infty).$$

Define the variable-order function $\Theta(s)$ by:

$$\Theta(s) = \begin{cases} \frac{3}{2}, & s \in J_1 := [0,1], \\ \frac{9}{5}, & s \in J_2 :=]1,2]. \end{cases} \tag{4.48}$$

Thus, we obtain

$$\begin{aligned} s^{\frac{1}{3}}|f(s,y_1,z_1) - f(s,y_2,z_2)| &= \left| \frac{e^{-s}}{(e^{e^{\frac{s^2}{1+s}}} + 4e^{2s} + 1)} \left(\frac{1}{1+y_1+z_1} - \frac{1}{1+y_2+z_2} \right) \right| \\ &\leq \frac{e^{-s}(|y_1-y_2| + |z_1-z_2|)}{(e^{e^{\frac{s^2}{1+s}}} + 4e^{2s} + 1)(1+y_1+z_1)(1+y_2+z_2)} \\ &\leq \frac{e^{-s}}{(e^{e^{\frac{s^2}{1+s}}} + 4e^{2s} + 1)}(|y_1-y_2| + |z_1-z_2|) \\ &\leq \frac{1}{(e+5)}|y_1-y_2| + \frac{1}{(e+5)}|z_1-z_2|. \end{aligned}$$

Therefore, the condition $(H2)$ holds true for $\delta = \frac{1}{3}$ and $K = L = \frac{1}{e+5}$.

By (4.48), the initial-type value problem (ITVP) (4.47) can be divided into two expressions as follows:

$$\begin{cases} {}^cF_{0+}^{\frac{3}{2}}\varkappa(s) = \dfrac{s^{-\frac{1}{3}}e^{-s}}{(e^{e^{\frac{s^2}{1+s}}}+4e^{2s}+1)(1+|\varkappa(s)|+|I_0^{\frac{3}{2}}\varkappa(s)|)}, & s \in J_1, \\ {}^cF_{1+}^{\frac{9}{5}}\varkappa(s) = \dfrac{s^{-\frac{1}{3}}e^{-s}}{(e^{e^{\frac{s^2}{1+s}}}+4e^{2s}+1)(1+|\varkappa(s)|+|I_0^{\frac{9}{5}}\varkappa(s)|)}, & s \in J_2. \end{cases}$$

For $s \in J_1$, we first consider the following initial-type value problem:

$$\begin{cases} {}^cF_{0+}^{\frac{3}{2}}\varkappa(s) = \dfrac{s^{-\frac{1}{3}}e^{-s}}{(e^{e^{\frac{s^2}{1+s}}}+4e^{2s}+1)(1+|\varkappa(s)|+|I_0^{\frac{3}{2}}\varkappa(s)|)}, & s \in J_1, \\ \varkappa(0) = 0, \quad \varkappa(1) = 0. \end{cases} \tag{4.49}$$

We are now in a position to check whether condition (4.44) is satisfied.

$$\frac{(\mathsf{T}_1{}^{1-\delta}-\mathsf{T}_0{}^{1-\delta})(\mathsf{T}_1-\mathsf{T}_0)^{\ltimes_1-1}}{(1-\delta)\Gamma(\ltimes_1)}\left(2K+\frac{2L(\mathsf{T}_1-\mathsf{T}_0)^{\ltimes_1}}{\Gamma(\ltimes_1+1)}\right) = \frac{2}{\frac{2}{3}(e+5)\Gamma(\frac{3}{2})}\left(1+\frac{1}{\Gamma(\frac{5}{2})}\right) \simeq 0.7685 < 1$$

Thus, condition (4.44) is satisfied.

Let $\vartheta(s) = s^{\frac{1}{2}}$. Consider the fractional integral:

$$\begin{aligned} I_{0+}^{\ltimes_1}\vartheta(s) &= \frac{1}{\Gamma(\frac{3}{2})}\int_0^s (s-t)^{\frac{1}{2}}t^{\frac{1}{2}}\,dt \\ &\leq \frac{1}{\Gamma(\frac{3}{2})}\int_0^s (s-t)^{\frac{1}{2}}\,dt \\ &\leq \frac{2}{3\Gamma(\frac{3}{2})}\vartheta(s) := \lambda_\vartheta\vartheta(s). \end{aligned}$$

Therefore, condition (H3) is satisfied with $\vartheta(s) = s^{\frac{1}{2}}$ and $\lambda_\vartheta = \frac{2}{3\Gamma(\frac{3}{2})}$.

In view of Theorem 32, the ITVP (4.49) admits a unique solution $\widetilde{\varkappa}_1 \in E_1$.

For $s \in J_2$, the ITVP (4.47) can be reformulated as follows:

$$\begin{cases} {}^cF_{1+}^{\frac{9}{5}}\varkappa(s) = \dfrac{s^{-\frac{1}{3}}e^{-s}}{(e^{e^{\frac{s^2}{1+s}}}+4e^{2s}+1)(1+|\varkappa(s)|+|I_0^{\frac{9}{5}}\varkappa(s)|)}, & s \in J_2, \\ \varkappa(1) = 0, \quad \varkappa(2) = 0. \end{cases} \tag{4.50}$$

We observe that

$$\frac{(\mathsf{T}_2{}^{1-\delta}-\mathsf{T}_1{}^{1-\delta})(\mathsf{T}_2-\mathsf{T}_1)^{\ltimes_2-1}}{(1-\delta)\Gamma(\ltimes_2)}\left(2K+\frac{2L(\mathsf{T}_2-\mathsf{T}_1)^{\ltimes_2}}{\Gamma(\ltimes_2+1)}\right) = \frac{2^{\frac{2}{3}}-1}{\frac{2}{3}\Gamma(\frac{9}{5})}\frac{2}{e+5}\left(1+\frac{1}{\Gamma(\frac{14}{5})}\right) \simeq 0.3913 < 1$$

Thus, condition (4.44) is satisfied.

Next, consider the fractional integral: Using the estimate $t^{\frac{1}{2}} \leq s^{\frac{1}{2}}$ for $t \in [1, s]$, we obtain:

$$\begin{aligned} I_{1^+}^{\ltimes_2}\vartheta(s) &= \frac{1}{\Gamma(\frac{9}{5})}\int_1^s (s-t)^{\frac{4}{5}} t^{\frac{1}{2}}\,dt \\ &\leq \frac{1}{\Gamma(\frac{9}{5})}\int_1^s (s-t)^{\frac{4}{5}}\,dt \\ &\leq \frac{5}{9\Gamma(\frac{9}{5})}\vartheta(s) := \lambda_\vartheta \vartheta(s). \end{aligned}$$

Hence, $I_{1^+}^{\Theta_2}\vartheta(s) \leq \lambda_{\vartheta\vartheta}(s)$, where $\lambda_\vartheta = \frac{5}{9\Gamma(\frac{9}{5})}$.

Therefore, condition (H3) is satisfied with $\vartheta(s) = s^{\frac{1}{2}}$ and $\lambda_\vartheta = \frac{5}{9\Gamma(\frac{9}{5})}$.

Taking Theorem 32 into account, the initial terminal value problem (4.50) admits a unique solution $\widetilde{\varkappa}_2 \in E_2$.

Thus, based on Theorem 33, the problem (4.47) admits a unique solution

$$\varkappa(s) = \begin{cases} \varkappa_1(s), & s \in J_1, \\ \varkappa_2(s), & s \in J_2, \end{cases}$$

where

$$\varkappa_2(s) = \begin{cases} 0, & s \in J_1, \\ \widetilde{\varkappa}_2(s), & s \in J_2. \end{cases}$$

By Theorem 34 the initial terminal value problem (4.47) possesses Ulam-Hyers-$\mathfrak{R}$assias stability with respect to ϑ.

4.4 Initial and Terminal Value Problem of a Fractional Thermostat Model of Variable Order via Piecewise Constant Function

In this section, we study the following fractional analog of the thermostat model:

$$\begin{cases} -{}^c\digamma^{\ltimes(s)}y(s) = f(s, y(s)), \ s \in J, \\ y'(0) = 0, \quad {}^c\digamma^{\ltimes(s)-1}y(\mathsf{T}) + y(0) = 0, \end{cases} \tag{4.51}$$

where $J = [0, \mathsf{T}]$, $0 < \mathsf{T} < \infty$, $\ltimes(s) : J \to (1, 2]$ denotes the variable order of the fractional derivatives, $f : J \times \mathbb{R} \to \mathbb{R}$ is a given function, and ${}^c\digamma^{\ltimes(s)}$ represents the left Caputo fractional derivative of variable-order $\ltimes(s)$.

This primary objective of this section is to investigate several qualitative properties of the solutions to the initial terminal value problem (ITVP) associated with the Caputo fractional thermostat model (4.51). To begin, we introduce a set of preliminary definitions and lemmas. An appropriate partition of the interval J is then constructed, guided by the Guo–Krasnoselskii fixed point theorem. From the ITVP corresponding to the variable-order Caputo fractional thermostat model (4.51), we derive suitable ITVPs for Caputo fractional differential equations of constant order. The existence of a solution to (4.51) is established. Finally, an illustrative example is presented to validate the theoretical results.

4.4.1 Existence Study

Let us state the premise as follows:

(H1) Let $n \in N$. Consider a finite sequence of points $\{\mathsf{T}_k\}_{k=0}^{n}$ such that $0 = \mathsf{T}_0 < \mathsf{T}_k < \mathsf{T}_n = \mathsf{T}$, $k = 1, \ldots, n-1$. Define subintervals $J_k := (\mathsf{T}_{k-1}, \mathsf{T}_k]$, $k = 1, 2, \ldots, n$. Then, the collection $\mathcal{P} = \cup_{k=1}^{n} J_k$ constitutes a partition of the interval $J = [0, \mathsf{T}]$.

For each $\jmath = 1, 2, \ldots, n$, let $E_\jmath = C(J_\jmath, \mathbb{R})$, denote the ßanach space of continuous functions $y : J_\jmath \to \mathbb{R}$, equipped with the norm $\|y\|_{E_i} = \sup_{s \in J_i} |y(s)|$.

Let $\ltimes(s) : J \to (1, 2]$ be a piecewise constant function with respect to the partition $\mathcal{P}$, i.e., $\ltimes(s) = \sum_{i=1}^{n} \ltimes_i I_i(s)$, where $1 < \ltimes_i \leq 2$ are constants, and I_i is the indicator function of the interval J_i defined by:

$$I_i(s) = \begin{cases} 1, & for\ s \in J_i, \\ 0, & otherwise. \end{cases}$$

For every s where J_i, for $i = 1, 2, \ldots, n$, the Caputo fractional derivative of variable order $\ltimes(s)$ for $y(s) \in C(J, \mathbb{R})$, according to (2.10), can be represented as a sum of left Caputo fractional derivatives of constant orders $\ltimes_j$, for $j = 1, 2, \ldots, i$, as follows:

$$^{c}\digamma_{0+}^{\ltimes(s)} y(s) = \sum_{j=1}^{i-1} \int_{\mathsf{T}_{j-1}}^{\mathsf{T}_j} \frac{(s-t)^{1-\ltimes_j}}{\Gamma(2-\ltimes_j)} y''(t)dt + \int_{\mathsf{T}_{i-1}}^{s} \frac{(s-t)^{1-\ltimes_i}}{\Gamma(2-\ltimes_i)} y''(t)dt. \quad (4.52)$$

Thus, the initial value problem (IVP) for the Caputo fractional thermostat model (4.51) can be formulated, for any $s \in J_i$, where $i = 1, 2, \ldots, n$, in the following manner:

$$-\sum_{j=1}^{i-1} \int_{\mathsf{T}_{j-1}}^{\mathsf{T}_j} \frac{(s-t)^{1-\ltimes_j}}{\Gamma(2-\ltimes_j)} y''(t)dt - \int_{\mathsf{T}_{i-1}}^{s} \frac{(s-t)^{1-\ltimes_i}}{\Gamma(2-\ltimes_i)} y''(t)dt = f(s, y(s)). \quad (4.53)$$

Suppose $\tilde{y} \in C(J_i, R)$ is a function such that $\tilde{y}(s) = 0$ for all $s \in [0, \mathsf{T}_{i-1}]$ and it satisfies the integral equation (4.53). Consequently, equation (4.53) reduces to:

$$-^{c}\digamma_{\mathsf{T}_{i-1}^{+}}^{\ltimes_i} \tilde{y}(s) = f(s, \tilde{y}(s)), \quad s \in J_i.$$

Considering the previously stated information, for any $i = 1, 2, \ldots, n$, we examine the following auxiliary initial value problem (IVP) for constant-order Caputo fractional differential equations.

$$\begin{cases} -{}^{c}F^{\ltimes_i}_{\mathsf{T}_{i-1}^{+}} y(s) = f(s, y(s)), \quad s \in J_i \\ y'(\mathsf{T}_{i-1}) = 0, \qquad {}^{c}F^{\ltimes_i - 1} y(\mathsf{T}_i) + y(\mathsf{T}_{i-1}) = 0. \end{cases} \tag{4.54}$$

Lemma 19 *A function $y \in C[\mathsf{T}_{i-1}, \mathsf{T}_i]$ is the solution of the initial and terminal value problem (4.54) if and only if it satisfies the integral equation*

$$y(s) = \int_{\mathsf{T}_{i-1}}^{\mathsf{T}_i} G_i(s,t) f(t) dt,$$

where $G_i(s,t)$ denotes the Green's function, given by

$$G_i(s,t) = \begin{cases} 1 - \frac{(s-t)^{\ltimes_i - 1}}{\Gamma(\ltimes_i)}, & t \leq s \\ \qquad 1, & t > s. \end{cases} \tag{4.55}$$

Proof: Using Lemma 2, we obtain that for some constants $c_0, c_1 \in \mathbb{R}$,, the solution $y(s)$ satisfies

$$y(s) = -I^{\ltimes_i} f(s) + c_0 + c_1 s = -\int_{\mathsf{T}_{i-1}}^{\mathsf{T}_i} \frac{(s-t)^{\ltimes_i - 1}}{\Gamma(\ltimes_i)} f(t) dt + c_0 + c_1 s. \tag{4.56}$$

In view of Lemma 2, we obtain

$$y'(s) = -\int_{\mathsf{T}_{i-1}}^{s} \frac{(s-t)^{\ltimes_i - 2}}{\Gamma(\ltimes_i - 1)} f(t) dt + c_1.$$

Since $y'(0) = 0$, we find that $c_1 = 0$.

It also follows that

$${}^{C}F^{\ltimes_i - 1} y(s) = -I^1 y(s).$$

Using the initial and terminal condition ${}^{C}F^{\ltimes - 1} y(\mathsf{T}_i) + u(\mathsf{T}_{i-1}) = 0$, we obtain

$$c_0 = \int_{\mathsf{T}_{i-1}}^{\mathsf{T}_i} f(t) dt.$$

Finally, substituting the values of c_0 and c_1 into equation (4.56), we obtain

$$y(s) = \int_{\mathsf{T}_{i-1}}^{\mathsf{T}_i} f(t) dt - \int_{\mathsf{T}_{i-1}}^{s} \frac{(s-t)^{\ltimes_i - 1}}{\Gamma(\ltimes_i)} f(t) dt = \int_{\mathsf{T}_{i-1}}^{\mathsf{T}_i} G_i(s,t) f(t) dt,$$

where $G_i(s,t)$ is given by equation (4.55).

Remark We observe that for each fixed point $t \in [\mathsf{T}_{i-1}, \mathsf{T}_i]$, the partial derivative $\frac{\partial G_i}{\partial s} = 0$ for $s \leq t$ and $\frac{\partial G_i}{\partial s} < 0$ for $s > t$, which implies that $G_i(s,t)$ is a decreasing function of s. It then follows that

$$\max_{s\in[\mathsf{T}_{i-1},\mathsf{T}_i]} G_i(s,t) = G_i(\mathsf{T}_{i-1},t) = \begin{cases} 1 & t > \mathsf{T}_{i-1}, \\ 1 + \frac{(\mathsf{T}_{i-1}-t)^{\ltimes_i -1}}{\Gamma(\ltimes_i)}, & t \leq \mathsf{T}_{i-1}, \end{cases}$$

and

$$\min_{s\in[\mathsf{T}_{i-1},\mathsf{T}_i]} G_i(s,t) = G_i(\mathsf{T}_i,t) = \begin{cases} 1 - \frac{(\mathsf{T}_i-t)^{\ltimes_i -1}}{\Gamma(\ltimes_i)} & t > \mathsf{T}_{i-1}, \\ 1 + \frac{(\mathsf{T}_{i-1}-t)^{\ltimes_i -1}-(\mathsf{T}_i-t)^{\ltimes_i -1}}{\Gamma(\ltimes_i)}, & t \leq \mathsf{T}_{i-1}, \end{cases}$$

Consequently, by examining the behavior of $G_i(s,t)$ with respect to t, we obtain

$$\min_{s,t\in[\mathsf{T}_{i-1},\mathsf{T}_i]} G_i(s,t) = 1 - \frac{(\mathsf{T}_i - \mathsf{T}_{i-1})^{\ltimes_i -1}}{\Gamma(\ltimes_i)}.$$

and

$$\max_{s,t\in[\mathsf{T}_{i-1},\mathsf{T}_i]} G_i(s,t) = 1 + \frac{\mathsf{T}_{i-1}^{\ltimes_i -1}}{\Gamma(\ltimes_i)}.$$

To establish the existence of positive solutions to the Caputo fractional thermostat model (4.51), we demonstrate that $G_i(s,t)$ satisfies the following property introduced by Lan and Webb in [66]:

(H2) There exists a measurable function $\phi : [\mathsf{T}_{i-1}, \mathsf{T}_i] \to [0,\infty)$, a subinterval $[a,b] \subseteq [\mathsf{T}_{i-1}, \mathsf{T}_i]$ and a constant $\lambda \in [0,1]$ such that

$$|G_i(s,t)| \leq \phi(t) \qquad \forall s,t \in [\mathsf{T}_{i-1}, \mathsf{T}_i]$$

and

$$G_i(s,t) \geq \lambda\phi(t) \qquad \forall s \in [a,b], \forall t \in [\mathsf{T}_{i-1}, \mathsf{T}_i].$$

Lemma 20 *If $\Gamma(\ltimes_i) > (\mathsf{T}_i - \mathsf{T}_{i-1})^{\ltimes_i -1}$, then $G_i(s,t) > 0$ for all $s,t \in [\mathsf{T}_{i-1}, \mathsf{T}_i]$, and $G_i(s,t)$ satisfies hypothesis (H2).*

Proof: We choose $[a,b] = [\mathsf{T}_{i-1}, \mathsf{T}_i]$, and observe that

$$|G_i(s,t)| = G_i(s,t) \leq 1 + \frac{\mathsf{T}_{i-1}^{\ltimes_i -1}}{\Gamma(\ltimes_i)} := \phi(t)$$

and

$$G_i(s,t) \geq \lambda\phi(t) \qquad \forall t,s \in [\mathsf{T}_{i-1}, \mathsf{T}_i]$$

where

$$\lambda = \frac{\Gamma(\ltimes_i) - (\mathsf{T}_i - \mathsf{T}_{i-1})^{\ltimes_i -1}}{\Gamma(\ltimes_i) + \mathsf{T}_{i-1}^{\ltimes_i -1}}. \tag{4.57}$$

Lemma 21 *If* $\Gamma(\ltimes_i) = (\mathsf{T}_i - \mathsf{T}_{i-1})^{\ltimes_i - 1}$, *then* $G_i(s,t) \geq 0$ *for all* $s, t \in [\mathsf{T}_{i-1}, \mathsf{T}_i]$, *and* $G_i(s,t)$ *satisfies hypothesis (H2).*

Proof: We choose $[a,b] = [\mathsf{T}_{i-1}, b]$ with $\mathsf{T}_{i-1} \leq b < \mathsf{T}_i$. Following the arguments in the previous lemma, we have

$$|G_i(s,t)| \leq 1 + \frac{\mathsf{T}_{i-1}^{\ltimes_i - 1}}{\Gamma(\ltimes_i)} := \phi(t), \qquad \forall s, t \in [\mathsf{T}_{i-1}, \mathsf{T}_i]$$

Also, by taking

$$\lambda = \frac{\Gamma(\ltimes_i) - (b - \mathsf{T}_{i-1})^{\ltimes_i - 1}}{\Gamma(\ltimes_i) + \mathsf{T}_{i-1}^{\ltimes_i - 1}},$$

we obtain

$$G(s,t) \geq \lambda \phi(t), \qquad \forall s \in [\mathsf{T}_{i-1}, b], \forall t \in [\mathsf{T}_{i-1}, \mathsf{T}_i].$$

Lemma 22 *If* $\Gamma(\ltimes_i) < (\mathsf{T}_i - \mathsf{T}_{i-1})^{\ltimes_i - 1}$, *then the kernel* $G(s,t)$ *changes sign on* $[\mathsf{T}_{i-1}, \mathsf{T}_i] \times [\mathsf{T}_{i-1}, \mathsf{T}_i]$, *and the function* $G_i(s,t)$ *satisfies condition (H2).*

Proof: We choose the interval $[a,b] = [\mathsf{T}_{i-1}, b]$, where $\mathsf{T}_{i-1} \leq b < \mathsf{T}_i$ such that $\Gamma(\ltimes_i) > (b - \mathsf{T}_{i-1})^{\ltimes_i - 1}$. Then, for all $s, t \in [\mathsf{T}_{i-1}, \mathsf{T}_i]$, we have

$$|G_i(s,t)| \leq max\Big\{1 + \frac{\mathsf{T}_{i-1}^{\ltimes_i - 1}}{\Gamma(\ltimes_i)}, \frac{(1 - \mathsf{T}_{i-1})^{\ltimes_i - 1}}{\Gamma(\ltimes_i)} - 1\Big\} := \phi(t).$$

and

$$G(s,t) \geq \lambda \phi(t) \qquad \forall s \in [\mathsf{T}_{i-1}, b], \forall t \in [\mathsf{T}_{i-1}, \mathsf{T}_i]$$

where

$$\lambda = min\Big\{\frac{\Gamma(\ltimes_i) - (b - \mathsf{T}_{i-1})^{\ltimes_i - 1}}{\Gamma(\ltimes_i) + \mathsf{T}_{i-1}^{\ltimes_i - 1}}, \frac{\Gamma(\ltimes_i) - -(b - \mathsf{T}_{i-1})^{\ltimes_i - 1}}{(\mathsf{T}_i - \mathsf{T}_{i-1})^{\ltimes_i - 1} - \Gamma(\ltimes_i)}\Big\}.$$

For the main results, we employ the well-known Guo-Krasnosel'skii fixed point theorem [47].

Main results

We define the following limits:

$$f_0 = \lim_{y \to 0+} \min_{s \in [\mathsf{T}_{i-1}, \mathsf{T}_i]} \frac{f(s,y)}{y}, \qquad f_0^\star = \lim_{y \to 0+} \max_{s \in [\mathsf{T}_{i-1}, \mathsf{T}_i]} \frac{f(s,y)}{y},$$

$$f_\infty = \lim_{y \to \infty} \min_{s \in [\mathsf{T}_{i-1}, \mathsf{T}_i]} \frac{f(s,y)}{y}, \qquad f_\infty^\star = \lim_{y \to \infty} \max_{s \in [\mathsf{T}_{i-1}, \mathsf{T}_i]} \frac{f(s,y)}{y}.$$

We now present the principal result of this study.

Theorem 35 *Let $f(t, y(t)) \in C([\mathsf{T}_{i-1}, \mathsf{T}_i] \times [0, \infty), [0, \infty))$. Assume that one of the following conditions holds:*

(i) *Sublinear case: $f_0 = \infty$ and $f_\infty = 0$.*

(ii) *Superlinear case: $f_0^\star = 0$ and $f_\infty^\star = \infty$.*

If $\Gamma(\ltimes_i) > (\mathsf{T}_i - \mathsf{T}_{i-1})^{\ltimes_i - 1}$, then the integral terminal value problem (ITVP) (4.54) admits at least one positive solution.

Proof: Let $C[\mathsf{T}_{i-1}, \mathsf{T}_i]$ denote the Banach space of all continuous real-valued functions on $[\mathsf{T}_{i-1}, \mathsf{T}_i]$, equipped with the usual supremum norm $\|.\|$.
We define the operator $\mathsf{T} : C[\mathsf{T}_{i-1}, \mathsf{T}_i] \to C[\mathsf{T}_{i-1}, \mathsf{T}_i]$ by

$$\mathsf{T}y(s) = \int_{\mathsf{T}_{i-1}}^{\mathsf{T}_i} G_i(s,t) f(t, y(t))dt,$$

where $G_i(s,t)$ is defined by equation (4.55).

It is clear from Lemma 19 that the fixed points of the operator T coincide with the solutions of problem (4.54).

We now define the cone

$$P = \{y \setminus y \in C[\mathsf{T}_{i-1}, \mathsf{T}_i], y(s) \geq 0, \min_{s \in [\mathsf{T}_{i-1}, \mathsf{T}_i]} y(s) \geq \lambda \|y\|\},$$

where λ is given by equation (4.57).

First, we show that $\mathsf{T}(P) \subset P$.

It follows from the continuity and the non-negativity of the functions G_i and f on their respective domains that if $y \in P$, then $\mathsf{T}y \in C[\mathsf{T}_{i-1}, \mathsf{T}_i]$ and $\mathsf{T}y(s) \geq 0$ for all $s \in [\mathsf{T}_{i-1}, \mathsf{T}_i]$. For a fixed $y \in P$ and for all $s \in [\mathsf{T}_{i-1}, \mathsf{T}_i]$, the fact that $G_i(s,t)$ satisfies condition (H2) yields the following chain of inequalities:

$$\begin{aligned}
\mathsf{T}y(s) &= \int_{\mathsf{T}_{i-1}}^{\mathsf{T}_i} G_i(s,t) f(t, y(t))dt \\
&\geq \lambda \int_{\mathsf{T}_{i-1}}^{\mathsf{T}_i} \phi(t) f(t, y(t))dt \\
&\geq \lambda \int_{\mathsf{T}_{i-1}}^{\mathsf{T}_i} \max_{s \in [\mathsf{T}_{i-1}, \mathsf{T}_i]} G_i(s,t) f(t, y(t))dt \\
&\geq \lambda \max_{s \in [\mathsf{T}_{i-1}, \mathsf{T}_i]} \int_{\mathsf{T}_{i-1}}^{\mathsf{T}_i} G_i(s,t) f(t, y(t))dt \\
&= \lambda \| \mathsf{T}\, y \|.
\end{aligned}$$

Hence, $\mathsf{T}(P) \subset P$.

We now show that the operator $\mathsf{T} : P \to P$ is completely continuous.

In view of the continuity of the functions G_i and f, it follows that the operator $\mathsf{T} : P \to P$ is continuous.

Let $\Omega \subset P$ be a bounded subset; that is, there exists a positive constant $M > 0$ such that $\|y\|_\infty \leq M$ for all $y \in \Omega$. Define

$$L = \max_{\mathsf{T}_{i-1} \leq s \leq \mathsf{T}_i, 0 \leq y \leq M} |f(s, y)| + 1.$$

Then for all $y \in \Omega$, we have

$$|\mathsf{T}\,y(s)| \leq \int_{\mathsf{T}_{i-1}}^{\mathsf{T}_i} G_i(s,t) f(t, y(t)) dt \leq L \int_{\mathsf{T}_{i-1}}^{\mathsf{T}_i} G_i(s,t) dt$$

for all $s \in [\mathsf{T}_{i-1}, \mathsf{T}_i]$. That is, the set $\mathsf{T}(\Omega)$ is bounded.

For each $y \in \Omega$ and $s_1, s_2 \in [\mathsf{T}_{i-1}, \mathsf{T}_i]$ with $s_1 < s_2$, we have

$$\begin{aligned}
|\mathsf{T}\,y(s_2) - \mathsf{T}y(s_1)| &= \Big| -\int_{\mathsf{T}_{i-1}}^{s_2} \frac{(s_2 - t)^{\ltimes_i - 1}}{\Gamma(\ltimes_i)} f(t, y(t)) dt + \int_{\mathsf{T}_{i-1}}^{s_1} \frac{(s_1 - t)^{\ltimes_i - 1}}{\Gamma(\ltimes_i)} f(t, y(t)) dt \Big| \\
&\leq \frac{1}{\Gamma(\ltimes_i)} \int_{\mathsf{T}_{i-1}}^{s_1} \Big((s_2 - t)^{\ltimes_i - 1} - (s_1 - t)^{\ltimes_i - 1} \Big) |f(t, y(t))| dt \\
&+ \frac{1}{\Gamma(\ltimes_i)} \int_{s_1}^{s_2} (s_2 - t)^{\ltimes_i - 1} |f(t, y(t))| dt \\
&\leq \frac{L}{\Gamma(\ltimes_i)} \Big(\int_{\mathsf{T}_{i-1}}^{s_1} \Big((s_2 - t)^{\ltimes_i - 1} - (s_1 - t)^{\ltimes_i - 1} \Big) dt + \int_{s_1}^{s_2} (s_2 - t)^{\ltimes_i - 1} dt \Big) \\
&= \frac{L}{\ltimes_i \Gamma(\ltimes_i)} \Big(- (s_2 - s_1)^{\ltimes_i} + s_2^{\ltimes_i} - s_1^{\ltimes_i} + (s_2 - s_1)^{\ltimes_i} \Big) \\
&= \frac{L}{\Gamma(\ltimes_i + 1)} (s_2^{\ltimes_i} - s_1^{\ltimes_i})
\end{aligned}$$

Clearly, the right-hand side of the above inequality tends to zero as $s_1 \to s_2$. Therefore, the set $\mathsf{T}(\Omega)$ is equicontinuous. It follows from the Arzela-Ascoli theorem that the operator $\mathsf{T} : P \to P$ is completely continuous.

We now consider the two cases:

(i) Sublinear case($f_0 = \infty$ and $f_\infty = 0$).

Since $f_0 = \infty$, there exists $\rho_1 > 0$ such that $f(s, y) \geq \delta_1 y$ for all $0 < y \leq \rho_1$, where δ_1 satisfies

$$\delta_1 \Big(1 - \frac{(\mathsf{T}_i - \mathsf{T}_{i-1})^{\ltimes_i - 1}}{\Gamma(\ltimes_i)} \Big) \geq 1. \tag{4.58}$$

Let $y \in P$ be such that $\|y\| = \rho_1$. Then we have the following inequalities:

$$\begin{aligned} \mathsf{T}y(s) &= \int_{\mathsf{T}_{i-1}}^{\mathsf{T}_i} G_i(s,t) f(t, y(t)) dt \\ &\geq \delta_1 \int_{\mathsf{T}_{i-1}}^{\mathsf{T}_i} G_i(s,t) y(t) dt \\ Hence, \mathsf{T}y(s) &\geq \delta_1 \|y\| \Big(1 - \frac{(\mathsf{T}_i - \mathsf{T}_{i-1})^{\ltimes_i - 1}}{\Gamma(\ltimes_i)}\Big) \\ &\geq \|y\|. \end{aligned}$$

This confirms that $\mathsf{T}y(s) \geq \|y\|$, supporting the lower bound condition in the sublinear case.

Let $\Omega_1 = \{y \in C[\mathsf{T}_{i-1}, \mathsf{T}_i] \setminus \|y\| < \rho_1\}$. Hence, for all $y \in P \cap \partial\Omega_1$, we have $\| \mathsf{T}\, y\| \geq \|y\|$.

Since $f(s, .)$ is continuous on $[0, \infty)$, we can define the function

$$\tilde{f}(s, y) = \max_{z \in [0,y]} \{f(s, z)\}.$$

It is clear that $\tilde{f}(s, y)$ is non-decreasing on $(0, \infty)$. Moreover, since $f_\infty = 0$, we have

$$\lim_{y \to \infty} \Big\{ \max_{s \in [\mathsf{T}_{i-1}, \mathsf{T}_i]} \frac{\tilde{f}(s, y)}{y} \Big\} = 0.$$

Therefore, there exists $\rho_2 > \rho_1 > 0$ such that $\tilde{f}(s, y) \leq \delta_2 y$ for all $y \geq \rho_2$, where δ_2 satisfies

$$\delta_2 \Big(1 + \frac{\mathsf{T}_{i-1}^{\ltimes_i - 1}}{\Gamma(\ltimes_i)}\Big) \leq 1. \tag{4.59}$$

Define $\Omega_2 = \{y \in C[\mathsf{T}_{i-1}, \mathsf{T}_i] \setminus \|y\| < \rho_2\}$ and let $y \in P$ be such that $\|y\| = \rho_2$. Then

$$\begin{aligned} \mathsf{T}y &= \int_{\mathsf{T}_{i-1}}^{\mathsf{T}_i} G_i(s,t) f(t, y(t)) dt \\ &\leq \int_{\mathsf{T}_{i-1}}^{\mathsf{T}_i} G_i(s,t) \tilde{f}(t, \|y\|) dt \\ &\leq \delta_2 \|y\| \Big(1 + \frac{\mathsf{T}_{i-1}^{\ltimes_i - 1}}{\Gamma(\ltimes_i)}\Big) \\ &\leq \|y\|. \end{aligned}$$

Hence, we have $\| \mathsf{T}\, y\| \leq \|y\|, y \in P \cap \partial\Omega_2$.

Thus, by applying the first part of the Guo-Krasnosel'skii fixed point theorem, we conclude that equation (4.54) admits at least one positive solution.

(ii) Superlinear case($f_0^\star = 0$ and $f_\infty^\star = \infty$).

Let $\delta_2 > 0$ be given as in equation (4.59).

Since $f_0^\star = 0$, there exists a constant $r_1 > 0$ such that $f(s, y) \leq \delta_2 y$ for $0 \leq y \leq r_1$. Take $y \in P$ such that $\|y\| = r_1$. Then we have

$$\begin{aligned}
\mathsf{T}y &= \int_{\mathsf{T}_{i-1}}^{\mathsf{T}_i} G_i(s,t)f(t,y(t))dt \\
&\leq \delta_2 \int_{\mathsf{T}_{i-1}}^{\mathsf{T}_i} G_i(s,t)y(t)dt \\
&\leq \delta_2\|y\|(1+\frac{\mathsf{T}_{i-1}^{\ltimes_i - 1}}{\Gamma(\ltimes_i)}) \\
&\leq \|y\|.
\end{aligned}$$

Let $\Omega_1 = \{y \in C[\mathsf{T}_{i-1}, \mathsf{T}_i] \backslash \|y\| < r_1\}$, we see that $\|\mathsf{T}y\| \leq \|y\|$ for $y \in P \cap \partial\Omega_1$.

Now, since $f_\infty^\star = \infty$, there exists $r > 0$. Then for all $y \in P \cap \partial\Omega_1$, we have$f(s, y) \geq \delta_1 u$ for all $y \geq r$, where δ_1 is as defined in equation (4.58).

Define $\Omega_2 = \{y \in C[\mathsf{T}_{i-1}, \mathsf{T}_i] \setminus \|y\| < r_2\}$, where $r_2 = \max(2r_1, \frac{r}{\lambda})$. Then for any $y \in P$ with $\|y\| = r_2$, it follows that

$$\min y(s) \geq \lambda\|y\| = \lambda r_2 \geq r,$$

and we obtain

$$\begin{aligned}
\mathsf{T}y &= \int_{\mathsf{T}_{i-1}}^{\mathsf{T}_i} G_i(s,t)f(t,y(t))dt \\
&\geq \delta_1 \int_{\mathsf{T}_{i-1}}^{\mathsf{T}_i} G_i(s,t)y(t)dt \\
&\geq \delta_1\|y\|\left(1-\frac{(\mathsf{T}_i - \mathsf{T}_{i-1})^{\ltimes_i - 1}}{\Gamma(\ltimes_i)}\right) \\
&\geq \|y\|.
\end{aligned}$$

This shows that $\|\mathsf{T}y\| \geq \|y\|$ for all $y \in P \cap \partial\Omega_2$. Hence, by the second part of the Guo-Krasnosel'skii fixed point theorem, we conclude that equation (4.54) admits at least one positive solution $y \in P \cap (\overline{\Omega_2} \setminus \Omega_1)$.

4.4.2 Uniqueness Study

The ßanach fixed-point theorem is fundamental to establishing the second conclusion.

We state the following hypothesis:

(H2) Let f be a continuous function on $J \times \mathbb{R}$, and let $\delta \in (0,1)$ be a fixed parameter such that the function δ in $(0,1)$ such that $s^{\delta} f(s,y)$ is also continuous on $J \times \mathbb{R}$. Furthermore, suppose there exists is a constant $K > 0$ such that

$$s^{\delta}|f(s,y_1) - f(s,y_2)| \leq K|y_1 - y_2|, \text{for any } y_1,\ y_2 \in \mathbb{R} \text{ and s} \in J.$$

Theorem 36 *Assume that conditions (H1) and (H2) hold, along with the inequality*

$$\frac{K(\mathsf{T}_i^{1-\delta} - \mathsf{T}_{i-1}^{1-\delta})}{1-\delta}\left(1 + \frac{\mathsf{T}_{i-1}^{\ltimes_i - 1}}{\Gamma(\ltimes_i)}\right) < 1 \tag{4.60}$$

for all $i \in \{1, 2, ..., n\}$.

If $\Gamma(\ltimes_i) \geq (\mathsf{T}_i - \mathsf{T}_{i-1})^{\ltimes_i - 1}$, *then the initial time-value problem (ITVP) given by equation (4.54) admits a unique solution in* E_i.

Proof: We shall apply the ßanach contraction principle to prove that the operator T possesses a unique fixed point.

For $\varkappa(s), y(s) \in E_i$, by Lemma(3.2) and Lemma(3.3), we obtain $|\mathsf{T}\varkappa(s) - \mathsf{T}y(s)|$

$$\begin{aligned}
&= \left|\int_{\mathsf{T}_{i-1}}^{\mathsf{T}_i} G_i(s,t) f(t,\varkappa(t))dt - \int_{\mathsf{T}_{i-1}}^{\mathsf{T}_i} G_i(s,t) f(t,y(t))dt\right| \\
&\leq \int_{\mathsf{T}_{i-1}}^{\mathsf{T}_i} G_i(s,t)\Big|f(t,\varkappa(t)) - f(t,y(t))\Big|dt \\
&\leq \left(1 + \frac{\mathsf{T}_{i-1}^{\ltimes_i - 1}}{\Gamma(\ltimes_i)}\right)\int_{\mathsf{T}_{i-1}}^{\mathsf{T}_i} t^{-\delta}\Big(K|\varkappa(t) - y(t)|\Big)dt \\
&\leq K\left(1 + \frac{\mathsf{T}_{i-1}^{\ltimes_i - 1}}{\Gamma(\ltimes_i)}\right)\|\varkappa - y\|_{E_i}\int_{\mathsf{T}_{i-1}}^{\mathsf{T}_i} t^{-\delta}dt \\
&\leq \frac{K(\mathsf{T}_i^{1-\delta} - \mathsf{T}_{i-1}^{1-\delta})}{1-\delta}\left(1 + \frac{\mathsf{T}_{i-1}^{\ltimes_i - 1}}{\Gamma(\ltimes_i)}\right)\|\varkappa - y\|_{E_i}.
\end{aligned}$$

Therefore, the operator T is a contraction by (4.60). Consequently, T has a unique fixed point $\widetilde{\varkappa}_i \in E_i$, which constitutes the exclusive solution to the initial and terminal value problem, based on ßanach's contraction mapping theorem (4.54).

We shall now demonstrate the existence result for Caputo fractional thermostat model (4.51).

Theorem 37 *Suppose that conditions (H1) and (H2) are satisfied for all indices $i \in \{1, 2, \ldots, n\}$.*

Then, the initial and terminal value problem associated with the Caputo fractional thermostat model (4.51) admits at least one solution in $C(J, \mathbb{R})$.

Proof: For all $i \in \{1, 2, \ldots, n\}$, based on Theorem 35, the initial and terminal value problem (4.54) admits at least one solution $\widetilde{y}_i \in E_i$.

For each $i \in \{1, 2, \ldots, n\}$, we define the function

$$y_i(s) = \begin{cases} 0, & s \in [0, \mathsf{T}_{i-1}], \\ \widetilde{y}_i(s), & s \in J_i, \end{cases}$$

where $J_i = [\mathsf{T}_{i-1}, \mathsf{T}_i]$ Therefore, the function y_i defined on $[0, \mathsf{T}_i]$ with values in $\mathbb{R}$ is a solution to the integral equation (4.53) for any $s \in J_i$.

Consequently, the function

$$y(s) = \begin{cases} y_1(s), & s \in J_1, \\ y_2(s), & s \in J_2, \\ \dots\dots\dots\dots\dots \\ y_n(s), & s \in J_n = [\mathsf{T}_{n-1}, \mathsf{T}] \end{cases}$$

is a solution of the initial and terminal value problem (ITVP) for the Caputo fractional thermostat model (4.51) in $C(J, \mathbb{R})$.

4.4.3 Ulam-Hyers Stability Conditions

Theorem 38 *Let conditions (H1), (H3), and inequality (4.60) be satisfied. Then, the Caputo fractional thermostat model (4.51) is Ulam-Hyers stable.*

Proof: Let $\epsilon > 0$ be an arbitrary number, and let the function $z(s)$ from $C(J, \mathbb{R})$ satisfy the inequality (2.28).

For any $i \in \{1, 2, ..., n\}$, we define the as follows: - For $i = 1$: $z_1(s) \equiv z(s), s \in [0, \mathsf{T}_1]$. - For $i = 2, 3, ..., n$;

$$z_i(s) = \begin{cases} 0, & s \in [0, \mathsf{T}_{i-1}], \\ z(s), & s \in J_i. \end{cases}$$

For each $i \in \{1, 2, \ldots, n\}$, based on equation (4.52), for s in J_i we obtain for $S \in J_i$:

$${}^{c}F_{0+}^{\ltimes(s)} z_i(s) = \int_{\mathsf{T}_{i-1}}^{s} \frac{(s-t)^{1-\ltimes_i}}{\Gamma(2-\ltimes_i)} z^{(2)}(t)dt.$$

Taking the fractional integral $I^{\ltimes_i}_{\mathsf{T}_{i-1}^+}$ of both sides of inequality (2.28), we obtain:

$$\begin{aligned}\Big| - z_i(s) &- \int_{\mathsf{T}_{i-1}}^{\mathsf{T}_i} G_i(s,t)f(t,z_i(t))dt\Big| \\ &\leq \epsilon \int_{\mathsf{T}_{i-1}}^{s} \frac{(s-t)^{\ltimes_i - 1}}{\Gamma(\ltimes_i)}dt \\ &\leq \epsilon \frac{(\mathsf{T}_i - \mathsf{T}_{i-1})^{\ltimes_i}}{\Gamma(\ltimes_i + 1)}.\end{aligned}$$

In accordance with Theorem 37, the Caputo fractional thermostat model described in (4.51) admits a solution $\varkappa in\ C(J,\mathbb{R})$, given by $\varkappa(s) = \varkappa_i(s)$, $s \in J_i,\ i = 1,2,...,n$, where

$$\varkappa_i = \begin{cases} 0, & s \in [0, \mathsf{T}_{i-1}], \\ \widetilde{\varkappa}_i, & s \in J_i, \end{cases} \tag{4.61}$$

and each $\widetilde{\varkappa}_i \in E_i$ is a solution of equation (4.54). Based on Lemma 19, the integral equation

$$\widetilde{\varkappa}_i(s) = \int_{\mathsf{T}_{i-1}}^{\mathsf{T}_i} G_i(s,t)f(t,\widetilde{\varkappa}_i(t))dt, \tag{4.62}$$

holds.

Let $s \in J_i,\ i = 1,2,...,n$. Then by equations (4.61) and (4.62) we get

$$\begin{aligned} & |z(s) - \varkappa(s)| = |z(s) - \varkappa_i(s)| = |z_i(s) - \widetilde{\varkappa}_i(s)| \\ = & \left|z_i(s) - \textstyle\int_{\mathsf{T}_{i-1}}^{\mathsf{T}_i} G_i(s,t)f(t,\widetilde{\varkappa}_i(t))dt\right| \\ \leq & \left|z_i(s) - \textstyle\int_{\mathsf{T}_{i-1}}^{\mathsf{T}_i} G_i(s,t)f(t,z_i(t))dt\right| \\ + & \left|\textstyle\int_{\mathsf{T}_{i-1}}^{\mathsf{T}_i} G_i(s,t)f(t,z_i(t))dt - \int_{\mathsf{T}_{i-1}}^{s} G_i(s,t)f(t,\widetilde{\varkappa}_i(t))dt\right| \\ \leq & \left| - z_i(s) - \textstyle\int_{\mathsf{T}_{i-1}}^{\mathsf{T}_i} G_i(s,t)f(t,z_i(t))dt\right| \\ + & \left|\textstyle\int_{\mathsf{T}_{i-1}}^{\mathsf{T}_i} G_i(s,t)f(t,z_i(t))dt - \int_{\mathsf{T}_{i-1}}^{s} G_i(s,t)f(t,\widetilde{\varkappa}_i(t))dt\right| \\ \leq & \ \epsilon \tfrac{(\mathsf{T}_i - \mathsf{T}_{i-1})^{\ltimes_i}}{\Gamma(\ltimes_i+1)} \\ + & \left(1 + \tfrac{\mathsf{T}_{i-1}^{\ltimes_i - 1}}{\Gamma(\ltimes_i)}\right) \textstyle\int_{\mathsf{T}_{i-1}}^{\mathsf{T}_i} |f(t,z_i(t))dt - f(t,\widetilde{\varkappa}_i(t))|dt \\ \leq & \ \epsilon \tfrac{(\mathsf{T}_i - \mathsf{T}_{i-1})^{\ltimes_i}}{\Gamma(\ltimes_i+1)} + \left(1 + \tfrac{\mathsf{T}_{i-1}^{\ltimes_i - 1}}{\Gamma(\ltimes_i)}\right) \textstyle\int_{\mathsf{T}_{i-1}}^{\mathsf{T}_i} t^{-\delta}(K|z_i(t) - \widetilde{\varkappa}_i(t)|dt \\ \leq & \ \epsilon \tfrac{(\mathsf{T}_i - \mathsf{T}_{i-1})^{\ltimes_i}}{\Gamma(\ltimes_i+1)} + \left(1 + \tfrac{\mathsf{T}_{i-1}^{\ltimes_i - 1}}{\Gamma(\ltimes_i)}\right)(K\|z_i - \widetilde{\varkappa}_i\|_{E_i}) \textstyle\int_{\mathsf{T}_{i-1}}^{\mathsf{T}_i} t^{-\delta}dt \\ \leq & \ \epsilon \tfrac{(\mathsf{T}_i - \mathsf{T}_{i-1})^{\ltimes_i}}{\Gamma(\ltimes_i+1)} + \tfrac{K({\mathsf{T}_i}^{1-\delta} - {\mathsf{T}_{i-1}}^{1-\delta})}{1-\delta}\left(1 + \tfrac{\mathsf{T}_{i-1}^{\ltimes_i - 1}}{\Gamma(\ltimes_i)}\right)\|z_i - \widetilde{\varkappa}_i\|_{E_i} \\ \leq & \ \epsilon \tfrac{(\mathsf{T}_i - \mathsf{T}_{i-1})^{\ltimes_i}}{\Gamma(\ltimes_i+1)} + \mu\|z - \varkappa\|, \end{aligned}$$

where

$$\mu = \max_{i=1,2,\ldots,n} \frac{K({\mathsf{T}_i}^{1-\delta} - {\mathsf{T}_{i-1}}^{1-\delta})}{1-\delta}\Big(1 + \frac{\mathsf{T}_{i-1}^{\ltimes_i - 1}}{\Gamma(\ltimes_i)}\Big).$$

Then,

$$\|z - \varkappa\|(1-\mu) \leq \frac{(\mathsf{T}_i - \mathsf{T}_{i-1})^{\ltimes_i}}{\Gamma(\ltimes_i + 1)}\epsilon.$$

We establish, for any $s \in J_i$

$$|z(s) - \varkappa(s)| \leq \|z - \varkappa\| \leq \frac{(\mathsf{T}_i - \mathsf{T}_{i-1})^{\ltimes_i}}{(1-\mu)\Gamma(\ltimes_i + 1)}\epsilon := c_f\epsilon.$$

Therefore, the Caputo fractional thermostat model (4.51) is Ulam-Hyers stable.

4.4.4 Approximate Application

The following fractional initial and terminal value problem warrants our attention:

$$\begin{cases} -{}^{c}\digamma^{\ltimes(s)} y(s) = \frac{e^s+1}{\sqrt{y(s)}}, & s \in J := [0,2], \\ y'(0) = 0, & {}^{c}\digamma^{\ltimes(s)-1} y(2) + y(0) = 0. \end{cases} \tag{4.63}$$

Let

$$f(s,y) = \frac{e^s + 1}{\sqrt{y}}, \; (s,y) \in [0,2] \times [0,+\infty).$$

$$\ltimes(s) = \begin{cases} 1.4, & s \in J_1 := [0,1], \\ 1.5, & s \in J_2 :=]1,2]. \end{cases} \tag{4.64}$$

In accordance with equation (4.54), we study two supplementary initial and terminal value problems for Caputo fractional differential equations of fixed order, as described by (4.64):

First, consider

$$\begin{cases} -{}^{c}\digamma_{0+}^{1.4} y(s) = \frac{e^s+1}{\sqrt{y(s)}}, & s \in J_1, \\ y'(0) = 0, & {}^{c}\digamma_{0+}^{0.4} y(1) + y(0) = 0. \end{cases} \tag{4.65}$$

Second, consider

$$\begin{cases} {}^{c}\digamma_{1+}^{1.5} y(s) = \frac{e^s+1}{\sqrt{y(s)}}, & s \in J_2, \\ y'(1) = 0, & {}^{c}\digamma_{1+}^{0.5} y(2) + y(1) = 0. \end{cases} \tag{4.66}$$

Clearly, $f_0 = \infty$ and $f_\infty = 0$.

For $i = 1$, we compute $\Gamma(1.4) - 1 \approx -0.11274 < 0$. We take

$$\lambda_1 = min\Big\{\frac{\Gamma(\ltimes_i) - (b - \mathsf{T}_{i-1})^{\ltimes_i - 1}}{\Gamma(\ltimes_i) + \mathsf{T}_{i-1}^{\ltimes_i - 1}}, \frac{\Gamma(\ltimes_i) - (b - \mathsf{T}_{i-1})^{\ltimes_i - 1}}{(\mathsf{T}_i - \mathsf{T}_{i-1})^{\ltimes_i - 1} - \Gamma(\ltimes_i)}\Big\}$$
$$= \frac{\Gamma(1.4) - b^{\ltimes_i - 1}}{\Gamma(1.4)}$$

We then consider the cone $P_1 = \{y \setminus y \in C[0,1], \min_{s \in [0,1]} y(s) \geq \lambda_1 \|y\|\}$.
By Theorem 35, we conclude that the initial and terminal value problem (ITVP) given in equation (4.65) admits a positive solution $\widetilde{y}_1 \in P_1$.

For $i = 2$, we compute $\Gamma(1.5) - 1 \approx -0.11377 < 0$. We define

$$\lambda_2 = min\Big\{\frac{\Gamma(\ltimes_i) - (b - \mathsf{T}_{i-1})^{\ltimes_i - 1}}{\Gamma(\ltimes_i) + \mathsf{T}_{i-1}^{\ltimes_i - 1}}, \frac{\Gamma(\ltimes_i) - (b - \mathsf{T}_{i-1})^{\ltimes_i - 1}}{(\mathsf{T}_i - \mathsf{T}_{i-1})^{\ltimes_i - 1} - \Gamma(\ltimes_i)}\Big\}$$
$$= \frac{\Gamma(1.5) - (b-1)^{1.5}}{\Gamma(1.5) + 1},$$

We then consider the cone $P_2 = \{y \setminus y \in C[1,2], \min_{s \in [1,2]} y(s) \geq \lambda_2 \|y\|\}$.
By Theorem 35, we conclude that the initial and terminal value problem (ITVP) given in equation (4.66) admits a positive solution $\widetilde{y}_2 \in P_2$.

According to Theorem 37, the ITVP for the Caputo fractional thermostat model described by equation (4.63) admits a solution

$$y(s) = \begin{cases} \widetilde{y}_1(s), & s \in J_1, \\ y_2(s), & s \in J_2, \end{cases}$$

where

$$y_2(s) = \begin{cases} 0, & s \in J_1, \\ \widetilde{y}_2(s), & s \in J_2. \end{cases}$$

In accordance with Theorem 38, the initial and terminal value problem (ITVP) for the Caputo fractional thermostat model given by equation (4.63) is Ulam-Hyers stable.

5

Initial and Terminal Value Problems of Hadamard Fractional Differential Equations of Variable Order

In the present chapter, for given initial and terminal value problems involving Hadamard's variable-order fractional differential equations, the existence properties are analyzed. To achieve this objective, we first formulate the necessary conditions to verify the existence and uniqueness of solutions. Subsequently, we examine the stability of the derived solutions through Ulam–Hyers and Ulam–Hyers–Rassias theories.

5.1 Qualitative Study on Solutions of a Hadamard Variable-Order Initial and Terminal Value Problem via Ulam-Hyers-$\Re$assias Stability

The following initial and terminal value problem (ITVP) of Hadamard variable-order type with terminal conditions is addressed in this section in from a qualitative perspective:

$$\begin{cases} ({}^{H}F_{1^+}^{\ltimes(s)}\varkappa)(s) = f(s, \varkappa(s)), \ s \in \mathcal{J} := [1, \mathsf{T}], \\ \varkappa(1) = 0, \quad \varkappa(\mathsf{T}) = 0, \end{cases} \tag{5.1}$$

where $1 < \mathsf{T} < +\infty$, $\ltimes(s) : [1, \mathsf{T}] \to (1, 2]$ is a variable-order function, and $f : \mathcal{J} \times \mathbb{R} \to \mathbb{R}$ is continuous. The operator ${}^{H}F_{1^+}^{\ltimes(s)}$ denotes the Hadamard fractional derivative of variable order $\ltimes(s)$.

Our research aims to establish novel criteria for the existence and uniqueness of solutions to the Hadamard variable-order initial and terminal value problem (ITVP) presented in equation 5.1. Furthermore, we investigate the Ulam-Hyers-$\Re$assias stability of the corresponding solution to the Hadamard variable-order ITVP 5.1.

DOI: 10.1201/9781003687900-5

5.1.1 Main Results on the Existence of Solutions

We denote the space of all continuous mappings by $C(\mathcal{J},\mathbb{R})$, equipped with the supremum norm

$$\|\varkappa\| = \sup\{|\varkappa(s)|,\ s \in \mathcal{J}\}.$$

Further, we introduce the following essential assumptions:

(H1) Let $n \in \mathbb{N}$, and define the partition

$$\mathcal{P} = \{\mathcal{J}_1 := [0, \mathsf{T}_1], \mathcal{J}_2 := (\mathsf{T}_1, \mathsf{T}_2], \mathcal{J}_3 := (\mathsf{T}_2, \mathsf{T}_3], \dots, \mathcal{J}_n := (\mathsf{T}_{n-1}, \mathsf{T}]\}$$

as a partition of the interval $\mathcal{J}$, and assume that the function $\ltimes(s) : \mathcal{J} \to (1, 2]$ is piecewise constant function with respect to $\mathcal{P}$; that is,

$$\ltimes(s) = \sum_{i=1}^{n} \ltimes_i \mathcal{I}_i(s) = \begin{cases} \ltimes_1, & \text{if } s \in \mathcal{J}_1, \\ \ltimes_2, & \text{if } s \in \mathcal{J}_2, \\ \vdots & \vdots \quad \vdots \\ \ltimes_n, & \text{if } s \in \mathcal{J}_n, \end{cases}$$

in which $1 < \ltimes_i \leq 2$ for each $i \in \mathcal{N}_n$, with $\triangleright iin\mathbb{R}$. Let $\mathcal{I}_i$ denote the indicator function of the subinterval $\mathcal{J}_i := (\mathsf{T}_{i-1}, \mathsf{T}_i]$, where $\mathsf{T}_0 := 0, and \mathsf{T}_n = \mathsf{T}$. Then,

$$\mathcal{I}_i(s) = \begin{cases} 1, & \text{for } s \in \mathcal{J}_i, \\ 0, & \text{for elsewhere.} \end{cases}$$

(H2) Let $i \in \{1, 2, \dots, n\}$, and let $f \in C(\mathcal{J} \times \mathbb{R}, \mathbb{R})$. Assume there exists a number $\delta \in (0, 1)$ such $(\ell ogs)^{\delta} f \in C(\mathcal{J} \times \mathbb{R}, \mathbb{R})$ and there exists a constant $K > 0$ satisfying

$$(\ell ogs)^{\delta} |f(s, \varkappa_1) - f(s, \varkappa_2)| \leq K|\varkappa_1 - \varkappa_2| \ for\ any\ \varkappa_1,\ \varkappa_2, \in \mathbb{R} \ and\ s \in \mathcal{J}_i.$$

Moreover, let $E_i = C(\mathcal{J}_i, \mathbb{R})$ denote the space of continuous functions on $\mathcal{J}_i$, which forms a ßanach space under the norm

$$\|\varkappa\|_{E_i} = \sup_{s \in \mathcal{J}_i} |\varkappa(s)|,$$

where $i \in \{1, 2, ..., n\}$.

We now undertake a critical examination of the variable-order Hadamard initial and terminal value problem (ITVP) given in equation (5.1), with the objective of establishing the principal results.

The differential equation of the variable order Hadamard ITVP (5.1) can be simplified by (2.12) as follows:

$$\frac{1}{\Gamma(2 - \ltimes(s))} (s\frac{d}{ds})^2 \int_1^s (\ell og\frac{s}{t})^{1-\ltimes(s)} \frac{\varkappa(t)}{t} dt = f(s, \varkappa(s)), \quad s \in \mathcal{J}. \tag{5.2}$$

According to assumption $(H1)$, the equation(5.2) can be represented on each subinterval $\mathcal{J}_i$ as follows:

$$(s\frac{d}{ds})^2\Big(\frac{1}{\Gamma(2-\ltimes_1)}\int_1^{\mathsf{T}_1}(\ell og\frac{s}{t})^{1-\ltimes_1}\frac{\varkappa(t)}{t}dt+\cdots+$$

$$\frac{1}{\Gamma(2-\ltimes_i)}\int_{\mathsf{T}_{i-1}}^{s}(\ell og\frac{s}{t})^{1-\ltimes_i}\frac{\varkappa(t)}{t}dt\Big)=f(s,\varkappa(s)),\ for\ \ s\in\mathcal{J}_i. \tag{5.3}$$

The definition of a solution to the variable-order Hadamard initial and terminal value problem (ITVP) given in equation (5.1), which is fundamental to the present study, is provided below.

Definition 51 *We say that the variable-order Hadamard initial and terminal value problem (ITVP) given in equation (5.1) admits a solution if there exist functions $\varkappa_i, i=1,2,...,n$, such that $\varkappa_i$ satisfies equation (5.3) and the boundary conditions $\varkappa_i(1)=0=\varkappa_i(\mathsf{T}_i)$.*

Based on the information provided, the variable-order Hadamard initial and terminal value problem (ITVP) described by equation (5.1) may be reformulated as equation (5.2). Consequently, within each subinterval $\mathcal{J}_i$ where $i\in\{1,2,...,n\}$, it can be expressed as equation (5.3). For $0\le s\le \mathsf{T}_{i-1}$, we assume $\varkappa(s)\equiv 0$. Under this assumption, equation (5.3) reduces to the following form:

$$({}^{H}\digamma^{\ltimes_i}_{\mathsf{T}^{+}_{i-1}}\varkappa)(s)=f(s,\varkappa(s)),\quad s\in\mathcal{J}_i.$$

We now proceed with our investigation of the corresponding standard Hadamard initial and terminal value problem (ITVP), which is formulated as follows:

$$\begin{cases}({}^{H}\digamma^{\ltimes_i}_{\mathsf{T}^{+}_{i-1}}\varkappa)(s)=f(s,\varkappa(s)), & s\in\mathcal{J}_i\\ \varkappa(\mathsf{T}_{i-1})=0, \quad \varkappa(\mathsf{T}_i)=0.\end{cases} \tag{5.4}$$

We may establish the existence criterion for solutions to the corresponding standard Hadamard initial and terminal value problem (ITVP) with the aid of the following auxiliary proposition, denoted as Proposition 5.4.

Proposition 7 *A function $\varkappa\in E_i$ is a solution of the equivalent standard Hadamard initial and terminal value problem (ITVP) given in equation (5.4) if and only if $\varkappa$ satisfies the following condition:*

$$\varkappa(s)=-(\ell og\frac{\mathsf{T}_i}{\mathsf{T}_{i-1}})^{1-\ltimes_i}(\ell og\frac{s}{\mathsf{T}_{i-1}})^{\ltimes_i-1}{}^{H}I^{\ltimes_i}_{\mathsf{T}^{+}_{i-1}}f(\mathsf{T}_i,\varkappa(\mathsf{T}_i))+{}^{H}I^{\ltimes_i}_{\mathsf{T}^{+}_{i-1}}f(s,\varkappa(s)). \tag{5.5}$$

Proof: The solution of the analogous standard Hadamard initial and terminal value problem (ITVP) given in equation (5.4) can be represented by $\varkappa\in E_i$.

Applying the Hadamard integral operator ${}^{H}I^{\ltimes_i}_{\mathsf{T}_{i-1}^+}$ to both sides of equation (5.4), and invoking Lemma 3, we obtain:

$$\varkappa(s) = \omega_1(\ell og\frac{s}{\mathsf{T}_{i-1}})^{\ltimes_i-1} + \omega_2(\ell og\frac{s}{\mathsf{T}_{i-1}})^{\ltimes_i-2} + {}^{H}I^{\ltimes_i}_{\mathsf{T}_{i-1}^+} f(s, \varkappa(s)), \ \ s \in \mathcal{J}_i.$$

Given that $\varkappa(\mathsf{T}_{i-1}) = 0$ and under the stated hypothesis on the function f, it follows that $\omega_2 = 0$. By imposing the condition $\varkappa(s)$ satisfying $\varkappa(\mathsf{T}_i) = 0$, we obtain

$$\omega_1 = -(\ell og\frac{\mathsf{T}_i}{\mathsf{T}_{i-1}})^{1-\ltimes_i}\, {}^{H}I^{\ltimes_i}_{\mathsf{T}_{i-1}^+} f(\mathsf{T}_i, \varkappa(\mathsf{T}_i)).$$

Therefore, the solution $\varkappa(s)$ on the interval $\mathcal{J}_i$ can be expressed as

$$\varkappa(s) = -(\ell og\frac{\mathsf{T}_i}{\mathsf{T}_{i-1}})^{1-\ltimes_i}(\ell og\frac{s}{\mathsf{T}_{i-1}})^{\ltimes_i-1}\, {}^{H}I^{\ltimes_i}_{\mathsf{T}_{i-1}^+} f(\mathsf{T}_i, \varkappa(\mathsf{T}_i)) + {}^{H}I^{\ltimes_i}_{\mathsf{T}_{i-1}^+} f(s, \varkappa(s)), \ \ s \in \mathcal{J}_i.$$

On the other hand, let $\varkappa \in E_i$ represent the expression obtained in equation (5.5). Then, based on the continuity of $(\ell ogs)^{\delta} f$ and the application of Lemma 3, it can be readily deduced that $\varkappa$ is a solution to the equivalent standard Hadamard initial and terminal value problem (ITVP) given in equation (5.4).

The following result is established using the Schauder fixed point theorem.

Theorem 39 *Assume that conditions (H1) and (H2) hold. Let*

$$\frac{K\Big((\ell og\mathsf{T}_i)^{1-\delta} - (\ell og\mathsf{T}_{i-1})^{1-\delta}\Big)}{(1-\delta)\Gamma(\ltimes_i)}(\ell og\frac{\mathsf{T}_i}{\mathsf{T}_{i-1}})^{\ltimes_i-1} < \frac{1}{2}.$$

Then the variable order Hadamard initial and terminal value problem (ITVP) given in equation (5.4) admits a solution on E_i.

Proof: First, we transform 5.4, the corresponding standard Hadamard initial value problem (ITVP), into a fixed-point formulation. Define

$$W : E_i \to E_i, \qquad i \in \mathbb{N}_1^n,$$

by

$$W\varkappa(s) = -\frac{1}{\Gamma(\ltimes_i)}(\ell og\frac{\mathsf{T}_i}{\mathsf{T}_{i-1}})^{1-\ltimes_i}(\ell og\frac{s}{\mathsf{T}_{i-1}})^{\ltimes_i-1}\int_{\mathsf{T}_{i-1}}^{\mathsf{T}_i}(\ell og\frac{\mathsf{T}_i}{t})^{\ltimes_i-1}\frac{f(t,\varkappa(t))}{t}dt$$

$$+\frac{1}{\Gamma(\ltimes_i)}\int_{\mathsf{T}_{i-1}}^{s}(\ell og\frac{s}{t})^{\ltimes_i-1}\frac{f(t,\varkappa(t))}{t}dt. \tag{5.6}$$

Given the specifications of Hadamard integrals and the continuity of $(\ell ogs)^{\delta} f$, we may conclude that the operator $W : E_i \to E_i$ defined above is well-defined. Let

$$R_i \geq \frac{\frac{2f^*}{\Gamma(\ltimes_i+1)}(\ell og\frac{\mathsf{T}_i}{\mathsf{T}_{i-1}})^{\ltimes_i}}{1-\frac{2K}{(1-\delta)\Gamma(\ltimes_i)}(\ell og\frac{\mathsf{T}_i}{\mathsf{T}_{i-1}})^{\ltimes_i-1}\Big((\ell og\mathsf{T}_i)^{1-\delta} - (\ell og\mathsf{T}_{i-1})^{1-\delta}\Big)},$$

where

$$f^* = \sup_{s\in\mathcal{J}_i} |f(s,0)|.$$

We now consider the set

$$B_{R_i} = \{\varkappa \in E_i,\ \|\varkappa\|_{E_i} \leq R_i\}.$$

It is evident that B_{R_i} is closed, convex, bounded, and nonempty.

We will now examine whether W satisfies the stated conditions of the Schauder fixed point theorem. The argument will be implemented in several stages.

STEP 1: $W(B_{R_i}) \subseteq (B_{R_i})$.
For $\varkappa \in B_{R_i}$ and by assumption (H2), we obtain:

$$\begin{aligned}
|W\varkappa(s)| &\leq \frac{1}{\Gamma(\ltimes_i)}(\ell og\frac{\mathsf{T}_i}{\mathsf{T}_{i-1}})^{1-\ltimes_i}(\ell og\frac{s}{\mathsf{T}_{i-1}})^{\ltimes_i-1}\int_{\mathsf{T}_{i-1}}^{\mathsf{T}_i}(\ell og\frac{\mathsf{T}_i}{t})^{\ltimes_i-1}|f(t,\varkappa(t))|\frac{dt}{t}\\
&+ \frac{1}{\Gamma(\ltimes_i)}\int_{\mathsf{T}_{i-1}}^{s}(\ell og\frac{s}{t})^{\ltimes_i-1}|f(t,\varkappa(t))|\frac{dt}{t}\\
&\leq \frac{2}{\Gamma(\ltimes_i)}\int_{\mathsf{T}_{i-1}}^{\mathsf{T}_i}(\ell og\frac{\mathsf{T}_i}{t})^{\ltimes_i-1}|f(t,\varkappa(t))|\frac{dt}{t}\\
&\leq \frac{2}{\Gamma(\ltimes_i)}\int_{\mathsf{T}_{i-1}}^{\mathsf{T}_i}(\ell og\frac{\mathsf{T}_i}{t})^{\ltimes_i-1}|f(t,\varkappa(t))-f(t,0)|\frac{dt}{t}\\
&+ \frac{2}{\Gamma(\ltimes_i)}\int_{\mathsf{T}_{i-1}}^{\mathsf{T}_i}(\ell og\frac{\mathsf{T}_i}{t})^{\ltimes_i-1}|f(t,0)|\frac{dt}{t}\\
&\leq \frac{2}{\Gamma(\ltimes_i)}\int_{\mathsf{T}_{i-1}}^{\mathsf{T}_i}(\ell og\frac{\mathsf{T}_i}{t})^{\ltimes_i-1}(\ell og t)^{-\delta}K|\varkappa(t))|\frac{dt}{t} + \frac{2f^*}{\Gamma(\ltimes_i)}\int_{\mathsf{T}_{i-1}}^{\mathsf{T}_i}(\ell og\frac{\mathsf{T}_i}{t})^{\ltimes_i-1}\frac{dt}{t}\\
&\leq \frac{2K}{\Gamma(\ltimes_i)}(\ell og\frac{\mathsf{T}_i}{\mathsf{T}_{i-1}})^{\ltimes_i-1}\|\varkappa\|_{E_i}\int_{\mathsf{T}_{i-1}}^{\mathsf{T}_i}(\ell og t)^{-\delta}\frac{dt}{t} + \frac{2f^*}{\Gamma(\ltimes_i+1)}(\ell og\frac{\mathsf{T}_i}{\mathsf{T}_{i-1}})^{\ltimes_i}\\
&\leq \frac{2K}{(1-\delta)\Gamma(\ltimes_i)}(\ell og\frac{\mathsf{T}_i}{\mathsf{T}_{i-1}})^{\ltimes_i-1}R_i\Big((\ell og\mathsf{T}_i)^{1-\delta} - (\ell og\mathsf{T}_{i-1})^{1-\delta}\Big)\\
&+ \frac{2f^*}{\Gamma(\ltimes_i+1)}(\ell og\frac{\mathsf{T}_i}{\mathsf{T}_{i-1}})^{\ltimes_i}\\
&\leq R_i.
\end{aligned}$$

which means that $W(B_{R_i}) \subseteq B_{R_i}$.

STEP 2: W is continuous.
Let $(\varkappa_n)$ be a sequence satisfying $\varkappa_n \to \varkappa$ in E_i. For $s \in \mathcal{J}_i$, we estimate

$$|(W\varkappa_n)(s) - (W\varkappa)(s)|$$

$$\leq \frac{1}{\Gamma(\ltimes_i)}(\ell og\frac{\mathsf{T}_i}{\mathsf{T}_{i-1}})^{1-\ltimes_i}(\ell og\frac{s}{\mathsf{T}_{i-1}})^{\ltimes_i-1}\int_{\mathsf{T}_{i-1}}^{\mathsf{T}_i}(\ell og\frac{\mathsf{T}_i}{t})^{\ltimes_i-1}|f(t,\varkappa_n(t)) - f(t,\varkappa(t))|\frac{dt}{t}$$

$$+ \frac{1}{\Gamma(\ltimes_i)}\int_{\mathsf{T}_{i-1}}^{s}(\ell og\frac{s}{t})^{\ltimes_i-1}|f(t,\varkappa_n(t)) - f(t,\varkappa(t))|\frac{dt}{t}$$

$$\leq \frac{1}{\Gamma(\ltimes_i)}(\ell og\frac{\mathsf{T}_i}{\mathsf{T}_{i-1}})^{1-\ltimes_i}(\ell og\frac{\mathsf{T}_i}{\mathsf{T}_{i-1}})^{\ltimes_i-1}\int_{\mathsf{T}_{i-1}}^{\mathsf{T}_i}(\ell og\frac{\mathsf{T}_i}{t})^{\ltimes_i-1}|f(t,\varkappa_n(t)) - f(t,\varkappa(t))|\frac{dt}{t}$$

$$+ \frac{1}{\Gamma(\ltimes_i)}\int_{\mathsf{T}_{i-1}}^{\mathsf{T}_i}(\ell og\frac{\mathsf{T}_i}{t})^{\ltimes_i-1}|f(t,\varkappa_n(t)) - f(t,\varkappa(t))|\frac{dt}{t}$$

$$\leq \frac{2}{\Gamma(\ltimes_i)}\int_{\mathsf{T}_{i-1}}^{\mathsf{T}_i}(\ell og\frac{\mathsf{T}_i}{t})^{\ltimes_i-1}|f(t,\varkappa_n(t)) - f(t,\varkappa(t))|\frac{dt}{t}$$

$$\leq \frac{2}{\Gamma(\ltimes_i)}\int_{\mathsf{T}_{i-1}}^{\mathsf{T}_i}(\ell og\frac{\mathsf{T}_i}{t})^{\ltimes_i-1}(\ell ogt)^{-\delta}K|\varkappa_n(t)) - \varkappa(t)|\frac{dt}{t}$$

$$\leq \frac{2K}{\Gamma(\ltimes_i)}(\ell og\frac{\mathsf{T}_i}{\mathsf{T}_{i-1}})^{\ltimes_i-1}\|\varkappa_n - \varkappa\|_{E_i}\int_{\mathsf{T}_{i-1}}^{\mathsf{T}_i}(\ell ogt)^{-\delta}\frac{dt}{t}$$

$$\leq \frac{2K[(\ell og\mathsf{T}_i)^{1-\delta} - (\ell og\mathsf{T}_{i-1})^{1-\delta}]}{(1-\delta)\Gamma(\ltimes_i)}(\ell og\frac{\mathsf{T}_i}{\mathsf{T}_{i-1}})^{\ltimes_i-1}\|\varkappa_n - \varkappa\|_{E_i}.$$

Therefore,

$$\|(W\varkappa_n) - (W\varkappa)\|_{E_i} \to 0 \quad as \quad n \to \infty.$$

Consequently, the operator W is continuous on E_i.

STEP 3: W is compact.
The goal of this step is to demonstrate the relative compactness of $W(B_{R_i})$, which in turn implies the compactness of W. Evidently, $W(B_{R_i})$ is uniformly bounded. From Step 1, we observed that

$$W(B_{R_i}) = \{W(\varkappa) : \varkappa \in B_{R_i}\} \subset B_{R_i}.$$

Hence for every $\varkappa \in B_{R_i}$, we obtain $\|W(\varkappa)\|_{E_i} \leq R_i$, which implies the uniform boundedness of $W(B_{R_i})$. For $s_1, s_2 \in \mathcal{J}_i$ with $s_1 < s_2$, and $\varkappa \in B_{R_i}$, we estimate

$$|(W\varkappa)(s_2) - (W\varkappa)(s_1)|$$

$$\leq \frac{1}{\Gamma(\ltimes_i)}(\ell og\frac{\mathsf{T}_i}{\mathsf{T}_{i-1}})^{1-\ltimes_i}\Big((\ell og\frac{s_2}{\mathsf{T}_{i-1}})^{\ltimes_i-1} - (\ell og\frac{s_1}{\mathsf{T}_{i-1}})^{\ltimes_i-1}\Big)$$

$$\int_{\mathsf{T}_{i-1}}^{\mathsf{T}_i} (\ell og\frac{\mathsf{T}_i}{t})^{\ltimes_i-1}|f(t,\varkappa(t))|\frac{dt}{t}$$

$$+ \frac{1}{\Gamma(\ltimes_i)}\int_{\mathsf{T}_{i-1}}^{s_1}\Big((\ell og\frac{s_2}{t})^{\ltimes_i-1} - (\ell og\frac{s_1}{t})^{\ltimes_i-1}\Big)|f(t,\varkappa(t))|\frac{dt}{t}$$

$$+ \frac{1}{\Gamma(\ltimes_i)}\int_{s_1}^{s_2}(\ell og\frac{s_2}{t})^{\ltimes_i-1}|f(t,\varkappa(t))|\frac{dt}{t}$$

$$\leq \frac{1}{\Gamma(\ltimes_i)}(\ell og\frac{\mathsf{T}_i}{\mathsf{T}_{i-1}})^{1-\ltimes_i}\Big((\ell og\frac{s_2}{\mathsf{T}_{i-1}})^{\ltimes_i-1} - (\ell og\frac{s_1}{\mathsf{T}_{i-1}})^{\ltimes_i-1}\Big)$$

$$\int_{\mathsf{T}_{i-1}}^{\mathsf{T}_i} (\ell og\frac{\mathsf{T}_i}{t})^{\ltimes_i-1}|f(t,\varkappa(t)) - f(t,0)|\frac{dt}{t}$$

$$+ \frac{1}{\Gamma(\ltimes_i)}(\ell og\frac{\mathsf{T}_i}{\mathsf{T}_{i-1}})^{1-\ltimes_i}\Big((\ell og\frac{s_2}{\mathsf{T}_{i-1}})^{\ltimes_i-1} - (\ell og\frac{s_1}{\mathsf{T}_{i-1}})^{\ltimes_i-1}\Big)\int_{\mathsf{T}_{i-1}}^{\mathsf{T}_i} (\ell og\frac{\mathsf{T}_i}{t})^{\ltimes_i-1}|f(t,0)|\frac{dt}{t}$$

$$+ \frac{1}{\Gamma(\ltimes_i)}\int_{\mathsf{T}_{i-1}}^{s_1}\Big((\ell og\frac{s_2}{t})^{\ltimes_i-1} - (\ell og\frac{s_1}{t})^{\ltimes_i-1}\Big)|f(t,\varkappa(t)) - f(t,0)|\frac{dt}{t}$$

$$+ \frac{1}{\Gamma(\ltimes_i)}\int_{\mathsf{T}_{i-1}}^{s_1}\Big((\ell og\frac{s_2}{t})^{\ltimes_i-1} - (\ell og\frac{s_1}{t})^{\ltimes_i-1}\Big)|f(t,0)|\frac{dt}{t}$$

$$+ \frac{1}{\Gamma(\ltimes_i)}\int_{s_1}^{s_2}(\ell og\frac{s_2}{t})^{\ltimes_i-1}|f(t,\varkappa(t)) - f(t,0)|\frac{dt}{t} + \frac{1}{\Gamma(\ltimes_i)}\int_{s_1}^{s_2}(\ell og\frac{s_2}{t})^{\ltimes_i-1}|f(t,0)|\frac{dt}{t}$$

$$\leq \frac{1}{\Gamma(\ltimes_i)}(\ell og\frac{\mathsf{T}_i}{\mathsf{T}_{i-1}})^{1-\ltimes_i}\Big((\ell og\frac{s_2}{\mathsf{T}_{i-1}})^{\ltimes_i-1} - (\ell og\frac{s_1}{\mathsf{T}_{i-1}})^{\ltimes_i-1}\Big)$$

$$\int_{\mathsf{T}_{i-1}}^{\mathsf{T}_i} (\ell og\frac{\mathsf{T}_i}{t})^{\ltimes_i-1}(\ell ogt)^{-\delta}K|\varkappa(t)|\frac{dt}{t}$$

$$+ \frac{f^*}{\Gamma(\ltimes_i)}(\ell og\frac{\mathsf{T}_i}{\mathsf{T}_{i-1}})^{1-\ltimes_i}\Big((\ell og\frac{s_2}{\mathsf{T}_{i-1}})^{\ltimes_i-1} - (\ell og\frac{s_1}{\mathsf{T}_{i-1}})^{\ltimes_i-1}\Big)\int_{\mathsf{T}_{i-1}}^{\mathsf{T}_i} (\ell og\frac{\mathsf{T}_i}{t})^{\ltimes_i-1}\frac{dt}{t}$$

$$+ \frac{1}{\Gamma(\ltimes_i)}\int_{\mathsf{T}_{i-1}}^{s_1}\Big((\ell og\frac{s_2}{t})^{\ltimes_i-1} - (\ell og\frac{s_1}{t})^{\ltimes_i-1}\Big)(\ell ogt)^{-\delta}K|\varkappa(t)|\frac{dt}{t}$$

$$+ \frac{f^*}{\Gamma(\ltimes_i)}\int_{\mathsf{T}_{i-1}}^{s_1}\Big((\ell og\frac{s_2}{t})^{\ltimes_i-1} - (\ell og\frac{s_1}{t})^{\ltimes_i-1}\Big)\frac{dt}{t}$$

$$+ \quad \frac{1}{\Gamma(\ltimes_i)}\int_{s_1}^{s_2}(\ell og\frac{s_2}{t})^{\ltimes_i-1}(\ell og t)^{-\delta}K|\varkappa(t)|\frac{dt}{t}$$

$$+ \quad \frac{f^*}{\Gamma(\ltimes_i)}\int_{s_1}^{s_2}(\ell og\frac{s_2}{t})^{\ltimes_i-1}\frac{dt}{t}$$

$$\leq \quad \frac{K[(\ell og\mathsf{T}_i)^{1-\delta}-(\ell og\mathsf{T}_{i-1})^{1-\delta}]\|\varkappa\|_{E_i}}{(1-\delta)\Gamma(\ltimes_i)}\Big((\ell og\frac{s_2}{\mathsf{T}_{i-1}})^{\ltimes_i-1}-(\ell og\frac{s_1}{\mathsf{T}_{i-1}})^{\ltimes_i-1}\Big)$$

$$+ \quad \frac{f^*}{\Gamma(\ltimes_i+1)}(\ell og\frac{\mathsf{T}_i}{\mathsf{T}_{i-1}})\Big((\ell og\frac{s_2}{\mathsf{T}_{i-1}})^{\ltimes_i-1}-(\ell og\frac{s_1}{\mathsf{T}_{i-1}})^{\ltimes_i-1}\Big)$$

$$+ \quad \frac{K[(\ell og s_1)^{1-\delta}-(\ell og\mathsf{T}_{i-1})^{1-\delta}]\|\varkappa\|_{E_i}}{(1-\delta)\Gamma(\ltimes_i)}(\ell og\frac{s_2}{s_1})^{\ltimes_i-1}$$

$$+ \quad \frac{f^*}{\Gamma(\ltimes_i+1)}[(\ell og\frac{s_2}{\mathsf{T}_{i-1}})^{\ltimes_i}-(\ell og\frac{s_2}{s_1})^{\ltimes_i}-(\ell og\frac{s_1}{\mathsf{T}_{i-1}})^{\ltimes_i}]$$

$$+ \quad \frac{K[(\ell og s_2)^{1-\delta}-(\ell og s_1)^{1-\delta}]\|\varkappa\|_{E_i}}{(1-\delta)\Gamma(\ltimes_i)}(\ell og\frac{s_2}{s_1})^{\ltimes_i-1}+\frac{f^*}{\Gamma(\ltimes_i+1)}(\ell og\frac{s_2}{s_1})^{\ltimes_i}$$

$$\leq \quad \Big(\frac{K[(\ell og\mathsf{T}_i)^{1-\delta}-(\ell og\mathsf{T}_{i-1})^{1-\delta}]\|\varkappa\|_{E_i}}{(1-\delta)\Gamma(\ltimes_i)}+\frac{f^*}{\Gamma(\ltimes_i+1)}(\ell og\frac{\mathsf{T}_i}{\mathsf{T}_{i-1}})\Big)$$

$$\Big((\ell og\frac{s_2}{\mathsf{T}_{i-1}})^{\ltimes_i-1}-(\ell og\frac{s_1}{\mathsf{T}_{i-1}})^{\ltimes_i-1}\Big)+\frac{f^*}{\Gamma(\ltimes_i+1)}[(\ell og\frac{s_2}{\mathsf{T}_{i-1}})^{\ltimes_i}-(\ell og\frac{s_1}{\mathsf{T}_{i-1}})^{\ltimes_i}]$$

$$+ \quad \frac{K[(\ell og s_2)^{1-\delta}-(\ell og\mathsf{T}_{i-1})^{1-\delta}]\|\varkappa\|_{E_i}}{(1-\delta)\Gamma(\ltimes_i)}(\ell og\frac{s_2}{s_1})^{\ltimes_i-1}.$$

Therefore, $|(W\varkappa)(s_2)-(W\varkappa)(s_1)|\to 0$ as $|s_2-s_1|\to 0$ which indicates that $W(B_{R_i})$ is equicontinuous.

Taking into account the first three phases and invoking the Arzela-Ascoli theorem, we conclude that W is completely continuous.

By the Schauder fixed point theorem, the conventional Hadamard initial value problem (ITVP) equivalent to equation (5.4) possesses at least one solution $\widetilde{\varkappa}_i$ in B_{R_i}.

5.1.2 Properties of Uniqueness

Theorem 40 *Consider hypotheses* $(H1)$ *and* $(H2)$. *If*

$$\frac{2K[(\ell og\mathsf{T}_i)^{1-\delta}-(\ell og\mathsf{T}_{i-1})^{1-\delta}]}{(1-\delta)\Gamma(\ltimes_i)}(\ell og\frac{\mathsf{T}_i}{\mathsf{T}_{i-1}})^{\ltimes_i-1}<1, \qquad (5.7)$$

then the variable order Hadamard ITVP (5.4) involves a solution in E_i uniquely.

Proof: We will utilize the ßanach fixed-point theorem to prove that W, as defined in equation (5.6), possesses a unique fixed point. For $\varkappa(s), \varkappa^*(s) \in E_i$, we may write:

$$|(W\varkappa)(s) - (W\varkappa^*)(s)|$$

$$\begin{aligned}
&\leq \frac{1}{\Gamma(\ltimes_i)}(\ell og\frac{\mathsf{T}_i}{\mathsf{T}_{i-1}})^{1-\ltimes_i}(\ell og\frac{s}{\mathsf{T}_{i-1}})^{\ltimes_i-1}\int_{\mathsf{T}_{i-1}}^{\mathsf{T}_i}(\ell og\frac{\mathsf{T}_i}{t})^{\ltimes_i-1}|f(t,\varkappa(t)) - f(t,\varkappa^*(t))|\frac{dt}{t}\\
&+ \frac{1}{\Gamma(\ltimes_i)}\int_{\mathsf{T}_{i-1}}^{s}(\ell og\frac{s}{t})^{\ltimes_i-1}|f(t,\varkappa(t)) - f(t,\varkappa^*(t))|\frac{dt}{t}\\
&\leq \frac{1}{\Gamma(\ltimes_i)}(\ell og\frac{\mathsf{T}_i}{\mathsf{T}_{i-1}})^{1-\ltimes_i}(\ell og\frac{\mathsf{T}_i}{\mathsf{T}_{i-1}})^{\ltimes_i-1}\int_{\mathsf{T}_{i-1}}^{\mathsf{T}_i}(\ell og\frac{\mathsf{T}_i}{t})^{\ltimes_i-1}|f(t,\varkappa(t)) - f(t,\varkappa^*(t))|\frac{dt}{t}\\
&+ \frac{1}{\Gamma(\ltimes_i)}\int_{\mathsf{T}_{i-1}}^{\mathsf{T}_i}(\ell og\frac{\mathsf{T}_i}{t})^{\ltimes_i-1}|f(t,\varkappa(t)) - f(t,\varkappa^*(t))|\frac{dt}{t}\\
&\leq \frac{2}{\Gamma(\ltimes_i)}\int_{\mathsf{T}_{i-1}}^{\mathsf{T}_i}(\ell og\frac{\mathsf{T}_i}{t})^{\ltimes_i-1}|f(t,\varkappa(t)) - f(t,\varkappa^*(t))|\frac{dt}{t}\\
&\leq \frac{2}{\Gamma(\ltimes_i)}\int_{\mathsf{T}_{i-1}}^{\mathsf{T}_i}(\ell og\frac{\mathsf{T}_i}{t})^{\ltimes_i-1}(\ell og t)^{-\delta}K|\varkappa(t)) - \varkappa^*(t)|\frac{dt}{t}\\
&\leq \frac{2K}{\Gamma(\ltimes_i)}(\ell og\frac{\mathsf{T}_i}{\mathsf{T}_{i-1}})^{\ltimes_i-1}\|\varkappa - \varkappa^*\|_{E_i}\int_{\mathsf{T}_{i-1}}^{\mathsf{T}_i}(\ell og t)^{-\delta}\frac{dt}{t}\\
&\leq \frac{2K[(\ell og\mathsf{T}_i)^{1-\delta} - (\ell og\mathsf{T}_{i-1})^{1-\delta}]}{(1-\delta)\Gamma(\ltimes_i)}(\ell og\frac{\mathsf{T}_i}{\mathsf{T}_{i-1}})^{\ltimes_i-1}\|\varkappa - \varkappa^*\|_{E_i}.
\end{aligned}$$

Consequently, by equation (5.7), the operator W is a contraction. Hence, W admits a unique fixed-point $\widetilde{\varkappa}_i \in E_i$, which coincides with the unique solution of the equivalent standard Hadamard initial value problem (ITVP) given in equation (5.4).

We now proceed with our investigation of the variable-order Hadamard fractional ITVP (5.1). The following premise is presented:

(H3) Suppose $f \in C(J \times \mathbb{R}, \mathbb{R})$ and there exists a number $0 \leq \delta \leq 1$ such that $(\ell ogs)^\delta f \in C(J \times \mathbb{R}, \mathbb{R})$, and there exist constants K and L such that $K, L > 0$ such that

$(\ell ogs)^\delta|f(s,\varkappa_1) - f(s,\varkappa_2)| \leq K|\varkappa_1 - \varkappa_2|$, for any $\varkappa_1,\ \varkappa_2, \in \mathbb{R}$ and $s \in J$.

Theorem 41 *Let us assume that the conditions (H1) and (H3), along with inequality (5.7), are satisfied for every $i \in \{1, 2, \ldots, n\}$. Then the initial value problem (ITVP) (5.1) admits a unique solution in $C(J, \mathbb{R})$.*

As stated in Theorem 40, the standard Hadamard initial value problem (ITVP) (5.4) admits a unique solution $\widetilde{\varkappa}_i \in E_i$ for every $i \in \{1, 2, \ldots, n\}$. We define the function $\varkappa_i$ as

$$\varkappa_i(s) = \begin{cases} 0, & s \in [1, \mathsf{T}_{i-1}], \\ \widetilde{\varkappa}_i, & s \in \mathcal{J}_i. \end{cases} \tag{5.8}$$

It is known that the function $\varkappa_i \in C([1, \mathsf{T}_i], \mathbb{R})$ given by (5.8) satisfies the corresponding equation

$$(s\frac{d}{ds})^2\Big(\frac{1}{\Gamma(2-\ltimes_1)}\int_1^{\mathsf{T}_1}(\ell og\frac{s}{t})^{1-\ltimes_1}\frac{\varkappa_i(t)}{t}dt + \cdots + \frac{1}{\Gamma(2-\ltimes_i)}\int_{\mathsf{T}_{i-1}}^{s}(\ell og\frac{s}{t})^{1-\ltimes_i}\frac{\varkappa_i(t)}{t}dt\Big) = f(s, \varkappa_i(s)),$$

For $s \in \mathcal{J}_i$, this implies that $\varkappa_i$ is a distinct solution of equation (5.3), satisfying the conditions $\varkappa_i(1) = 0$, $\varkappa_i(\mathsf{T}_i) = \widetilde{\varkappa}_i(\mathsf{T}_i) = 0$.
Then, define the function $\varkappa(s)$ as

$$\varkappa(s) = \begin{cases} \varkappa_1(s), & s \in \mathcal{J}_1, \\ \varkappa_2(s) = \begin{cases} 0, & s \in \mathcal{J}_1, \\ \widetilde{\varkappa}_2, & s \in \mathcal{J}_2 \end{cases} \\ \cdot \\ \cdot \\ \cdot \\ \varkappa_i(s) = \begin{cases} 0, & s \in [1, \mathsf{T}_{i-1}], \\ \widetilde{\varkappa}_i, & s \in \mathcal{J}_i, \end{cases} \end{cases} .$$

This function $\varkappa(s)$ constitutes a unique solution of the variable-order Hadamard initial value problem (ITVP) (5.1), and our argument is thus complete.

5.1.3 Criteria of Stability

In the sequel, we aim to investigate the Ulam-Hyers-$\mathfrak{R}$assias stability of solutions to the postulated variable-order problem, since stability constitutes one of the principal qualitative criteria for solutions to specified initial value problems (ITVPs), particularly the Hadamard-type ITVP (5.1).

Theorem 42 *Consider the hypotheses (H1), (H3), and the inequality (5.7). Assume the following:*

(H4) *There exists a function $\phi \in C(\mathcal{J}_i, \mathbb{R}_+)$ which is increasing, and a constant $\lambda_\phi > 0$ such that for all $s \in \mathcal{J}_i$,*

$$^H I^{\ltimes_i}_{\mathsf{T}_{i-1}^+}\phi(s) \le \lambda_{\phi(s)}\phi(s).$$

Then, the variable-order Hadamard initial value problem (ITVP) (5.1) is Ulam-Hyers-$\Re$assias stable with respect to ϕ.

Proof: Assume that for every $\forall\,\epsilon > 0$, the function $\mathrm{r} \in C(\mathcal{J}_i, \mathbb{R})$ satisfies the inequality

$$|{}^{H}F^{\ltimes_i}_{\mathsf{T}_{i-1}^{+}}\mathrm{r}(s) - f(s, \mathrm{r}(s))| \leq \epsilon\phi(s), s \in \mathcal{J}_i. \tag{5.9}$$

For any $i \in \{1, 2, ..., n\}$, define the functions $r_1(s) \equiv r(s), s \in [1, \mathsf{T}_1]$ and for $i = 2, 3, ..., n$:

$$r_i(s) = \begin{cases} 0, & s \in [0, \mathsf{T}_{i-1}], \\ r(s), & s \in \mathcal{J}_i, \end{cases}$$

Applying the fractional integral operator ${}^{H}I^{\ltimes_i}_{\mathsf{T}_{i-1}^{+}}$ to both sides of inequality (5.9), we obtain for all $s \in \mathcal{J}_i$:

$$|\mathrm{r_i}(s) + \frac{1}{\Gamma(\ltimes_i)}(\ell og\frac{\mathsf{T}_i}{\mathsf{T}_{i-1}})^{1-\ltimes_i}(\ell og\frac{s}{\mathsf{T}_{i-1}})^{\ltimes_i-1}\int_{\mathsf{T}_{i-1}}^{\mathsf{T}_i}(\ell og\frac{\mathsf{T}_i}{t})^{\ltimes_i-1}\frac{f(t,\mathrm{r_i}(t))}{t}dt$$

$$-\frac{1}{\Gamma(\ltimes_i)}\int_{\mathsf{T}_{i-1}}^{s}(\ell og\frac{s}{t})^{\ltimes_i-1}\frac{f(t,\mathrm{r_i}(t))}{t}dt|$$

$$\leq \epsilon\frac{1}{\Gamma(\ltimes_i)}\int_{\mathsf{T}_{i-1}}^{s}\frac{1}{t}(\ell og\frac{s}{t})^{\ltimes_i-1}\phi(t)dt$$

$$\leq \epsilon\lambda_{\phi(s)}\phi(s).$$

According to the argument above, the initial value problem (ITVP) 5.1 admits a solution $\varkappa$ defined by $\varkappa(s) = \varkappa_i(s)$ for $s \in \mathcal{J}_i$, $i = 1, 2, ..., n$, where

$$\varkappa_i(s) = \begin{cases} 0, & s \in [1, \mathsf{T}_{i-1}], \\ \widetilde{\varkappa}_i, & s \in \mathcal{J}_i, \end{cases} \tag{5.10}$$

and $\widetilde{\varkappa}_i \in E_i$ is a solution of equation (5.4). Based on Proposition (7), the integral equation

$$\widetilde{\varkappa}_i(s) = -\frac{1}{\Gamma(\ltimes_i)}(\ell og\frac{\mathsf{T}_i}{\mathsf{T}_{i-1}})^{1-\ltimes_i}(\ell og\frac{s}{\mathsf{T}_{i-1}})^{\ltimes_i-1}\int_{\mathsf{T}_{i-1}}^{\mathsf{T}_i}(\ell og\frac{\mathsf{T}_i}{t})^{\ltimes_i-1}\frac{f(t,\widetilde{\varkappa}_i(t))}{t}dt$$

$$+\frac{1}{\Gamma(\ltimes_i)}\int_{\mathsf{T}_{i-1}}^{s}(\ell og\frac{s}{t})^{\ltimes_i-1}\frac{f(t,\widetilde{\varkappa}_i(t))}{t}dt.$$

holds.

Then, we have, for each $s \in \mathcal{J}_i$:

$$
\begin{aligned}
& |\mathrm{r}(s) - \varkappa(s)| = |\mathrm{r}(s) - \varkappa_{\mathrm{i}}(s)| = |\mathrm{r}_{\mathrm{i}}(s) - \widetilde{\varkappa}_{\mathrm{i}}(s)| \\
= \ & |\mathrm{r}_{\mathrm{i}}(s) + \frac{1}{\Gamma(\ltimes_i)}(\ell og\frac{\mathsf{T}_i}{\mathsf{T}_{i-1}})^{1-\ltimes_i}(\ell og\frac{s}{\mathsf{T}_{i-1}})^{\ltimes_i - 1}\int_{\mathsf{T}_{i-1}}^{\mathsf{T}_i}(\ell og\frac{\mathsf{T}_i}{t})^{\ltimes_i - 1}\frac{f(t,\widetilde{\varkappa}_{\mathrm{i}}(t))}{t}dt \\
- \ & \frac{1}{\Gamma(\ltimes_i)}\int_{\mathsf{T}_{i-1}}^{s}(\ell og\frac{s}{t})^{\ltimes_i - 1}\frac{f(t,\widetilde{\varkappa}_{\mathrm{i}}(t))}{t}dt| \\
\leq \ & |\mathrm{r}_{\mathrm{i}}(s) + \frac{1}{\Gamma(\ltimes_i)}(\ell og\frac{\mathsf{T}_i}{\mathsf{T}_{i-1}})^{1-\ltimes_i}(\ell og\frac{s}{\mathsf{T}_{i-1}})^{\ltimes_i - 1}\int_{\mathsf{T}_{i-1}}^{\mathsf{T}_i}(\ell og\frac{\mathsf{T}_i}{t})^{\ltimes_i - 1}\frac{f(t,\mathrm{r}_{\mathrm{i}}(t))}{t}dt \\
- \ & \frac{1}{\Gamma(\ltimes_i)}\int_{\mathsf{T}_{i-1}}^{s}(\ell og\frac{s}{t})^{\ltimes_i - 1}\frac{f(t,\mathrm{r}_{\mathrm{i}}(t))}{t}dt| \\
+ \ & \frac{1}{\Gamma(\ltimes_i)}(\ell og\frac{\mathsf{T}_i}{\mathsf{T}_{i-1}})^{1-\ltimes_i}(\ell og\frac{s}{\mathsf{T}_{i-1}})^{\ltimes_i - 1}\int_{\mathsf{T}_{i-1}}^{\mathsf{T}_i}(\ell og\frac{\mathsf{T}_i}{t})^{\ltimes_i - 1}\frac{|f(t,\mathrm{r}_{\mathrm{i}}(t)) - f(t,\widetilde{\varkappa}_{\mathrm{i}}(t))|}{t}dt \\
+ \ & \frac{1}{\Gamma(\ltimes_i)}\int_{\mathsf{T}_{i-1}}^{s}(\ell og\frac{s}{t})^{\ltimes_i - 1}\frac{|f(t,\mathrm{r}_{\mathrm{i}}(t)) - f(t,\widetilde{\varkappa}_{\mathrm{i}}(t))|}{t}dt \\
\leq \ & \lambda_{\phi(s)}\epsilon\phi(s) + \frac{1}{\Gamma(\ltimes_i)}(\ell og\frac{\mathsf{T}_i}{\mathsf{T}_{i-1}})^{1-\ltimes_i}(\ell og\frac{s}{\mathsf{T}_{i-1}})^{\ltimes_i - 1}\int_{\mathsf{T}_{i-1}}^{\mathsf{T}_i}(\ell og\frac{\mathsf{T}_i}{t})^{\ltimes_i - 1}(\ell ogt)^{-\delta}\frac{K|\mathrm{r}_{\mathrm{i}}(t) - \widetilde{\varkappa}_{\mathrm{i}}(t)|}{t}dt \\
+ \ & \frac{1}{\Gamma(\ltimes_i)}\int_{\mathsf{T}_{i-1}}^{s}(\ell og\frac{s}{t})^{\ltimes_i - 1}(\ell ogt)^{-\delta}\frac{K|\mathrm{r}_{\mathrm{i}}(t) - \widetilde{\varkappa}_{\mathrm{i}}(t)|}{t}dt \\
\leq \ & \lambda_{\phi(s)}\epsilon\phi(s) + \frac{1}{\Gamma(\ltimes_i)}(\ell og\frac{s}{\mathsf{T}_{i-1}})^{\ltimes_i - 1}\int_{\mathsf{T}_{i-1}}^{\mathsf{T}_i}(\ell ogt)^{-\delta}\frac{K|\mathrm{r}_{\mathrm{i}}(t) - \widetilde{\varkappa}_{\mathrm{i}}(t)|}{t}dt \\
+ \ & \frac{1}{\Gamma(\ltimes_i)}(\ell og\frac{s}{\mathsf{T}_{i-1}})^{\ltimes_i - 1}\int_{\mathsf{T}_{i-1}}^{s}(\ell ogt)^{-\delta}\frac{K|\mathrm{r}_{\mathrm{i}}(t) - \widetilde{\varkappa}_{\mathrm{i}}(t)|}{t}dt \\
\leq \ & \lambda_{\phi(s)}\epsilon\phi(s) + \frac{(\ell og\mathsf{T}_i)^{1-\delta} - (\ell og\mathsf{T}_{i-1})^{1-\delta}}{(1-\delta)\Gamma(\ltimes_i)}(\ell og\frac{s}{\mathsf{T}_{i-1}})^{\ltimes_i - 1}K\|\mathrm{r}_{\mathrm{i}} - \widetilde{\varkappa}_{\mathrm{i}}\|_{E_i} \\
+ \ & \frac{(\ell ogs)^{1-\delta} - (\ell og\mathsf{T}_{i-1})^{1-\delta}}{(1-\delta)\Gamma(\ltimes_i)}(\ell og\frac{s}{\mathsf{T}_{i-1}})^{\ltimes_i - 1}K\|\mathrm{r}_{\mathrm{i}} - \widetilde{\varkappa}_{\mathrm{i}}\|_{E_i} \\
\leq \ & \lambda_{\phi(s)}\epsilon\phi(s) + \frac{2K[(\ell og\mathsf{T}_i)^{1-\delta} - (\ell og\mathsf{T}_{i-1})^{1-\delta}]}{(1-\delta)\Gamma(\ltimes_i)}(\ell og\frac{\mathsf{T}_i}{\mathsf{T}_{i-1}})^{\ltimes_i - 1}\|\mathrm{r}_{\mathrm{i}} - \widetilde{\varkappa}_{\mathrm{i}}\|_{E_i} \\
\leq \ & \lambda_{\phi(s)}\epsilon\phi(s) + \mu\|\mathrm{r} - \varkappa\|
\end{aligned}
$$

where

$$\mu = \max_{i=1,2,\ldots,n} \frac{2K[(\ell og \mathsf{T}_i)^{1-\delta} - (\ell og \mathsf{T}_{i-1})^{1-\delta}]}{(1-\delta)\Gamma(\ltimes_i)} (\ell og \frac{\mathsf{T}_i}{\mathsf{T}_{i-1}})^{\ltimes_i - 1}.$$

Then, in the context of nonlinear fractional differential equations of variable order, define

$$\|\mathrm{r} - \varkappa\|(1-\mu) \leq \lambda_{\phi(s)} \epsilon \phi(s).$$

It follows, for each $s \in \mathcal{J}$, that

$$|\mathrm{r}(s) - \varkappa(s)| \leq \|\mathrm{r} - \varkappa\| \leq \frac{\lambda_{\phi(s)}}{(1-\mu)} \epsilon \phi(s) := c_f \epsilon \phi(s).$$

Then, with respect to the kernel ϕ, the variable-order Hadamard-type initial value problem (ITVP) defined in (5.1) is Ulam-Hyers-$\mathfrak{R}$assias stable.

5.1.4 Illustrative Approach

Example 3 *In accordance with Section 5.1, we formulate the variable-order Hadamard initial terminal value problem (ITVP) in the following structure:*

$$\begin{cases} {}^{H}F_{1^+}^{\ltimes(s)} \varkappa(s) = \dfrac{7}{5\sqrt{\pi}} (\ell ogs)^{\ltimes(s)} + \dfrac{(\ell ogs)^{-\frac{1}{3}}}{s+1} \varkappa(s), \\ \varkappa(1) = 0, \;\; \varkappa(e) = 0, \end{cases} \tag{5.11}$$

for $s \in \mathcal{J} := [1, e]$*, where*

$$f(s, \varkappa) = \frac{7}{5\sqrt{\pi}} (\ell ogs)^{\ltimes(s)} + \frac{(\ell ogs)^{-\frac{1}{3}}}{s+1} \varkappa(s), (s, \varkappa) \in [1, e] \times [0, +\infty),$$

and

$$\ltimes(s) = \begin{cases} 1.3, & s \in \mathcal{J}_1 := [1, 2], \\ 1.7, & s \in \mathcal{J}_2 := (2, e]. \end{cases} \tag{5.12}$$

The graphs of the function f*, corresponding to two distinct values of the variable order* $\ltimes(s)$ *over the subintervals* $\mathcal{J}_1$ *and* $\mathcal{J}_2$*, are presented in Figures 5.1 and 5.2. Accordingly, we obtain*

$$\begin{aligned} (\ell ogs)^{\frac{1}{3}} |f(s, \varkappa) - f(s, \varkappa^*)| &= \left| \frac{7}{5\sqrt{\pi}} (\ell ogs)^{\ltimes(s)+\frac{1}{3}} + \frac{1}{s+1} \varkappa(s) - \frac{7}{5\sqrt{\pi}} (\ell ogs)^{\ltimes(s)+\frac{1}{3}} - \frac{1}{s+1} \varkappa^*(s) \right| \\ &\leq \frac{1}{s+1} |\varkappa(s) - \varkappa^*(s)| \\ &\leq \frac{1}{2} |\varkappa(s) - \varkappa^*(s)|. \end{aligned}$$

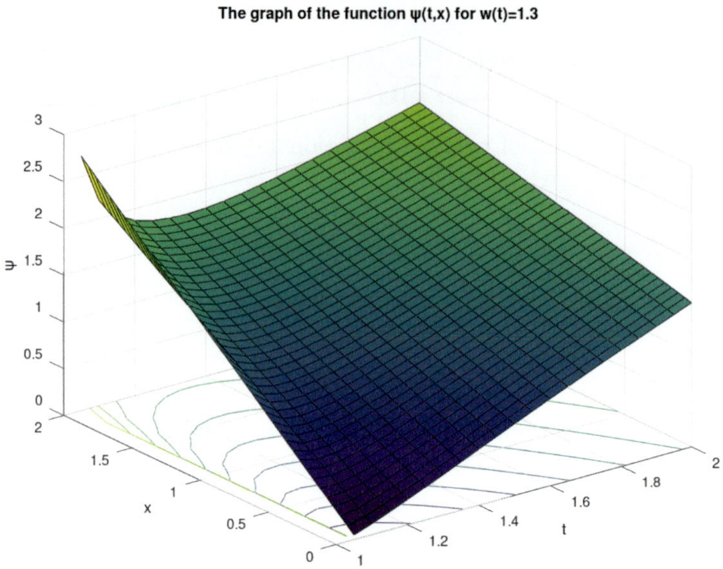

FIGURE 5.1: Graph of the function ψ corresponding to $w(t) = 1.3$ over the interval $\mathbb{I}_1 = [1, 2]$.

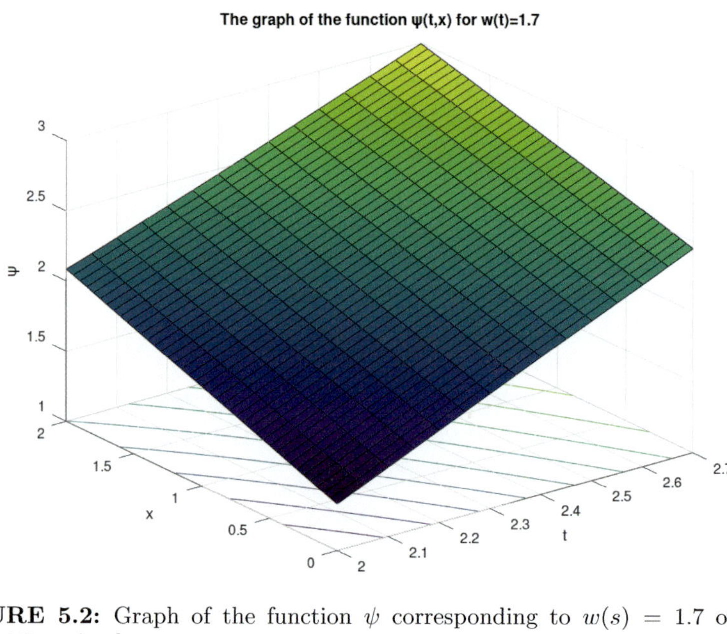

FIGURE 5.2: Graph of the function ψ corresponding to $w(s) = 1.7$ over the interval $\mathbb{I}_1 = (2, e]$.

Thus the hypothesis (H3) is validated with $\delta = \dfrac{1}{3}$ *and* $K = \dfrac{1}{2}$.
By equation (5.12)*, the differential equation of the variable order Hadamard ITVP (5.11) is decomposed into two distinct differential equations as follows:*

$$\begin{cases} {}^{H}\mathcal{F}_{1^+}^{1.3}\varkappa(s) = \dfrac{7}{5\sqrt{\pi}}(\ell ogs)^{1.3} + \dfrac{(\ell ogs)^{-\frac{1}{3}}}{s+1}\varkappa(s), & s \in \mathcal{J}_1, \\ {}^{H}\mathcal{F}_{1^+}^{1.7}\varkappa(s) = \dfrac{7}{5\sqrt{\pi}}(\ell ogs)^{1.7} + \dfrac{(\ell ogs)^{-\frac{1}{3}}}{s+1}\varkappa(s), & s \in \mathcal{J}_2. \end{cases}$$

For $s \in \mathcal{J}_1$*, the variable-order Hadamard ITVP (5.11) corresponds to the equivalent standard Hadamard ITVP.*

$$\begin{cases} {}^{H}\mathcal{F}_{1^+}^{1.3}\varkappa(s) = \dfrac{7}{5\sqrt{\pi}}(\ell ogs)^{1.3} + \dfrac{(\ell ogs)^{-\frac{1}{3}}}{s+1}\varkappa(s), \quad s \in \mathcal{J}_1, \\ \varkappa(1) = 0, \ \varkappa(2) = 0. \end{cases} \tag{5.13}$$

Clearly one can verify that condition (5.7) is satisfied. Indeed,

$$\frac{2K[(\ell og\mathsf{T}_1)^{1-\delta} - (\ell og\mathsf{T}_0)^{1-\delta}]}{(1-\delta)\Gamma(\ltimes_1)}(\ell og\frac{\mathsf{T}_1}{\mathsf{T}_0})^{\ltimes_1 - 1} = \frac{(\ell og2)^{\frac{2}{3}}}{(\frac{2}{3})\Gamma(1.3)}(\ell og2)^{0.3} \simeq 0.5279 < 1.$$

Let $\phi(s) = (\ell ogs)^{\frac{1}{2}}$*. Hence*

$$\begin{aligned} {}^{H}I_{1^+}^{\ltimes_1}\phi(s) &= \frac{1}{\Gamma(1.3)}\int_1^s (\ell og\frac{s}{t})^{1.3-1}\frac{(\ell ogt)^{\frac{1}{2}}}{t}dt \\ &\leq \frac{1}{\Gamma(1.3)}\int_1^s (\ell og\frac{s}{t})^{0.3}\frac{1}{t}dt \\ &\leq \frac{0.75}{\Gamma(2.3)}(\ell ogs)^{\frac{1}{2}} := \lambda_{\phi(s)}\phi(s). \end{aligned}$$

Therefore the condition (H4) is satisfied with $\phi(s) = (\ell ogs)^{\frac{1}{2}}$ *and* $\lambda_{\phi(s)} = \dfrac{0.75}{\Gamma(2.3)}$.

By Theorem 40, the equivalent standard Hadamard ITVP (5.13) admits a unique solution $\varkappa_1 \in E_1$.

On the other side, for $s \in \mathcal{J}_2$, the variable order Hadamard ITVP (5.11) can be reformulated as follows:

$$\begin{cases} {}^{H}\digamma_{1^+}^{1.7}\varkappa(s) = \dfrac{7}{5\sqrt{\pi}}(\ell ogs)^{1.7} + \dfrac{(\ell ogs)^{-\frac{1}{3}}}{s+1}\varkappa(s), \quad s \in \mathcal{J}_2 \\ \varkappa(2) = 0, \quad \varkappa(e) = 0. \end{cases} \tag{5.14}$$

Evidently,

$$\frac{2K[(\ell og\mathsf{T}_2)^{1-\delta} - (\ell og\mathsf{T}_1)^{1-\delta}]}{(1-\delta)\Gamma(\ltimes_2)}(\ell og\frac{\mathsf{T}_2}{\mathsf{T}_1})^{\ltimes_2 - 1} = \frac{1-(\ell og2)^{\frac{2}{3}}}{(\frac{2}{3})\Gamma(1.7)}(\ell og\frac{e}{2})^{0.7} \simeq 0.0742 < 1.$$

Thus, condition (5.7) is satisfied, and

$$\begin{aligned} {}^{H}I_{2^+}^{\ltimes_2}\phi(s) &= \frac{1}{\Gamma(1.7)}\int_2^s (\ell og\frac{s}{t})^{1.7-1}\frac{(\ell ogt)^{\frac{1}{2}}}{t}dt \\ &\leq \frac{1}{\Gamma(1.7)}\int_2^s (\ell og\frac{s}{t})^{0.7}\frac{1}{t}dt \\ &\leq \frac{1}{\Gamma(2.7)}(\ell ogs)^{\frac{1}{2}} := \lambda_{\phi(s)}\phi(s). \end{aligned}$$

This confirms that condition (H4) is satisfied with $\phi(s) = (\ell ogs)^{\frac{1}{2}}$ and $\lambda_{\phi(s)} = \dfrac{1}{\Gamma(2.7)}$.

By Theorem 40, the equivalent standard Hadamard ITVP (5.14) admits a unique solution $\varkappa_2 \in E_2$.

Then, by Theorem 41, the variable order Hadamard ITVP (5.11) admits a unique solution of the form

$$\varkappa(s) = \begin{cases} \widetilde{\varkappa}_1(s), \quad s \in \mathcal{J}_1, \\ \varkappa_2(s) = \begin{cases} 0, \quad s \in \mathcal{J}_1, \\ \widetilde{\varkappa}_2(s), \quad s \in \mathcal{J}_2, \end{cases} \end{cases}$$

where

$$\varkappa_2(s) = \begin{cases} 0, \quad s \in \mathcal{J}_1, \\ \widetilde{\varkappa}_2(s), \quad s \in \mathcal{J}_2. \end{cases}$$

On the other hand, by Theorem 42, the variable-order Hadamard ITVP (5.11) is Ulam-Hyers-$\mathfrak{R}$assias stable.

5.2 Darbo Fixed Point Criterion for Solutions of a Nonlinear Variable-Order Hadamard Problem and Ulam-Hyers-ℜassias Stability

This section investigates the existence of solutions to the nonlinear variable-order Hadamard fractional ınitial and terminal υalue problem:

$$\begin{cases} {}^{H}F_{1^+}^{\ltimes(s)} \varkappa(s) + f(s, \varkappa(s)) = 0, & s \in \mathcal{J} := [1, \mathsf{T}], \\ \varkappa(1) = \varkappa(\mathsf{T}) = 0, \end{cases} \tag{5.15}$$

where T satisfies $1 < \mathsf{T} < +\infty$, $\ltimes(s)$ is a variable-order function with $1 < \ltimes(s) \leq 2$, $f : \mathcal{J} \times X \to X$ is a continuous mapping (with X being a ßanach space), and ${}^{H}F_{1^+}^{\ltimes(s)}$ denotes the Hadamard fractional derivative of variable order $\ltimes(s)$.

In this chapter, we investigate—for the first time—a fractional ınitial and terminal υalue problem in the Hadamard framework with variable order. By integrating existing concepts related to Kuratowski's measure of noncompactness with Darbo's fixed point criterion, we establish the conditions required for the existence of solutions to the system within specified generalized subintervals. To reformulate the variable-order Hadamard fractional ınitial and terminal υalue problem presented in equation (5.15) into a standard Hadamard fractional framework, we employ piecewise constant approximations for the variable-order function. To ensure the coherence and completeness of our results, we further examine the behavior of solutions under an additional criterion—namely, Ulam-Hyers-ℜassias stability—and support our findings with a numerical example.

5.2.1 Investigation of the Existence of Solutions

Let's state the underlying presumptions:

(H1) Let $n \in \mathbb{N}$ be a fixed integer, and consider a finite sequence of points $\{\mathsf{T}_k\}_{k=0}^{n}$ such that $1 = \mathsf{T}_0 < \mathsf{T}_k < \mathsf{T}_n = \mathsf{T}$ for $k = 1, \ldots, n-1$. For each $k = 1, 2, \ldots, n$, define the subintervals. Then, the collection $\mathcal{P} := \bigcup_{k=1}^{n} \mathcal{J}_k$ forms a partition of the interval $\mathcal{J} := [1, \mathsf{T}]$. Let $\ltimes(s) : \mathcal{J} \to (1, 2]$ be a piecewise constant function with respect to the partition $\mathcal{P}$, defined as follows:

$$\ltimes(s) = \sum_{i=1}^{n} \ltimes_i I_i(s) = \begin{cases} \ltimes_1, & if\ s \in \mathcal{J}_1, \\ \ltimes_2, & if\ s \in \mathcal{J}_2, \\ \cdot \\ \cdot \\ \cdot \\ \ltimes_n, & if\ s \in \mathcal{J}_n, \end{cases}$$

where $1 < \ltimes_i \leq 2$ are constants, and I_i denotes the indicator function for the interval $\mathcal{J}_i$, defined for each $i = 1, 2, \ldots, n$ as:

$$I_i(s) = \begin{cases} 1, & for\ s \in \mathcal{J}_i, \\ 0, & otherwise. \end{cases}$$

(H2) Let $(\ell ogs)^{\delta} f : \mathcal{J} \times X \to X$ be a continuous function, where $0 < \delta < 1$. Then there exists a constant $K > 0$ such that

$$(\ell ogs)^{\delta} \|f(s, y_1) - f(s, y_2)\| \leq K \|y_1 - y_2\|,$$

holds for all $y_1, y_2 \in X$ and $s \in \mathcal{J}$.

Additionally, consider the set U of all mappings $u : \mathcal{J} \to X$. Define:

- For each $t \in J, U(t) = \{u(t), u \in U\}$ item Over the entire interval $J, U(\mathcal{J}) = \{U(s) : u \in U,\ s \in \mathcal{J}\}$.

Now, let us apply Krasnoselskii's measure of noncompactness (KMNC) and Darbo's fixed point criterion to establish the existence of a solution to the Hadamard fractional initial and terminal value problem of variable order given in equation (5.15).

Consider any $s \in \mathcal{J}_i$, with $i = 1, 2, \ldots, n$. The left Hadamard fractional derivative of the function $\varkappa(s) \in C(\mathcal{J}, X)$ with variable order $\ltimes(s)$ can be expressed as a summation of the left Hadamard fractional derivatives of constant orders $\ltimes_i$, where i ranges from 1 to n:

$$\begin{aligned} {}^{H}F_{1^+}^{\ltimes(s)} \varkappa(s) &= (s\frac{d}{ds})^2 \Big(\frac{1}{\Gamma(2 - \ltimes_1)} \int_1^{\mathsf{T}_1} (\ell og\frac{s}{t})^{1-\ltimes_1} \frac{\varkappa(t)}{t} dt + \ldots \\ &+ \frac{1}{\Gamma(2 - \ltimes_i)} \int_{\mathsf{T}_{i-1}}^{s} (\ell og\frac{s}{t})^{1-\ltimes_i} \frac{\varkappa(t)}{t} dt\Big) dt. \end{aligned} \tag{5.16}$$

Given equation (5.16), the Hadamard fractional ınitial and terminal value problem of variable order (5.15) can be reformulated for each $s \in \mathcal{J}_i$, with $i = 1, 2, \ldots, n$ in the following manner:

$$\begin{aligned} &(s\frac{d}{ds})^2 \Big(\frac{1}{\Gamma(2 - \ltimes_1)} \int_1^{\mathsf{T}_1} (\ell og\frac{s}{t})^{1-\ltimes_1} \frac{\varkappa(t)}{t} dt + \ldots \\ &+ \frac{1}{\Gamma(2 - \ltimes_i)} \int_{\mathsf{T}_{i-1}}^{s} (\ell og\frac{s}{t})^{1-\ltimes_i} \frac{\varkappa(t)}{t} dt\Big) + f(s, \varkappa(s)) = 0. \end{aligned} \tag{5.17}$$

Definition 52 *To ensure the existence of a solution to the Hadamard fractional ınitial and terminal value problem of variable order (5.15), it is necessary that there exist functions $\varkappa_i$, where $i = 1, 2, \ldots, n$, such that $\varkappa_i \in C([1, \mathsf{T}_i], X)$ and that these functions satisfy equation (5.17) along with the ınitial and terminal conditions: $\varkappa_i(1) = 0$ and $\varkappa_i(\mathsf{T}_i) = 0$.*

When $\varkappa \in C(\mathcal{J}, X)$ vanishes on the interval $[1, \mathsf{T}_{i-1}]$ and satisfies the integral equation (5.17), the equation can be reduced to the following simplified form for $s \in \mathcal{J}_i$:

$$ {}^{H}F^{\ltimes_i}_{\mathsf{T}_{i-1}^+}\varkappa(s) + f(s, \varkappa(s)) = 0, \quad s \in \mathcal{J}_i. $$

In the subsequent analysis, we apply our results to the standard Hadamard initial and terminal value problem with constant fractional order, formulated as follows:

$$ \begin{cases} {}^{H}F^{\ltimes_i}_{\mathsf{T}_{i-1}^+}\varkappa(s) + f(s, \varkappa(s)) = 0, \quad s \in \mathcal{J}_i, \\ \varkappa(\mathsf{T}_{i-1}) = 0, \quad \varkappa(\mathsf{T}_i) = 0. \end{cases} \tag{5.18} $$

In the following section, we present a detailed discussion of the central principal findings from our analysis concerning the solutions of the Hadamard constant-order fractional initial and terminal value problem (5.18).

Lemma 23 *To be a solution of the Hadamard constant-order fractional initial and terminal value problem (5.18), a function $\varkappa \in E_i$ must satisfy the integral equation given by:*

$$ \varkappa(s) = \int_{\mathsf{T}_{i-1}}^{\mathsf{T}_i} \frac{1}{t} G_i(s,t) f(t, \varkappa(t))dt, s \in \mathcal{J}_i, \tag{5.19} $$

In this context, $G_i(s,t)$ denotes the Green function defined by:

$$ G_i(s,t) = \frac{1}{\Gamma(\ltimes_i)} \begin{cases} \left(\ell og\frac{\mathsf{T}_i}{\mathsf{T}_{i-1}}\right)^{1-\ltimes_i} \left[\left(\ell og\frac{s}{\mathsf{T}_{i-1}}\right)\left(\ell og\frac{\mathsf{T}_i}{t}\right)\right]^{\ltimes_i - 1} - \left(\ell og\frac{s}{t}\right)^{\ltimes_i - 1}, \\ \qquad \mathsf{T}_{i-1} \le t \le s \le \mathsf{T}_i, \\ \left(\ell og\frac{\mathsf{T}_i}{\mathsf{T}_{i-1}}\right)^{1-\ltimes_i} \left[\left(\ell og\frac{s}{\mathsf{T}_{i-1}}\right)\left(\ell og\frac{\mathsf{T}_i}{t}\right)\right]^{\ltimes_i - 1}, \\ \qquad \mathsf{T}_{i-1} \le s \le t \le \mathsf{T}_i, \end{cases} \tag{5.20} $$

where $i \in \{1, 2, ..., n\}$.

Proof: Assume that the Hadamard constant-order fractional initial and terminal value problem (5.18) is satisfied by $\varkappa \in E_i$. By applying the operator ${}^{H}I^{\ltimes_i}_{\mathsf{T}_{i-1}^+}$ to both sides of equation (5.18) and invoking Lemma 3, we obtain:

$$ \varkappa(s) = \omega_1(\ell og\frac{s}{\mathsf{T}_{i-1}})^{\ltimes_i - 1} + \omega_2(\ell og\frac{s}{\mathsf{T}_{i-1}})^{\ltimes_i - 2} - {}^{H}I^{\ltimes_i}_{\mathsf{T}_{i-1}^+} f(s, \varkappa(s)), \quad s \in \mathcal{J}_i, \ i \in \{1, 2, ..., n\}. $$

Using the definition of f and the condition $\varkappa(\mathsf{T}_{i-1}) = 0$, it follows that $\omega_2 = 0$.

Assuming further that $\varkappa(\mathsf{T}_i) = 0$, we then obtain:

$$ \omega_1 = (\ell og\frac{\mathsf{T}_i}{\mathsf{T}_{i-1}})^{1-\ltimes_i} \, {}^{H}I^{\ltimes_i}_{\mathsf{T}_{i-1}^+} f(\mathsf{T}_i, \varkappa(\mathsf{T}_i)). $$

Thus, for $s \in \mathcal{J}_i$, we have:

$$\varkappa(s) = (\ell og\frac{\mathsf{T}_i}{\mathsf{T}_{i-1}})^{1-\ltimes_i}(\ell og\frac{s}{\mathsf{T}_{i-1}})^{\ltimes_i-1}\ {}^{H}I^{\ltimes_i}_{\mathsf{T}_{i-1}^+} f(\mathsf{T}_i, \varkappa(\mathsf{T}_i)) - {}^{H}I^{\ltimes_i}_{\mathsf{T}_{i-1}^+} f(s, \varkappa(s)).$$

Thus, the solution to the Hadamard constant-order fractional initial and terminal value problem (5.18) can be obtained through the following procedure:

$$\begin{aligned}
\varkappa(s) &= (\ell og\frac{\mathsf{T}_i}{\mathsf{T}_{i-1}})^{1-\ltimes_i}(\ell og\frac{s}{\mathsf{T}_{i-1}})^{\ltimes_i-1}\frac{1}{\Gamma(\ltimes_i)}\int_{\mathsf{T}_{i-1}}^{\mathsf{T}_i}(\ell og\frac{\mathsf{T}_i}{t})^{\ltimes_i-1}\frac{f(t,\varkappa(t))}{t}dt\\
&- \frac{1}{\Gamma(\ltimes_i)}\int_{\mathsf{T}_{i-1}}^{s}(\ell og\frac{s}{t})^{\ltimes_i-1}\frac{f(t,\varkappa(t))}{t}dt\\
&= \frac{1}{\Gamma(\ltimes_i)}\Big[\int_{\mathsf{T}_{i-1}}^{s}\Big[(\ell og\frac{\mathsf{T}_i}{\mathsf{T}_{i-1}})^{1-\ltimes_i}(\ell og\frac{s}{\mathsf{T}_{i-1}})^{\ltimes_i-1}(\ell og\frac{\mathsf{T}_i}{t})^{\ltimes_i-1} - (\ell og\frac{s}{t})^{\ltimes_i-1}\Big]\frac{f(t,\varkappa(t))}{t}dt\\
&+ \int_{s}^{\mathsf{T}_i}(\ell og\frac{\mathsf{T}_i}{\mathsf{T}_{i-1}})^{1-\ltimes_i}(\ell og\frac{s}{\mathsf{T}_{i-1}})^{\ltimes_i-1}(\ell og\frac{\mathsf{T}_i}{t})^{\ltimes_i-1}\frac{f(t,\varkappa(t))}{t}dt\Big].
\end{aligned}$$

Moreover, the continuity property of the Green function implies that

$$\varkappa(s) = \int_{\mathsf{T}_{i-1}}^{\mathsf{T}_i}\frac{1}{t}G_i(s,t)f(t,\varkappa(t))dt,\ \ s \in \mathcal{J}_i.$$

On the other hand, assume that $\varkappa \in E_i$ is a solution to the integral equation (5.19). By considering the continuity of $(\ell ogs)^{\delta} f$ and invoking Lemma 3, it is readily confirmed that $\varkappa$ satisfies the Hadamard constant-order fractional initial and terminal value problem (5.18).

Proposition 8 *For $0 < \delta < 1$ and with $(\ell ogs)^{\delta} f : \mathcal{J} \times X \to X$ being continuous, and $\ltimes(s) : \mathcal{J} \to (1, 2]$ fulfilling condition $(H1)$, the Green's function as given by (5.20) possesses the following characteristics:*

(1) $G_i(s,t) \geq 0$ *for all* $\mathsf{T}_{i-1} \leq s,\ t \leq \mathsf{T}_i$,

(2) $\max_{s\in\mathcal{J}_i} G_i(s,t) = G_i(t,t),\ t \in \mathcal{J}_i$,

(3) $G_i(t,t)$ *shows a single unique maximum, represented by:*

$$\max_{t\in\mathcal{J}_i} G_i(t,t) = \frac{1}{\Gamma(\ltimes_i)}\Big(\frac{\ell og\mathsf{T}_i - \ell og\mathsf{T}_{i-1}}{4}\Big)^{\ltimes_i-1},$$

where $i = 1, 2, ..., n$.

Proof: Let $\varphi(s,t) = \Big(\ell og\frac{\mathsf{T}_i}{\mathsf{T}_{i-1}}\Big)^{1-\ltimes_i}\Big[\Big(\ell og\frac{s}{\mathsf{T}_{i-1}}\Big)\Big(\ell og\frac{\mathsf{T}_i}{t}\Big)\Big]^{\ltimes_i-1} - \Big(\ell og\frac{s}{t}\Big)^{\ltimes_i-1}.$

Clearly, it follows that

$$\begin{aligned}
\varphi_s(s,t) &= \Big(\frac{\ltimes_i - 1}{s}\Big)\Big(\ell og\frac{\mathsf{T}_i}{\mathsf{T}_{i-1}}\Big)^{1-\ltimes_i}\Big(\ell og\frac{\mathsf{T}_i}{t}\Big)^{\ltimes_i-1}\Big(\ell og\frac{s}{\mathsf{T}_{i-1}}\Big)^{\ltimes_i-2} - \Big(\frac{\ltimes_i - 1}{s}\Big)\Big(\ell og\frac{s}{t}\Big)^{\ltimes_i-2} \\
&\leq \Big(\frac{\ltimes_i - 1}{s}\Big)\Big(\ell og\frac{\mathsf{T}_i}{t}\Big)^{1-\ltimes_i}\Big(\ell og\frac{\mathsf{T}_i}{t}\Big)^{\ltimes_i-1}\Big(\ell og\frac{s}{t}\Big)^{\ltimes_i-2} - \Big(\frac{\ltimes_i - 1}{s}\Big)\Big(\ell og\frac{s}{t}\Big)^{\ltimes_i-2} = 0,
\end{aligned}$$

This implies that $\varphi(s,t)$ is non-increasing in s. Hence, $\varphi(s,t) \geq \varphi(\mathsf{T}_i, t) = 0$ for $\mathsf{T}_{i-1} \leq t \leq s \leq \mathsf{T}_i$.

As a result, $G_i(s,t) \geq 0$ holds for every $\mathsf{T}_{i-1} \leq s,t \leq \mathsf{T}_i$ and for $i = 1,\ldots,n$.

Moreover, since $\varphi(s,t)$ does not increase with respect to s, we have $\varphi(s,t) \leq \varphi(t,t)$ for all $\mathsf{T}_{i-1} \leq t \leq s \leq \mathsf{T}_i$.

Alternatively, when $\mathsf{T}_{i-1} \leq s \leq t \leq \mathsf{T}_i$, it follows that:

$$\Big(\ell og\Big(\frac{\mathsf{T}_i}{\mathsf{T}_{i-1}}\Big)\Big)^{1-\ltimes_i}\Big(\ell og\Big(\frac{s}{\mathsf{T}_{i-1}}\Big)\ell og\Big(\frac{T_i}{t}\Big)\Big)^{\ltimes_i-1} \leq \Big(\ell og\Big(\frac{\mathsf{T}_i}{\mathsf{T}_{i-1}}\Big)\Big)^{1-\ltimes_i}\Big(\ell og\Big(\frac{t}{\mathsf{T}_{i-1}}\Big)\ell og\Big(\frac{\mathsf{T}_i}{t}\Big)\Big)^{\ltimes_i-1}.$$

These observations confirm that

$$\max_{s\in[\mathsf{T}_{i-1},\mathsf{T}_i]} G_i(s,t) = G_i(t,t),\ t \in [\mathsf{T}_{i-1},\mathsf{T}_i],\ i = 1, \dots, n.$$

Moreover, we verify condition (3) from Proposition 8. Clearly, $G_i(t,t)$ does not attain its maximum at T_{i-1} and T_i for $i = 1,\ldots,n$.

For $t \in [\mathsf{T}_{i-1},\mathsf{T}_i]$, $i = 1,\dots,n$, we have:

$$\begin{aligned}
\frac{dG_i(t,t)}{dt} &= \Big(\frac{\ltimes_i - 1}{t}\Big)\Big(\ell og\frac{\mathsf{T}_i}{\mathsf{T}_{i-1}}\Big)^{1-\ltimes_i}\Big(\ell og\frac{t}{\mathsf{T}_{i-1}}\Big)^{\ltimes_i-2}\Big(\ell og\frac{\mathsf{T}_i}{t}\Big)^{\ltimes_i-2}\Big[\Big(\ell og\frac{\mathsf{T}_i}{t}\Big) - \Big(\ell og\frac{t}{\mathsf{T}_{i-1}}\Big)\Big], \\
&= \Big(\frac{\ltimes_i - 1}{t}\Big)\Big(\ell og\frac{\mathsf{T}_i}{\mathsf{T}_{i-1}}\Big)^{1-\ltimes_i}\Big(\ell og\frac{t}{\mathsf{T}_{i-1}}\Big)^{\ltimes_i-2}\Big(\ell og\frac{\mathsf{T}_i}{t}\Big)^{\ltimes_i-2}\Big[\ell og(\mathsf{T}_i\mathsf{T}_{i-1}) - \ell og(t^2)\Big],
\end{aligned}$$

This indicates that the maximum points of $G_i(t,t)$ occur at $t = \sqrt{\mathsf{T}_i\mathsf{T}_{i-1}}$ for $i = 1,\ldots,n$. Consequently, for $i = 1,\ldots,n$,

$$\begin{aligned}
\max_{t\in[\mathsf{T}_{i-1},\mathsf{T}_i]} G_i(t,t) &= G_i\Big(\sqrt{\mathsf{T}_i\mathsf{T}_{i-1}}, \sqrt{\mathsf{T}_i\mathsf{T}_{i-1}}\Big) \\
&= \frac{1}{\Gamma(\ltimes_i)}\Big(\frac{1}{4}\ell og\frac{\mathsf{T}_i}{\mathsf{T}_{\,i-1}}\Big)^{\ltimes_i-1} \\
&= \frac{1}{\Gamma(\ltimes_i)}\Big(\frac{\ell og\mathsf{T}_i - \ell og\mathsf{T}_{i-1}}{4}\Big)^{\ltimes_i-1}.
\end{aligned}$$

The applicability of Darbo's fixed point theorem, as examined in this section, is crucial for establishing the existence of solutions to the Hadamard constant-order fractional initial and terminal value problem (5.18).

Theorem 43 *Provided that both conditions (H1) and (H2) are satisfied, and*

$$\frac{K\Big((\ell og\mathsf{T}_i)^{1-\delta}-(\ell og\mathsf{T}_{i-1})^{1-\delta}\Big)\Big(\ell og\mathsf{T}_i-\ell og\mathsf{T}_{i-1}\Big)^{\ltimes_i-1}}{4^{\ltimes_i-1}(1-\delta)\Gamma(\ltimes_i)}<1, \qquad (5.21)$$

then there exists at least one solution to the Hadamard constant-order fractional initial and terminal value problem (5.18) on E_i.

Proof: We construct the operator using the approach outlined below:

$$W: E_i \to E_i$$

defined as

$$W\varkappa(s)=\int_{\mathsf{T}_{i-1}}^{\mathsf{T}_i}\frac{1}{t}G_i(s,t)f(t,\varkappa(t))dt,\quad s\in\mathcal{J}_i. \qquad (5.22)$$

Assuming the continuity of the function $(\ell og s)^{\delta}f$ and the properties of fractional integrals, we may conclude that the operator $W: E_i \to E_i$, as defined in (5.22) is well-defined.

Let

$$R_i\geq\frac{\dfrac{f^{\star}(\ell og\mathsf{T}_i-\ell og\mathsf{T}_{i-1})^{\ltimes_i}}{4^{\ltimes_i-1}\Gamma(\ltimes_i)}}{1-\dfrac{K\Big((\ell og\mathsf{T}_i)^{1-\delta}-(\ell og\mathsf{T}_{i-1})^{1-\delta}\Big)\Big(\ell og\mathsf{T}_i-\ell og\mathsf{T}_{i-1}\Big)^{\ltimes_i-1}}{4^{\ltimes_i-1}(1-\delta)\Gamma(\ltimes_i)}},$$

where

$$f^{\star}=\sup_{s\in\mathcal{J}_i}\|f(s,0)\|.$$

Let us consider the following set:

$$B_{R_i}=\{\varkappa\in E_i,\ \|\varkappa\|_{E_i}\leq R_i\}.$$

It is evident that B_{R_i} is closed, bounded, convex, and nonempty.

In the subsequent four steps, we shall demonstrate that the operator W satisfies the hypotheses of Darbo's fixed point theorem.

STEP 1: $W(B_{R_i}) \subseteq (B_{R_i})$.
For $\varkappa \in B_{R_i}$, by Proposition 8 and assumption (H2), we obtain:

$$\begin{aligned}
\|W\varkappa(s)\| &= \Big\| \int_{\mathsf{T}_{i-1}}^{\mathsf{T}_i} \frac{1}{t} G_i(s,t) f(t,\varkappa(t))dt \Big\| \\
&\leq \int_{\mathsf{T}_{i-1}}^{\mathsf{T}_i} \frac{1}{t} G_i(s,t)\|f(t,\varkappa(t))\|dt \\
&\leq \frac{1}{\Gamma(\ltimes_i)}\Big(\frac{\ell og\mathsf{T}_i - \ell og\mathsf{T}_{i-1}}{4}\Big)^{\ltimes_i - 1} \int_{\mathsf{T}_{i-1}}^{\mathsf{T}_i} \frac{1}{t}\|f(t,\varkappa(t))\|dt \\
&\leq \frac{1}{\Gamma(\ltimes_i)}\Big(\frac{\ell og\mathsf{T}_i - \ell og\mathsf{T}_{i-1}}{4}\Big)^{\ltimes_i - 1} \int_{\mathsf{T}_{i-1}}^{\mathsf{T}_i} \frac{1}{t}\Big\|f(t,\varkappa(t)) - f(t,0)\Big\|dt \\
&+ \frac{1}{\Gamma(\ltimes_i)}\Big(\frac{\ell og\mathsf{T}_i - \ell og\mathsf{T}_{i-1}}{4}\Big)^{\ltimes_i - 1} \int_{\mathsf{T}_{i-1}}^{\mathsf{T}_i} \frac{1}{t}\|f(t,0)\|dt \\
&\leq \frac{1}{\Gamma(\ltimes_i)}\Big(\frac{\ell og\mathsf{T}_i - \ell og\mathsf{T}_{i-1}}{4}\Big)^{\ltimes_i - 1} \int_{\mathsf{T}_{i-1}}^{\mathsf{T}_i} \frac{1}{t}(\ell ogt)^{-\delta}(K\|\varkappa(t)\|)dt \\
&+ \frac{f^\star\Big(\ell og\mathsf{T}_i - \ell og\mathsf{T}_{i-1}\Big)^{\ltimes_i}}{4^{\ltimes_i - 1}\Gamma(\ltimes_i)} \\
&\leq \frac{K}{\Gamma(\ltimes_i)}\Big(\frac{\ell og\mathsf{T}_i - \ell og\mathsf{T}_{i-1}}{4}\Big)^{\ltimes_i - 1} \|\varkappa\|_{E_i} \int_{\mathsf{T}_{i-1}}^{\mathsf{T}_i} \frac{1}{t}(\ell ogt)^{-\delta}dt \\
&+ \frac{f^\star\Big(\ell og\mathsf{T}_i - \ell og\mathsf{T}_{i-1}\Big)^{\ltimes_i}}{4^{\ltimes_i - 1}\Gamma(\ltimes_i)} \\
&\leq \frac{K}{\Gamma(\ltimes_i)}\Big(\frac{\ell og\mathsf{T}_i - \ell og\mathsf{T}_{i-1}}{4}\Big)^{\ltimes_i - 1} R_i\Big(\frac{(\ell og\mathsf{T}_i)^{1-\delta} - (\ell og\mathsf{T}_{i-1})^{1-\delta}}{1-\delta}\Big) \\
&+ \frac{f^\star\Big(\ell og\mathsf{T}_i - \ell og\mathsf{T}_{i-1}\Big)^{\ltimes_i}}{4^{\ltimes_i - 1}\Gamma(\ltimes_i)} \\
&\leq \frac{K\Big((\ell og\mathsf{T}_i)^{1-\delta} - (\ell og\mathsf{T}_{i-1})^{1-\delta}\Big)\Big(\ell og\mathsf{T}_i - \ell og\mathsf{T}_{i-1}\Big)^{\ltimes_i - 1}}{4^{\ltimes_i - 1}(1-\delta)\Gamma(\ltimes_i)} R_i \\
&+ \frac{f^\star\Big(\ell og\mathsf{T}_i - \ell og\mathsf{T}_{i-1}\Big)^{\ltimes_i}}{4^{\ltimes_i - 1}\Gamma(\ltimes_i)} \\
&\leq R_i.
\end{aligned}$$

This implies that $W(B_{R_i}) \subseteq B_{R_i}$.

STEP 2: W is continuous.

Assume that the sequence $(\varkappa_n)$ converges to $\varkappa$ in E_i and let $s \in \mathcal{J}_i$. Then,

$$\|(W\varkappa_n)(s) - (W\varkappa)(s)\|$$

$$\begin{aligned}
&\leq \int_{\mathsf{T}_{i-1}}^{\mathsf{T}_i} \frac{1}{t} G_i(s,t) \Big\| f(t, \varkappa_n(t)) - f(t, \varkappa(t)) \Big\| dt \\
&\leq \frac{1}{\Gamma(\ltimes_i)} \Big(\frac{\ell og \mathsf{T}_i - \ell og \mathsf{T}_{i-1}}{4} \Big)^{\ltimes_i - 1} \int_{\mathsf{T}_{i-1}}^{\mathsf{T}_i} \frac{1}{t} \Big\| f(t, \varkappa_n(t)) - f(t, \varkappa(t)) \Big\| dt \\
&\leq \frac{1}{\Gamma(\ltimes_i)} \Big(\frac{\ell og \mathsf{T}_i - \ell og \mathsf{T}_{i-1}}{4} \Big)^{\ltimes_i - 1} \int_{\mathsf{T}_{i-1}}^{\mathsf{T}_i} \frac{1}{t} (\ell og t)^{-\delta} (K \|\varkappa_n(t) - \varkappa(t)\|) dt \\
&\leq \frac{1}{\Gamma(\ltimes_i)} \Big(\frac{\ell og \mathsf{T}_i - \ell og \mathsf{T}_{i-1}}{4} \Big)^{\ltimes_i - 1} (K \|\varkappa_n - \varkappa\|_{E_i}) \int_{\mathsf{T}_{i-1}}^{\mathsf{T}_i} \frac{1}{t} (\ell og t)^{-\delta} dt \\
&\leq \frac{K \Big((\ell og \mathsf{T}_i)^{1-\delta} - (\ell og \mathsf{T}_{i-1})^{1-\delta} \Big) \Big(\ell og \mathsf{T}_i - \ell og \mathsf{T}_{i-1} \Big)^{\ltimes_i - 1}}{4^{\ltimes_i - 1} (1-\delta) \Gamma(\ltimes_i)} \|\varkappa_n - \varkappa\|_{E_i},
\end{aligned}$$

That is, we obtain

$$\|(W\varkappa_n) - (W\varkappa)\|_{E_i} \to 0 \text{ as } n \to \infty.$$

Therefore, the operator W is continuous in the space E_i.

STEP 3: $W(B_{R_i})$ is confined and consistent in continuity.

From Step 1, we have $W(B_{R_i}) = \{W(\varkappa) : \varkappa \in B_{R_i}\} \subset B_{R_i}$. Thus, for each $\varkappa \in B_{R_i}$, it follows that $\|W(\varkappa)\|_{E_i} \leq R_i$. In other words, this implies that $W(B_{R_i})$ is bounded. It remains to verify the equicontinuity of $W(B_{R_i})$.

Now, for all $s_1 < s_2 \in \mathcal{J}_i,\ s_1 < s_2$ and $\varkappa \in B_{R_i}$, we express:

$$\begin{aligned}
\|(W\varkappa)(s_2) - (W\varkappa)(s_1)\| &= \Big\|\int_{\mathsf{T}_{i-1}}^{\mathsf{T}_i} \frac{1}{t} G_i(s_2,t) f(t,\varkappa(t))dt - \int_{\mathsf{T}_{i-1}}^{\mathsf{T}_i} \frac{1}{t} G_i(s_1,t) f(t,\varkappa(t))dt\Big\| \\
&\leq \int_{\mathsf{T}_{i-1}}^{\mathsf{T}_i} \frac{1}{t}\Big\|\Big(G_i(s_2,t) - G_i(s_1,t)\Big) f(t,\varkappa(t))\Big\| dt \\
&\leq \int_{\mathsf{T}_{i-1}}^{\mathsf{T}_i} \frac{1}{t}\Big\|G_i(s_2,t) - G_i(s_1,t)\Big\| \|f(t,\varkappa(t))\| dt \\
&\leq \int_{\mathsf{T}_{i-1}}^{\mathsf{T}_i} \frac{1}{t}\Big\|G_i(s_2,t) - G_i(s_1,t)\Big\|\Big(\Big\|f(t,\varkappa(t)) - f(t,0)\Big\| + \|f(t,0)\|\Big) dt \\
&\leq \int_{\mathsf{T}_{i-1}}^{\mathsf{T}_i} \frac{1}{t}\Big\|G_i(s_2,t) - G_i(s_1,t)\Big\|\Big[(\ell og t)^{-\delta}(K\|\varkappa(t)\|) + f^\star\Big] dt
\end{aligned}$$

$$\begin{aligned}
&\leq \int_{\mathsf{T}_{i-1}}^{\mathsf{T}_i} \Big\|G_i(s_2,t) - G_i(s_1,t)\Big\|\Big[\frac{1}{t}(\ell og t)^{-\delta}\Big(K\|\varkappa\|_{E_i}\Big) + \frac{1}{t} f^\star\Big] dt \\
&\leq \frac{K(\ell og \mathsf{T}_{i-1})^{-\delta}}{\mathsf{T}_{i-1}} \|\varkappa\|_{E_i} R_i \int_{\mathsf{T}_{i-1}}^{\mathsf{T}_i} \Big\|G_i(s_2,t) - G_i(s_1,t)\Big\| dt \\
&+ \frac{f^\star}{\mathsf{T}_{i-1}} \int_{\mathsf{T}_{i-1}}^{\mathsf{T}_i} \Big\|G_i(s_2,t) - G_i(s_1,t)\Big\| dt,
\end{aligned}$$

By virtue of the continuity of Green's function G_i, it follows that $\|(W\varkappa)(s_2) - (W\varkappa)(s_1)\|_{E_i} \to 0$ as $|s_2 - s_1| \to 0$. Consequently, $W(B_{R_i})$ is equicontinuous.

Remark 15 *[23]. Note that the inequality*

$$\mu\Big((\ell og s)^\delta \|f(s, B_1)\|\Big) \leq K\mu(B_1),$$

conforms to assumption (H2) for each bounded subset $B_1 \subset X$ *and for all* $s \in \mathcal{J}$, *where* B_1 *is bounded.*

STEP 4: W is a k-set contraction mapping.
For $U \in B_{R_i}$ and $s \in \mathcal{J}_i$, we deduce,

$$\begin{aligned}
\mu(W(U)(s)) &= \mu((W\varkappa)(s), \varkappa \in U) \\
&\leq \Big\{\int_{\mathsf{T}_{i-1}}^{\mathsf{T}_i} \frac{1}{t} G_i(s,t)\mu f(t,\varkappa(t))dt, \varkappa \in U\Big\}.
\end{aligned}$$

Remark 15. It follows that for $U \in B_{R_i}$ and $s \in \mathcal{J}_i$, we have:

$$\begin{aligned}
\mu(W(U)(s)) &\leq \Big\{\int_{\mathsf{T}_{i-1}}^{\mathsf{T}_i} \frac{1}{t} G_i(s,t)[K\mu(\{\varkappa(t), \varkappa \in U\})]\Big\} \\
&\leq \Big\{\frac{1}{\Gamma(\ltimes_i)}\Big(\frac{\ell og\mathsf{T}_i - \ell og\mathsf{T}_{i-1}}{4}\Big)^{\ltimes_i - 1}\Big[K\widehat{\mu}(U)\int_{\mathsf{T}_{i-1}}^{\mathsf{T}_i} \frac{1}{t}(\ell ogt)^{-\delta}dt\Big], \varkappa \in U\Big\} \\
&\leq \frac{K\Big((\ell og\mathsf{T}_i{}^{1-\delta}) - (\ell og\mathsf{T}_{i-1}{}^{1-\delta})\Big)(\ell og\mathsf{T}_i - \ell og\mathsf{T}_{i-1})^{\ltimes_i - 1}}{4^{\ltimes_i - 1}(1-\delta)\Gamma(\ltimes_i)}\widehat{\mu}(U).
\end{aligned}$$

Therefore, the measure of the image set satisfies:

$$\widehat{\mu}(WU) \leq \frac{K\Big((\ell og\mathsf{T}_i{}^{1-\delta}) - (\ell og\mathsf{T}_{i-1}{}^{1-\delta})\Big)(\ell og\mathsf{T}_i - \ell og\mathsf{T}_{i-1})^{\ltimes_i - 1}}{4^{\ltimes_i - 1}(1-\delta)\Gamma(\ltimes_i)}\widehat{\mu}(U).$$

We then conclude that W is a k-set contraction by (5.21).

Consequently, the Hadamard constant-order fractional ınitial and terminal υalue problem (5.18) admits at least one solution, $\widetilde{\varkappa}_i \in B_{R_i}$, since all criteria of the Darbo fixed point theorem are satisfied. As $B_{R_i} \subset E_i$, the Theorem 43 step is demonstrated.

We now proceed to establish the existence theorem for the variable order Hadamard fractional ınitial and terminal υalue problem (5.15).

Theorem 44 *Assuming that conditions (H1), (H2), and inequality (5.21) hold for all* $i \in \{1, 2, ..., n\}$, *the variable-order Hadamard fractional ınitial and terminal* υ*alue problem (5.15) admits at least one solution in* $C(\mathcal{J}, X)$.

Proof: As stated in Theorem 43, the Hadamard constant-order fractional ınitial and terminal υalue problem (5.18) admits at least one solution $\widetilde{\varkappa}_i \in E_i$ for each index $i \in \{1, 2, ..., n\}$.

Let us define the function $\varkappa_i$ for every $i \in \{1, 2, ..., n\}$ as follows:

$$\varkappa_i = \begin{cases} 0, & s \in [1, \mathsf{T}_{i-1}], \\ \widetilde{\varkappa}_i, & s \in \mathcal{J}_i. \end{cases}$$

Accordingly, the function $\varkappa_i \in C([1, \mathsf{T}_i], X)$ satisfies the integral equation (5.17) for all $s \in \mathcal{J}_i$ and meets the boundary conditions $\varkappa_i(1) = 0$ and $\varkappa_i(\mathsf{T}_i) = 0$.

Thus, the mapping $\varkappa(s)$ defined by

$$\varkappa(s) = \begin{cases} \varkappa_1(s), & s \in \mathcal{J}_1, \\ \varkappa_2(s) = \begin{cases} 0, & s \in \mathcal{J}_1, \\ \widetilde{\varkappa}_2, & s \in \mathcal{J}_2, \end{cases} \\ \vdots \\ \varkappa_n(s) = \begin{cases} 0, & s \in [1, \mathsf{T}_{n-1}], \\ \widetilde{\varkappa}_n, & s \in \mathcal{J}_n, \end{cases} \end{cases} .$$

provides a solution to the variable-order Hadamard fractional ınitial and terminal υalue problem (5.15).

5.2.2 Study of Stability

Theorem 45 *Assume (H1), (H2), (5.21), and (H3).*

Let $\vartheta \in C(\mathcal{J}, X)$ be an increasing function. Then, there exists a positive constant $\lambda_\vartheta > 0$ such that for each $s \in \mathcal{J}_i$, the following inequality holds:

$${}^H I^{\ltimes_i}_{\mathsf{T}_{i-1}{}^+} \vartheta(s) \le \lambda_{\vartheta(s)} \vartheta(s),$$

where ${}^H I^{\ltimes_i}_{\mathsf{T}_{i-1}{}^+} + \mathsf{T}_i{}^+$ denotes the Hadamard fractional integral of variable order.

In this case, the Hadamard fractional ınitial and terminal υalue problem of variable order (5.15) exhibits Ulam-Hyers-$\mathfrak{R}$assias stability with respect to ϑ.

Proof: Let $\epsilon > 0$ be an arbitrary number, and let the function $z(s) \in C(\mathcal{J}, X)$ satisfy the inequality (2.29).

For any $i \in \{1, 2, ..., n\}$ we define the functions $z_1(s) \equiv z(s)$, for $s \in [1, \mathsf{T}_1]$ and for $i = 2, 3, ..., n$:

$$z_i(s) = \begin{cases} 0, & s \in [1, \mathsf{T}_{i-1}], \\ z(s), & s \in \mathcal{J}_i. \end{cases}$$

For any $i \in \{1, 2, ..., n\}$, according to equality (5.16), for $s \in \mathcal{J}_i$, we obtain

$${}^H \digamma^{\ltimes(s)}_{1+} z_i(s) = \frac{1}{\Gamma(2 - \ltimes(s))} (s\frac{d}{ds})^2 \int_{\mathsf{T}_{i-1}}^{s} (\ell og\frac{s}{t})^{1-\ltimes_i} \frac{z(t)}{t} dt,$$

where $HD^1 + \varrho(s)$ denotes the Hadamard fractional derivative of variable order $\varrho(s)$, and $\Gamma(\cdot)$ is the Gamma function.

Applying the Hadamard fractional integral ${}^{H}I^{\ltimes_i}_{\mathsf{T}^{+}_{i-1}}$ to both sides of inequality (2.29), we obtain

$$\Big\|z(s)+\int_{\mathsf{T}_{i-1}}^{\mathsf{T}_i}\frac{1}{t}G_i(s,t)f(t,z(t))dt\Big\| \le \frac{\epsilon}{\Gamma(\ltimes_i)}\int_{\mathsf{T}_{i-1}}^{s}\frac{1}{t}(\ell og\frac{s}{t})^{\ltimes_i-1}\vartheta(t)dt$$

$$\le \epsilon\lambda_{\vartheta(s)}\vartheta(s).$$

Theorem 44 guarantees that the variable-order Hadamard fractional initial and terminal value problem (5.15) admits a solution $\varkappa\in C(\mathcal{J},X)$ defined by $\varkappa(s)=\varkappa_i(s)$ for $s\in\mathcal{J}_i$, where $i=1,2,...,n$, and where

$$\varkappa_i(s)=\begin{cases}0, & s\in[1,\mathsf{T}_{i-1}],\\ \widetilde{\varkappa}_i(s), & s\in\mathcal{J}_i,\end{cases} \tag{5.23}$$

with $\widetilde{\varkappa}_i\in E_i$ representing a solution to the Hadamard constant-order fractional initial and terminal value problem (5.18). Lemma 23 establishes that the integral equation

$$\widetilde{\varkappa}_i(s)=-\frac{(\mathsf{T}_i-\mathsf{T}_{i-1})^{-1}(s-\mathsf{T}_{i-1})}{\Gamma(\ltimes_i)}\int_{\mathsf{T}_{i-1}}^{\mathsf{T}_i}(\mathsf{T}_i-t)^{\ltimes_i-1}f(t,\widetilde{\varkappa}_i(t))dt$$

$$+\frac{1}{\Gamma(\ltimes_i)}\int_{\mathsf{T}_{i-1}}^{s}(s-t)^{\ltimes_i-1}f(t,\widetilde{\varkappa}_i(t))dt,\quad s\in\mathcal{J}_i. \tag{5.24}$$

holds.

Let $s\in\mathcal{J}_i$, where $i\in\{1,2,...,n\}$. Then, by equations (5.23) and (5.24), we obtain:

$$\|z(s)-\varkappa(s)\|=\|z(s)-\varkappa_i(s)\|=\|z_i(s)-\widetilde{\varkappa}_i(s)\|$$

$$=\|z_i(s)-\int_{\mathsf{T}_{i-1}}^{\mathsf{T}_i}\frac{1}{t}G_i(s,t)f(t,\widetilde{\varkappa}_i(t))dt\|$$

$$\le\Big\|z_i(s)-\int_{\mathsf{T}_{i-1}}^{\mathsf{T}_i}\frac{1}{t}G_i(s,t)f(t,z_i(t))dt\Big\|+\int_{\mathsf{T}_{i-1}}^{\mathsf{T}_i}\frac{1}{t}G_i(s,t)\Big\|f(t,z_i(t))-f(t,\widetilde{\varkappa}_i(t))\Big\|dt$$

$$\le\Big\|z_i(s)+\int_{\mathsf{T}_{i-1}}^{\mathsf{T}_i}\frac{1}{t}G_i(s,t)f(t,z_i(t))dt\Big\|+\int_{\mathsf{T}_{i-1}}^{\mathsf{T}_i}\frac{1}{t}G_i(s,t)\Big\|f(t,z_i(t))-f(t,\widetilde{\varkappa}_i(t))\Big\|dt$$

$$\le\lambda_{\vartheta(s)}\epsilon\vartheta(s)+\frac{1}{\Gamma(\ltimes_i)}\Big(\frac{\ell og\mathsf{T}_i-\ell og\mathsf{T}_{i-1}}{4}\Big)^{\ltimes_i-1}\int_{\mathsf{T}_{i-1}}^{\mathsf{T}_i}(\ell ogt)^{-\delta}\frac{K\|z_i(t)-\widetilde{\varkappa}_i(t)\|}{t}dt$$

$$\leq \lambda_{\vartheta(s)}\epsilon\vartheta(s) + \frac{K}{\Gamma(\ltimes_i)}\Big(\frac{\ell og\mathsf{T}_i - \ell og\mathsf{T}_{i-1}}{4}\Big)^{\ltimes_i - 1}\|z_i - \widetilde{\varkappa}_i\|_{E_i}\int_{\mathsf{T}_{i-1}}^{\mathsf{T}_i}\frac{1}{t}(\ell ogt)^{-\delta}dt$$

$$\leq \lambda_{\vartheta(s)}\epsilon\vartheta(s) + \frac{K\Big((\ell og\mathsf{T}_i)^{1-\delta} - (\ell og\mathsf{T}_{i-1})^{1-\delta}\Big)\Big(\ell og\mathsf{T}_i - \ell og\mathsf{T}_{i-1}\Big)^{\ltimes_i - 1}}{(1-\text{ß})4^{\ltimes_i - 1}\Gamma(\ltimes_i)}\|z_i - \widetilde{\varkappa}_i\|_{E_i}$$

$$\leq \lambda_{\vartheta(s)}\epsilon\vartheta(s) + \mu\|z - \varkappa\|,$$

where

$$\mu = \max_{i=1,2,\dots,n}\frac{K\Big((\ell og\mathsf{T}_i)^{1-\delta} - (\ell og\mathsf{T}_{i-1})^{1-\delta}\Big)\Big(\ell og\mathsf{T}_i - \ell og\mathsf{T}_{i-1}\Big)^{\ltimes_i - 1}}{(1-\delta)4^{\ltimes_i - 1}\Gamma(\ltimes_i)}.$$

Then

$$\|z - \varkappa\|(1-\mu) \leq \lambda_{\vartheta(s)}\epsilon\vartheta(s),$$

and so by assuming $c_f := \frac{\lambda_{\vartheta(s)}}{(1-\mu)}$,

$$\|z(s) - \varkappa(s)\| \leq \frac{\lambda_{\vartheta(s)}\vartheta(s)}{(1-\mu)}\epsilon := c_f\epsilon\vartheta(s).$$

Thus, the Hadamard fractional ınitial and terminal υalue problem of variable order (5.15) is stable in the Ulam-Hyers-$\mathfrak{R}$assias sense with respect to ϑ.

5.2.3 Applied Approach

We examine the variable-order Hadamard fractional ınitial and terminal υalue problem:

$$\begin{cases} {}^{H}\digamma_{1^+}^{\ltimes(s)}\varkappa_n(s) + \frac{1}{(1+\ell ogs)}\Big(1 + \frac{\|\varkappa_n(s)\|}{1+\|\varkappa_n(s)\|}\Big) = 0, \quad s \in \mathcal{J} = [1, e], \\ \varkappa_n(1) = 0, \quad \varkappa_n(e) = 0. \end{cases} \tag{5.25}$$

Let

$$E = l^1 = \{\varkappa = (\varkappa_1, \varkappa_2, ..., \varkappa_n, ...), \sum_{n=1}^{\infty}|\varkappa_n| < \infty\},$$

then E denotes a ßanach space, and its norm is given by $\|\varkappa\| = \sum_{n=1}^{\infty}|\varkappa_n|$.

Define the function

$$f(s, \varkappa) = \frac{1}{(1+\ell ogs)}\Big(1 + \frac{\|\varkappa\|}{1+\|\varkappa\|}\Big), \quad s \in [1, e], \ \varkappa \in E,$$

and let the variable-order function $\ltimes(s)$ be defined piecewise as

$$\ltimes(s) = \begin{cases} 1.2, & s \in \mathcal{J}_1 := [1, 2], \\ 1.6, & s \in \mathcal{J}_2 :=]2, e]. \end{cases} \tag{5.26}$$

Then, we obtain

$$\begin{aligned}(\ell ogs)^{\frac{1}{4}}|f(s,\varkappa)-f(s,y)| &= \frac{(\ell ogs)^{\frac{1}{4}}}{(1+\ell ogs)}\left|\frac{\|\varkappa\|}{1+\|\varkappa\|}-\frac{\|y\|}{1+\|y\|}\right| \\ &\leq \frac{1}{2}\|\varkappa-y\|.\end{aligned}$$

Thus, condition (H2) holds with $\delta=\frac{1}{4}$ and $K=\frac{1}{2}$.

The variable-order Hadamard fractional initial and terminal value problem (5.25) can be classified into the following categories based on the piecewise definition in (5.26):

$$\begin{cases} {}^{H}F_{1^+}^{1.2}\varkappa_n(s)+\frac{1}{(1+\ell ogs)}\left(1+\frac{|\varkappa_n(s)|}{1+|\varkappa_n(s)|}\right)=0, & s\in\mathcal{J}_1, \\ {}^{H}F_{2^+}^{1.6}\varkappa_n(s)+\frac{1}{(1+\ell ogs)}\left(1+\frac{|\varkappa_n(s)|}{1+|\varkappa_n(s)|}\right)=0, & s\in\mathcal{J}_2. \end{cases}$$

When $s\in\mathcal{J}_1$, the variable-order Hadamard fractional initial and terminal value problem (5.25) reduces to the constant-order Hadamard constant-order fractional initial and terminal value problem:

$$\begin{cases} {}^{H}F_{1^+}^{1.2}\varkappa_n(s)+\frac{1}{(1+\ell ogs)}\left(1+\frac{|\varkappa_n(s)|}{1+|\varkappa_n(s)|}\right)=0, & s\in\mathcal{J}_1, \\ \varkappa_n(1)=0, \quad \varkappa_n(2)=0. \end{cases} \tag{5.27}$$

Now, let us demonstrate that inequality (5.21) is satisfied. Evidently, the resulting value is:

$$\frac{K\Big((\ell og\mathsf{T}_1)^{1-\delta}-(\ell og\mathsf{T}_0)^{1-\delta}\Big)\Big(\ell og\mathsf{T}_1-\ell og\mathsf{T}_0\Big)^{\ltimes_1-1}}{4^{\ltimes_1-1}(1-\delta)\Gamma(\ltimes_1)}=\frac{\frac{1}{2}(\ell og2)^{\frac{3}{4}}(\ell og2)^{0.2}}{(4^{0.2})\frac{3}{4}\Gamma(1.2)}\simeq 0.1941<1.$$

In contrast, let $\vartheta(s)=(\ell ogs)^{\frac{1}{2}}$. In this case,

$$\begin{aligned}{}^{H}I_{1^+}^{\ltimes_1}\vartheta(s) &= \frac{1}{\Gamma(1.2)}\int_1^2(\ell og\frac{s}{t})^{1.2-1}\frac{(\ell ogs)^{\frac{1}{2}}}{t}dt \\ &\leq \frac{(\ell ogs)^{\frac{1}{2}}}{\Gamma(1.2)}\int_1^2(\ell og\frac{2}{t})^{0.2}\frac{dt}{t} \\ &\leq \frac{(\ell og2)^{1.2}}{\Gamma(2.2)}(\ell ogs)^{\frac{1}{2}}:=\lambda_{\vartheta(s)}\vartheta(s).\end{aligned}$$

Hence, hypothesis (H3) is fulfilled with $\vartheta(s)=logs$ and $\lambda_{\vartheta(s)}=\dfrac{(\ell og2)^{1.2}}{\Gamma(2.2)}\in\mathbb{R}$.

Theorem (43) guarantees that a solution $\widetilde{\varkappa}_1 \in E_1$ exists for the Hadamard constant-order fractional initial and terminal value problem (5.27).

In the interval $s \in \mathcal{J}_2$, the Hadamard fractional initial and terminal value problem of variable order (5.25) is equivalent to the following constant-order fractional initial and terminal value problem, namely:

$$\begin{cases} {}^H F_{2^+}^{1.6} \varkappa_n(s) + \frac{1}{(1+\ell ogs)}\left(1 + \frac{|\varkappa_n(s)|}{1+|\varkappa_n(s)|}\right) = 0, \quad s \in \mathcal{J}_2, \\ \varkappa_n(2) = 0, \quad \varkappa_n(e) = 0. \end{cases} \tag{5.28}$$

We observe that

$$\frac{K\Big((\ell og\mathsf{T}_2)^{1-\delta} - (\ell og\mathsf{T}_1)^{1-\delta}\Big)\Big(\ell og\mathsf{T}_2 - \ell og\mathsf{T}_1\Big)^{\ltimes_2 - 1}}{4^{\ltimes_2-1}(1-\delta)\Gamma(\ltimes_2)} = \frac{\frac{1}{2}\Big(1 - (\ell og2)^{\frac{3}{4}}\Big)(1-\ell og2)^{0.6}}{(4^{0.6})\frac{3}{4}\Gamma(1.6)} \simeq 0.0191 < 1.$$

Thus, condition 5.21 is satisfied on the subinterval $\mathcal{J}_2$. Additionally, we have:

$$\begin{aligned} {}^H I_{2^+}^{\ltimes_1}\vartheta(s) &= \frac{1}{\Gamma(1.6)}\int_2^e (\ell og\frac{s}{t})^{1.6-1}\frac{(\ell ogs)^{\frac{1}{2}}}{t}dt \\ &\leq \frac{(\ell ogs)^{\frac{1}{2}}}{\Gamma(1.6)}\int_2^e (\ell og\frac{e}{t})^{0.6}\frac{dt}{t} \\ &\leq \frac{(\ell og\frac{e}{2})^{1.6}}{\Gamma(2.6)}(\ell ogs)^{\frac{1}{2}} := \lambda_{\vartheta(s)}\vartheta(t). \end{aligned}$$

Thus, (H3) is also applicable with $\vartheta(s) = logs$ and $\lambda_{\vartheta(s)} = \dfrac{(\ell og\frac{e}{2})^{1.6}}{\Gamma(2.6)} \in \mathbb{R}$.

The existence of a solution $\widetilde{\varkappa}_2 \in E_2$ for the Hadamard constant-order fractional initial and terminal value problem (5.28) follows from Theorem (43).

Therefore, by Theorem 44, the Hadamard fractional initial and terminal value problem of variable order (5.25) admits a solution:

$$\varkappa(s) = \begin{cases} \widetilde{\varkappa}_1(s), & s \in \mathcal{J}_1, \\ \varkappa_2(s), & s \in \mathcal{J}_2, \end{cases}$$

where

$$\varkappa_2(s) = \begin{cases} 0, & s \in \mathcal{J}_1, \\ \widetilde{\varkappa}_2(s), & s \in \mathcal{J}_2. \end{cases}$$

As demonstrated by Theorem 45, the Hadamard fractional initial and terminal value problem of variable order (5.25) is Ulam-Hyers-$\mathfrak{R}$assias stable with respect to ϑ.

5.3 Initial and Terminal Value Problems for Hadamard Fractional Differential Equations of Variable Order via the Kuratowski Measure of Noncompactness Technique

The initial and terminal value issue (ITVP) under consideration is given by

$$\begin{cases} {}^{H}\digamma_{1^+}^{\ltimes(s)}\varkappa(s) = f(s,\varkappa(s),{}^{H}I_{1^+}^{\ltimes(s)}\varkappa(s)), \ s\in\mathcal{J}:=[1,\intercal], \\ \varkappa(1)=\varkappa(\intercal)=0, \end{cases} \tag{5.29}$$

where $1<\ltimes(s)\leq 2$, $f:\mathcal{J}\times X\times X\to X$ is a continuous function, and ${}^{H}\digamma_{1^+}^{\ltimes(s)}$, ${}^{H}I_{1^+}^{\ltimes(s)}$ are denote the Hadamard fractional derivative and integral of variable order, respectively.

In this section, we investigate the existence of a solution to equation (5.29). Furthermore, we examine the Ulam-Hyers-$\mathfrak{R}$assias stability of the solution obtained.

5.3.1 Existence Properties

We begin by stating the foundational assumptions underlying our analysis:

(H1) Let $n\in\mathbb{N}$, and let $\mathcal{P}=\{\mathcal{J}_1:=[1,\intercal_1],\mathcal{J}_2:=(\intercal_1,\intercal_2],\mathcal{J}_3:=(\intercal_2,\intercal_3],\ldots,\mathcal{J}_n:=(\intercal_{n-1},\intercal]\}$ be a partition of the interval $\mathcal{J}:=[1,\intercal]$. Define the function $\ltimes(s):\mathcal{J}\to(1,2]$ as a piecewise constant function with respect to the partition $\mathcal{P}$, such that

$$\ltimes(s)=\sum_{i=1}^{n}\ltimes_i I_i(s)=\begin{cases} \ltimes_1, & if\ s\in\mathcal{J}_1, \\ \ltimes_2, & if\ s\in\mathcal{J}_2, \\ \cdot & \\ \cdot & \\ \cdot & \\ \ltimes_n, & if\ s\in\mathcal{J}_n, \end{cases}$$

where $1<\ltimes_i\leq 2$ are constants for each $i=1,2,\ldots,n$, I_i denotes the indicator function of the interval $\mathcal{J}_i:=(\intercal_{i-1},\intercal_i]$, with $\intercal_0:=1$ and $\intercal_n:=\intercal$), defined by

$$I_i(s)=\begin{cases} 1, & if\ s\in\mathcal{J}_i, \\ 0, & otherwise. \end{cases}$$

(H2) Let $i\in\{1,2,\ldots,n\}$ be a fixed integer, and let $f:\mathcal{J}_i\times\mathbb{R}\times\mathbb{R}\to\mathbb{R}$ be a continuous function. Assume there exists a constant $\delta\in[0,\min_{s\in\mathcal{J}}|\ltimes(s)|]$ such that the function $(\ell ogs)^{\delta}f$ is continuous on $\mathcal{J}_i\times\mathbb{R}\times\mathbb{R}$. Furthermore, let $K>0$ and $L>0$ be constants satisfying:

Remark 16 *According to the remark in [[23], p. 20], it can be readily shown that condition (H2) is equivalent to the following inequality:*

$$\zeta(s^{\delta}\|f(s,B_1,B_2)\|) \leq K\zeta(B_1) + L\zeta(B_2),$$

for any bounded sets $B_1, B_2 \subset X$ and for each $s \in \mathcal{J}$.

Further, for a given set U of functions $u : \mathcal{J} \to X$, we define

$$U(s) = \{u(s), u \in U\}, \ s \in \mathcal{J}$$

and

$$U(\mathcal{J}) = \{u(s) : v \in U, \ s \in \mathcal{J}\}.$$

We can now establish the existence of a solution to the initial value problem (5.29) by applying Darbo's fixed point theorem in conjunction with Kuratowski's measure of noncompactness.

For each $i \in \{1, 2, \ldots, n\}$, let $E_i = C(\mathcal{J}_i, X)$ denote the ßanach space of continuous functions $\varkappa : \mathcal{J}_i \to X$, equipped with the norm

$$\|\varkappa\|_{E_i} = \sup_{s \in \mathcal{J}_i} \|\varkappa(s)\|.$$

To begin with, we examine the ınitial and terminal υalue problem (ITVP) specified in equation (5.29).

In light of identity (2.12), the ITVP (5.29) can be reformulated as

$$\frac{1}{\Gamma(2-\ltimes(s))}(s\frac{d}{ds})^2 \int_1^s (\ell og\frac{s}{t})^{1-\ltimes(s)}\frac{\varkappa(t)}{t}dt = f(s, \varkappa(s), {}^H I_{1^+}^{\ltimes(s)}\varkappa(s)), \quad s \in \mathcal{J}. \tag{5.30}$$

Taking assumption $(H1)$ into account, the equation(5.30) on the interval $\mathcal{J}_i$, for $i = 1, 2, ..., n$, can be expressed as

$$(s\frac{d}{ds})^2\Big(\frac{1}{\Gamma(2-\ltimes_1)}\int_1^{\mathsf{T}_1} (\ell og\frac{s}{t})^{1-\ltimes_1}\frac{\varkappa(t)}{t}dt + ...$$

$$+\frac{1}{\Gamma(2-\ltimes_i)}\int_{\mathsf{T}_{i-1}}^s (\ell og\frac{s}{t})^{1-\ltimes_i}\frac{\varkappa(t)}{t}dt\Big) = f(s, \varkappa(s), {}^H I_{1^+}^{\ltimes(s)}\varkappa(s)), \quad s \in \mathcal{J}_i. \tag{5.31}$$

We shall present the solution to the ITVP solution (5.29) in the following.

Definition 53 *The initial and terminal value problem (ITVP) (5.29) admits a solution if there exist mappings $\varkappa_i$, $i = 1, 2, ..., n$, such that, $\varkappa_i \in C([1, \mathsf{T}_i], X)$ satisfy equation (5.31) and the boundary conditions $\varkappa_i(1) = 0 = \varkappa_i(\mathsf{T}_i)$.*

Based on the previous observation, ITVP (5.29) can be reformulated as shown in equation (5.30), with the subintervals $\mathcal{J}_i$, for $i \in \{1, 2, \dots, n\}$ represented as in equation (5.31).

For $1 \leq s \leq \mathsf{T}_{i-1}$, we set $\varkappa(s) \equiv 0$. Then equation (5.31) reduces to

$$^{H}F^{\ltimes_i}_{\mathsf{T}^+_{i-1}}\varkappa(s) = f(s, \varkappa(s), {}^{H}I^{\ltimes_i}_{\mathsf{T}^+_{i-1}}\varkappa(s)), \quad s \in \mathcal{J}_i.$$

We will address the ensuing initial and terminal value problem (ITVP):

$$\begin{cases} ^{H}F^{\ltimes_i}_{\mathsf{T}^+_{i-1}}\varkappa(s) = f(s, \varkappa(s), {}^{H}I^{\ltimes_i}_{\mathsf{T}^+_{i-1}}\varkappa(s)), \quad s \in \mathcal{J}_i \\ \varkappa(\mathsf{T}_{i-1}) = 0, \varkappa(\mathsf{T}_i) = 0. \end{cases} \tag{5.32}$$

The following lemma will serve as the foundation for our solution to equation (5.32).

Lemma 24 *A function $\varkappa \in E_i$ is a solution of equation* (5.32) *if and only if $\varkappa$ it satisfies the corresponding equation.*

$$\varkappa(s) = -(\ell og\frac{\mathsf{T}_i}{\mathsf{T}_{i-1}})^{1-\ltimes_i}(\ell og\frac{s}{\mathsf{T}_{i-1}})^{\ltimes_i-1}\, {}^{H}I^{\ltimes_i}_{\mathsf{T}^+_{i-1}} f(\mathsf{T}_i, \varkappa(\mathsf{T}_i), {}^{H}I^{\ltimes_i}_{\mathsf{T}^+_{i-1}}\varkappa(\mathsf{T}_i))$$
$$+ {}^{H}I^{\ltimes_i}_{\mathsf{T}^+_{i-1}} f(s, \varkappa(s), {}^{H}I^{\ltimes_i}_{\mathsf{T}^+_{i-1}}\varkappa(s)). \tag{5.33}$$

Proof: Assume that $\varkappa \in E_i$ is a solution to the ITVP (5.32). Applying the operator ${}^{H}I^{\ltimes_i}_{\mathsf{T}^+_{i-1}}$ to both sides of equation (5.32), and invoking Lemma 3, we obtain:

$$\varkappa(s) = \omega_1(\ell og\frac{s}{\mathsf{T}_{i-1}})^{\ltimes_i-1} + \omega_2(\ell og\frac{s}{\mathsf{T}_{i-1}})^{\ltimes_i-2} + {}^{H}I^{\ltimes_i}_{\mathsf{T}^+_{i-1}} f(s, \varkappa(s), {}^{H}I^{\ltimes_i}_{\mathsf{T}^+_{i-1}}\varkappa(s)), \quad s \in \mathcal{J}_i.$$

Given that f is continuous and $\varkappa(\mathsf{T}_{i-1}) = 0$, it follows that $\omega_2 = 0$.

Given that $\varkappa(s)$ satisfies $\varkappa(\mathsf{T}_i) = 0$, it follows that

$$\omega_1 = -(\ell og\frac{\mathsf{T}_i}{\mathsf{T}_{i-1}})^{1-\ltimes_i}\, {}^{H}I^{\ltimes_i}_{\mathsf{T}^+_{i-1}} f(\mathsf{T}_i, \varkappa(\mathsf{T}_i), {}^{H}I^{\ltimes_i}_{\mathsf{T}^+_{i-1}}\varkappa(\mathsf{T}_i))$$

Therefore, we obtain the explicit representation:

$$\varkappa(s) = -(\ell og\frac{\mathsf{T}_i}{\mathsf{T}_{i-1}})^{1-\ltimes_i}(\ell og\frac{s}{\mathsf{T}_{i-1}})^{\ltimes_i-1}\, {}^{H}I^{\ltimes_i}_{\mathsf{T}^+_{i-1}} f(\mathsf{T}_i, \varkappa(\mathsf{T}_i), {}^{H}I^{\ltimes_i}_{\mathsf{T}^+_{i-1}}\varkappa(\mathsf{T}_i))$$
$$+ {}^{H}I^{\ltimes_i}_{\mathsf{T}^+_{i-1}} f(s, \varkappa(s), {}^{H}I^{\ltimes_i}_{\mathsf{T}^+_{i-1}}\varkappa(s)), \quad s \in \mathcal{J}_i.$$

In contrast, consider the integral equation (5.33) whose solution is $\varkappa \in E_i$. By applying Lemma 3 and utilizing the continuity of the function $(\ell ogs)^{\delta} f$, we conclude that $\varkappa$ also satisfies the ITVP (5.32).

The Darbo fixed point theorem forms the foundation for our first existence result.

Theorem 46 *Assume that conditions (H1) and (H2) hold. If*

$$\frac{(\ell og \mathsf{T}_i)^{1-\delta} - (\ell og \mathsf{T}_{i-1})^{1-\delta}}{(1-\delta)\Gamma(\ltimes_i)}(\ell og\frac{\mathsf{T}_i}{\mathsf{T}_{i-1}})^{\ltimes_i - 1}\Big(K + \frac{L}{\Gamma(\ltimes_i + 1)}(\ell og\frac{\mathsf{T}_i}{\mathsf{T}_{i-1}})^{\ltimes_i}\Big) < \frac{1}{2}. \tag{5.34}$$

Then there exists at least one solution to the ITVP (5.32) in E_i.

Proof: We construct the operator

$$W : E_i \to E_i,$$

as detailed below

$$W\varkappa(s) = -\frac{1}{\Gamma(\ltimes_i)}(\ell og\frac{\mathsf{T}_i}{\mathsf{T}_{i-1}})^{1-\ltimes_i}(\ell og\frac{s}{\mathsf{T}_{i-1}})^{\ltimes_i-1}\int_{\mathsf{T}_{i-1}}^{\mathsf{T}_i}(\ell og\frac{\mathsf{T}_i}{t})^{\ltimes_i-1}\frac{f(t,\varkappa(t),{}^H I^{\ltimes_i}_{\mathsf{T}_{i-1}^+}\varkappa(t))}{t}dt$$

$$+\frac{1}{\Gamma(\ltimes_i)}\int_{\mathsf{T}_{i-1}}^{s}(\ell og\frac{s}{t})^{\ltimes_i-1}\frac{f(t,\varkappa(t),{}^H I^{\ltimes_i}_{\mathsf{T}_{i-1}^+}\varkappa(t))}{t}dt. \tag{5.35}$$

Assuming the continuity of the function $(\ell ogs)^{\delta} f$ and the properties of fractional integrals, we conclude that the operator $W : E_i \to E_i$ defined in equation (5.35), is well-defined.

Let

$$R_i \geq \frac{\frac{2f^*}{\Gamma(\ltimes_i+1)}(\ell og\frac{\mathsf{T}_i}{\mathsf{T}_{i-1}})^{\ltimes_i}}{1-\frac{2[(\ell og\mathsf{T}_i)^{1-\delta}-(\ell og\mathsf{T}_{i-1})^{1-\delta}]}{(1-\delta)\Gamma(\ltimes_i)}(\ell og\frac{\mathsf{T}_i}{\mathsf{T}_{i-1}})^{\ltimes_i-1}\Big(K+\frac{L}{\Gamma(\ltimes_i+1)}(\ell og\frac{\mathsf{T}_i}{\mathsf{T}_{i-1}})^{\ltimes_i}\Big)}$$

where

$$f^\star = \sup_{s\in\mathcal{J}_i}|f(s,0,0)|$$

We consider the set

$$B_{R_i} = \{\varkappa \in E_i,\ \|\varkappa\|_{E_i} \leq R_i\}$$

It is evident that B_{R_i} is bounded, closed, convex, and nonempty.

We now proceed to verify that the operator W satisfies the assumptions of the Darbo fixed point theorem. This will be demonstrated in four stages.

STEP 1: Claim: $W(B_{R_i}) \subseteq (B_{R_i})$.

For $\varkappa \in B_{R_i}$ and by assumption (H2), we obtain:

$$\|W\varkappa(s)\|$$

$$
\begin{aligned}
&\leq \frac{1}{\Gamma(\ltimes_i)}(\ell og\frac{\mathsf{T}_i}{\mathsf{T}_{i-1}})^{1-\ltimes_i}(\ell og\frac{s}{\mathsf{T}_{i-1}})^{\ltimes_i-1}\int_{\mathsf{T}_{i-1}}^{\mathsf{T}_i}(\ell og\frac{\mathsf{T}_i}{t})^{\ltimes_i-1}\|f(t,\varkappa(t),{}^H I^{\ltimes_i}_{\mathsf{T}^+_{i-1}}\varkappa(t))\|\frac{dt}{t}\\
&+ \frac{1}{\Gamma(\ltimes_i)}\int_{\mathsf{T}_{i-1}}^{s}(\ell og\frac{s}{t})^{\ltimes_i-1}\|f(t,\varkappa(t),{}^H I^{\ltimes_i}_{\mathsf{T}^+_{i-1}}\varkappa(t))\|\frac{dt}{t}\\
&\leq \frac{2}{\Gamma(\ltimes_i)}\int_{\mathsf{T}_{i-1}}^{\mathsf{T}_i}(\ell og\frac{\mathsf{T}_i}{t})^{\ltimes_i-1}\|f(t,\varkappa(t),{}^H I^{\ltimes_i}_{\mathsf{T}^+_{i-1}}\varkappa(t))\|\frac{dt}{t}
\end{aligned}
$$

$$
\begin{aligned}
&\leq \frac{2}{\Gamma(\ltimes_i)}\int_{\mathsf{T}_{i-1}}^{\mathsf{T}_i}(\ell og\frac{\mathsf{T}_i}{t})^{\ltimes_i-1}\|f(t,\varkappa(t),{}^H I^{\ltimes_i}_{\mathsf{T}^+_{i-1}}\varkappa(t))-f(t,0,0)\|\frac{dt}{t}\\
&+ \frac{2}{\Gamma(\ltimes_i)}\int_{\mathsf{T}_{i-1}}^{\mathsf{T}_i}(\ell og\frac{\mathsf{T}_i}{t})^{\ltimes_i-1}\|f(t,0,0)\|\frac{dt}{t}\\
&\leq \frac{2}{\Gamma(\ltimes_i)}\int_{\mathsf{T}_{i-1}}^{\mathsf{T}_i}(\ell og\frac{\mathsf{T}_i}{t})^{\ltimes_i-1}(\ell og t)^{-\delta}\Big(K\|\varkappa(t)\|+L\|{}^H I^{\ltimes_i}_{\mathsf{T}^+_{i-1}}\varkappa(t)\|\Big)\frac{dt}{t}\\
&+ \frac{2f^\star}{\Gamma(\ltimes_i)}\int_{\mathsf{T}_{i-1}}^{\mathsf{T}_i}(\ell og\frac{\mathsf{T}_i}{t})^{\ltimes_i-1}\frac{dt}{t}\\
&\leq \frac{2}{\Gamma(\ltimes_i)}(\ell og\frac{\mathsf{T}_i}{\mathsf{T}_{i-1}})^{\ltimes_i-1}\Big(K\|\varkappa\|_{E_i}+L\|{}^H I^{\ltimes_i}_{\mathsf{T}^+_{i-1}}\varkappa\|_{E_i}\Big)\int_{\mathsf{T}_{i-1}}^{\mathsf{T}_i}(\ell og t)^{-\delta}\frac{dt}{t}\\
&+ \frac{2f^\star}{\Gamma(\ltimes_i+1)}(\ell og\frac{\mathsf{T}_i}{\mathsf{T}_{i-1}})^{\ltimes_i}\\
&\leq \frac{2}{\Gamma(\ltimes_i)}(\ell og\frac{\mathsf{T}_i}{\mathsf{T}_{i-1}})^{\ltimes_i-1}\Big(K+\frac{L}{\Gamma(\ltimes_i+1)}(\ell og\frac{\mathsf{T}_i}{\mathsf{T}_{i-1}})^{\ltimes_i}\Big)\|\varkappa\|_{E_i}\int_{\mathsf{T}_{i-1}}^{\mathsf{T}_i}(\ell og t)^{-\delta}\frac{dt}{t}\\
&+ \frac{2f^\star}{\Gamma(\ltimes_i+1)}(\ell og\frac{\mathsf{T}_i}{\mathsf{T}_{i-1}})^{\ltimes_i}\\
&\leq \frac{2[(\ell og\mathsf{T}_i)^{1-\delta}-(\ell og\mathsf{T}_{i-1})^{1-\delta}]}{(1-\delta)\Gamma(\ltimes_i)}(\ell og\frac{\mathsf{T}_i}{\mathsf{T}_{i-1}})^{\ltimes_i-1}\Big(K+\frac{L}{\Gamma(\ltimes_i+1)}(\ell og\frac{\mathsf{T}_i}{\mathsf{T}_{i-1}})^{\ltimes_i}\Big)R_i\\
&+ \frac{2f^\star}{\Gamma(\ltimes_i+1)}(\ell og\frac{\mathsf{T}_i}{\mathsf{T}_{i-1}})^{\ltimes_i}\\
&\leq R_i,
\end{aligned}
$$

Since $\|\varkappa\| \leq E_i R_i$, it follows that $W(B_{R_i}) \in (B_{R_i})$.

STEP 2: Claim: The operator W is continuous.

Let $(\varkappa_n)$ be a sequence in E_i such that $\varkappa_n \to \varkappa \in inE_i$. Then,

$$\|(W\varkappa_n)(s) - (W\varkappa)(s)\|$$

$$\begin{aligned}
\leq\ & \frac{1}{\Gamma(\ltimes_i)}(\ell og\frac{\mathsf{T}_i}{\mathsf{T}_{i-1}})^{1-\ltimes_i}(\ell og\frac{s}{\mathsf{T}_{i-1}})^{\ltimes_i-1}\int_{\mathsf{T}_{i-1}}^{\mathsf{T}_i}(\ell og\frac{\mathsf{T}_i}{t})^{\ltimes_i-1} \\
& \|f(t,\varkappa_n(t),{}^H I^{\ltimes_i}_{\mathsf{T}^+_{i-1}}\varkappa_n(t)) - f(t,\varkappa(t),{}^H I^{\ltimes_i}_{\mathsf{T}^+_{i-1}}\varkappa(t))\|\frac{dt}{t} \\
+\ & \frac{1}{\Gamma(\ltimes_i)}\int_{\mathsf{T}_{i-1}}^{s}(\ell og\frac{s}{t})^{\ltimes_i-1}\|f(t,\varkappa_n(t),{}^H I^{\ltimes_i}_{\mathsf{T}^+_{i-1}}\varkappa_n(t)) - f(t,\varkappa(t),{}^H I^{\ltimes_i}_{\mathsf{T}^+_{i-1}}\varkappa(t))\|\frac{dt}{t} \\
\leq\ & \frac{2}{\Gamma(\ltimes_i)}(\ell og\frac{s}{\mathsf{T}_{i-1}})^{\ltimes_i-1}\int_{\mathsf{T}_{i-1}}^{\mathsf{T}_i}\|f(t,\varkappa_n(t),{}^H I^{\ltimes_i}_{\mathsf{T}^+_{i-1}}\varkappa_n(t)) - f(t,\varkappa(t),{}^H I^{\ltimes_i}_{\mathsf{T}^+_{i-1}}\varkappa(t))\|\frac{dt}{t} \\
\leq\ & \frac{2}{\Gamma(\ltimes_i)}(\ell og\frac{s}{\mathsf{T}_{i-1}})^{\ltimes_i-1}\int_{\mathsf{T}_{i-1}}^{\mathsf{T}_i}(\ell ogt)^{-\delta}\Big(K\|\varkappa_n(t)-\varkappa(t)\| + L\ {}^H I^{\ltimes_i}_{\mathsf{T}^+_{i-1}}\|\varkappa_n(t)-\varkappa(t)\|\Big)\frac{dt}{t}
\end{aligned}$$

$$\begin{aligned}
\leq\ & \frac{2}{\Gamma(\ltimes_i)}\Big(K\|\varkappa_n-\varkappa\|_{E_i} + L\ \|{}^H I^{\ltimes_i}_{\mathsf{T}^+_{i-1}}(\varkappa_n-\varkappa)\|_{E_i}\Big)(\ell og\frac{s}{\mathsf{T}_{i-1}})^{\ltimes_i-1}\int_{\mathsf{T}_{i-1}}^{\mathsf{T}_i}(\ell ogt)^{-\delta}\frac{dt}{t} \\
\leq\ & \frac{2[(\ell og\mathsf{T}_i)^{1-\delta} - (\ell og\mathsf{T}_{i-1})^{1-\delta}]}{(1-\delta)\Gamma(\ltimes_i)}(\ell og\frac{\mathsf{T}_i}{\mathsf{T}_{i-1}})^{\ltimes_i-1}\Big(K + \frac{L}{\Gamma(\ltimes_i+1)}(\ell og\frac{\mathsf{T}_i}{\mathsf{T}_{i-1}})^{\ltimes_i}\Big)\|\varkappa_n-\varkappa\|_{E_i},
\end{aligned}$$

i.e., we obtain

$$\|(W\varkappa_n) - (W\varkappa)\|_{E_i} \to 0 \quad as \quad n \to \infty.$$

Hence, the continuity of W is established on E_i.

STEP 3: Claim: W is characterized by boundedness and equicontinuity.

From Step 1, we conclude that $W(B_{R_i}) = \{W(\varkappa) : \varkappa \in B_{R_i}\} \subset B_{R_i}$. Therefore, $\|W(\varkappa)\|_{E_i} \leq R_i$ for all $\varkappa \in B_{R_i}$, implying that $W(B_{R_i})$ is bounded. Next, we must show that $W(B_{R_i})$ is equicontinuous.

For $s_1, s_2 \in \mathcal{J}_i,\ s_1 < s_2$ and $\varkappa \in B_{R_i}$, we have

$$\|(W\varkappa)(s_2) - (W\varkappa)(s_1)\|$$

$$\begin{aligned}
= \ & \Big\| - \frac{1}{\Gamma(\ltimes_i)}(\ell og\frac{\mathsf{T}_i}{\mathsf{T}_{i-1}})^{1-\ltimes_i}(\ell og\frac{s_2}{\mathsf{T}_{i-1}})^{\ltimes_i-1}\int_{\mathsf{T}_{i-1}}^{\mathsf{T}_i}(\ell og\frac{\mathsf{T}_i}{t})^{\ltimes_i-1}f(t,\varkappa(t),{}^H I_{\mathsf{T}_{i-1}^+}^{\ltimes_i}\varkappa(t))\frac{dt}{t} \\
+ \ & \frac{1}{\Gamma(\ltimes_i)}(\ell og\frac{\mathsf{T}_i}{\mathsf{T}_{i-1}})^{1-\ltimes_i}(\ell og\frac{s_1}{\mathsf{T}_{i-1}})^{\ltimes_i-1}\int_{\mathsf{T}_{i-1}}^{\mathsf{T}_i}(\ell og\frac{\mathsf{T}_i}{t})^{\ltimes_i-1}f(t,\varkappa(t),{}^H I_{\mathsf{T}_{i-1}^+}^{\ltimes_i}\varkappa(t))\frac{dt}{t} \\
+ \ & \frac{1}{\Gamma(\ltimes_i)}\int_{\mathsf{T}_{i-1}}^{s_2}(\ell og\frac{s_2}{t})^{\ltimes_i-1}f(t,\varkappa(t),{}^H I_{\mathsf{T}_{i-1}^+}^{\ltimes_i}\varkappa(t))\frac{dt}{t} \\
- \ & \frac{1}{\Gamma(\ltimes_i)}\int_{\mathsf{T}_{i-1}}^{s_1}(\ell og\frac{s_1}{t})^{\ltimes_i-1}f(t,\varkappa(t),{}^H I_{\mathsf{T}_{i-1}^+}^{\ltimes_i}\varkappa(t))\frac{dt}{t}\Big\| \\
\leq \ & \frac{1}{\Gamma(\ltimes_i)}(\ell og\frac{\mathsf{T}_i}{\mathsf{T}_{i-1}})^{1-\ltimes_i}\Big((\ell og\frac{s_2}{\mathsf{T}_{i-1}})^{\ltimes_i-1} - (\ell og\frac{s_1}{\mathsf{T}_{i-1}})^{\ltimes_i-1}\Big) \\
& \int_{\mathsf{T}_{i-1}}^{\mathsf{T}_i}(\ell og\frac{\mathsf{T}_i}{t})^{\ltimes_i-1}\|f(t,\varkappa(t),{}^H I_{\mathsf{T}_{i-1}^+}^{\ltimes_i}\varkappa(t))\|\frac{dt}{t} \\
+ \ & \frac{1}{\Gamma(\ltimes_i)}\int_{\mathsf{T}_{i-1}}^{s_1}\Big((\ell og\frac{s_2}{t})^{\ltimes_i-1} - (\ell og\frac{s_1}{t})^{\ltimes_i-1}\Big)\|f(t,\varkappa(t),{}^H I_{\mathsf{T}_{i-1}^+}^{\ltimes_i}\varkappa(t))\|\frac{dt}{t} \\
+ \ & \frac{1}{\Gamma(\ltimes_i)}\int_{s_1}^{s_2}(\ell og\frac{s_2}{t})^{\ltimes_i-1}\|f(t,\varkappa(t),{}^H I_{\mathsf{T}_{i-1}^+}^{\ltimes_i}\varkappa(t))\|\frac{dt}{t} \\
\leq \ & \frac{1}{\Gamma(\ltimes_i)}(\ell og\frac{\mathsf{T}_i}{\mathsf{T}_{i-1}})^{1-\ltimes_i}\Big((\ell og\frac{s_2}{\mathsf{T}_{i-1}})^{\ltimes_i-1} - (\ell og\frac{s_1}{\mathsf{T}_{i-1}})^{\ltimes_i-1}\Big) \\
& \int_{\mathsf{T}_{i-1}}^{\mathsf{T}_i}(\ell og\frac{\mathsf{T}_i}{t})^{\ltimes_i-1}\|f(t,\varkappa(t),{}^H I_{\mathsf{T}_{i-1}^+}^{\ltimes_i}\varkappa(t)) - f(t,0,0)\|\frac{dt}{t} \\
+ \ & \frac{1}{\Gamma(\ltimes_i)}(\ell og\frac{\mathsf{T}_i}{\mathsf{T}_{i-1}})^{1-\ltimes_i}\Big((\ell og\frac{s_2}{\mathsf{T}_{i-1}})^{\ltimes_i-1} - (\ell og\frac{s_1}{\mathsf{T}_{i-1}})^{\ltimes_i-1}\Big)\int_{\mathsf{T}_{i-1}}^{\mathsf{T}_i}(\ell og\frac{\mathsf{T}_i}{t})^{\ltimes_i-1}\|f(t,0,0)\|\frac{dt}{t}
\end{aligned}$$

$$
\begin{aligned}
&+ \frac{1}{\Gamma(\ltimes_i)} \int_{\mathsf{T}_{i-1}}^{s_1} (\ell og\frac{s_2}{s_1})^{\ltimes_i - 1} \| f(t, \varkappa(t), {}^H I^{\ltimes_i}_{\mathsf{T}^+_{i-1}} \varkappa(t)) - f(t,0,0) \| \frac{dt}{t} \\
&+ \frac{1}{\Gamma(\ltimes_i)} \int_{\mathsf{T}_{i-1}}^{s_1} (\ell og\frac{s_2}{s_1})^{\ltimes_i - 1} \| f(t,0,0) \| \frac{dt}{t} \\
&+ \frac{1}{\Gamma(\ltimes_i)} \int_{s_1}^{s_2} (\ell og\frac{s_2}{t})^{\ltimes_i - 1} \| f(t, \varkappa(t), {}^H I^{\ltimes_i}_{\mathsf{T}^+_{i-1}} \varkappa(t)) - f(t,0,0) \| \frac{dt}{t} \\
&+ \frac{1}{\Gamma(\ltimes_i)} \int_{s_1}^{s_2} (\ell og\frac{s_2}{t})^{\ltimes_i - 1} \| f(t,0,0) \| \frac{dt}{t} \\
&\le \frac{1}{\Gamma(\ltimes_i)} (\ell og\frac{\mathsf{T}_i}{\mathsf{T}_{i-1}})^{1-\ltimes_i} \Big((\ell og\frac{s_2}{\mathsf{T}_{i-1}})^{\ltimes_i - 1} - (\ell og\frac{s_1}{\mathsf{T}_{i-1}})^{\ltimes_i - 1} \Big) \\
&\quad \int_{\mathsf{T}_{i-1}}^{\mathsf{T}_i} (\ell og\frac{\mathsf{T}_i}{t})^{\ltimes_i - 1} (\ell og t)^{-\delta} (K \|\varkappa(t)\| + L\ {}^H I^{\ltimes_i}_{\mathsf{T}^+_{i-1}} \|\varkappa(t)\|) \frac{dt}{t} \\
&+ \frac{f^\star}{\Gamma(\ltimes_i)} (\ell og\frac{\mathsf{T}_i}{\mathsf{T}_{i-1}})^{1-\ltimes_i} \Big((\ell og\frac{s_2}{\mathsf{T}_{i-1}})^{\ltimes_i - 1} - (\ell og\frac{s_1}{\mathsf{T}_{i-1}})^{\ltimes_i - 1} \Big) \int_{\mathsf{T}_{i-1}}^{\mathsf{T}_i} (\ell og\frac{\mathsf{T}_i}{t})^{\ltimes_i - 1} \frac{dt}{t} \\
&+ \frac{1}{\Gamma(\ltimes_i)} \int_{\mathsf{T}_{i-1}}^{s_1} (\ell og\frac{s_2}{s_1})^{\ltimes_i - 1} (\ell og t)^{-\delta} (K \|\varkappa(t)\| + L\ {}^H I^{\ltimes_i}_{\mathsf{T}^+_{i-1}} \|\varkappa(t)\|) \frac{dt}{t} \\
&+ \frac{f^\star}{\Gamma(\ltimes_i)} \int_{\mathsf{T}_{i-1}}^{s_1} (\ell og\frac{s_2}{s_1})^{\ltimes_i - 1} \frac{dt}{t} \\
&+ \frac{1}{\Gamma(\ltimes_i)} \int_{s_1}^{s_2} (\ell og\frac{s_2}{t})^{\ltimes_i - 1} (\ell og t)^{-\delta} (K \|\varkappa(t)\| + L\ {}^H I^{\ltimes_i}_{\mathsf{T}^+_{i-1}} \|\varkappa(t)\|) \frac{dt}{t} \\
&+ \frac{f^\star}{\Gamma(\ltimes_i)} \int_{s_1}^{s_2} (\ell og\frac{s_2}{t})^{\ltimes_i - 1} \frac{dt}{t} \\
&\le \frac{1}{\Gamma(\ltimes_i)} (\ell og\frac{\mathsf{T}_i}{\mathsf{T}_{i-1}})^{1-\ltimes_i} (\ell og\frac{\mathsf{T}_i}{\mathsf{T}_{i-1}})^{\ltimes_i - 1} \Big((\ell og\frac{s_2}{\mathsf{T}_{i-1}})^{\ltimes_i - 1} - (\ell og\frac{s_1}{\mathsf{T}_{i-1}})^{\ltimes_i - 1} \Big) \\
&\quad \Big(K \|\varkappa\|_{E_i} + L \| {}^H I^{\ltimes_i}_{\mathsf{T}^+_{i-1}} \varkappa \|_{E_i} \Big) \int_{\mathsf{T}_{i-1}}^{\mathsf{T}_i} (\ell og t)^{-\delta} \frac{dt}{t} + \frac{f^\star}{\Gamma(\ltimes_i + 1)} (\ell og\frac{\mathsf{T}_i}{\mathsf{T}_{i-1}})^{1-\ltimes_i} \\
&\quad \Big((\ell og\frac{s_2}{\mathsf{T}_{i-1}})^{\ltimes_i - 1} - (\ell og\frac{s_1}{\mathsf{T}_{i-1}})^{\ltimes_i - 1} \Big) (\ell og\frac{s_i}{s_{i-1}})^{\ltimes_i} \\
&+ \frac{(\ell og s_2)^{1-\delta} - (\ell og s_1)^{1-\delta}}{(1-\delta)\Gamma(\ltimes_i)} (\ell og\frac{s_2}{s_1})^{\ltimes_i - 1} \Big(K \|\varkappa\|_{E_i} + L \| {}^H I^{\ltimes_i}_{\mathsf{T}^+_{i-1}} \varkappa \|_{E_i} \Big) \\
&+ \frac{f^\star}{\Gamma(\ltimes_i)} (\ell og\frac{s_2}{s_1})^{\ltimes_i - 1} (\ell og s_1 - \ell og \mathsf{T}_{i-1}) \\
&+ \frac{(\ell og s_2)^{1-\delta} - (\ell og s_1)^{1-\delta}}{(1-\delta)\Gamma(\ltimes_i)} (\ell og\frac{s_2}{s_1})^{\ltimes_i - 1} \Big(K \|\varkappa\|_{E_i} + L \| {}^H I^{\ltimes_i}_{\mathsf{T}^+_{i-1}} \varkappa \|_{E_i} \Big) + \frac{f^\star}{\Gamma(\ltimes_i + 1)} (\ell og\frac{s_2}{s_1})^{\ltimes_i} \\
&\le \frac{(\ell og \mathsf{T}_i)^{1-\delta} - (\ell og \mathsf{T}_{i-1})^{1-\delta}}{(1-\delta)\Gamma(\ltimes_i)} \Big(K + \frac{L}{\Gamma(\ltimes_i + 1)} (\ell og\frac{\mathsf{T}_i}{\mathsf{T}_{i-1}})^{\ltimes_i} \Big) \|\varkappa\|_{E_i} \\
&\quad \Big((\ell og\frac{s_2}{\mathsf{T}_{i-1}})^{\ltimes_i - 1} - (\ell og\frac{s_1}{\mathsf{T}_{i-1}})^{\ltimes_i - 1} \Big)
\end{aligned}
$$

$$
\begin{aligned}
&+ \frac{f^\star}{\Gamma(\ltimes_i+1)}(\ell og\frac{\mathsf{T}_i}{\mathsf{T}_{i-1}})\Big((\ell og\frac{s_2}{\mathsf{T}_{i-1}})^{\ltimes_i-1}-(\ell og\frac{s_1}{\mathsf{T}_{i-1}})^{\ltimes_i-1}\Big)\\
&+ \frac{2((\ell og s_2)^{1-\delta}-(\ell og s_1)^{1-\delta})}{(1-\delta)\Gamma(\ltimes_i)}(\ell og\frac{s_2}{s_1})^{\ltimes_i-1}\Big(K+\frac{L}{\Gamma(\ltimes_i+1)}(\ell og\frac{\mathsf{T}_i}{\mathsf{T}_{i-1}})^{\ltimes_i}\Big)\|\varkappa\|_{E_i}\\
&+ \frac{f^\star}{\Gamma(\ltimes_i)}(\ell og\frac{s_2}{s_1})^{\ltimes_i-1}(\ell og s_1-\ell og\mathsf{T}_{i-1})+\frac{f^\star}{\Gamma(\ltimes_i+1)}(\ell og\frac{s_2}{s_1})^{\ltimes_i}\\
&\le \Big[\frac{(\ell og\mathsf{T}_i)^{1-\delta}-(\ell og\mathsf{T}_{i-1})^{1-\delta}}{(1-\delta)\Gamma(\ltimes_i)}\Big(K+\frac{L}{\Gamma(\ltimes_i+1)}(\ell og\frac{\mathsf{T}_i}{\mathsf{T}_{i-1}})^{\ltimes_i}\Big)\|\varkappa\|_{E_i}+\frac{f^\star}{\Gamma(\ltimes_i+1)}(\ell og\frac{\mathsf{T}_i}{\mathsf{T}_{i-1}})\Big]\\
&\quad \Big((\ell og\frac{s_2}{\mathsf{T}_{i-1}})^{\ltimes_i-1}-(\ell og\frac{s_1}{\mathsf{T}_{i-1}})^{\ltimes_i-1}\Big)\\
&+ \Big[\frac{2((\ell og s_2)^{1-\delta}-(\ell og s_1)^{1-\delta})}{(1-\delta)\Gamma(\ltimes_i)}\Big(K+\frac{L}{\Gamma(\ltimes_i+1)}(\ell og\frac{\mathsf{T}_i}{\mathsf{T}_{i-1}})^{\ltimes_i}\Big)\|\varkappa\|_{E_i}\\
&+ \frac{f^\star}{\Gamma(\ltimes_i)}(\ell og s_1-\ell og\mathsf{T}_{i-1})\Big](\ell og\frac{s_2}{s_1})^{\ltimes_i-1}+\frac{f^\star}{\Gamma(\ltimes_i+1)}(\ell og\frac{s_2}{s_1})^{\ltimes_i}
\end{aligned}
$$

Hence, $\|(W\varkappa)(s_2)-(W\varkappa)(s_1)\| \to 0$ as $|s_2-s_1| \to 0$. This implies that $W(B_{R_i})$ is equicontinuous.

STEP 4: The operator W is a k-set contraction.

For $U \in B_{R_i}$ and $s \in \mathcal{J}_i$, we obtain:

$$
\begin{aligned}
\zeta(W(U)(s)) &= \zeta((Wy)(s), y\in U)\\
&\le \Big\{\tfrac{1}{\Gamma(\ltimes_i)}(\ell og\tfrac{\mathsf{T}_i}{\mathsf{T}_{i-1}})^{1-\ltimes_i}(\ell og\tfrac{s}{\mathsf{T}_{i-1}})^{\ltimes_i-1}\int_{\mathsf{T}_{i-1}}^{\mathsf{T}_i}(\ell og\tfrac{\mathsf{T}_i}{t})^{\ltimes_i-1}\zeta f(t,\varkappa(t),{}^H I^{\ltimes_i}_{\mathsf{T}_{i-1}^+}\varkappa(t))dt\\
&\quad +\tfrac{1}{\Gamma(\ltimes_i)}\int_{\mathsf{T}_{i-1}}^{s}(\ell og\tfrac{s}{t})^{\ltimes_i-1}\zeta f(t,\varkappa(t),{}^H I^{\ltimes_i}_{\mathsf{T}_{i-1}^+}\varkappa(t))dt,\ y\in U\Big\}.
\end{aligned}
$$

Then, Remark 16 implies that for each $t \in \mathcal{J}_i$,

$$
\begin{aligned}
&\zeta(W(U)(s))\\
&\le \Big\{\tfrac{1}{\Gamma(\ltimes_i)}(\ell og\tfrac{\mathsf{T}_i}{\mathsf{T}_{i-1}})^{1-\ltimes_i}(\ell og\tfrac{s}{\mathsf{T}_{i-1}})^{\ltimes_i-1}\int_{\mathsf{T}_{i-1}}^{\mathsf{T}_i}(\ell og\tfrac{\mathsf{T}_i}{t})^{\ltimes_i-1}\Big[K\widehat{\zeta}(U)\int_{\mathsf{T}_{i-1}}^{\mathsf{T}_i}t^{-\delta}dt\\
&\quad +\tfrac{L}{\Gamma(\ltimes_i+1)}(\ell og\tfrac{\mathsf{T}_i}{\mathsf{T}_{i-1}})^{\ltimes_i}\widehat{\zeta}(U)\int_{\mathsf{T}_{i-1}}^{\mathsf{T}_i}t^{-\delta}dt\Big]+\tfrac{1}{\Gamma(\ltimes_i)}\int_{\mathsf{T}_{i-1}}^{s}(\ell og\tfrac{s}{t})^{\ltimes_i-1}\Big[K\widehat{\zeta}(U)\int_{\mathsf{T}_{i-1}}^{s}t^{-\delta}dt\\
&\quad +\tfrac{L}{\Gamma(\ltimes_i+1)}(\ell og\tfrac{s}{\mathsf{T}_{i-1}})^{\ltimes_i}\widehat{\zeta}(U)\int_{s_{i-1}}^{s}t^{-\delta}dt\Big],\ \varkappa\in U\Big\}.\\
&\le \Big\{\tfrac{2}{\Gamma(\ltimes_i)}(\ell og\tfrac{\mathsf{T}_i}{\mathsf{T}_{i-1}})^{\ltimes_i-1}\int_{\mathsf{T}_{i-1}}^{\mathsf{T}_i}\Big[K\widehat{\zeta}(U)\int_{\mathsf{T}_{i-1}}^{\mathsf{T}_i}t^{-\delta}dt+\tfrac{L}{\Gamma(\ltimes_i+1)}(\ell og\tfrac{\mathsf{T}_i}{\mathsf{T}_{i-1}})^{\ltimes_i}\widehat{\zeta}(U)\int_{\mathsf{T}_{i-1}}^{\mathsf{T}_i}t^{-\delta}dt\Big],\ \varkappa\in U\Big\}.\\
&\le \tfrac{2((\ell og\mathsf{T}_i)^{1-\delta}-(\ell og\mathsf{T}_{i-1})^{1-\delta})}{(1-\delta)\Gamma(\ltimes_i)}(\ell og\tfrac{\mathsf{T}_i}{\mathsf{T}_{i-1}})^{\ltimes_i-1}\Big(K+\tfrac{L}{\Gamma(\ltimes_i+1)}(\ell og\tfrac{\mathsf{T}_i}{\mathsf{T}_{i-1}})^{\ltimes_i}\Big)\widehat{\zeta}(U)
\end{aligned}
$$

Thus,

$$\widehat{\varsigma}(WU) \leq \frac{2((\ell og\mathsf{T}_i)^{1-\delta} - (\ell og\mathsf{T}_{i-1})^{1-\delta})}{(1-\delta)\Gamma(\ltimes_i)}(\ell og\frac{\mathsf{T}_i}{\mathsf{T}_{i-1}})^{\ltimes_i - 1}\Big(K + \frac{L}{\Gamma(\ltimes_i + 1)}(\ell og\frac{\mathsf{T}_i}{\mathsf{T}_{i-1}})^{\ltimes_i}\Big)\widehat{\varsigma}(U).$$

We then conclude that W forms a strict contraction by (5.34). According to the Darbo fixed point theorem, $W(\widetilde{\varkappa}_i) = \widetilde{\varkappa}_i$ is a solution of the initial terminal value problem (ITVP) (5.32), and hence, there exists $\widetilde{\varkappa}_i \in B_{R_i}$.

We shall now demonstrate that there are solutions to the ITVP (5.29).

Present the premise that follows:

(H3) Assume f is a continuous function on $\mathcal{J} \times \mathbb{R} \times \mathbb{R}$. Suppose there exists a δ with $0 \leq \delta \leq \min_{s \in \mathcal{J}} |\ltimes(s)|$ such that the function $(\ell ogs)^{\delta} f$ is continuous on $\mathcal{J} \times \mathbb{R} \times \mathbb{R}$. Additionally, assume there exist constants $K > 0$ and $L > 0$ such that

$$(\ell ogs)^{\delta}|f(s, \varkappa_1, y_1) - f(s, \varkappa_2, y_2)| \leq K|\varkappa_1 - \varkappa_2| + L|y_1 - y_2|,$$

for all $\varkappa_1, \varkappa_2, y_1, y_2 \in \mathbb{R}$ and $s \in \mathcal{J}$.

Theorem 47 *Assume that conditions (H1), (H3) and inequality (5.34) hold for all* $i \in 1, 2, ..., n$. *Then, the initial and terminal value problem (ITVP) (5.29) admits a solution in* $C(\mathcal{J}, \mathbb{R})$.

Proof: In accordance with Theorem 46, the initial and terminal value problem for Hadamard fractional differential equations of constant order (5.32) admits a solution $\widetilde{\varkappa}_i \in E_i$ for each $i \in \{1, 2, \ldots, n\}$. We define the function

$$\varkappa_i = \begin{cases} 0, & s \in [1, \mathsf{T}_{i-1}], \\ \widetilde{\varkappa}_i, & s \in \mathcal{J}_i, \end{cases} \tag{5.36}$$

so that the mapping $\varkappa_i \in C([1, \mathsf{T}_i], \mathbb{R})$ satisfies the integral relation (5.31) for $s \in \mathcal{J}_i$, with $\varkappa_i(1) = 0$ and $\varkappa_i(\mathsf{T}_i) = \widetilde{\varkappa}_i(\mathsf{T}_i) = 0$, in accordance with the terminal conditions.

Thus, the mapping

$$\varkappa(s)=\begin{cases}\varkappa_1(s), & s\in\mathcal{J}_1,\\ \varkappa_2(s)=\begin{cases}0, & s\in\mathcal{J}_1,\\ \widetilde{\varkappa}_2, & s\in\mathcal{J}_2\end{cases} & .\\ \cdot\\ \cdot\\ \cdot\\ \varkappa_n(s)=\begin{cases}0, & s\in[1,\mathsf{T}_{i-1}],\\ \widetilde{\varkappa}_i, & s\in\mathcal{J}_i\end{cases}\end{cases}$$

defines a piecewise function $\varkappa\in C(J,\mathcal{R})$ that constitutes a solution to the initial and terminal value problem (ITVP) (5.29).

5.3.2 Stability Results

Theorem 48 *Assume that conditions (H1), (H3), and inequality (5.34) hold, and additionally:*

(H4) *The mapping $\vartheta\in C(\mathcal{J}_i,\mathbb{R}_+)$ is non-decreasing and there exists $\lambda_\vartheta>0$ such that, for each $s\in\mathcal{J}_i$,*

$$^{H}I^{\ltimes_i}_{\mathsf{T}_{i-1}^+}\vartheta(s)\le\lambda_{\vartheta(s)}\vartheta(s),$$

then, the initial and terminal value problem (ITVP) (5.29) *is Ulam-Hyers-$\mathfrak{R}$assias stable with respect to ϑ.*

Proof: Let $\epsilon>0$ be an arbitrary constant, and let the function $z(s)\in C(\mathcal{J},\mathbb{R})$ satisfy the inequality (2.30).

For each $i\in\{1,2,\ldots,n\}$, define the functions $z_i(s)$ as follows: - For $i=1$, set $z_1(s)\equiv z(s)$ for $s\in[1,\mathsf{T}_1]$. - For $i=2,3,\ldots,n$, define

$$z_i(s)=\begin{cases}0, & s\in[1,\mathsf{T}_{i-1}],\\ z(s), & s\in\mathcal{J}_i.\end{cases}$$

For $s\in\mathcal{J}_i,\ i=1,2,\ldots,n$ we obtain

$$^{H}F^{u(s)}_{1^+}z_i(s)=\frac{s^2}{\Gamma(2-u(s))}\frac{d^2}{ds^2}\int_{\mathsf{T}_{i-1}}^{s}(\ell og\frac{s}{t})^{1-u_i}\frac{z(t)}{t}dt.$$

We compute the Hadamard fractional integral ${}^{HI}_{\mathsf{T}_{i-1}^{+}} I^{\ltimes_i}$ on both sides of inequality (2.30), resulting in

$$\begin{aligned}
\|z_i(s) \quad &+ \quad \frac{1}{\Gamma(\ltimes_i)}(\ell og\frac{\mathsf{T}_i}{\mathsf{T}_{i-1}})^{1-\ltimes_i}(\ell og\frac{s}{\mathsf{T}_{i-1}})^{\ltimes_i-1}\int_{\mathsf{T}_{i-1}}^{\mathsf{T}_i}(\ell og\frac{\mathsf{T}_i}{t})^{\ltimes_i-1}\frac{f(t,z_i(t),{}^{H}I^{\ltimes_i}_{\mathsf{T}_{i-1}^{+}}z_i(t))}{t}dt \\
&- \quad \frac{1}{\Gamma(\ltimes_i)}\int_{\mathsf{T}_{i-1}}^{s}(\ell og\frac{s}{t})^{\ltimes_i-1}\frac{f(t,z_i(t),{}^{H}I^{\ltimes_i}_{\mathsf{T}_{i-1}^{+}}z_i(t))}{t}dt\| \\
&\leq \quad \epsilon\frac{1}{\Gamma(\ltimes_i)}\int_{\mathsf{T}_{i-1}}^{s}\frac{1}{t}(\ell og\frac{s}{t})^{\ltimes_i-1}\vartheta(t)dt \\
&\leq \quad \epsilon\lambda_{\vartheta(s)}\vartheta(\mathsf{T}).
\end{aligned}$$

Based on Theorem 47, the initial and terminal value problem (ITVP) (5.29) admits a solution $\varkappa \in C(\mathcal{J},\mathbb{R})$, expressed as $\varkappa(s) = \varkappa_i(s)$ for $s \in \mathcal{J}_i$, $i = 1,2,\ldots,n$, where

$$\varkappa_i = \begin{cases} 0, & s \in [1,\mathsf{T}_{i-1}], \\ \widetilde{\varkappa}_i, & s \in \mathcal{J}_i, \end{cases} \tag{5.37}$$

and $\widetilde{\varkappa}_i \in E_i$ solves the initial and terminal value problem (ITVP) (5.32). By Lemma 24, the integral equation

$$\begin{aligned}
\widetilde{\varkappa}_i(s) \quad &= \quad - \quad \frac{1}{\Gamma(\ltimes_i)}(\ell og\frac{\mathsf{T}_i}{\mathsf{T}_{i-1}})^{1-\ltimes_i}(\ell og\frac{s}{\mathsf{T}_{i-1}})^{\ltimes_i-1}\int_{\mathsf{T}_{i-1}}^{\mathsf{T}_i}(\ell og\frac{\mathsf{T}_i}{t})^{\ltimes_i-1}\frac{f(t,\widetilde{\varkappa}_i(t),{}^{H}I^{\ltimes_i}_{\mathsf{T}_{i-1}^{+}}\widetilde{\varkappa}_i(t))}{t}dt \\
&+ \quad \frac{1}{\Gamma(\ltimes_i)}\int_{\mathsf{T}_{i-1}}^{s}(\ell og\frac{s}{t})^{\ltimes_i-1}\frac{f(t,\widetilde{\varkappa}_i(t),{}^{H}I^{\ltimes_i}_{\mathsf{T}_{i-1}^{+}}\widetilde{\varkappa}_i(t))}{t}dt.
\end{aligned}$$

holds.

Hence, for $s \in \mathcal{J}_i$ with $i = 1,2,\ldots,n$, we obtain

$$\|z(s) - \varkappa(s)\| = \|z(s) - \varkappa_i(s)\| = \|z_i(s) - \widetilde{\varkappa}_i(s)\|$$

$$
\begin{aligned}
= & \ \|z_i(s) + \frac{1}{\Gamma(\ltimes_i)}(\ell og\frac{\mathsf{T}_i}{\mathsf{T}_{i-1}})^{1-\ltimes_i}(\ell og\frac{s}{\mathsf{T}_{i-1}})^{\ltimes_i-1}\int_{\mathsf{T}_{i-1}}^{\mathsf{T}_i}(\ell og\frac{\mathsf{T}_i}{t})^{\ltimes_i-1}\frac{f(t,\widetilde{\varkappa}_i(t),{}^H I_{\mathsf{T}_{i-1}^+}^{\ltimes_i}\widetilde{\varkappa}_i(t))}{t}dt \\
- & \ \frac{1}{\Gamma(\ltimes_i)}\int_{\mathsf{T}_{i-1}}^{s}(\ell og\frac{s}{t})^{\ltimes_i-1}\frac{f(t,\widetilde{\varkappa}_i(t),{}^H I_{\mathsf{T}_{i-1}^+}^{\ltimes_i}\widetilde{\varkappa}_i(t))}{t}dt\| \\
\leq & \ \|z_i(s) + \frac{1}{\Gamma(\ltimes_i)}(\ell og\frac{\mathsf{T}_i}{\mathsf{T}_{i-1}})^{1-\ltimes_i}(\ell og\frac{s}{\mathsf{T}_{i-1}})^{\ltimes_i-1}\int_{\mathsf{T}_{i-1}}^{\mathsf{T}_i}(\ell og\frac{\mathsf{T}_i}{t})^{\ltimes_i-1}\frac{f(t,z_i(t),{}^H I_{\mathsf{T}_{i-1}^+}^{\ltimes_i}z_i(t))}{t}dt \\
- & \ \frac{1}{\Gamma(\ltimes_i)}\int_{\mathsf{T}_{i-1}}^{s}(\ell og\frac{s}{t})^{\ltimes_i-1}\frac{f(t,z_i(t),{}^H I_{\mathsf{T}_{i-1}^+}^{\ltimes_i}z_i(t))}{t}dt\| \\
+ & \ \frac{1}{\Gamma(\ltimes_i)}(\ell og\frac{\mathsf{T}_i}{\mathsf{T}_{i-1}})^{1-\ltimes_i}(\ell og\frac{s}{\mathsf{T}_{i-1}})^{\ltimes_i-1}\int_{\mathsf{T}_{i-1}}^{\mathsf{T}_i}(\ell og\frac{\mathsf{T}_i}{t})^{\ltimes_i-1} \\
& \ \frac{\|f(t,z_i(t),{}^H I_{\mathsf{T}_{i-1}^+}^{\ltimes_i}z_i(t)) - f(t,\widetilde{\varkappa}_i(t),{}^H I_{\mathsf{T}_{i-1}^+}^{\ltimes_i}\widetilde{\varkappa}_i(t))\|}{t}dt \\
+ & \ \frac{1}{\Gamma(\ltimes_i)}\int_{\mathsf{T}_{i-1}}^{s}(\ell og\frac{s}{t})^{\ltimes_i-1}\frac{\|f(t,z_i(t),{}^H I_{\mathsf{T}_{i-1}^+}^{\ltimes_i}z_i(t)) - f(t,\widetilde{\varkappa}_i(t),{}^H I_{\mathsf{T}_{i-1}^+}^{\ltimes_i}\widetilde{\varkappa}_i(t))\|}{t}dt \\
\leq & \ \lambda_{\vartheta(s)}\epsilon\vartheta(s) + \frac{1}{\Gamma(\ltimes_i)}(\ell og\frac{\mathsf{T}_i}{\mathsf{T}_{i-1}})^{1-\ltimes_i}(\ell og\frac{s}{\mathsf{T}_{i-1}})^{\ltimes_i-1} \\
& \ \int_{\mathsf{T}_{i-1}}^{\mathsf{T}_i}(\ell og\frac{\mathsf{T}_i}{t})^{\ltimes_i-1}\frac{1}{t}(\ell ogt)^{-\delta}(K\|z_i(t)-\widetilde{\varkappa}_i(t)\| + L\ {}^H I_{\mathsf{T}_{i-1}^+}^{\ltimes_i}\|z_i(t)-\widetilde{\varkappa}_i(t)\|)dt \\
+ & \ \frac{1}{\Gamma(\ltimes_i)}\int_{\mathsf{T}_{i-1}}^{s}(\ell og\frac{s}{t})^{\ltimes_i-1}\frac{1}{t}(\ell ogt)^{-\delta}(K\|z_i(t)-\widetilde{\varkappa}_i(t)\| + L\ {}^H I_{\mathsf{T}_{i-1}^+}^{\ltimes_i}\|z_i(t)-\widetilde{\varkappa}_i(t)\|)dt \\
\leq & \ \lambda_{\vartheta(s)}\epsilon\vartheta(s) + \frac{1}{\Gamma(\ltimes_i)}(\ell og\frac{s}{\mathsf{T}_{i-1}})^{\ltimes_i-1} \\
& \ \int_{\mathsf{T}_{i-1}}^{\mathsf{T}_i}\frac{1}{t}(\ell ogt)^{-\delta}(K\|z_i(t)-\widetilde{\varkappa}_i(t)\| + L\ {}^H I_{\mathsf{T}_{i-1}^+}^{\ltimes_i}\|z_i(t)-\widetilde{\varkappa}_i(t)\|)dt \\
+ & \ \frac{1}{\Gamma(\ltimes_i)}(\ell og\frac{s}{\mathsf{T}_{i-1}})^{\ltimes_i-1}\int_{\mathsf{T}_{i-1}}^{s}\frac{1}{t}(\ell ogt)^{-\delta}(K\|z_i(t)-\widetilde{\varkappa}_i(t)\| + L\ {}^H I_{\mathsf{T}_{i-1}^+}^{\ltimes_i}\|z_i(t)-\widetilde{\varkappa}_i(t)\|)dt \\
\leq & \ \lambda_{\vartheta(s)}\epsilon\vartheta(s) + \frac{(\ell og\mathsf{T}_i)^{1-\delta} - (\ell og\mathsf{T}_{i-1})^{1-\delta}}{(1-\delta)\Gamma(\ltimes_i)}(\ell og\frac{s}{\mathsf{T}_{i-1}})^{\ltimes_i-1} \\
& \ (K\|z_i-\widetilde{\varkappa}_i\|_{E_i} + L\ {}^H I_{\mathsf{T}_{i-1}^+}^{\ltimes_i}\|z_i-\widetilde{\varkappa}_i\|_{E_i}) \\
+ & \ \frac{(\ell ogs)^{1-\delta} - (\ell og\mathsf{T}_{i-1})^{1-\delta}}{(1-\delta)\Gamma(\ltimes_i)}(\ell og\frac{s}{\mathsf{T}_{i-1}})^{\ltimes_i-1}(K\|z_i-\widetilde{\varkappa}_i\|_{E_i} + L\ {}^H I_{\mathsf{T}_{i-1}^+}^{\ltimes_i}\|z_i-\widetilde{\varkappa}_i\|_{E_i}) \\
\leq & \ \lambda_{\vartheta(s)}\epsilon\vartheta(s) + \frac{2[(\ell og\mathsf{T}_i)^{1-\delta} - (\ell og\mathsf{T}_{i-1})^{1-\delta}]}{(1-\delta)\Gamma(\ltimes_i)}(\ell og\frac{\mathsf{T}_i}{\mathsf{T}_{i-1}})^{\ltimes_i-1} \\
& \ (K + \frac{L}{\Gamma(\ltimes_i+1)}(\ell og\frac{\mathsf{T}_i}{\mathsf{T}_{i-1}})^{\ltimes_i})\|z_i-\widetilde{\varkappa}_i\|_{E_i} \\
\leq & \ \lambda_{\vartheta(s)}\epsilon\vartheta(s) + \mu\|z-\varkappa\|,
\end{aligned}
$$

where

$$\mu = \max_{i=1,2,\ldots,n} \frac{2[(\ell og\mathsf{T}_i)^{1-\delta} - (\ell og\mathsf{T}_{i-1})^{1-\delta}]}{(1-\delta)\Gamma(\ltimes_i)}(\ell og\frac{\mathsf{T}_i}{\mathsf{T}_{i-1}})^{\ltimes_i - 1}(K + \frac{L}{\Gamma(\ltimes_i + 1)}(\ell og\frac{\mathsf{T}_i}{\mathsf{T}_{i-1}})^{\ltimes_i}).$$

Then,

$$\|z - \varkappa\|(1-\mu) \leq \lambda_{\vartheta(s)}\epsilon\vartheta(s).$$

We find, for any $s \in \mathcal{J}$,

$$|z(s) - \varkappa(s)| \leq \|z - \varkappa\| \leq \frac{\lambda_{\vartheta(s)}\vartheta(s)}{1-\mu}\epsilon := c_f\epsilon.$$

Therefore, the ITVP (5.29) is Ulam-Hyers-$\mathfrak{R}$assias stable with respect to ϑ.

5.3.3 Applied Representation

We address the fractional ınitial and terminal value problem in this case:

$$\begin{cases} F_{1^+}^{\ltimes(s)}\varkappa(s) = \frac{7}{5\sqrt{\pi}}(\ell ogs)^{\ltimes(s)} + \frac{(\ell ogs)^{-\frac{1}{3}}}{s+3}\varkappa(s) + \frac{\sqrt{\pi}\ell ogs}{s^2+1}\,{}^H I_{1^+}^{\ltimes(s)}\varkappa(s), \quad s \in \mathcal{J} := [1,e], \\ \varkappa(1) = 0, \quad \varkappa(e) = 0. \end{cases} \tag{5.38}$$

Let

$$f(s,\varkappa,y) = \frac{7}{5\sqrt{\pi}}(\ell ogs)^{\ltimes(s)} + \frac{(\ell ogs)^{-\frac{1}{3}}}{s+3}\varkappa(s) + \frac{\sqrt{\pi}\ell ogs}{s^2+1}y(s),$$

$$(s,\varkappa,y) \in [1,e] \times [0,+\infty) \times [0,+\infty).$$

$$where \qquad \ltimes(s) = \begin{cases} 1.3, & s \in \mathcal{J}_1 := [1,2], \\ 1.7, & s \in \mathcal{J}_2 :=]2,e]. \end{cases} \tag{5.39}$$

Accordingly, we find

$$(\ell ogs)^{\frac{1}{3}}\Big|f(s,\varkappa_1,y_1) - f(s,\varkappa_2,y_2)\Big|$$

$$\begin{aligned} &= \left|\frac{1}{s+3}\varkappa_1(s) + \frac{\sqrt{\pi}(\ell ogs)^{\frac{4}{3}}}{s^2+1}y_1(s) - \frac{1}{s+3}\varkappa_2(s) - \frac{\sqrt{\pi}(\ell ogs)^{\frac{4}{3}}}{s^2+1}y_2(s)\right| \\ &\leq \frac{1}{s+3}|\varkappa_1(s) - \varkappa_2(s)| + \frac{\sqrt{\pi}(\ell ogs)^{\frac{4}{3}}}{s^2+1}|y_1(s) - y_2(s)| \\ &\leq \frac{1}{4}|\varkappa_1(s) - \varkappa_2(s)| + \frac{\sqrt{\pi}}{2}|y_1(s) - y_2(s)|. \end{aligned}$$

Ergo, hypothesis (H3) holds with $\delta = \frac{1}{3}$, $K = \frac{1}{4}, L = \frac{\sqrt{\pi}}{2}$.

Based on (5.39), the initial and terminal value problem (ITVP) (5.38) can be decomposed into two expressions as follows:

$$\begin{cases} F_{1^+}^{1.3}\varkappa(s) = \frac{7}{5\sqrt{\pi}}(\ell ogs)^{1.3} + \frac{(\ell ogs)^{-\frac{1}{3}}}{s+3}\varkappa(s) + \frac{\sqrt{\pi}\ell ogs}{s^2+1}\, {}^{H}I_{1^+}^{1.3}\varkappa(s), & s \in \mathcal{J}_1, \\ F_{2^+}^{1.7}\varkappa(s) = \frac{7}{5\sqrt{\pi}}(\ell ogs)^{1.7} + \frac{(\ell ogs)^{-\frac{1}{3}}}{s+3}\varkappa(s) + \frac{\sqrt{\pi}\ell ogs}{s^2+1}\, {}^{H}I_{1^+}^{1.7}\varkappa(s), & s \in \mathcal{J}_2. \end{cases}$$

For $s \in \mathcal{J}_1$, the ITVP (5.38) is equivalent to the following initial and terminal value problem:

$$\begin{cases} F_{1^+}^{1.3}\varkappa(s) = \frac{7}{5\sqrt{\pi}}(\ell ogs)^{1.3} + \frac{(\ell ogs)^{-\frac{1}{3}}}{s+3}\varkappa(s) + \frac{\sqrt{\pi}\ell ogs}{s^2+1}\, {}^{H}I_{1^+}^{1.3}\varkappa(s), & s \in \mathcal{J}_1, \\ \varkappa(1) = 0, \quad \varkappa(2) = 0. \end{cases} \tag{5.40}$$

We then demonstrate that the given condition (5.34) is satisfied:

$$\begin{aligned} &\frac{(\ell og\mathrm{T}_1)^{1-\delta} - (\ell og\mathrm{T}_0)^{1-\delta}}{(1-\delta)\Gamma(\ltimes_1)}(\ell og\frac{\mathrm{T}_1}{\mathrm{T}_0})^{\ltimes_1 - 1}\Big(K + \frac{L}{\Gamma(\ltimes_1 + 1)}(\ell og\frac{\mathrm{T}_1}{\mathrm{T}_0})^{\ltimes_1}\Big) \\ &= \frac{(\ell og2)^{\frac{2}{3}}}{(\frac{2}{3})\Gamma(1.3)}(\ell og2)^{0.3}\Big(\frac{1}{4} + \frac{\frac{\sqrt{\pi}}{2}}{\Gamma(2.3)}(\ell og2)^{1.3}\Big) \\ &\simeq 0.3809 < \frac{1}{2}. \end{aligned}$$

Let $\vartheta(s) = (\ell ogs)^{\frac{1}{2}}$. Then

$$\begin{aligned} {}^{H}I_{1^+}^{\ltimes_1}\vartheta(s) &= \frac{1}{\Gamma(1.3)}\int_1^s (\ell og\frac{s}{t})^{1.3-1}\frac{(\ell ogt)^{\frac{1}{2}}}{t}dt \\ &\le \frac{1}{\Gamma(1.3)}\int_1^s (\ell og\frac{s}{t})^{0.3}\frac{1}{t}dt \\ &\le \frac{0.75}{\Gamma(2.3)}(\ell ogs)^{\frac{1}{2}} := \lambda_{\vartheta(s)}\vartheta(s). \end{aligned}$$

Thus, condition (H4) is satisfied with $\vartheta(s) = (\ell ogs)^{\frac{1}{2}}$ and $\lambda_{\vartheta(s)} = \frac{0.75}{\Gamma(2.3)}$.

By Theorem 46, the ITVP (5.40) admits a solution $\widetilde{\varkappa}_1 \in E_1$.

For $s \in \mathcal{J}_2$, the ITVP (5.38) can been written as following initial and terminal value problem:

$$\begin{cases} F_{2^+}^{1.7}\varkappa(s) = \frac{7}{5\sqrt{\pi}}(\ell ogs)^{1.7} + \frac{(\ell ogs)^{-\frac{1}{3}}}{s+3}\varkappa(s) + \frac{\sqrt{\pi}\ell ogs}{s^2+1}\, {}^{H}I_{1^+}^{1.7}\varkappa(s), & s \in \mathcal{J}_2, \\ \varkappa(2) = 0, \quad \varkappa(e) = 0. \end{cases} \tag{5.41}$$

We observe that

$$\frac{(\ell og\mathsf{T}_2)^{1-\delta}-(\ell og\mathsf{T}_1)^{1-\delta}}{(1-\delta)\Gamma(\ltimes_2)}(\ell og\frac{\mathsf{T}_2}{\mathsf{T}_1})^{\ltimes_2-1}\Big(K+\frac{L}{\Gamma(\ltimes_2+1)}(\ell og\frac{\mathsf{T}_2}{\mathsf{T}_1})^{\ltimes_2}\Big)$$
$$=\frac{1-(\ell og2)^{\frac{2}{3}}}{(\frac{2}{3})\Gamma(1.7)}(\ell og\frac{e}{2})^{0.7}\Big(\frac{1}{4}+\frac{\frac{\sqrt{\pi}}{2}}{\Gamma(2.7)}(\ell og\frac{e}{2})^{1.7}\Big)$$
$$\simeq 0.0242<\frac{1}{2}.$$

Accordingly the condition (5.34) is satisfied. Moreover,

$$\begin{aligned}
{}^{H}I_{2^+}^{\ltimes_2}\vartheta(s) &= \frac{1}{\Gamma(1.7)}\int_2^s(\ell og\frac{s}{t})^{1.7-1}\frac{(\ell ogt)^{\frac{1}{2}}}{t}dt\\
&\leq \frac{1}{\Gamma(1.7)}\int_2^s(\ell og\frac{s}{t})^{0.7}\frac{1}{t}dt\\
&\leq \frac{1}{\Gamma(2.7)}(\ell ogs)^{\frac{1}{2}}:=\lambda_{\vartheta(s)}\vartheta(s).
\end{aligned}$$

Thus condition (H4) is fulfilled with $\vartheta(s)=(\ell ogs)^{\frac{1}{2}}$, $\lambda_{\vartheta(s)}=\frac{1}{\Gamma(2.7)}$.

By virtue of Theorem 46, the initial and terminal value problem (5.41) admits a solution $\widetilde{\varkappa}_2\in E_2$.

Based on Theorem 47, the initial and terminal value problem (5.38) admits a solution of the form

$$\varkappa(s)=\begin{cases}\widetilde{\varkappa}_1(s), & s\in\mathcal{J}_1,\\ \varkappa_2(s), & s\in\mathcal{J}_2,\end{cases}$$

where

$$\varkappa_2(s)=\begin{cases}0, & s\in\mathcal{J}_1\\ \widetilde{\varkappa}_2(s), & s\in\mathcal{J}_2.\end{cases}$$

As stated in Theorem 48, the initial and terminal value problem (ITVP) (5.38) is stable in the sense of Ulam-Hyers-$\mathfrak{R}$assias.

5.4 Multiterm Initial and Terminal Value Problems of Hadamard Fractional Differential Equations of Variable Order

We address the initial and terminal value problem (ITVP) for multiterm Hadamard fractional differential equations of variable order.

$$\begin{cases}{}^{H}\digamma_{1^+}^{\ltimes(s)}\varkappa(s)=f(s,\varkappa(s),{}^{H}I_{1^+}^{\ltimes(s)}\varkappa(s)), \ s\in\mathcal{J}:=[1,\mathsf{T}],\\ \varkappa(1)=\varkappa(\mathsf{T})=0,\end{cases}\tag{5.42}$$

where $1 < \mathsf{T} < +\infty$, $1 < \ltimes(s) \leq 2$, and $f : \mathcal{J}\times\mathbb{R} \to \mathbb{R}$ is a given function. Here, ${}^{H}\digamma_{1^+}^{\ltimes(s)}$ and ${}^{H}I_{1^+}^{\ltimes(s)}$ denote the Hadamard fractional derivative and integral of order $\ltimes(s)$, respectively.

In this section, we seek a solution to (5.42) and investigate its Ulam-Hyers-$\mathfrak{R}$assias stability.

5.4.1 Existence Criteria

We begin by stating the underlying assumptions:

(H1) Let $n \in \mathbb{N}$ be a fixed integer. Consider a partition $\mathcal{P} = \{\mathcal{J}_1 := [1, \mathsf{T}_1], \mathcal{J}_2 := (\mathsf{T}_1, \mathsf{T}_2], \mathcal{J}_3 := (\mathsf{T}_2, \mathsf{T}_3], ...\mathcal{J}_n := (\mathsf{T}_{n-1}, \mathsf{T}]\}$ of the interval $\mathcal{J} := [1, \mathsf{T}]$, where $1 < \mathsf{T}_1 < \mathsf{T}_2 < \cdots < \mathsf{T}_n = \mathsf{T}$. Define a piecewise constant function $\ltimes(s) : \mathcal{J} \to (1, 2]$ with respect to the partition $\mathcal{P}$, given by

$$\ltimes(s) = \sum_{i=1}^{n} \ltimes_i I_i(s) = \begin{cases} \ltimes_1, & if\ s \in \mathcal{J}_1, \\ \ltimes_2, & if\ s \in \mathcal{J}_2, \\ \quad . & \\ \quad . & \\ \quad . & \\ \ltimes_n, & if\ s \in \mathcal{J}_n, \end{cases}$$

where $1 < \ltimes_i \leq 2$ are constants, and I_i specifies the interval $\mathcal{J}_i := (\mathsf{T}_{i-1}, \mathsf{T}_i]$, for $i = 1, 2, \ldots, n$ (with $\mathsf{T}_0 = 1$ and $\mathsf{T}_n = \mathsf{T}$), such that

$$I_i(s) = \begin{cases} 1, & for\ s \in \mathcal{J}_i, \\ 0, & otherwise. \end{cases}$$

The ßanach space of continuous functions is denoted by $E_i = C(\mathcal{J}_i, \mathbb{R})$ for each $i \in \{1, 2, \ldots, n\}$. For a function $h : \mathcal{J}_i \to \mathbb{R}$, the norm is defined by

$$\|\varkappa\|_{E_i} = \sup_{s\in\mathcal{J}_i} |\varkappa(s)|.$$

We begin by examining the initial terminal value problem (ITVP) as specified in equation (5.42).

Next, the left Hadamard fractional derivative of variable order $\ltimes(s)$ for a function $\varkappa(s) \in C(\mathcal{J}_i, \mathbb{R})$, as defined in equation (2.12), for $s \in \mathcal{J}_i$, $i \in \{1, 2, ..., n\}$, may be expressed as a sum of left Hadamard fractional derivatives of constant orders $\ltimes_k$, where $k \in 1, 2, ..., i$.

$$ {}^{H}\digamma_{1^+}^{\ltimes(s)}\varkappa(s) = \frac{1}{\Gamma(2-\ltimes(s))}(s\frac{d}{ds})^2 \int_1^s (\ell og\frac{s}{t})^{1-\ltimes(s)} \frac{\varkappa(t)}{t}dt.$$

$$= (s\frac{d}{ds})^2 \Big(\frac{1}{\Gamma(2-\ltimes_1)} \int_1^{\mathsf{T}_1} (\ell og\frac{s}{t})^{1-\ltimes_1} \frac{\varkappa(t)}{t}dt + ... + \frac{1}{\Gamma(2-\ltimes_i)} \int_{\mathsf{T}_{i-1}}^{s} (\ell og\frac{s}{t})^{1-\ltimes_i} \frac{\varkappa(t)}{t}dt \Big). \tag{5.43}$$

Then, equation(5.42) on the interval $\mathcal{J}_i$, for $i = 1, 2, ..., n$ can be expressed as follows:

$$(s\frac{d}{ds})^2\Big(\frac{1}{\Gamma(2-\ltimes_1)}\int_1^{\mathsf{T}_1}(\ell og\frac{s}{t})^{1-\ltimes_1}\frac{\varkappa(t)}{t}dt + ...$$

$$+\frac{1}{\Gamma(2-\ltimes_i)}\int_{\mathsf{T}_{i-1}}^{s}(\ell og\frac{s}{t})^{1-\ltimes_i}\frac{\varkappa(t)}{t}dt\Big) = f(s, \varkappa(s), {}^H I_{1+}^{\ltimes(s)}\varkappa(s)), \quad s \in \mathcal{J}_i. \tag{5.44}$$

We now introduce the solution to the initial and terminal value problem (5.42). This formulation constitutes the essential component of the present work.

Definition 54 *The initial and terminal value problem (ITVP) given in equation (5.42) admits a solution if there exist functions $\varkappa_i$, $i = 1, 2, \ldots, n$, such that each $\varkappa_i$ satisfies the equation (5.44) along with the initial and terminal conditions $\varkappa_i(1) = 0$ and $\varkappa_i(\mathsf{T}_i) = 0$.*

According the observation above, for $s \in \mathcal{J}_i$, where $i \in \{1, 2, ..., n\}$, the ITVP (5.42) can be reformulated as equation (5.44).

For $1 \leq s \leq \mathsf{T}_{i-1}$, we take $\varkappa(s) \equiv 0$. Then, equation (5.44) becomes

$${}^H F_{\mathsf{T}_{i-1}^+}^{\ltimes_i}\varkappa(s) = f(s, \varkappa(s), {}^H I_{\mathsf{T}_{i-1}^+}^{\ltimes_i}\varkappa(s)), \quad s \in \mathcal{J}_i.$$

We now consider the next auxiliary initial and terminal value problem (ITVP) associated with constant-order Hadamard fractional differential equations.

$$\begin{cases} {}^H F_{\mathsf{T}_{i-1}^+}^{\ltimes_i}\varkappa(s) = f(s, \varkappa(s), {}^H I_{\mathsf{T}_{i-1}^+}^{\ltimes_i}\varkappa(s)), \quad s \in \mathcal{J}_i, \\ \varkappa(\mathsf{T}_{i-1}) = 0, \varkappa(\mathsf{T}_i) = 0. \end{cases} \tag{5.45}$$

In the forthcoming analysis, the next lemma will serve as a key instrument in establishing the solution of equation (5.45).

Lemma 25 *Let $i \in \{1, 2, \ldots, n\}$ be a natural number, and let $f : \mathcal{J}_i \times \mathbb{R} \times \mathbb{R} \to \mathbb{R}$ be a continuous function. Suppose there exists a number $0 \leq \delta \leq \min_{s\in\mathcal{J}} |\ltimes(s)|$ such that $(\ln s)^\delta f \in C(\mathcal{J}_i \times \mathbb{R} \times \mathbb{R}, \mathbb{R})$.*

The function $\varkappa \in E_i$ solves equation (5.45) if and only if $\varkappa$ satisfies the integral equation

$$\varkappa(s) = -(\ell og\frac{\mathsf{T}_i}{\mathsf{T}_{i-1}})^{1-\ltimes_i}(\ell og\frac{s}{\mathsf{T}_{i-1}})^{\ltimes_i-1}\, {}^H I_{\mathsf{T}_{i-1}^+}^{\ltimes_i} f(\mathsf{T}_i, \varkappa(\mathsf{T}_i), {}^H I_{\mathsf{T}_{i-1}^+}^{\ltimes_i}\varkappa(\mathsf{T}_i))$$

$$+ {}^H I_{\mathsf{T}_{i-1}^+}^{\ltimes_i} f(s, \varkappa(s), {}^H I_{\mathsf{T}_{i-1}^+}^{\ltimes_i}\varkappa(s)). \tag{5.46}$$

Proof: We presume that $\varkappa \in E_i$ is a solution of the ITVP (5.45). Applying the operator ${}^H I^{\ltimes_i}_{\mathsf{T}^+_{i-1}}$ to both sides of (5.45) and invoking Lemma 3, we obtain

$$\varkappa(s) = \omega_1(\ell og\frac{s}{\mathsf{T}_{i-1}})^{\ltimes_i-1} + \omega_2(\ell og\frac{s}{\mathsf{T}_{i-1}})^{\ltimes_i-2} + {}^H I^{\ltimes_i}_{\mathsf{T}^+_{i-1}} f(s, \varkappa(s), {}^H I^{\ltimes_i}_{\mathsf{T}^+_{i-1}} \varkappa(s)), \quad s \in \mathcal{J}_i.$$

Given the assumption of function f, together with $\varkappa(\mathsf{T}_{i-1}) = 0$, we conclude that $\omega_2 = 0$.

Let $\varkappa(s)$ satisfy $\varkappa(\mathsf{T}_i) = 0$. Then, we observe that

$$\omega_1 = -(\ell og\frac{\mathsf{T}_i}{\mathsf{T}_{i-1}})^{1-\ltimes_i} \, {}^H I^{\ltimes_i}_{\mathsf{T}^+_{i-1}} f(\mathsf{T}_i, \varkappa(\mathsf{T}_i), {}^H I^{\ltimes_i}_{\mathsf{T}^+_{i-1}} \varkappa(\mathsf{T}_i)).$$

Thus, we derive

$$\varkappa(s) = -(\ell og\frac{\mathsf{T}_i}{\mathsf{T}_{i-1}})^{1-\ltimes_i}(\ell og\frac{s}{\mathsf{T}_{i-1}})^{\ltimes_i-1} \, {}^H I^{\ltimes_i}_{\mathsf{T}^+_{i-1}} f(\mathsf{T}_i, \varkappa(\mathsf{T}_i), {}^H I^{\ltimes_i}_{\mathsf{T}^+_{i-1}} \varkappa(\mathsf{T}_i))$$
$$+ {}^H I^{\ltimes_i}_{\mathsf{T}^+_{i-1}} f(s, \varkappa(s), {}^H I^{\ltimes_i}_{\mathsf{T}^+_{i-1}} \varkappa(s)), \quad s \in \mathcal{J}_i.$$

In contrast, consider the integral equation (5.46), whose solution is $\varkappa \in E_i$. We determine that $\varkappa$ is the solution to initial and terminal value problem (5.45), based on the continuity of function $(\ell og s)^{\delta} f$ and Lemma 3.

We shall demonstrate the existence and uniqueness of solutions to the ITVP (5.45). The Banach contraction principle and the Krasnoselskii fixed point theorem serve as the foundation for this conclusion.

Theorem 49 *Let the conditions of Lemma 25 be satisfied and there exist constants $K,\ L > 0$, such that,*

$(\ell og s)^{\delta}|f(s, \varkappa_1, y_1) - f(s, \varkappa_2, y_2)| \leq K|\varkappa_1 - \varkappa_2| + L|y_1 - y_2|$, *for any* $\varkappa_1,\ \varkappa_2,\ y_1,\ y_2, \in \mathbb{R}$, $s \in \mathcal{J}_i$, *and the inequality*

$$\frac{(\ell og\mathsf{T}_i)^{1-\delta} - (\ell og\mathsf{T}_{i-1})^{1-\delta}}{(1-\delta)\Gamma(\ltimes_i)}(\ell og\frac{\mathsf{T}_i}{\mathsf{T}_{i-1}})^{\ltimes_i-1}\Big(K + \frac{L}{\Gamma(\ltimes_i+1)}(\ell og\frac{\mathsf{T}_i}{\mathsf{T}_{i-1}})^{\ltimes_i}\Big) < \frac{1}{2}. \tag{5.47}$$

Then, the ITVP (5.45) has a unique solution on E_i.

Proof: We address the fixed point problem (5.45) and formulate the operators

$$W_1,\ W_2 : E_i \to E_i,$$

as follows:

$$W_1\varkappa(s) = -\frac{1}{\Gamma(\Bbbk_i)}(\ell og\frac{\mathsf{T}_i}{\mathsf{T}_{i-1}})^{1-\Bbbk_i}(\ell og\frac{s}{\mathsf{T}_{i-1}})^{\Bbbk_i-1}\int_{\mathsf{T}_{i-1}}^{\mathsf{T}_i}(\ell og\frac{\mathsf{T}_i}{t})^{\Bbbk_i-1}\frac{f(t,\varkappa(t),{}^H I_{\mathsf{T}_{i-1}^+}^{\Bbbk_i}\varkappa(t))}{t}dt \tag{5.48}$$

$$W_2\varkappa(s) = \frac{1}{\Gamma(\Bbbk_i)}\int_{\mathsf{T}_{i-1}}^{s}(\ell og\frac{s}{t})^{\Bbbk_i-1}\frac{f(t,\varkappa(t),{}^H I_{\mathsf{T}_{i-1}^+}^{\Bbbk_i}\varkappa(t))}{t}dt. \tag{5.49}$$

The mappings $W_1, W_2 : E_i \to E_i$ given by (5.48), (5.49) are properly defined, as may be inferred from the continuity of the function $(\ell ogs)^{\delta} f$ and the properties of fractional integrals.

Let

$$R_i \geq \frac{\frac{2f^*}{\Gamma(\Bbbk_i+1)}(\ell og\frac{\mathsf{T}_i}{\mathsf{T}_{i-1}})^{\Bbbk_i}}{1-\frac{2[(\ell og\mathsf{T}_i)^{1-\delta}-(\ell og\mathsf{T}_{i-1})^{1-\delta}]}{(1-\delta)\Gamma(\Bbbk_i)}(\ell og\frac{\mathsf{T}_i}{\mathsf{T}_{i-1}})^{\Bbbk_i-1}\left(K+\frac{L}{\Gamma(\Bbbk_i+1)}(\ell og\frac{\mathsf{T}_i}{\mathsf{T}_{i-1}})^{\Bbbk_i}\right)},$$

where

$$f^* = \sup_{s\in\mathcal{J}_i}|f(s,0,0)|.$$

We now examine the set

$$B_{R_i} = \{\varkappa \in E_i,\ \|\varkappa\|_{E_i} \leq R_i\}.$$

It is evident that B_{R_i} is closed, convex, bounded, and nonempty.

Initially, we will demonstrate the existence of four phases of solutions to the ITVP (5.45).

STEP 1: Claim: $W_1(B_{R_i}) + W_2(B_{R_i}) \subseteq (B_{R_i})$.
For $\varkappa \in B_{R_i}$, we get

$$|W_1\varkappa(t) + W_2\varkappa(s)|$$

$$\begin{aligned}
&\leq \frac{1}{\Gamma(\ltimes_i)}(\ell og\frac{T_i}{T_{i-1}})^{1-\ltimes_i}(\ell og\frac{s}{T_{i-1}})^{\ltimes_i-1}\int_{T_{i-1}}^{T_i}(\ell og\frac{T_i}{t})^{\ltimes_i-1}|f(t,\varkappa(t),{}^H I^{\ltimes_i}_{T^+_{i-1}}\varkappa(t))|\frac{dt}{t}\\
&+ \frac{1}{\Gamma(\ltimes_i)}\int_{T_{i-1}}^{s}(\ell og\frac{s}{t})^{\ltimes_i-1}|f(t,\varkappa(t),{}^H I^{\ltimes_i}_{T^+_{i-1}}\varkappa(t))|\frac{dt}{t}\\
&\leq \frac{2}{\Gamma(\ltimes_i)}\int_{T_{i-1}}^{T_i}(\ell og\frac{T_i}{t})^{\ltimes_i-1}|f(t,\varkappa(t),{}^H I^{\ltimes_i}_{T^+_{i-1}}\varkappa(t))|\frac{dt}{t}\\
&\leq \frac{2}{\Gamma(\ltimes_i)}\int_{T_{i-1}}^{T_i}(\ell og\frac{T_i}{t})^{\ltimes_i-1}|f(t,\varkappa(t),{}^H I^{\ltimes_i}_{T^+_{i-1}}\varkappa(t))-f(t,0,0)|\frac{dt}{t}\\
&+ \frac{2}{\Gamma(\ltimes_i)}\int_{T_{i-1}}^{T_i}(\ell og\frac{T_i}{t})^{\ltimes_i-1}|f(t,0,0)|\frac{dt}{t}\\
&\leq \frac{2}{\Gamma(\ltimes_i)}\int_{T_{i-1}}^{T_i}(\ell og\frac{T_i}{t})^{\ltimes_i-1}(\ell ogt)^{-\delta}\Big(K|\varkappa(t)|+L|{}^H I^{\ltimes_i}_{T^+_{i-1}}\varkappa(t)|\Big)\frac{dt}{t}+\frac{2f^*}{\Gamma(\ltimes_i)}\int_{T_{i-1}}^{T_i}(\ell og\frac{T_i}{t})^{\ltimes_i-1}\frac{dt}{t}\\
&\leq \frac{2}{\Gamma(\ltimes_i)}(\ell og\frac{T_i}{T_{i-1}})^{\ltimes_i-1}\Big(K\|\varkappa\|_{E_i}+L\|{}^H I^{\ltimes_i}_{T^+_{i-1}}\varkappa\|_{E_i}\Big)\int_{T_{i-1}}^{T_i}(\ell ogt)^{-\delta}\frac{dt}{t}+\frac{2f^*}{\Gamma(\ltimes_i+1)}(\ell og\frac{T_i}{T_{i-1}})^{\ltimes_i}\\
&\leq \frac{2}{\Gamma(\ltimes_i)}(\ell og\frac{T_i}{T_{i-1}})^{\ltimes_i-1}\Big(K+\frac{L}{\Gamma(\ltimes_i+1)}(\ell og\frac{T_i}{T_{i-1}})^{\ltimes_i}\Big)\|\varkappa\|_{E_i}\int_{T_{i-1}}^{T_i}(\ell ogt)^{-\delta}\frac{dt}{t}\\
&+ \frac{2f^*}{\Gamma(\ltimes_i+1)}(\ell og\frac{T_i}{T_{i-1}})^{\ltimes_i}\\
&\leq \frac{2[(\ell og T_i)^{1-\delta}-(\ell og T_{i-1})^{1-\delta}]}{(1-\delta)\Gamma(\ltimes_i)}(\ell og\frac{T_i}{T_{i-1}})^{\ltimes_i-1}\Big(K+\frac{L}{\Gamma(\ltimes_i+1)}(\ell og\frac{T_i}{T_{i-1}})^{\ltimes_i}\Big)R_i\\
&+ \frac{2f^*}{\Gamma(\ltimes_i+1)}(\ell og\frac{T_i}{T_{i-1}})^{\ltimes_i}\\
&\leq R_i,
\end{aligned}$$

which means that $W_1(B_{R_i})+W_2(B_{R_i})\subseteq(B_{R_i})$.

STEP 2: Assertion: W_1 is continuous.
We presuppose that the sequence $(\varkappa_n)$ converges to $\varkappa$ in the ßanach space E_i. Then,

$$\|(W_1\varkappa_n)-(W_1\varkappa)\|_{E_i}\to 0.$$

For $s\in\mathcal{J}_i$, we have

$$|(W_1\varkappa_n)(s)-(W_1\varkappa)(s)|$$

$$\begin{aligned}
&\leq \frac{1}{\Gamma(\ltimes_i)}(\ell og\frac{T_i}{T_{i-1}})^{1-\ltimes_i}(\ell og\frac{s}{T_{i-1}})^{\ltimes_i-1}\int_{T_{i-1}}^{T_i}(\ell og\frac{T_i}{t})^{\ltimes_i-1}\\
&\quad |f(t,\varkappa_n(t),{}^H I^{\ltimes_i}_{T^+_{i-1}}\varkappa_n(t))-f(t,\varkappa(t),{}^H I^{\ltimes_i}_{T^+_{i-1}}\varkappa(t))|\frac{dt}{t}\\
&\leq \frac{1}{\Gamma(\ltimes_i)}(\ell og\frac{s}{T_{i-1}})^{\ltimes_i-1}\int_{T_{i-1}}^{T_i}|f(t,\varkappa_n(t),{}^H I^{\ltimes_i}_{T^+_{i-1}}\varkappa_n(t))-f(t,\varkappa(t),{}^H I^{\ltimes_i}_{T^+_{i-1}}\varkappa(t))|\frac{dt}{t}
\end{aligned}$$

$$
\begin{aligned}
&\leq \frac{1}{\Gamma(\ltimes_i)}(\ell og\frac{s}{\mathsf{T}_{i-1}})^{\ltimes_i-1}\int_{\mathsf{T}_{i-1}}^{\mathsf{T}_i}(\ell ogt)^{-\delta}\Big(K|\varkappa_n(t)-\varkappa(t)|+L\ {}^HI^{\ltimes_i}_{\mathsf{T}^+_{i-1}}|\varkappa_n(t)-\varkappa(t)|\Big)\frac{dt}{t}\\
&\leq \frac{1}{\Gamma(\ltimes_i)}\Big(K\|\varkappa_n-\varkappa\|_{E_i}+L\ \|{}^HI^{\ltimes_i}_{\mathsf{T}^+_{i-1}}(\varkappa_n-\varkappa)\|_{E_i}\Big)(\ell og\frac{s}{\mathsf{T}_{i-1}})^{\ltimes_i-1}\int_{\mathsf{T}_{i-1}}^{\mathsf{T}_i}(\ell ogt)^{-\delta}\frac{dt}{t}\\
&\leq \frac{(\ell og\mathsf{T}_i)^{1-\delta}-(\ell og\mathsf{T}_{i-1})^{1-\delta}}{(1-\delta)\Gamma(\ltimes_i)}(\ell og\frac{\mathsf{T}_i}{\mathsf{T}_{i-1}})^{\ltimes_i-1}\Big(K+\frac{L}{\Gamma(\ltimes_i+1)}(\ell og\frac{\mathsf{T}_i}{\mathsf{T}_{i-1}})^{\ltimes_i}\Big)\|\varkappa_n-\varkappa\|_{E_i},
\end{aligned}
$$

and

$$\|(W_1\varkappa_n)-(W_1\varkappa)\|_{E_i}\to 0\quad as\quad n\to\infty.$$

Consequently, the operator W_1 is continuous over E_i.

STEP 3: Claim: $W_1(B_{R_i})$ is relatively compact.
We aim to show that $W_1(B_{R_i})$ is equicontinuous. For $s_1,s_2\in\mathcal{J}_i$ with $s_1<s_2$ and $\varkappa\in B_{R_i}$, we have: 0

$$|(W_1\varkappa)(s_2)-(W_1\varkappa)(s_1)|\leq\|W_1\varkappa\|_{E_i}\ \text{(using the definition of the operator } W_1).$$

To demonstrate equicontinuity, it is sufficient to show that the right-hand side of the inequality is uniformly bounded by some function that depends on s_1 and s_2 but not on $\varkappa$.

By ensuring that the variation of $(W_1\varkappa)(s)$ is controlled uniformly for all $\varkappa\in B_{R_i}$, we can establish that $W_1(B_{R_i})$ is indeed equicontinuous. 0

$$
\begin{aligned}
&= \Big|-\frac{1}{\Gamma(\ltimes_i)}(\ell og\frac{\mathsf{T}_i}{\mathsf{T}_{i-1}})^{1-\ltimes_i}(\ell og\frac{s_2}{\mathsf{T}_{i-1}})^{\ltimes_i-1}\int_{\mathsf{T}_{i-1}}^{\mathsf{T}_i}(\ell og\frac{\mathsf{T}_i}{t})^{\ltimes_i-1}f(t,\varkappa(t),{}^HI^{\ltimes_i}_{\mathsf{T}^+_{i-1}}\varkappa(t))\frac{dt}{t}\\
&+ \frac{1}{\Gamma(\ltimes_i)}(\ell og\frac{\mathsf{T}_i}{\mathsf{T}_{i-1}})^{1-\ltimes_i}(\ell og\frac{s_1}{\mathsf{T}_{i-1}})^{\ltimes_i-1}\int_{\mathsf{T}_{i-1}}^{\mathsf{T}_i}(\ell og\frac{\mathsf{T}_i}{t})^{\ltimes_i-1}f(t,\varkappa(t),{}^HI^{\ltimes_i}_{\mathsf{T}^+_{i-1}}\varkappa(t))\frac{dt}{t}\Big|\\
&\leq \frac{1}{\Gamma(\ltimes_i)}(\ell og\frac{\mathsf{T}_i}{\mathsf{T}_{i-1}})^{1-\ltimes_i}\Big((\ell og\frac{s_2}{\mathsf{T}_{i-1}})^{\ltimes_i-1}-(\ell og\frac{s_1}{\mathsf{T}_{i-1}})^{\ltimes_i-1}\Big)\\
&\quad\int_{\mathsf{T}_{i-1}}^{\mathsf{T}_i}(\ell og\frac{\mathsf{T}_i}{t})^{\ltimes_i-1}|f(t,\varkappa(t),{}^HI^{\ltimes_i}_{\mathsf{T}^+_{i-1}}\varkappa(t))|\frac{dt}{t}\\
&\leq \frac{1}{\Gamma(\ltimes_i)}(\ell og\frac{\mathsf{T}_i}{\mathsf{T}_{i-1}})^{1-\ltimes_i}\Big((\ell og\frac{s_2}{\mathsf{T}_{i-1}})^{\ltimes_i-1}-(\ell og\frac{s_1}{\mathsf{T}_{i-1}})^{\ltimes_i-1}\Big)\\
&\quad\int_{\mathsf{T}_{i-1}}^{\mathsf{T}_i}(\ell og\frac{\mathsf{T}_i}{t})^{\ltimes_i-1}|f(t,\varkappa(t),{}^HI^{\ltimes_i}_{\mathsf{T}^+_{i-1}}\varkappa(t))-f(t,0,0)|\frac{dt}{t}\\
&+ \frac{1}{\Gamma(\ltimes_i)}(\ell og\frac{\mathsf{T}_i}{\mathsf{T}_{i-1}})^{1-\ltimes_i}\Big((\ell og\frac{s_2}{\mathsf{T}_{i-1}})^{\ltimes_i-1}-(\ell og\frac{s_1}{\mathsf{T}_{i-1}})^{\ltimes_i-1}\Big)\int_{\mathsf{T}_{i-1}}^{\mathsf{T}_i}(\ell og\frac{\mathsf{T}_i}{t})^{\ltimes_i-1}|f(t,0,0)|\frac{dt}{t}\\
&\leq \frac{1}{\Gamma(\ltimes_i)}(\ell og\frac{\mathsf{T}_i}{\mathsf{T}_{i-1}})^{1-\ltimes_i}\Big((\ell og\frac{s_2}{\mathsf{T}_{i-1}})^{\ltimes_i-1}-(\ell og\frac{s_1}{\mathsf{T}_{i-1}})^{\ltimes_i-1}\Big)\\
&\quad\int_{\mathsf{T}_{i-1}}^{\mathsf{T}_i}(\ell og\frac{\mathsf{T}_i}{t})^{\ltimes_i-1}(\ell ogt)^{-\delta}(K|\varkappa(t)|+L\ {}^HI^{\ltimes_i}_{\mathsf{T}^+_{i-1}}|\varkappa(t)|)\frac{dt}{t}\\
&+ \frac{f^*}{\Gamma(\ltimes_i)}(\ell og\frac{\mathsf{T}_i}{\mathsf{T}_{i-1}})^{1-\ltimes_i}\Big((\ell og\frac{s_2}{\mathsf{T}_{i-1}})^{\ltimes_i-1}-(\ell og\frac{s_1}{\mathsf{T}_{i-1}})^{\ltimes_i-1}\Big)\int_{\mathsf{T}_{i-1}}^{\mathsf{T}_i}(\ell og\frac{\mathsf{T}_i}{t})^{\ltimes_i-1}\frac{dt}{t}
\end{aligned}
$$

$$\leq \frac{1}{\Gamma(\ltimes_i)}(\ell og\frac{T_i}{T_{i-1}})^{1-\ltimes_i}(\ell og\frac{T_i}{T_{i-1}})^{\ltimes_i-1}\Big((\ell og\frac{s_2}{T_{i-1}})^{\ltimes_i-1}-(\ell og\frac{s_1}{T_{i-1}})^{\ltimes_i-1}\Big)$$
$$\Big(K\|\varkappa\|_{E_i}+L\|^H I^{\ltimes_i}_{T^+_{i-1}}\varkappa\|_{E_i}\Big)\int_{T_{i-1}}^{T_i}(\ell ogt)^{-\delta}\frac{dt}{t}$$
$$+\frac{f^*}{\Gamma(\ltimes_i)}(\ell og\frac{T_i}{T_{i-1}})^{1-\ltimes_i}\Big((\ell og\frac{s_2}{T_{i-1}})^{\ltimes_i-1}-(\ell og\frac{s_1}{T_{i-1}})^{\ltimes_i-1}\Big)\int_{T_{i-1}}^{T_i}(\ell og\frac{T_i}{t})^{\ltimes_i-1}\frac{dt}{t}$$
$$\leq \frac{(\ell ogT_i)^{1-\delta}-(\ell ogT_{i-1})^{1-\delta}}{(1-\delta)\Gamma(\ltimes_i)}\Big(K+\frac{L}{\Gamma(\ltimes_i+1)}(\ell og\frac{T_i}{T_{i-1}})^{\ltimes_i}\Big)\|\varkappa\|_{E_i}$$
$$\Big((\ell og\frac{s_2}{T_{i-1}})^{\ltimes_i-1}-(\ell og\frac{s_1}{T_{i-1}})^{\ltimes_i-1}\Big)$$
$$+\frac{f^*}{\Gamma(\ltimes_i+1)}(\ell og\frac{T_i}{T_{i-1}})\Big((\ell og\frac{s_2}{T_{i-1}})^{\ltimes_i-1}-(\ell og\frac{s_1}{T_{i-1}})^{\ltimes_i-1}\Big)$$
$$\leq \Big[\frac{(\ell ogT_i)^{1-\delta}-(\ell ogT_{i-1})^{1-\delta}}{(1-\delta)\Gamma(\ltimes_i)}\Big(K+\frac{L}{\Gamma(\ltimes_i+1)}(\ell og\frac{T_i}{T_{i-1}})^{\ltimes_i}\Big)\|\varkappa\|_{E_i}+\frac{f^*}{\Gamma(\ltimes_i+1)}(\ell og\frac{T_i}{T_{i-1}})\Big]$$
$$\Big((\ell og\frac{s_2}{T_{i-1}})^{\ltimes_i-1}-(\ell og\frac{s_1}{T_{i-1}})^{\ltimes_i-1}\Big).$$

Hence $|(W_1\varkappa)(s_2)-(W_1\varkappa)(s_1)|\to 0$ as $|s_2-s_1|\to 0$. It implies that $W_1(B_{R_i})$ is equicontinuous.

STEP 4: Assertion: The operator W_2 constitutes a strict contraction on B_{R_i}. For $\varkappa(s),\overline{\varkappa}(s)\in E_i$, we obtain:

$$|(W_2\varkappa)(s)-(W_2\overline{\varkappa})(s)|$$

$$\leq \frac{1}{\Gamma(\ltimes_i)}\int_{T_{i-1}}^{s}(\ell og\frac{s}{t})^{\ltimes_i-1}|f(t,\varkappa(t),^H I^{\ltimes_i}_{T^+_{i-1}}\varkappa(t))-f(t,\overline{\varkappa}(t),^H I^{\ltimes_i}_{T^+_{i-1}}\overline{\varkappa}(t))|\frac{dt}{t}$$
$$\leq \frac{1}{\Gamma(\ltimes_i)}\int_{T_{i-1}}^{T_i}(\ell og\frac{T_i}{t})^{\ltimes_i-1}|f(t,\varkappa(t),^H I^{\ltimes_i}_{T^+_{i-1}}\varkappa(t))-f(t,\overline{\varkappa}(t),^H I^{\ltimes_i}_{T^+_{i-1}}\overline{\varkappa}(t))|\frac{dt}{t}$$
$$\leq \frac{1}{\Gamma(\ltimes_i)}\int_{T_{i-1}}^{T_i}(\ell og\frac{T_i}{t})^{\ltimes_i-1}(\ell ogt)^{-\delta}\Big(K|\varkappa(t)-\overline{\varkappa}(t)|+L\ {}^H I^{\ltimes_i}_{T^+_{i-1}}|\varkappa_n(t)-\overline{\varkappa}(t)|\Big)\frac{dt}{t}$$
$$\leq \frac{1}{\Gamma(\ltimes_i)}\Big(K\|\varkappa-\overline{\varkappa}\|_{E_i}+L\ \|^H I^{\ltimes_i}_{T^+_{i-1}}(\varkappa-\overline{\varkappa})\|_{E_i}\Big)(\ell og\frac{T_i}{T_{i-1}})^{\ltimes_i-1}\int_{T_{i-1}}^{T_i}(\ell ogt)^{-\delta}\frac{dt}{t}$$
$$\leq \frac{1}{\Gamma(\ltimes_i)}\Big(K+\frac{L}{\Gamma(\ltimes_i+1)}(\ell og\frac{T_i}{T_{i-1}})^{\ltimes_i}\Big)\|\varkappa-\overline{\varkappa}\|_{E_i}(\ell og\frac{T_i}{T_{i-1}})^{\ltimes_i-1}\int_{T_{i-1}}^{T_i}(\ell ogt)^{-\delta}\frac{dt}{t}$$
$$\leq \frac{(\ell ogT_i)^{1-\delta}-(\ell ogT_{i-1})^{1-\delta}}{(1-\delta)\Gamma(\ltimes_i)}(\ell og\frac{T_i}{T_{i-1}})^{\ltimes_i-1}\Big(K+\frac{L}{\Gamma(\ltimes_i+1)}(\ell og\frac{T_i}{T_{i-1}})^{\ltimes_i}\Big)\|\varkappa-\overline{\varkappa}\|_{E_i}.$$

Thus, we conclude that W_2 is a strict contraction, as established by (5.47). Hence, by the Krasnoselskii fixed point theorem, there exists $\widetilde{\varkappa}_i \in B_{R_i}$ such that

$$W_1(\widetilde{\varkappa}_i) + W_2(\widetilde{\varkappa}_i) = \widetilde{\varkappa}_i,$$

which is a solution to the initial and terminal value problem (5.45). Since $B_{R_i} \subset E_i$, it follows that $\widetilde{\varkappa}_i \in E_i$.

5.4.2 Main Uniqueness Results

We now establish the uniqueness of solutions to the initial and terminal value problem (5.45) using the ßanach contraction principle. Consider the operator

$$W : E_i \to E_i,$$

We define the operator W as follows:

$$(W\varkappa)(s) = (W_1\varkappa)(s) + (W_2\varkappa)(s), \quad for \quad \varkappa(s) \in E_i.$$

STEP 5: Assertion: W is a contraction on the space E_i.
For $\varkappa(s), \overline{\varkappa}(s) \in E_i$, we obtain:

$$|W\varkappa(s) - W\overline{\varkappa}(s)|$$

$$\begin{aligned}
\leq\ & \frac{1}{\Gamma(\ltimes_i)}(\ell og\frac{\mathsf{T}_i}{\mathsf{T}_{i-1}})^{1-\ltimes_i}(\ell og\frac{s}{\mathsf{T}_{i-1}})^{\ltimes_i-1}\int_{\mathsf{T}_{i-1}}^{\mathsf{T}_i}(\ell og\frac{\mathsf{T}_i}{t})^{\ltimes_i-1} \\
& |f(t,\varkappa(t),{}^H I^{\ltimes_i}_{\mathsf{T}^+_{i-1}}\varkappa(t)) - f(t,\overline{\varkappa}(t),{}^H I^{\ltimes_i}_{\mathsf{T}^+_{i-1}}\overline{\varkappa}(t))|\frac{dt}{t} \\
+\ & \frac{1}{\Gamma(\ltimes_i)}\int_{\mathsf{T}_{i-1}}^{s}(\ell og\frac{s}{t})^{\ltimes_i-1}|f(t,\varkappa(t),{}^H I^{\ltimes_i}_{\mathsf{T}^+_{i-1}}\varkappa(t)) - f(t,\overline{\varkappa}(t),{}^H I^{\ltimes_i}_{\mathsf{T}^+_{i-1}}\overline{\varkappa}(t))|\frac{dt}{t} \\
\leq\ & \frac{2}{\Gamma(\ltimes_i)}\int_{\mathsf{T}_{i-1}}^{\mathsf{T}_i}(\ell og\frac{\mathsf{T}_i}{t})^{\ltimes_i-1}|f(t,\varkappa(t),{}^H I^{\ltimes_i}_{\mathsf{T}^+_{i-1}}\varkappa(t)) - f(t,\overline{\varkappa}(t),{}^H I^{\ltimes_i}_{\mathsf{T}^+_{i-1}}\overline{\varkappa}(t))|\frac{dt}{t} \\
\leq\ & \frac{2}{\Gamma(\ltimes_i)}\int_{\mathsf{T}_{i-1}}^{\mathsf{T}_i}(\ell og\frac{\mathsf{T}_i}{t})^{\ltimes_i-1}(\ell ogt)^{-\delta}\Big(K|\varkappa(t)-\overline{\varkappa}(t)| + L|{}^H I^{\ltimes_i}_{\mathsf{T}^+_{i-1}}(\varkappa(t)-\overline{\varkappa}(t))|\Big)\frac{dt}{t} \\
\leq\ & \frac{2}{\Gamma(\ltimes_i)}(\ell og\frac{\mathsf{T}_i}{\mathsf{T}_{i-1}})^{\ltimes_i-1}\Big(K\|\varkappa-\overline{\varkappa}\|_{E_i} + L\|{}^H I^{\ltimes_i}_{\mathsf{T}^+_{i-1}}(\varkappa-\overline{\varkappa})\|_{E_i}\Big)\int_{\mathsf{T}_{i-1}}^{\mathsf{T}_i}(\ell ogt)^{-\delta}\frac{dt}{t} \\
\leq\ & \frac{2}{\Gamma(\ltimes_i)}(\ell og\frac{\mathsf{T}_i}{\mathsf{T}_{i-1}})^{\ltimes_i-1}\Big(K + \frac{L}{\Gamma(\ltimes_i+1)}(\ell og\frac{\mathsf{T}_i}{\mathsf{T}_{i-1}})^{\ltimes_i}\Big)\|\varkappa-\overline{\varkappa}\|_{E_i}\int_{\mathsf{T}_{i-1}}^{\mathsf{T}_i}(\ell ogt)^{-\delta}\frac{dt}{t} \\
\leq\ & \frac{2[(\ell og\mathsf{T}_i)^{1-\delta} - (\ell og\mathsf{T}_{i-1})^{1-\delta}]}{(1-\delta)\Gamma(\ltimes_i)}(\ell og\frac{\mathsf{T}_i}{\mathsf{T}_{i-1}})^{\ltimes_i-1}\Big(K + \frac{L}{\Gamma(\ltimes_i+1)}(\ell og\frac{\mathsf{T}_i}{\mathsf{T}_{i-1}})^{\ltimes_i}\Big)\|\varkappa-\overline{\varkappa}\|_{E_i}
\end{aligned}$$

Thus, we infer that the mapping W forms a contraction by (5.47). Therefore, W has a unique fixed point $\widetilde{\varkappa}_i \in E_i$, according to ßanach's contraction principle. This unique fixed point is the unique solution to the ITVP (5.45), thereby validating the assertion of Theorem 49.

We will address the existence and uniqueness of solutions to the initial and terminal value problem (5.42) in the following discussion.

We now present the premise that follows:

(H2) Suppose $f \in C(\mathcal{J} \times \mathbb{R} \times \mathbb{R}, \mathbb{R})$ and there exists a number $0 \leq \delta \leq \min_{s\in\mathcal{J}} |\ltimes(s)|$ such that $(\ell ogs)^{\delta} f$ remains continuous on $\mathcal{J} \times \mathbb{R} \times \mathbb{R}$. Furthermore, assume the existence of positive constants K and L such that

$$(\ell ogs)^{\delta}|f(s, \varkappa_1, y_1) - f(s, \varkappa_2, y_2)| \leq K|\varkappa_1 - \varkappa_2| + L|y_1 - y_2|,$$

for all $\varkappa_1, \varkappa_2, y_1, y_2 \in \mathbb{R}$ and $s \in \mathcal{J}$.

Theorem 50 *If conditions (H1), (H2), and the inequality (5.47) hold for every $i \in \{1, 2, \ldots, n\}$, then the initial and terminal value problem (5.42) has a unique solution in the space $C(\mathcal{J}, \mathbb{R})$.*

Proof: Theorem 49 asserts that, for the initial and terminal value problem (5.45) in the context of nonlinear fractional differential equations of variable order involving Hadamard fractional differential equations of constant order, there exists a unique solution $\widetilde{\varkappa}_i \in E_i$ for every $i \in \{1, 2, \ldots, n\}$. We now introduce the function

$$\varkappa_i = \begin{cases} 0, & s \in [1, \mathsf{T}_{i-1}], \\ \widetilde{\varkappa}_i, & s \in \mathcal{J}_i, \end{cases} \tag{5.50}$$

where $J_i \subseteq [1, \mathsf{T}_i]$ denotes the subinterval associated with the i-th component.

Consequently, the function $\varkappa_i \in C([1, \mathsf{T}_i], \mathbb{R})$ is a solution to the integral equation (5.44) for $s \in \mathcal{J}_i$, subject to the conditions $\varkappa_i(1) = 0$ and $\varkappa_i(\mathsf{T}_i) = \widetilde{\varkappa}_i(\mathsf{T}_i) = 0$.

Define the function $\varkappa : [1, \mathsf{T}_n] \to \mathcal{R}$ by

$$\varkappa(s) = \begin{cases} \varkappa_1(s), \quad s \in \mathcal{J}_1, \\ \varkappa_2(s) = \begin{cases} 0, & s \in \mathcal{J}_1, \\ \widetilde{\varkappa}_2, & s \in \mathcal{J}_2, \end{cases} \\ \vdots \\ \varkappa_n(s) = \begin{cases} 0, & s \in [1, \mathsf{T}_{n-1}], \\ \widetilde{\varkappa}_n, & s \in \mathcal{J}_n, \end{cases} \end{cases}.$$

Then $\varkappa(s)$ is the unique solution of the ITVP (5.42).

5.4.3 Conditions of Ulam-Hyers Stability

Theorem 51 *Let conditions (H1), (H2) and the inequality (5.47) be satisfied for any* $i \in 1, 2, ..., n$*. Then, the initial and terminal value problem ITVP (5.42) is Ulam-Hyers stable.*

Proof: Let $\epsilon > 0$ be an arbitrary number and the function $z(t) \in C(\mathcal{J}, \mathbb{R})$ satisfy the inequality (2.31).

For $i \in \{1, 2, \ldots, n\}$, we define the functions as follows: $z_1(s) \equiv z(s),\ s \in [1, \mathsf{T}_1]$ and for $i = 2, 3, \ldots, n$:

$$z_i(s) = \begin{cases} 0, & s \in [1, \mathsf{T}_{i-1}], \\ z(s), & s \in \mathcal{J}_i. \end{cases}$$

Based on equality (5.43), for $s \in \mathcal{J}_i$ and $i \in \{1, 2, \ldots, n\}$, we have

$$^{H}F_{1+}^{\ltimes(s)} z_i(s) = \frac{s^2}{\Gamma(2 - \ltimes(s))} \frac{d^2}{ds^2} \int_{\mathsf{T}_{i-1}}^{s} (\ln \frac{s}{t})^{1-\ltimes_i} \frac{z(t)}{t} dt. \tag{5.51}$$

Applying the Hadamard fractional integral $^{H}I^{\ltimes_i}_{\mathsf{T}_{i-1}^{+}}$ to both sides of inequality (2.31), we derive:

$$\begin{aligned} |z(s) \ &+ \ \frac{1}{\Gamma(\ltimes_i)}(\ell og\frac{\mathsf{T}_i}{\mathsf{T}_{i-1}})^{1-\ltimes_i}(\ell og\frac{s}{\mathsf{T}_{i-1}})^{\ltimes_i-1}\int_{\mathsf{T}_{i-1}}^{\mathsf{T}_i}(\ell og\frac{\mathsf{T}_i}{t})^{\ltimes_i-1}\frac{f(t, z(t), ^{H}I^{\ltimes_i}_{\mathsf{T}_{i-1}^{+}} z(t))}{t}dt \\ &- \ \frac{1}{\Gamma(\ltimes_i)}\int_{\mathsf{T}_{i-1}}^{s}(\ell og\frac{s}{t})^{\ltimes_i-1}\frac{f(t, z(t), ^{H}I^{\ltimes_i}_{\mathsf{T}_{i-1}^{+}} z(t))}{t}dt| \\ &\leq \ \epsilon\frac{1}{\Gamma(\ltimes_i)}\int_{\mathsf{T}_{i-1}}^{s}\frac{1}{t}(\ell og\frac{s}{t})^{\ltimes_i-1}dt \\ &\leq \ \epsilon\frac{1}{\Gamma(\ltimes_i+1)}(\ell og\frac{\mathsf{T}_i}{\mathsf{T}_{i-1}})^{\ltimes_i}. \end{aligned}$$

According to Theorem 50, there exists a unique solution $\varkappa \in C(\mathcal{J}, \mathbb{R})$ for the initial and terminal value problem (5.42), given by $\varkappa(s) = \varkappa_i(s)$ for $s \in \mathcal{J}_i,\ i = 1, 2, \ldots, n$, where

$$\varkappa_i = \begin{cases} 0, & s \in [1, \mathsf{T}_{i-1}], \\ \widetilde{\varkappa}_i, & s \in \mathcal{J}_i, \end{cases} \tag{5.52}$$

and $\widetilde{\varkappa}_i \in E_i$ is the unique solution of (5.45). By Lemma 25, the integral equation

$$\widetilde{\varkappa}_i(s) = -\frac{1}{\Gamma(\ltimes_i)}(\ell og\frac{\mathsf{T}_i}{\mathsf{T}_{i-1}})^{1-\ltimes_i}(\ell og\frac{s}{\mathsf{T}_{i-1}})^{\ltimes_i-1}\int_{\mathsf{T}_{i-1}}^{\mathsf{T}_i}(\ell og\frac{\mathsf{T}_i}{t})^{\ltimes_i-1}\frac{f(t,\varkappa(t),{}^H I_{\mathsf{T}_{i-1}^+}^{\ltimes_i}\varkappa(t))}{t}dt$$
$$+\frac{1}{\Gamma(\ltimes_i)}\int_{\mathsf{T}_{i-1}}^{s}(\ell og\frac{s}{t})^{\ltimes_i-1}\frac{f(t,\varkappa(t),{}^H I_{\mathsf{T}_{i-1}^+}^{\ltimes_i}\varkappa(t))}{t}dt. \tag{5.53}$$

holds.

Then, for $s \in \mathcal{J}_i$, $i = 1, 2, \ldots, n$, we obtain

$$|z(s) - \varkappa(s)| = |z(s) - \varkappa_i(s)| = |z_i(s) - \widetilde{\varkappa}_i(s)|$$

$$= \left|z_i(s) + \frac{1}{\Gamma(\ltimes_i)}(\ell og\frac{\mathsf{T}_i}{\mathsf{T}_{i-1}})^{1-\ltimes_i}(\ell og\frac{s}{\mathsf{T}_{i-1}})^{\ltimes_i-1}\int_{\mathsf{T}_{i-1}}^{\mathsf{T}_i}(\ell og\frac{\mathsf{T}_i}{t})^{\ltimes_i-1}\frac{f(t,\widetilde{\varkappa}_i(t),{}^H I_{\mathsf{T}_{i-1}^+}^{\ltimes_i}\widetilde{\varkappa}_i(t))}{t}dt\right.$$
$$\left.-\frac{1}{\Gamma(\ltimes_i)}\int_{\mathsf{T}_{i-1}}^{s}(\ell og\frac{s}{t})^{\ltimes_i-1}\frac{f(t,\widetilde{\varkappa}_i(t),{}^H I_{\mathsf{T}_{i-1}^+}^{\ltimes_i}\widetilde{\varkappa}_i(t))}{t}dt\right|$$
$$\leq \left|z_i(s) + \frac{1}{\Gamma(\ltimes_i)}(\ell og\frac{\mathsf{T}_i}{\mathsf{T}_{i-1}})^{1-\ltimes_i}(\ell og\frac{s}{\mathsf{T}_{i-1}})^{\ltimes_i-1}\int_{\mathsf{T}_{i-1}}^{\mathsf{T}_i}(\ell og\frac{\mathsf{T}_i}{t})^{\ltimes_i-1}\frac{f(t,z_i(t),{}^H I_{\mathsf{T}_{i-1}^+}^{\ltimes_i}z_i(t))}{t}dt\right.$$
$$\left.-\frac{1}{\Gamma(\ltimes_i)}\int_{\mathsf{T}_{i-1}}^{s}(\ell og\frac{s}{t})^{\ltimes_i-1}\frac{f(t,z_i(t),{}^H I_{\mathsf{T}_{i-1}^+}^{\ltimes_i}z_i(t))}{t}dt\right|$$

$$
\begin{aligned}
&+ \frac{1}{\Gamma(\ltimes_i)}(\ell og\frac{\mathsf{T}_i}{\mathsf{T}_{i-1}})^{1-\ltimes_i}(\ell og\frac{s}{\mathsf{T}_{i-1}})^{\ltimes_i-1} \\
&\quad \int_{\mathsf{T}_{i-1}}^{\mathsf{T}_i}(\ell og\frac{\mathsf{T}_i}{t})^{\ltimes_i-1}\frac{|f(t,z_i(t),{}^H I^{\ltimes_i}_{\mathsf{T}^+_{i-1}}z_i(t))-f(t,\widetilde{\varkappa}_i(t),{}^H I^{\ltimes_i}_{\mathsf{T}^+_{i-1}}\widetilde{\varkappa}_i(t))|}{t}dt \\
&+ \frac{1}{\Gamma(\ltimes_i)}\int_{\mathsf{T}_{i-1}}^{s}(\ell og\frac{s}{t})^{\ltimes_i-1}\frac{|f(t,z_i(t),{}^H I^{\ltimes_i}_{\mathsf{T}^+_{i-1}}z_i(t))-f(t,\widetilde{\varkappa}_i(t),{}^H I^{\ltimes_i}_{\mathsf{T}^+_{i-1}}\widetilde{\varkappa}_i(t))|}{t}dt \\
&\le \epsilon\frac{1}{\Gamma(\ltimes_i+1)}(\ell og\frac{\mathsf{T}_i}{\mathsf{T}_{i-1}})^{\ltimes_i}+\frac{1}{\Gamma(\ltimes_i)}(\ell og\frac{\mathsf{T}_i}{\mathsf{T}_{i-1}})^{1-\ltimes_i}(\ell og\frac{s}{\mathsf{T}_{i-1}})^{\ltimes_i-1} \\
&\quad \int_{\mathsf{T}_{i-1}}^{\mathsf{T}_i}(\ell og\frac{\mathsf{T}_i}{t})^{\ltimes_i-1}\frac{1}{t}(\ell ogt)^{-\delta}(K|z_i(t)-\widetilde{\varkappa}_i(t)|+L\ {}^H I^{\ltimes_i}_{\mathsf{T}^+_{i-1}}|z_i(t)-\widetilde{\varkappa}_i(t)|)dt \\
&+ \frac{1}{\Gamma(\ltimes_i)}\int_{\mathsf{T}_{i-1}}^{s}(\ell og\frac{s}{t})^{\ltimes_i-1}\frac{1}{t}(\ell ogt)^{-\delta}(K|z_i(t)-\widetilde{\varkappa}_i(t)|+L\ {}^H I^{\ltimes_i}_{\mathsf{T}^+_{i-1}}|z_i(t)-\widetilde{\varkappa}_i(t)|)dt \\
&\le \epsilon\frac{1}{\Gamma(\ltimes_i+1)}(\ell og\frac{\mathsf{T}_i}{\mathsf{T}_{i-1}})^{\ltimes_i}+\frac{1}{\Gamma(\ltimes_i)}(\ell og\frac{s}{\mathsf{T}_{i-1}})^{\ltimes_i-1}\int_{\mathsf{T}_{i-1}}^{\mathsf{T}_i}\frac{1}{t}(\ell ogt)^{-\delta}(K|z_i(t)-\widetilde{\varkappa}_i(t)| \\
&+ L\ {}^H I^{\ltimes_i}_{\mathsf{T}^+_{i-1}}|z_i(t)-\widetilde{\varkappa}_i(t)|)dt \\
&+ \frac{1}{\Gamma(\ltimes_i)}(\ell og\frac{s}{\mathsf{T}_{i-1}})^{\ltimes_i-1}\int_{\mathsf{T}_{i-1}}^{s}\frac{1}{t}(\ell ogt)^{-\delta}(K|z_i(t)-\widetilde{\varkappa}_i(t)|+L\ {}^H I^{\ltimes_i}_{\mathsf{T}^+_{i-1}}|z_i(t)-\widetilde{\varkappa}_i(t)|)dt \\
&\le \epsilon\frac{1}{\Gamma(\ltimes_i+1)}(\ell og\frac{\mathsf{T}_i}{\mathsf{T}_{i-1}})^{\ltimes_i}+\frac{(\ell og\mathsf{T}_i)^{1-\delta}-(\ell og\mathsf{T}_{i-1})^{1-\delta}}{(1-\delta)\Gamma(\ltimes_i)}(\ell og\frac{s}{\mathsf{T}_{i-1}})^{\ltimes_i-1}(K\|z_i-\widetilde{\varkappa}_i\|_{E_i} \\
&+ L\ {}^H I^{\ltimes_i}_{\mathsf{T}^+_{i-1}}\|z_i-\widetilde{\varkappa}_i\|_{E_i}) \\
&+ \frac{(\ell ogs)^{1-\delta}-(\ell og\mathsf{T}_{i-1})^{1-\delta}}{(1-\delta)\Gamma(\ltimes_i)}(\ell og\frac{s}{\mathsf{T}_{i-1}})^{\ltimes_i-1}(K\|z_i-\widetilde{\varkappa}_i\|_{E_i}+L\ {}^H I^{\ltimes_i}_{\mathsf{T}^+_{i-1}}\|z_i-\widetilde{\varkappa}_i\|_{E_i}) \\
&\le \epsilon\frac{1}{\Gamma(\ltimes_i+1)}(\ell og\frac{\mathsf{T}_i}{\mathsf{T}_{i-1}})^{\ltimes_i}+\frac{2[(\ell og\mathsf{T}_i)^{1-\delta}-(\ell og\mathsf{T}_{i-1})^{1-\delta}]}{(1-\delta)\Gamma(\ltimes_i)}(\ell og\frac{\mathsf{T}_i}{\mathsf{T}_{i-1}})^{\ltimes_i-1} \\
&\quad (K+\frac{L}{\Gamma(\ltimes_i+1)}(\ell og\frac{\mathsf{T}_i}{\mathsf{T}_{i-1}})^{\ltimes_i})\|z_i-\widetilde{\varkappa}_i\|_{E_i} \\
&\le \epsilon\frac{1}{\Gamma(\ltimes_i+1)}(\ell og\frac{\mathsf{T}_i}{\mathsf{T}_{i-1}})^{\ltimes_i}+\mu\|z-\varkappa\|,
\end{aligned}
$$

where

$$
\mu=\max_{i=1,2,\dots,n}\frac{2[(\ell og\mathsf{T}_i)^{1-\delta}-(\ell og\mathsf{T}_{i-1})^{1-\delta}]}{(1-\delta)\Gamma(\ltimes_i)}(\ell og\frac{\mathsf{T}_i}{\mathsf{T}_{i-1}})^{\ltimes_i-1}(K+\frac{L}{\Gamma(\ltimes_i+1)}(\ell og\frac{\mathsf{T}_i}{\mathsf{T}_{i-1}})^{\ltimes_i}).
$$

Then,

$$\|z - \varkappa\|(1-\mu) \leq \epsilon \frac{1}{\Gamma(\ltimes_i + 1)} (\ell og \frac{\mathsf{T}_i}{\mathsf{T}_{i-1}})^{\ltimes_i}.$$

We obtain, for each $s \in \mathcal{J}$,

$$|z(s) - \varkappa(s)| \leq \|z - \varkappa\| \leq \frac{1}{(1-\mu)\Gamma(\ltimes_i + 1)} (\ell og \frac{\mathsf{T}_i}{\mathsf{T}_{i-1}})^{\ltimes_i} \epsilon := c_f \epsilon.$$

Therefore, the initial and terminal value problem (5.42) is Ulam-Hyers stable.

5.4.4 Illustrative Representation

We address the fractional initial and terminal value problem in this case:

$$\begin{cases} F_{1^+}^{\ltimes(s)} \varkappa(s) = \frac{7}{5\sqrt{\pi}} (\ell ogs)^{\ltimes(s)} + \frac{(\ell ogs)^{-\frac{1}{3}}}{s+3} \varkappa(s) + \frac{\sqrt{\pi} \ell ogs}{s^2+1} \, {}^H I_{1^+}^{\ltimes(s)} \varkappa(s), \quad s \in \mathcal{J} := [1, e], \\ \varkappa(1) = 0, \quad \varkappa(e) = 0. \end{cases} \tag{5.54}$$

Let

$$f(s, \varkappa, y) = \frac{7}{5\sqrt{\pi}} (\ell ogs)^{\ltimes(s)} + \frac{(\ell ogs)^{-\frac{1}{3}}}{s+3} \varkappa(s) + \frac{\sqrt{\pi} \ell ogs}{s^2+1} y(s), \ (s, \varkappa, y) \in [1, e] \times [0, +\infty) \times [0, +\infty).$$

Define the operator $\ltimes(s)$ piecewise as:

$$\ltimes(s) = \begin{cases} 1.3, & s \in \mathcal{J}_1 := [1, 2], \\ 1.7, & s \in \mathcal{J}_2 :=]2, e]. \end{cases} \tag{5.55}$$

Hence, we derive

$$\begin{aligned} & (\ell ogs)^{\frac{1}{3}} |f(s, \varkappa_1, y_1) - f(s, \varkappa_2, y_2)| \\ = & \left| \frac{1}{s+3} \varkappa_1(s) + \frac{\sqrt{\pi} (\ell ogs)^{\frac{4}{3}}}{s^2+1} y_1(s) - \frac{1}{s+3} \varkappa_2(s) - \frac{\sqrt{\pi} (\ell ogs)^{\frac{4}{3}}}{s^2+1} y_2(s) \right| \\ \leq & \frac{1}{s+3} |\varkappa_1(s) - \varkappa_2(s)| + \frac{\sqrt{\pi} (\ell ogs)^{\frac{4}{3}}}{s^2+1} |y_1(s) - y_2(s)| \\ \leq & \frac{1}{4} |\varkappa_1(s) - \varkappa_2(s)| + \frac{\sqrt{\pi}}{2} |y_1(s) - y_2(s)|. \end{aligned}$$

Consequently, $(H2)$ is satisfied with $\delta = \frac{1}{3}$, $K = \frac{1}{4}, L = \frac{\sqrt{\pi}}{2}$.

The problem equation (5.54) is split into two expressions according to (5.55)

$$\begin{cases} F^{1.3}_{1^+}\varkappa(s) = \frac{7}{5\sqrt{\pi}}(\ell ogs)^{1.3} + \frac{(\ell ogs)^{-\frac{1}{3}}}{s+3}\varkappa(s) + \frac{\sqrt{\pi}\ell ogs}{s^2+1}\ {}^{H}I^{1.3}_{1^+}\varkappa(s), & s \in \mathcal{J}_1, \\ F^{1.7}_{2^+}\varkappa(s) = \frac{7}{5\sqrt{\pi}}(\ell ogs)^{1.7} + \frac{(\ell ogs)^{-\frac{1}{3}}}{s+3}\varkappa(s) + \frac{\sqrt{\pi}\ell ogs}{s^2+1}\ {}^{H}I^{1.7}_{1^+}\varkappa(s), & s \in \mathcal{J}_2. \end{cases}$$

For $s \in \mathcal{J}_1$, the problem (5.54) is equivalent to the following formulation:

$$\begin{cases} F^{1.3}_{1^+}\varkappa(s) = \frac{7}{5\sqrt{\pi}}(\ell ogs)^{1.3} + \frac{(\ell ogs)^{-\frac{1}{3}}}{s+3}\varkappa(s) + \frac{\sqrt{\pi}\ell ogs}{s^2+1}\ {}^{H}I^{1.3}_{1^+}\varkappa(s), & s \in \mathcal{J}_1, \\ \varkappa(1) = 0, \quad \varkappa(2) = 0. \end{cases} \tag{5.56}$$

We then demonstrate the fulfillment of condition (5.47).

$$\begin{aligned} & \frac{(\ell og\mathsf{T}_1)^{1-\delta} - (\ell og\mathsf{T}_0)^{1-\delta}}{(1-\delta)\Gamma(\ltimes_1)}(\ell og\frac{\mathsf{T}_1}{\mathsf{T}_0})^{\ltimes_1 - 1}\Big(K + \frac{L}{\Gamma(\ltimes_1 + 1)}(\ell og\frac{\mathsf{T}_1}{\mathsf{T}_0})^{\ltimes_1}\Big) \\ = \ & \frac{(\ell og2)^{\frac{2}{3}}}{(\frac{2}{3})\Gamma(1.3)}(\ell og2)^{0.3}\Big(\frac{1}{4} + \frac{\frac{\sqrt{\pi}}{2}}{\Gamma(2.3)}(\ell og2)^{1.3}\Big) \\ \simeq \ & 0.3809 < \frac{1}{2} \end{aligned}$$

By applying Theorem 49, we conclude that the problem (5.56) admits a unique solution $\varkappa_1$ within E_1.

For $s \in \mathcal{J}_2$, the problem (5.54) can been written as follows:

$$\begin{cases} F^{1.7}_{2^+}\varkappa(s) = \frac{7}{5\sqrt{\pi}}(\ell ogs)^{1.7} + \frac{(\ell ogs)^{-\frac{1}{3}}}{s+3}\varkappa(s) + \frac{\sqrt{\pi}\ell ogs}{s^2+1}\ {}^{H}I^{1.7}_{1^+}\varkappa(s), & s \in \mathcal{J}_2, \\ \varkappa(2) = 0, \quad \varkappa(e) = 0. \end{cases} \tag{5.57}$$

Hence, we find that

$$\begin{aligned} & \frac{(\ell og\mathsf{T}_2)^{1-\delta} - (\ell og\mathsf{T}_1)^{1-\delta}}{(1-\delta)\Gamma(\ltimes_2)}(\ell og\frac{\mathsf{T}_2}{\mathsf{T}_1})^{\ltimes_2 - 1}\Big(K + \frac{L}{\Gamma(\ltimes_2 + 1)}(\ell og\frac{\mathsf{T}_2}{\mathsf{T}_1})^{\ltimes_2}\Big) \\ = \ & \frac{1 - (\ell og2)^{\frac{2}{3}}}{(\frac{2}{3})\Gamma(1.7)}(\ell og\frac{e}{2})^{0.7}\Big(\frac{1}{4} + \frac{\frac{\sqrt{\pi}}{2}}{\Gamma(2.7)}(\ell og\frac{e}{2})^{1.7}\Big) \\ \simeq \ & 0.0242 < \frac{1}{2}. \end{aligned}$$

Consequently, the requirement (5.47) is fulfilled.

Theorem 49 ensures that the problem (5.57) has a unique solution $\widetilde{\varkappa}_2 \in E_2$.

Based on Theorem 50, the initial and terminal value problem (5.54) admits a unique solution

$$\varkappa(s) = \begin{cases} \varkappa_1(s), & s \in \mathcal{J}_1, \\ \varkappa_2(s), & s \in \mathcal{J}_2, \end{cases}$$

where

$$\varkappa_2(s) = \begin{cases} 0, & s \in \mathcal{J}_1 \\ \widetilde{\varkappa}_2(s), & s \in \mathcal{J}_2. \end{cases}$$

Furthermore, Theorem 51 establishes that the initial and terminal value problem (5.54) is Ulam-Hyers stable.

6

Problems for Caputo-Hadamard Fractional Differential Equations with Variable Order

This chapter is devoted to the study of three types of variable-order fractional differential equation problems. The first and second concern an impulsive initial and terminal value problem, and an initial value problem involving the fractional Caputo–Hadamard derivative. The third addresses a multiterm initial and terminal value problem involving the fractional Riemann–Liouville derivative in the sense of Atangana–Baleanu of variable order. In addition to fixed point theorems, we employ the upper-lower solutions technique.

6.1 Multiterm Impulsive Caputo–Hadamard-Type Differential Equations of Fractional Variable Order

For the following impulsive initial and terminal value problem (ITVP), we address the existence, uniqueness, and stability of the solution in the Ulam-Hyers sense:

$$ {}^{CH}F^{\ltimes(s)}_{\mathsf{T}_i^+}\varkappa(s) = f(s, \varkappa(s), {}^{H}I^{\ltimes(s)}_{\mathsf{T}_i^+}\varkappa(s)),\ for\ s \in \mathcal{J} := [1, \mathsf{T}],\ s \neq \mathsf{T}_i,\ i = 1, ..., n, \tag{6.1} $$

$$ \Delta\varkappa|_{s=\mathsf{T}_i} = L_i(\varkappa(\mathsf{T}_i^-)),\ i = 1, ..., n, \tag{6.2} $$

$$ a\varkappa(1) + b\varkappa(\mathsf{T}) = c, \tag{6.3} $$

where $\mathsf{T} > e$, $0 < \ltimes(s) < 1$, $f : \mathcal{J}\times\mathbb{R}\times\mathbb{R} \longrightarrow \mathbb{R}$, $L_i : \mathbb{R} \longrightarrow \mathbb{R}$, $i = 1, ..., n$ are continuous functions. The operators ${}^{CH}F^{\ltimes(s)}_{\mathsf{T}_i^+}$ and ${}^{H}I^{\ltimes(s)}_{\mathsf{T}_i^+}$ denote the Caputo-Hadamard derivative and the Hadamard integral operators of variable order $\ltimes(s)$, respectively. The constants $a, b,$ and c are real numbers with $a + b \neq 0$. The impulse points satisfy $1 = \mathsf{T}_0 < \mathsf{T}_1 < ... < \mathsf{T}_n < \mathsf{T}_{n+1} = \mathsf{T}$. The jump condition is defined as: $\Delta\varkappa|_{s=\mathsf{T}_i} = \varkappa(\mathsf{T}_i^+) - \varkappa(\mathsf{T}_i^-)$, where $\varkappa(\mathsf{T}_i^+) = \lim_{\mathsf{h}\longrightarrow 0^+}\varkappa(\mathsf{T}_i + \mathsf{h})$ and $\varkappa(\mathsf{T}_i^-) = \lim_{\mathsf{h}\longrightarrow 0^-}\varkappa(\mathsf{T}_i + \mathsf{h})$ represent the right-hand and left-hand limits of $\varkappa(s)$ at $s = \mathsf{T}_i$ respectively, for $i = 1, ..., n$.

DOI: 10.1201/9781003687900-6

6.1.1 Existence Analysis

Let us state the underlying presumption:

(H1) Let $P = \{\mathcal{J}_0 := [\mathsf{T}_0, \mathsf{T}_1], \mathcal{J}_1 := (\mathsf{T}_1, \mathsf{T}_2], \mathcal{J}_2 := (\mathsf{T}_2, \mathsf{T}_3], ..., \mathcal{J}_n := (\mathsf{T}_n, \mathsf{T}_{n+1}]\}$ be a partition of the interval $\mathcal{J}$, where $\mathsf{T}_0 = 1$, $\mathsf{T}_{n+1} = \mathsf{T}$. Let $\ltimes^*(s) : \mathcal{J} \to (0,1)$ be a function that is piecewise constant with respect to P, and define $\ltimes^* = \sup_{s\in\mathcal{J}} \ltimes(s),$. That is,

$$\ltimes(s) = \sum_{i=0}^{n} \ltimes_i I_i(s) = \begin{cases} \ltimes_0, & if \ s \in \mathcal{J}_0, \\ \ltimes_1, & if \ s \in \mathcal{J}_1, \\ . \\ . \\ . \\ \ltimes_n, & if \ s \in \mathcal{J}_n, \end{cases}$$

where $0 < \ltimes_i \leq \ltimes^* < 1$, are constants, and

$$I_i(s) = \begin{cases} 1, & for \ s \in \mathcal{J}_i, \quad i = 1, ..., n \\ 0, & eleswhere. \end{cases}$$

We also provide the definition of the solution to the ITVP corresponding to equations (6.1)–(6.3).

Definition 55 *The function* $\varkappa \in PC(\mathcal{J}, \mathbb{R})$ *is said to be a solution of the ITVP (6.1)–(6.3) if* $\varkappa$ *satisfies the equation* ${}^{CH}F_{\mathsf{T}_i^+}^{\ltimes(s)} \varkappa(s) = f(s, \varkappa(s), {}^{H}I_{\mathsf{T}_i^+}^{\ltimes(s)} \varkappa(s))$ *on* $\mathcal{J}_i$ *along with the conditions*

$$\Delta\varkappa|_{s=\mathsf{T}_i} = L_i(\varkappa(\mathsf{T}_i^-)), \ \ i = 1, ..., n,$$

and

$$a\varkappa(1) + b\varkappa(\mathsf{T}) = c.$$

We now introduce the following auxiliary proposition.

Proposition 9 *Let* $h : \mathcal{J} \to \mathbb{R}$ *be continuous. Then the solution of the following impulsive ITVP*

$${}^{CH}F_{\mathsf{T}_i^+}^{\ltimes_i} \varkappa(s) = h(s), \quad s \in \mathcal{J}_i \ \ i = 1, ..., n, \tag{6.4}$$

$$\Delta\,\varkappa|_{s=\mathsf{T}_i} = L_i(\varkappa(\mathsf{T}_i^-)), i = 1, ..., n, \tag{6.5}$$

$$a\varkappa(1) + b\varkappa(\mathsf{T}) = c, \tag{6.6}$$

is given by

$$
\varkappa(s)=\begin{cases}
\dfrac{-1}{a+b}\left[b\displaystyle\sum_{j=1}^{n}\frac{1}{\Gamma(\ltimes_{j-1})}\int_{\mathsf{T}_{j-1}}^{\mathsf{T}_j}\left(\ell og\frac{\mathsf{T}_j}{t}\right)^{\ltimes_{j-1}-1}h(t)\frac{dt}{t}\right.\\
+\dfrac{b}{\Gamma(\ltimes_n)}\displaystyle\int_{\mathsf{T}_n}^{\mathsf{T}}\left(\ell og\frac{\mathsf{T}}{t}\right)^{\ltimes_n-1}h(t)\frac{dt}{t}\\
\left.+b\displaystyle\sum_{j=1}^{n}L_j(\varkappa(\mathsf{T}_j^-))-c\right]+\dfrac{1}{\Gamma(\ltimes_0)}\displaystyle\int_1^s\left(\ell og\frac{s}{t}\right)^{\ltimes_0-1}h(t)\frac{dt}{t}, \quad s\in[\mathsf{T}_0,\mathsf{T}_1],\\
\dfrac{-1}{a+b}\left[b\displaystyle\sum_{j=1}^{n}\frac{1}{\Gamma(\ltimes_{j-1})}\int_{\mathsf{T}_{j-1}}^{\mathsf{T}_j}\left(\ell og\frac{\mathsf{T}_j}{t}\right)^{\ltimes_{j-1}-1}h(t)\frac{dt}{t}\right.\\
+\dfrac{b}{\Gamma(\ltimes_n)}\displaystyle\int_{\mathsf{T}_n}^{\mathsf{T}}\left(\ell og\frac{\mathsf{T}}{t}\right)^{\ltimes_n-1}h(t)\frac{dt}{t}\\
\left.+b\displaystyle\sum_{j=1}^{n}L_j(\varkappa(\mathsf{T}_j^-))-c\right]+\displaystyle\sum_{j=1}^{i}\frac{1}{\Gamma(\ltimes_{j-1})}\int_{\mathsf{T}_{j-1}}^{\mathsf{T}_j}\left(\ell og\frac{\mathsf{T}_j}{t}\right)^{\ltimes_{j-1}-1}h(t)\frac{dt}{t}\\
+\dfrac{1}{\Gamma(\ltimes_i)}\displaystyle\int_{\mathsf{T}_i}^{s}\left(\ell og\frac{s}{t}\right)^{\ltimes_i-1}h(t)\frac{dt}{t}+\sum_{j=1}^{i}L_j(\varkappa(\mathsf{T}_j^-)), \quad s\in(\mathsf{T}_i,\mathsf{T}_{i+1}],\\
i=1,...,n.
\end{cases}
\tag{6.7}
$$

Proof: Let $\varkappa$ be a solution of the ITVP (6.4)-(6.6). If $s\in[\mathsf{T}_0,\mathsf{T}_1]$, then by Lemma 4, we obtain

$$
\varkappa(s)=\omega_0+\frac{1}{\Gamma(\ltimes_0)}\int_1^s\left(\ell og\frac{s}{t}\right)^{\ltimes_0-1}h(t)\frac{dt}{t}, \quad \omega_0\in\mathbb{R}.
$$

If $s\in(\mathsf{T}_1,\mathsf{T}_2]$, then Lemma 4 implies

$$
\begin{aligned}
\varkappa(s) &= \varkappa(\mathsf{T}_1^+)+\frac{1}{\Gamma(\ltimes_1)}\int_{\mathsf{T}_1}^s\left(\ell og\frac{s}{t}\right)^{\ltimes_1-1}h(t)\frac{dt}{t}\\
&= \Delta\varkappa|_{s=\mathsf{T}_1}+\varkappa(\mathsf{T}_1^-)+\frac{1}{\Gamma(\ltimes_1)}\int_{\mathsf{T}_1}^s\left(\ell og\frac{s}{t}\right)^{\ltimes_1-1}h(t)\frac{dt}{t}\\
&= \omega_0+L_1(\varkappa(\mathsf{T}_1^-))+\frac{1}{\Gamma(\ltimes_0)}\int_1^{\mathsf{T}_1}\left(\ell og\frac{\mathsf{T}_1}{t}\right)^{\ltimes_0-1}h(t)\frac{dt}{t}\\
&\quad+\frac{1}{\Gamma(\ltimes_1)}\int_{\mathsf{T}_1}^s\left(\ell og\frac{s}{t}\right)^{\ltimes_1-1}h(t)\frac{dt}{t}.
\end{aligned}
$$

If $s \in (\mathsf{T}_2, \mathsf{T}_3]$, then Lemma 4 implies

$$\begin{aligned}
\varkappa(s) &= \varkappa(\mathsf{T}_2^+) + \frac{1}{\Gamma(\ltimes_2)} \int_{\mathsf{T}_2}^{s} \left(\ell og\frac{s}{t}\right)^{\ltimes_2 - 1} h(t)\frac{dt}{t} \\
&= \Delta\varkappa|_{s=\mathsf{T}_2} + \varkappa(\mathsf{T}_2^-) + \frac{1}{\Gamma(\ltimes_2)} \int_{\mathsf{T}_2}^{s} \left(\ell og\frac{s}{t}\right)^{\ltimes_2 - 1} h(t)\frac{dt}{t} \\
&= \omega_0 + L_2(\varkappa(\mathsf{T}_2^-)) + L_1(\varkappa(\mathsf{T}_1^-)) + \frac{1}{\Gamma(\ltimes_0)} \int_{1}^{\mathsf{T}_1} \left(\ell og\frac{\mathsf{T}_1}{t}\right)^{\ltimes_0 - 1} h(t)\frac{dt}{t} \\
&+ \frac{1}{\Gamma(\ltimes_1)} \int_{\mathsf{T}_1}^{\mathsf{T}_2} \left(\ell og\frac{\mathsf{T}_2}{t}\right)^{\ltimes_1 - 1} h(t)\frac{dt}{t}\,\frac{1}{\Gamma(\ltimes_2)} \int_{\mathsf{T}_2}^{s} \left(\ell og\frac{s}{t}\right)^{\ltimes_2 - 1} h(t)\frac{dt}{t}\,.
\end{aligned}$$

Then, the solution $\varkappa(s)$ for $s \in (\mathsf{T}_i, \mathsf{T}_{i+1}]$, $i = 1, ..., n$, can be written as

$$\begin{aligned}
\varkappa(s) &= \omega_0 + \sum_{j=1}^{i} \frac{1}{\Gamma(\ltimes_{j-1})} \int_{\mathsf{T}_{j-1}}^{\mathsf{T}_j} \left(\ell og\frac{\mathsf{T}_j}{t}\right)^{\ltimes_{j-1} - 1} h(t)\frac{dt}{t} \\
&+ \frac{1}{\Gamma(\ltimes_i)} \int_{\mathsf{T}_i}^{s} \left(\ell og\frac{s}{t}\right)^{\ltimes_i - 1} h(t)\frac{dt}{t} + \sum_{j=1}^{i} L_j(\varkappa(\mathsf{T}_j^-)).
\end{aligned}$$

Applying the initial and terminal conditions $a\varkappa(1) + b\varkappa(\mathsf{T}) = c$, we obtain

$$\begin{aligned}
c &= \omega_0(a+b) + b\sum_{j=1}^{n} \frac{1}{\Gamma(\ltimes j - 1)} \int_{\mathsf{T}_{j-1}}^{\mathsf{T}_j} \left(\ell og\frac{\mathsf{T}_j}{t}\right)^{\ltimes j - 1 - 1} h(t)\frac{dt}{t} \\
&+ \frac{b}{\Gamma(\ltimes_n)} \int_{\mathsf{T}_n}^{M} \left(\ell og\frac{\mathsf{T}}{t}\right)^{\ltimes_n - 1} h(t)\frac{dt}{t} + b\sum_{j=1}^{n} L_j(\varkappa(\mathsf{T}_j^-)).
\end{aligned}$$

Then,

$$\begin{aligned}
\omega_0 &= \frac{-1}{a+b}\Big[b\sum_{j=1}^{n} \frac{1}{\Gamma(\ltimes_{j-1})} \int_{\mathsf{T}_{j-1}}^{\mathsf{T}_j} (\ell og\frac{\mathsf{T}_j}{t})^{\ltimes_{j-1}-1} h(t)\frac{dt}{t} \\
&+ \frac{b}{\Gamma(\ltimes_n)} \int_{\mathsf{T}_n}^{\mathsf{T}} (\ell og\frac{\mathsf{T}}{t})^{\ltimes_n - 1} h(t)\frac{dt}{t} + b\sum_{j=1}^{n} L_j(\varkappa(\mathsf{T}_j^-)) - c\Big].
\end{aligned}$$

Thus,

$$\begin{aligned}\varkappa(s) &= \frac{-1}{a+b}\Big[b\sum_{j=1}^{n}\frac{1}{\Gamma(\ltimes_{j-1})}\int_{\mathsf{T}_{j-1}}^{\mathsf{T}_j}(\ell og\frac{\mathsf{T}_j}{t})^{\ltimes_{j-1}-1}h(t)\frac{dt}{t}\\ &+ \frac{b}{\Gamma(\ltimes_n)}\int_{\mathsf{T}_n}^{\mathsf{T}}(\ell og\frac{\mathsf{T}}{t})^{\ltimes_n-1}h(t)\frac{dt}{t}+b\sum_{j=1}^{n}L_j(\varkappa(\mathsf{T}_j^-))-c\Big]\\ &+ \sum_{j=1}^{i}\frac{1}{\Gamma(\ltimes_{j-1})}\int_{\mathsf{T}_{j-1}}^{\mathsf{T}_j}(\ell og\frac{\mathsf{T}_j}{t})^{\ltimes_{j-1}-1}h(t)\frac{dt}{t}\\ &+ \frac{1}{\Gamma(\ltimes_i)}\int_{\mathsf{T}_i}^{s}(\ell og\frac{s}{t})^{\ltimes_i-1}h(t)\frac{dt}{t}+\sum_{j=1}^{i}L_j(\varkappa(\mathsf{T}_j^-)).\end{aligned}$$

On the other hand, it is simple to verify that $\varkappa$ satisfies the initial and terminal value problem (6.4)–(6.5).

Now, we present our first result, under the assumption that the following conditions are satisfied:

(H2) There exist constants $\digamma_1, \digamma_2 > 0$, such that

$$|f(s,\varkappa_1,y_1)-f(s,\varkappa_2,y_2)| \le \digamma_1|\varkappa_1-\varkappa_2|+\digamma_2|y_1-y_2|,$$

for any $\varkappa_1, \varkappa_2, y_1, y_2 \in \mathbb{R}$ and $s \in \mathcal{J}$.

(H3) For any $i = 1, \ldots, n$, $\varkappa \in \mathbb{R}$ and $s \in \mathcal{J}$, there exists $\digamma_3 > 0$ such that

$$|L_i(\varkappa(s))| \le \digamma_3|\varkappa(s)|.$$

Theorem 52 *Let conditions (H1)–(H3) hold, and suppose that*

$$\left(\frac{|b|}{|a+b|}+1\right)\left[\frac{(n+1)(\ell og\mathsf{T})^{\ltimes^*}}{2(\sqrt{2}-1)}\left(\digamma_1+\digamma_2\frac{(\ell og\mathsf{T})^{\ltimes^*}}{2(\sqrt{2}-1)}\right)+n\digamma_3\right]<1. \quad (6.8)$$

Then, the initial and terminal value problem (6.1)–(6.3) possesses a solution in $PC(\mathcal{J},\mathbb{R})$.

Proof: We define the operator

$$\mathcal{S} : PC(\mathcal{J}, \mathbb{R}) \rightarrow PC(\mathcal{J}, \mathbb{R}),$$

as follows:

$$\begin{aligned}
\mathcal{S}\varkappa(s) &= \frac{-1}{a+b}\Bigg[b\sum_{j=1}^{n} \frac{1}{\Gamma(\ltimes_{j-1})} \int_{\mathsf{T}_{j-1}}^{\mathsf{T}_j} \left(\ell og\frac{\mathsf{T}_j}{t}\right)^{\ltimes_{j-1}-1} f\left(t, \varkappa(t), I_{\mathsf{T}_{j-1}^{+}}^{\ltimes_{j-1}} \varkappa(t)\right) \frac{dt}{t} \\
&\quad + \frac{b}{\Gamma(\ltimes_n)} \int_{\mathsf{T}_n}^{\mathsf{T}} \left(\ell og\frac{\mathsf{T}}{t}\right)^{\ltimes_n - 1} f\left(t, \varkappa(t), I_{\mathsf{T}_n^{+}}^{\ltimes_n} \varkappa(t)\right) \frac{dt}{t} + b\sum_{j=1}^{n} L_j\left(\varkappa\left(\mathsf{T}_j^{-}\right)\right) - c\Bigg] \\
&\quad + \sum_{0<\mathsf{T}_j<s} \frac{1}{\Gamma(\ltimes_{j-1})} \int_{\mathsf{T}_{j-1}}^{\mathsf{T}_j} \left(\ell og\frac{\mathsf{T}_j}{t}\right)^{\ltimes_{j-1}-1} f\left(t, \varkappa(t), I_{\mathsf{T}_{j-1}^{+}}^{\ltimes_{j-1}} \varkappa(t)\right) \frac{dt}{t} \\
&\quad + \frac{1}{\Gamma(\ltimes_i)} \int_{\mathsf{T}_i}^{s} \left(\ell og\frac{s}{t}\right)^{\ltimes_i - 1} f\left(t, \varkappa(t), I_{\mathsf{T}_i^{+}}^{\ltimes_i} \varkappa(t)\right) \frac{dt}{t} \\
&\quad + \sum_{0<\mathsf{T}_j<s} L_j\left(\varkappa\left(\mathsf{T}_j^{-}\right)\right).
\end{aligned} \tag{6.9}$$

The operator $\mathcal{S}$ defined in (6.9) is well defined due to the continuity of function g and from the properties of fractional integrals.

Let the set

$$B_R = \{\varkappa \in PC(\mathcal{J}, \mathbb{R}), \|\varkappa\| \leq R\},$$

where

$$R \geq \frac{\left(\frac{|b|}{|a+b|} + 1\right) \frac{(n+1)(\ell og\mathsf{T})^{\ltimes^*}}{2(\sqrt{2}-1)} f^{\star} + \frac{|c|}{|a+b|}}{1 - \left(\frac{|b|}{|a+b|} + 1\right)\left[\frac{(n+1)(\ell og\mathsf{T})^{\ltimes^*}}{2(\sqrt{2}-1)}\left(F_1 + F_2 \frac{(\ell og\mathsf{T})^{\ltimes *}}{2(\sqrt{2}-1)}\right) + nF_3\right]},$$

and

$$f^{\star} = \sup_{s\in\mathcal{J}} |f(s, 0, 0)|.$$

Clearly, B_R is non-empty, closed, convex, and bounded.

STEP 1: Claim: $\mathcal{S}(B_R) \subseteq (B_R)$.
For $\varkappa \in B_R$, we obtain

$$
\begin{aligned}
&|\mathcal{S}\varkappa(s)| \\
\leq\ & \frac{1}{|a+b|}\Bigg[|b|\sum_{j=1}^{n}\frac{1}{\Gamma(\ltimes_{j-1})}\int_{\mathsf{T}_{j-1}}^{\mathsf{T}_j}\left(\ell og\frac{\mathsf{T}_j}{t}\right)^{\ltimes_{j-1}-1} \\
&|f(t,\varkappa(t),I^{\ltimes_{j-1}}_{\mathsf{T}^+_{j-1}}\varkappa(t)) - f(t,0,0)|\frac{dt}{t} \\
+\ & |b|\sum_{j=1}^{n}\frac{1}{\Gamma(\ltimes_{j-1})}\int_{\mathsf{T}_{j-1}}^{\mathsf{T}_j}\left(\ell og\frac{\mathsf{T}_j}{t}\right)^{\ltimes_{j-1}-1}|f(t,0,0)|\frac{dt}{t} \\
+\ & \frac{|b|}{\Gamma(\ltimes_n)}\int_{\mathsf{T}_n}^{\mathsf{T}}\left(\ell og\frac{\mathsf{T}}{t}\right)^{\ltimes_n-1}|f(t,\varkappa(t),I^{\ltimes_n}_{\mathsf{T}^+_n}\varkappa(t)) - f(t,0,0)|\frac{dt}{t} \\
+\ & \frac{|b|}{\Gamma(\ltimes_n)}\int_{\mathsf{T}_n}^{\mathsf{T}}\left(\ell og\frac{\mathsf{T}}{t}\right)^{\ltimes_n-1}|f(t,0,0)|\frac{dt}{t} + |b|\sum_{j=1}^{n}|L_j(\varkappa(\mathsf{T}_j^-))| + |c|\Bigg] \\
+\ & \sum_{0<\mathsf{T}_j<s}\frac{1}{\Gamma(\ltimes_{j-1})}\int_{\mathsf{T}_{j-1}}^{\mathsf{T}_j}\left(\ell og\frac{\mathsf{T}_j}{t}\right)^{\ltimes_{j-1}-1}|f(t,\varkappa(t),I^{\ltimes_{j-1}}_{\mathsf{T}^+_{j-1}}\varkappa(t)) - f(t,0,0)|\frac{dt}{t} \\
+\ & \sum_{0<\mathsf{T}_j<s}\frac{1}{\Gamma(\ltimes_{j-1})}\int_{\mathsf{T}_{j-1}}^{\mathsf{T}_j}\left(\ell og\frac{\mathsf{T}_j}{t}\right)^{\ltimes_{j-1}-1}|f(t,0,0)|\frac{dt}{t} \\
+\ & \frac{1}{\Gamma(\ltimes_i)}\int_{\mathsf{T}_i}^{s}\left(\ell og\frac{s}{t}\right)^{\ltimes_i-1}|f(t,\varkappa(t),I^{\ltimes_i}_{\mathsf{T}^+_i}\varkappa(t)) - f(t,0,0)|\frac{dt}{t} \\
+\ & \frac{1}{\Gamma(\ltimes_i)}\int_{\mathsf{T}_i}^{s}\left(\ell og\frac{s}{t}\right)^{\ltimes_i-1}|f(t,0,0)|\frac{dt}{t} + \sum_{0<\mathsf{T}_j<s}|L_j(\varkappa(\mathsf{T}_j^-))| \\
\leq\ & \frac{1}{|a+b|}\Bigg[|b|\sum_{j=1}^{n}\frac{1}{\Gamma(\ltimes_{j-1})}\int_{\mathsf{T}_{j-1}}^{\mathsf{T}_j}\left(\ell og\frac{\mathsf{T}_j}{t}\right)^{\ltimes_{j-1}-1}\left(\digamma_1|\varkappa(t)| + \digamma_2|I^{\ltimes_{j-1}}_{\mathsf{T}^+_{j-1}}\varkappa(t)|\right)\frac{dt}{t} \\
+\ & |b|f^*\sum_{j=1}^{n}\frac{1}{\Gamma(\ltimes_{j-1})}\int_{\mathsf{T}_{j-1}}^{\mathsf{T}_j}\left(\ell og\frac{\mathsf{T}_j}{t}\right)^{\ltimes_{j-1}-1}\frac{dt}{t} \\
+\ & \frac{|b|}{\Gamma(\ltimes_n)}\int_{\mathsf{T}_n}^{\mathsf{T}}\left(\ell og\frac{\mathsf{T}}{t}\right)^{\ltimes_n-1}\left(\digamma_1|\varkappa(t)| + \digamma_2|I^{\ltimes_n}_{\mathsf{T}^+_n}\varkappa(t)|\right)\frac{dt}{t} \\
+\ & \frac{|b|f^*}{\Gamma(\ltimes_n)}\int_{\mathsf{T}_n}^{\mathsf{T}}\left(\ell og\frac{\mathsf{T}}{t}\right)^{\ltimes_n-1}\frac{dt}{t} + |b|\sum_{j=1}^{n}\digamma_3|\varkappa(\mathsf{T}_j^-)| + |c|\Bigg] \\
+\ & \sum_{0<\mathsf{T}_j<s}\frac{1}{\Gamma(\ltimes_{j-1})}\int_{\mathsf{T}_{j-1}}^{\mathsf{T}_j}\left(\ell og\frac{\mathsf{T}_j}{t}\right)^{\ltimes_{j-1}-1}\left(\digamma_1|\varkappa(t)| + \digamma_2|I^{\ltimes_{j-1}}_{\mathsf{T}^+_{j-1}}\varkappa(t)|\right)\frac{dt}{t} \\
+\ & \sum_{0<\mathsf{T}_j<s}\frac{f^*}{\Gamma(\ltimes_{j-1})}\int_{\mathsf{T}_{j-1}}^{\mathsf{T}_j}\left(\ell og\frac{\mathsf{T}_j}{t}\right)^{\ltimes_{j-1}-1}\frac{dt}{t}
\end{aligned}
$$

$$
\begin{aligned}
&+ \frac{1}{\Gamma(\ltimes_i)}\int_{\mathsf{T}_i}^{s}\left(\ell og\frac{s}{t}\right)^{\ltimes_i-1}\left(\digamma_1|\varkappa(t)|+\digamma_2|I_{\mathsf{T}_i^+}^{\ltimes_i}\varkappa(t)|\right)\frac{dt}{t}\\
&+ \frac{f^*}{\Gamma(\ltimes_i)}\int_{\mathsf{T}_i}^{s}\left(\ell og\frac{s}{t}\right)^{\ltimes_i-1}\frac{dt}{t}+\sum_{0<\mathsf{T}_j<s}\digamma_3|\varkappa(\mathsf{T}_j^-)|\\
\leq\ & \frac{1}{|a+b|}\left[|b|\sum_{j=1}^{n}\frac{1}{\Gamma(\ltimes_{j-1})}\int_{\mathsf{T}_{j-1}}^{\mathsf{T}_j}\left(\ell og\frac{\mathsf{T}_j}{t}\right)^{\ltimes_{j-1}-1}\right.\\
&\left(\digamma_1\|\varkappa\|+\digamma_2\|\frac{(\ell og\mathsf{T})^{\ltimes_{j-1}}}{\Gamma(\ltimes_{j-1}+1)}\varkappa\|\right)\frac{dt}{t}\\
&+ |b|f^*\sum_{j=1}^{n}\frac{1}{\Gamma(\ltimes_{j-1})}\int_{\mathsf{T}_{j-1}}^{\mathsf{T}_j}\left(\ell og\frac{\mathsf{T}_j}{t}\right)^{\ltimes_{j-1}-1}\frac{dt}{t}\\
&+ \frac{|b|}{\Gamma(\ltimes_n)}\int_{\mathsf{T}_n}^{\mathsf{T}}\left(\ell og\frac{\mathsf{T}}{t}\right)^{\ltimes_n-1}\left(\digamma_1\|\varkappa\|+\digamma_2\|\frac{(\ell og\mathsf{T})^{\ltimes_n}}{\Gamma(\ltimes_n+1)}\varkappa\|\right)\frac{dt}{t}\\
&\left.+ \frac{|b|f^*}{\Gamma(\ltimes_n)}\int_{\mathsf{T}_n}^{\mathsf{T}}\left(\ell og\frac{\mathsf{T}}{t}\right)^{\ltimes_n-1}\frac{dt}{t}+|b|\sum_{j=1}^{n}\digamma_3\|\varkappa\|+|c|\right]\\
&+ \sum_{0<\mathsf{T}_j<s}\frac{1}{\Gamma(\ltimes_{j-1})}\int_{\mathsf{T}_{j-1}}^{\mathsf{T}_j}\left(\ell og\frac{\mathsf{T}_j}{t}\right)^{\ltimes_{j-1}-1}\left(\digamma_1\|\varkappa\|+\digamma_2\|\frac{(\ell og\mathsf{T})^{\ltimes_{j-1}}}{\Gamma(\ltimes_{j-1}+1)}\varkappa\|\right)\frac{dt}{t}\\
&+ \sum_{0<\mathsf{T}_j<s}\frac{f^*}{\Gamma(\ltimes_{j-1})}\int_{\mathsf{T}_{j-1}}^{\mathsf{T}_j}\left(\ell og\frac{\mathsf{T}_j}{t}\right)^{\ltimes_{j-1}-1}\frac{dt}{t}\\
&+ \frac{1}{\Gamma(\ltimes_i)}\int_{\mathsf{T}_i}^{s}\left(\ell og\frac{s}{t}\right)^{\ltimes_i-1}\left(\digamma_1\|\varkappa\|+\digamma_2\|\frac{(\ell og\mathsf{T})^{\ltimes_i}}{\Gamma(\ltimes_i+1)}\varkappa\|\right)\frac{dt}{t}\\
&+ \frac{f^*}{\Gamma(\ltimes_i)}\int_{\mathsf{T}_i}^{s}\left(\ell og\frac{s}{t}\right)^{\ltimes_i-1}\frac{dt}{t}+\sum_{0<\mathsf{T}_j<s}\digamma_3\|\varkappa\|
\end{aligned}
$$

$$
\begin{aligned}
\leq\ & \frac{1}{|a+b|}\left[|b|\frac{n(\ell og\mathsf{T})^{\ltimes_{j-1}}}{\Gamma(\ltimes^{j-1}+1)}\left(\digamma_1+\digamma_2\frac{(\ell og\mathsf{T})^{\ltimes_{j-1}}}{\Gamma(\ltimes_{j-1}+1)}\right)\|\varkappa\|+|b|f^*\frac{n(\ell og\mathsf{T})^{\ltimes_{j-1}}}{\Gamma(\ltimes_{j-1}+1)}\right.\\
&\left.+ \frac{|b|(\ell og\mathsf{T})^{\ltimes_n}}{\Gamma(\ltimes_n+1)}\left(\digamma_1+\digamma_2\frac{(\ell og\mathsf{T})^{\ltimes_n}}{\Gamma(\ltimes_n+1)}\right)\|\varkappa\|+\frac{|b|f^*(\ell og\mathsf{T})^{\ltimes_n}}{\Gamma(\ltimes_n+1)}\ |b|n\digamma_3\|\varkappa\|+|c|\right]\\
&+ \frac{n(\ell og\mathsf{T})^{\ltimes_{j-1}}}{\Gamma(\ltimes_{j-1}+1)}\left(\digamma_1+\digamma_2\frac{(\ell og\mathsf{T})^{\ltimes_{j-1}}}{\Gamma(\ltimes_{j-1}+1)}\right)\|\varkappa\|+\frac{nf^*(\ell og\mathsf{T})^{\ltimes_{j-1}}}{\Gamma(\ltimes_{j-1}+1)}\\
&+ \frac{(\ell og\mathsf{T})^{\ltimes_i}}{\Gamma(\ltimes_i+1)}\left(\digamma_1+\digamma_2\frac{(\ell og\mathsf{T})^{\ltimes_i}}{\Gamma(\ltimes_i+1)}\right)\|\varkappa\|+\frac{f^*(\ell og\mathsf{T})^{\ltimes_i}}{\Gamma(\ltimes_i+1)}+n\digamma_3\|\varkappa\|
\end{aligned}
$$

$$\leq \left(\frac{|b|}{|a+b|}+1\right)\left[\frac{(n+1)(\ell og \mathsf{T})^{\ltimes^*}}{2(\sqrt{2}-1)}\left(F_1+F_2\frac{(\ell og \mathsf{T})^{\ltimes^*}}{2(\sqrt{2}-1)}\right)+nF_3\right]\|\varkappa\|$$
$$+ \left(\frac{|b|}{|a+b|}+1\right)\frac{(n+1)(\ell og \mathsf{T})^{\ltimes^*}}{2(\sqrt{2}-1)}f^*+\frac{|c|}{|a+b|}$$
$$\leq R.$$

STEP 2: Claim: The operator $\mathcal{S}$ is continuous.
Let the sequence $(\varkappa_m)$ converge to $\varkappa$ in $PC(\mathcal{J},\mathbb{R})$. We will prove that

$$\|\mathcal{S}\varkappa_m - \mathcal{S}\varkappa\| \to 0.$$

For $s \in \mathcal{J}$, we have

$$\begin{aligned}
&|\mathcal{S}\varkappa_m(s)-\mathcal{S}\varkappa(s)|\\
\leq\ & \frac{1}{|a+b|}\Bigg[|b|\sum_{j=1}^{n}\frac{1}{\Gamma(\ltimes_{j-1})}\int_{\mathsf{T}_{j-1}}^{\mathsf{T}_j}\left(\ell og\frac{\mathsf{T}_j}{t}\right)^{\ltimes_{j-1}-1}\\
& |f(t,\varkappa_m(t),I^{\ltimes_{j-1}}_{\mathsf{T}_{j-1}^+}\varkappa_m(t))-f(t,\varkappa(t),I^{\ltimes_{j-1}}_{\mathsf{T}_{j-1}^+}\varkappa(t))|\frac{dt}{t}\\
+\ & \frac{|b|}{\Gamma(\ltimes_n)}\int_{\mathsf{T}_n}^{\mathsf{T}}\left(\ell og\frac{\mathsf{T}}{t}\right)^{\ltimes_n-1}|f(t,\varkappa_m(t),I^{\ltimes_n}_{\mathsf{T}_n^+}\varkappa_m(t))-f(t,\varkappa(t),I^{\ltimes_n}_{\mathsf{T}_n^+}\varkappa(t))|\frac{dt}{t}\\
+\ & |b|\sum_{j=1}^{n}|L_j(\varkappa_m(\mathsf{T}_j^-))-L_j(\varkappa(\mathsf{T}_j^-))|\Bigg]\\
+\ & \sum_{0<\mathsf{T}_j<s}\frac{1}{\Gamma(\ltimes_{j-1})}\int_{\mathsf{T}_{j-1}}^{\mathsf{T}_j}\left(\ell og\frac{\mathsf{T}_j}{t}\right)^{\ltimes_{j-1}-1}\\
& |f(t,\varkappa_m(t),I^{\ltimes_{j-1}}_{\mathsf{T}_{j-1}^+}\varkappa_m(t))-f(t,\varkappa(t),I^{\ltimes_{j-1}}_{\mathsf{T}_{j-1}^+}\varkappa(t))|\frac{dt}{t}
\end{aligned}$$

$$
\begin{aligned}
&+ \frac{1}{\Gamma(\ltimes_i)}\int_{\mathsf{T}_i}^{s}\left(\ell og\frac{s}{t}\right)^{\ltimes_i-1}|f(t,\varkappa_m(t),I_{\mathsf{T}_i^+}^{\ltimes_i}\varkappa_m(t))-f(t,\varkappa(t),I_{\mathsf{T}_i^+}^{\ltimes_i}\varkappa(t))|\frac{dt}{t}\\
&+ \sum_{0<\mathsf{T}_j<s}|L_j(\varkappa_m(\mathsf{T}_j^-))-L_j(\varkappa(\mathsf{T}_j^-))|\\
\leq\ & \frac{1}{|a+b|}\Bigg[|b|\sum_{j=1}^{n}\frac{1}{\Gamma(\ltimes_{j-1})}\int_{\mathsf{T}_{j-1}}^{\mathsf{T}_j}\left(\ell og\frac{\mathsf{T}_j}{t}\right)^{\ltimes_{j-1}-1}\\
&\left(F_1|\varkappa_m(t)-\varkappa(t)|+F_2|I_{\mathsf{T}_{j-1}^+}^{\ltimes_{j-1}}(\varkappa_m(t)-\varkappa(t))|\right)\frac{dt}{t}\\
&+ \frac{|b|}{\Gamma(\ltimes_n)}\int_{\mathsf{T}_n}^{\mathsf{T}}\left(\ell og\frac{\mathsf{T}}{t}\right)^{\ltimes_n-1}\left(F_1|\varkappa_m(t)-\varkappa(t)|+F_2|I_{\mathsf{T}_n^+}^{\ltimes_n}(\varkappa_m(t)-\varkappa(t))|\right)\frac{dt}{t}\\
&+ \sum_{0<\mathsf{T}_j<s}\frac{1}{\Gamma(\ltimes_{j-1})}\int_{\mathsf{T}_{j-1}}^{\mathsf{T}_j}\left(\ell og\frac{\mathsf{T}_j}{t}\right)^{\ltimes_{j-1}-1}\\
&\left(F_1|\varkappa_m(t)-\varkappa(t)|+F_2|I_{\mathsf{T}_{j-1}^+}^{\ltimes_{j-1}}(\varkappa_m(t)-\varkappa(t))|\right)\frac{dt}{t}\\
&+ \frac{1}{\Gamma(\ltimes_i)}\int_{\mathsf{T}_i}^{s}\left(\ell og\frac{s}{t}\right)^{\ltimes_i-1}\left(F_1|\varkappa_m(t)-\varkappa(t)|+F_2|I_{\mathsf{T}_i^+}^{\ltimes_i}(\varkappa_m(t)-\varkappa(t)|\right)\frac{dt}{t}\\
&+ \left(\frac{|b|}{|a+b|}+1\right)\sum_{j=1}^{n}|L_j(\varkappa_m(\mathsf{T}_j^-))-L_j(\varkappa(\mathsf{T}_j^-))|\\
\leq\ & \frac{1}{|a+b|}\left[|b|\frac{n(\ell og\mathsf{T})^{\ltimes_{j-1}}}{\Gamma(\ltimes_{j-1}+1)}\left(F_1+F_2\frac{(\ell og\mathsf{T})^{\ltimes_{j-1}}}{\Gamma(\ltimes_{j-1}+1)}\right)\|\varkappa_m-\varkappa\|\right.\\
&\left.+ \frac{|b|(\ell og\mathsf{T})^{\ltimes_n}}{\Gamma(\ltimes_n+1)}\left(F_1+F_2\frac{(\ell og\mathsf{T})^{\ltimes_n}}{\Gamma(\ltimes_n+1)}\right)\|\varkappa_m-\varkappa\|\right]\\
&+ \frac{n(\ell og\mathsf{T})^{\ltimes_{j-1}}}{\Gamma(\ltimes_{j-1}+1)}\left(F_1+F_2\frac{(\ell og\mathsf{T})^{\ltimes_{j-1}}}{\Gamma(\ltimes_{j-1}+1)}\right)\|\varkappa_m-\varkappa\|\\
&+ \frac{(\ell og\mathsf{T})^{\ltimes_i}}{\Gamma(\ltimes_i+1)}\left(F_1+F_2\frac{(\ell og\mathsf{T})^{\ltimes_i}}{\Gamma(\ltimes_i+1)}\right)\|\varkappa_m-\varkappa\|\\
&+ \left(\frac{|b|}{|a+b|}+1\right)\sum_{j=1}^{n}|L_j(\varkappa_m(\mathsf{T}_j^-))-L_j(\varkappa(\mathsf{T}_j^-))|\\
\leq\ & \left(\frac{|b|}{|a+b|}+1\right)\left(\frac{(n+1)(\ell og\mathsf{T})^{\ltimes^*}}{2(\sqrt{2}-1)}\right)\left(F_1+F_2\frac{(\ell og\mathsf{T})^{\ltimes^*}}{2(\sqrt{2}-1)}\right)\|\varkappa_m-\varkappa\|\\
&+ \left(\frac{|b|}{|a+b|}+1\right)\sum_{j=1}^{n}|L_j(\varkappa_m(\mathsf{T}_j^-))-L_j(\varkappa(\mathsf{T}_j^-))|.
\end{aligned}
$$

Since L_j is continuous, it follows that

$$\|\mathcal{S}\varkappa_m - \mathcal{S}\varkappa\| \to 0 \ \ as \ \ m \to \infty.$$

Therefore, the operator $\mathcal{S}$ is continuous.

STEP 3: Claim. The operator $\mathcal{S}$ is compact.
By Step 1, we have $\|\mathcal{S}(\varkappa)\| \le R$ for each $\varkappa \in B_R$, which implies that $\mathcal{S}(B_R)$ is bounded.

We now proceed to show that $\mathcal{S}(B_R)$ is equicontinuous.

For $s_1,\ s_2 \in \mathcal{J},\ s_1 < s_2$ and $\varkappa \in B_R$, we have

$$|\mathcal{S}\varkappa(s_2) - \mathcal{S}\varkappa(s_1)|$$

$$\begin{aligned}
\le\ & \frac{1}{\Gamma(\ltimes_i)}\int_1^{s_1}\left|\left(\ell og\frac{s_2}{t}\right)^{\ltimes_i-1} - \left(\ell og\frac{s_1}{t}\right)^{\ltimes_i-1}\right|\left|f\left(t,\varkappa(t),I^{\ltimes_i}_{\mathsf{T}_i^+}\varkappa(t)\right)\right|\frac{dt}{t}\\
&+\frac{1}{\Gamma(\ltimes_i)}\int_{s_1}^{s_2}\left(\ell og\frac{s_2}{t}\right)^{\ltimes_i-1}\left|f\left(t,\varkappa(t),I^{\ltimes_i}_{\mathsf{T}_i^+}\varkappa(t)\right)\right|\frac{dt}{t}+\sum_{0<\mathsf{T}_j<s_2-s_1}\left|L_j\left(\varkappa\left(\mathsf{T}_j^-\right)\right)\right|\\
\le\ & \frac{1}{\Gamma(\ltimes_i)}\int_1^{s_1}\left|\left(\ell og\frac{s_2}{t}\right)^{\ltimes_i-1} - \left(\ell og\frac{s_1}{t}\right)^{\ltimes_i-1}\right|\left|f\left(t,\varkappa(t),I^{\ltimes_i}_{\mathsf{T}_i^+}\varkappa(t)\right)-f(t,0,0)\right|\frac{dt}{t}\\
&+\frac{1}{\Gamma(\ltimes_i)}\int_1^{s_1}\left|\left(\ell og\frac{s_2}{t}\right)^{\ltimes_i-1} - \left(\ell og\frac{s_1}{t}\right)^{\ltimes_i-1}\right||f(t,0,0)|\frac{dt}{t}\\
&+\frac{1}{\Gamma(\ltimes_i)}\int_{s_1}^{s_2}\left(\ell og\frac{s_2}{t}\right)^{\ltimes_i-1}\left|f\left(t,\varkappa(t),I^{\ltimes_i}_{\mathsf{T}_i^+}\varkappa(t)\right)-f(t,0,0)\right|\frac{dt}{t}\\
&+\frac{1}{\Gamma(\ltimes_i)}\int_{s_1}^{s_2}\left(\ell og\frac{s_2}{t}\right)^{\ltimes_i-1}|f(t,0,0)|\frac{dt}{t}+\sum_{0<\mathsf{T}_j<s_2-s_1}\left|L_j\left(\varkappa\left(\mathsf{T}_j^-\right)\right)\right|\\
\le\ & \frac{1}{\Gamma(\ltimes_i)}\int_1^{s_1}\left|\left(\ell og\frac{s_2}{t}\right)^{\ltimes_i-1} - \left(\ell og\frac{s_1}{t}\right)^{\ltimes_i-1}\right|\left(\digamma_1|\varkappa(t)|+\digamma_2\left|I^{\ltimes_i}_{\mathsf{T}_i^+}\varkappa(t)\right|\right)\frac{dt}{t}\\
&+\frac{f^\star}{\Gamma(\ltimes_i)}\int_1^{s_1}\left|\left(\ell og\frac{s_2}{t}\right)^{\ltimes_i-1} - \left(\ell og\frac{s_1}{t}\right)^{\ltimes_i-1}\right|\frac{dt}{t}\\
&+\frac{1}{\Gamma(\ltimes_i)}\int_{s_1}^{s_2}\left(\ell og\frac{s_2}{t}\right)^{\ltimes_i-1}\left(\digamma_1|\varkappa(t)|+\digamma_2\left|I^{\ltimes_i}_{\mathsf{T}_i^+}\varkappa(t)\right|\right)\frac{dt}{t}
\end{aligned}$$

$$+\frac{f^{\star}}{\Gamma(\ltimes_i)}\int_{s_1}^{s_2}\left(\ell og\frac{s_2}{t}\right)^{\ltimes_i-1}\frac{dt}{t}+\sum_{0<\tau_j<s_2-s_1}\left|L_j\left(\varkappa\left(\tau_j^-\right)\right)\right|$$

$$\leq \frac{1}{\Gamma(\ltimes_i)}\int_1^{s_1}\left|\left(\ell og\frac{s_2}{t}\right)^{\ltimes_i-1}-\left(\ell og\frac{s_1}{t}\right)^{\ltimes_i-1}\right|\left(F_1\|\varkappa\|+F_2\left\|\frac{(\ell og\tau)^{\ltimes_i}}{\Gamma(\ltimes_i+1)}\varkappa\right\|\right)\frac{dt}{t}$$

$$+\frac{f^{\star}}{\Gamma(\ltimes_i)}\int_1^{s_1}\left|\left(\ell og\frac{s_2}{t}\right)^{\ltimes_i-1}-\left(\ell og\frac{s_1}{t}\right)^{\ltimes_i-1}\right|\frac{dt}{t}$$

$$+\frac{1}{\Gamma(\ltimes_i)}\int_{s_1}^{s_2}\left(\ell og\frac{s_2}{t}\right)^{\ltimes_i-1}\left(F_1\|\varkappa\|+F_2\left\|\frac{(\ell og\tau)^{\ltimes_i}}{\Gamma(\ltimes_i+1)}\varkappa\right\|\right)\frac{dt}{t}$$

$$+\frac{f^{\star}}{\Gamma(\ltimes_i)}\int_{s_1}^{s_2}\left(\ell og\frac{s_2}{t}\right)^{\ltimes_i-1}\frac{dt}{t}+\sum_{0<\tau_j<s_2-s_1}\left|L_j\left(\varkappa\left(\tau_j^-\right)\right)\right|$$

$$\leq \frac{1}{\Gamma(\ltimes_i)}\left[\left(F_1+F_2\frac{(\ell og\tau)^{\ltimes_i}}{\Gamma(\ltimes_i+1)}\right)\|\varkappa\|+f^{\star}\right]\int_1^{s_1}\left|\left(\ell og\frac{s_2}{t}\right)^{\ltimes_i-1}-\left(\ell og\frac{s_1}{t}\right)^{\ltimes_i-1}\right|\frac{dt}{t}$$

$$+\frac{1}{\Gamma(\ltimes_i)}\left[\left(F_1+F_2\frac{(\ell og\tau)^{\ltimes_i}}{\Gamma(\ltimes_i+1)}\right)\|\varkappa\|+f^{\star}\right]\int_{s_1}^{s_2}\left(\ell og\frac{s_2}{t}\right)^{\ltimes_i-1}\frac{dt}{t}$$

$$+\sum_{0<\tau_j<s_2-s_1}\left|L_j\left(\varkappa\left(\tau_j^-\right)\right)\right|.$$

As $s_1 \to s_2$, the right-hand side of the inequality tends to zero. Hence, $|\mathcal{S}\varkappa(s_2)-\mathcal{S}\varkappa(s_1)|\to 0$. It implies that $\mathcal{S}(B_R)$ is equicontinuous.

Thus, by the Schauder fixed point theorem, the initial and terminal value problem (6.1)–(6.3) possesses a solution in B_R. Since $B_R \subset PC(\mathcal{J},\mathbb{R})$, the claim of Theorem 52 is proved.

6.1.2 Uniqueness of Solution

We will now apply the ⋋anach contraction principle to confirm that the solutions of the ITVPs (6.1)–(6.3) are unique.

Theorem 53 *In addition to (H1) and (H2), assume that:*

(H4) For $j=1,\ldots,n$, there exists $F_4>0$ such that, for any $\varkappa, y\in\mathbb{R}$ and $s\in\mathcal{J}$,

$$|L_j(\varkappa(s))-L_j(y(s))|\leq F_4|\varkappa(s)-y(s)|.$$

Then, if

$$\left(\frac{|b|}{|a+b|}+1\right)\left[\frac{(n+1)(\ell og\mathsf{T})^{\ltimes^*}}{2(\sqrt{2}-1)}\left(F_1+F_2\frac{(\ell og\mathsf{T})^{\ltimes^*}}{2(\sqrt{2}-1)}\right)+nF_4\right]<1, \tag{6.10}$$

the condition (6.10) holds.

The ITVPs (6.1)–(6.3) possesses a solution uniquely determined on $PC(\mathcal{J},\mathbb{R})$.

Proof: For $s\in\mathcal{J}$ and $\varkappa, y\in PC(\mathcal{J},\mathbb{R})$, we have

$$
\begin{aligned}
&|\mathcal{S}\varkappa(s)-\mathcal{S}y(s)|\\
\le\ & \frac{1}{|a+b|}\left[|b|\sum_{j=1}^{n}\frac{1}{\Gamma(\ltimes_{j-1})}\int_{\mathsf{T}_{j-1}}^{\mathsf{T}_j}\left(\ell og\frac{\mathsf{T}_j}{t}\right)^{\ltimes_{j-1}-1}\right.\\
&\left|f\left(t,\varkappa(t),I_{\mathsf{T}_{j-1}^+}^{\ltimes_{j-1}}\varkappa(t)\right)-f\left(t,y(t),I_{\mathsf{T}_{j-1}^+}^{\ltimes_{j-1}}y(t)\right)\right|\frac{dt}{t}\\
+\ & \frac{|b|}{\Gamma(\ltimes_n)}\int_{\mathsf{T}_n}^{\mathsf{T}}\left(\ell og\frac{\mathsf{T}}{t}\right)^{\ltimes_n-1}\left|f\left(t,\varkappa(t),I_{\mathsf{T}_n^+}^{\ltimes_n}\varkappa(t)\right)-f\left(t,y(t),I_{\mathsf{T}_n^+}^{\ltimes_n}y(t)\right)\right|\frac{dt}{t}\\
+\ & \left.|b|\sum_{j=1}^{n}\left|L_j\left(\varkappa\left(\mathsf{T}_j^-\right)\right)-L_j\left(y\left(\mathsf{T}_j^-\right)\right)\right|\right]+\sum_{0<\mathsf{T}_j<s}\frac{1}{\Gamma(\ltimes_{j-1})}\int_{\mathsf{T}_{j-1}}^{\mathsf{T}_j}\left(\ell og\frac{\mathsf{T}_j}{t}\right)^{\ltimes_{j-1}-1}\\
&\left|f\left(t,\varkappa(t),I_{\mathsf{T}_{j-1}^+}^{\ltimes_{j-1}}\varkappa(t)\right)-f\left(t,y(t),I_{\mathsf{T}_{j-1}^+}^{\ltimes_{j-1}}y(t)\right)\right|\frac{dt}{t}\\
+\ & \frac{1}{\Gamma(\ltimes_i)}\int_{\mathsf{T}_i}^{s}\left(\ell og\frac{s}{t}\right)^{\ltimes_i-1}\left|f\left(t,\varkappa(t),I_{\mathsf{T}_i^+}^{\ltimes_i}\varkappa(t)\right)-f\left(t,y(t),I_{\mathsf{T}_i^+}^{\ltimes_i}y(t)\right)\right|\frac{dt}{t}\\
+\ & \sum_{0<\mathsf{T}_j<s}\left|L_j\left(\varkappa\left(\mathsf{T}_j^-\right)\right)-L_j\left(y\left(\mathsf{T}_j^-\right)\right)\right|\\
\le\ & \frac{1}{|a+b|}\left[|b|\sum_{j=1}^{n}\frac{1}{\Gamma(\ltimes_{j-1})}\int_{\mathsf{T}_{j-1}}^{\mathsf{T}_j}\left(\ell og\frac{\mathsf{T}_j}{t}\right)^{\ltimes_{j-1}-1}\right.\\
&\left(F_1|\varkappa(t)-y(t)|+F_2\left|I_{\mathsf{T}_{j-1}^+}^{\ltimes_{j-1}}(\varkappa(t)-y(t))\right|\right)\frac{dt}{t}+\frac{|b|}{\Gamma(\ltimes_n)}\int_{\mathsf{T}_n}^{\mathsf{T}}\left(\ell og\frac{\mathsf{T}}{t}\right)^{\ltimes_n-1}\\
&\left.\left(F_1|\varkappa(t)-y(t)|+F_2\left|I_{\mathsf{T}_n^+}^{\ltimes_n}(\varkappa(t)-y(t))\right|\right)\frac{dt}{t}+|b|\sum_{j=1}^{n}F_4\left|\varkappa\left(\mathsf{T}_j^-\right)-y\left(\mathsf{T}_j^-\right)\right|\right]
\end{aligned}
$$

$$+ \sum_{j=1}^{n} \frac{1}{\Gamma(\ltimes_{j-1})} \int_{\mathsf{T}_{j-1}}^{\mathsf{T}_j} \left(\ell og \frac{\mathsf{T}_j}{t}\right)^{\ltimes_{j-1}-1}$$

$$\left(F_1|\varkappa(t) - y(t)| + F_2 \left| I_{\mathsf{T}_{j-1}^+}^{\ltimes_{j-1}}(\varkappa(t) - y(t))\right|\right) \frac{dt}{t} + \frac{1}{\Gamma(\ltimes_i)} \int_{\mathsf{T}_i}^{s} \left(\ell og \frac{s}{t}\right)^{\ltimes_i - 1}$$

$$\left(F_1|\varkappa(t) - y(t)| + F_2 \left| I_{\mathsf{T}_i^+}^{\ltimes_i}(\varkappa(t) - y(t))\right|\right) \frac{dt}{t} + \sum_{j=1}^{n} F_4 \left|\varkappa\left(\mathsf{T}_j^-\right) - y\left(\mathsf{T}_j^-\right)\right|$$

$$\leq \frac{1}{|a+b|}\left[|b| \frac{n(\ell og\mathsf{T})^{\ltimes_{j-1}}}{\Gamma(\ltimes_{j-1}+1)}\left(F_1 + F_2 \frac{(\ell og\mathsf{T})^{\ltimes_{j-1}}}{\Gamma(\ltimes_{j-1}+1)}\right)\|\varkappa - y\|\right.$$

$$\left.\frac{|b|(\ell og\mathsf{T})^{\ltimes_n}}{\Gamma(\ltimes_n+1)}\left(F_1 + F_2 \frac{(\ell og\mathsf{T})^{\ltimes_n}}{\Gamma(\ltimes_n+1)}\right)\|\varkappa - y\|\right]$$

$$+ \frac{n(\ell og\mathsf{T})^{\ltimes_{j-1}}}{\Gamma(\ltimes_{j-1}+1)}\left(F_1 + F_2 \frac{(\ell og\mathsf{T})^{\ltimes_{j-1}}}{\Gamma(\ltimes_{j-1}+1)}\right)\|\varkappa - y\|$$

$$+ \frac{(\ell og\mathsf{T})^{\ltimes_i}}{\Gamma(\ltimes_i+1)}\left(F_1 + F_2 \frac{(\ell og\mathsf{T})^{\ltimes_i}}{\Gamma(\ltimes_i+1)}\right)\|\varkappa - y\| + \left(\frac{|b|}{|a+b|} + 1\right)\sum_{j=1}^{n} F_4\|\varkappa - y\|$$

$$\leq \left(\frac{|b|}{|a+b|} + 1\right)\left[\frac{(n+1)(\ell og\mathsf{T})^{\ltimes^*}}{2(\sqrt{2}-1)}\left(F_1 + F_2 \frac{(\ell og\mathsf{T})^{\ltimes^*}}{2(\sqrt{2}-1)}\right) + nF_4\right]\|\varkappa - y\|.$$

6.1.3 Study of the Solution's Behavior

To apply the obtained existence and uniqueness results, in this section we will consider the stability of the Ulam-Hyers solutions to the ITVPs (6.1)–(6.3).

Theorem 54 *Consider the hypotheses of Theorem 53. Consequently, the ITVPs (6.1)–(6.3) exhibit Ulam-Hyers stability.*

Proof: Assume $z(s)$ satisfies the inequality (2.32). Then, the integral inequality

$$z(s) + \frac{1}{a+b}\left[b\sum_{j=1}^{n} \frac{1}{\Gamma(\ltimes_{j-1})} \int_{\mathsf{T}_{j-1}}^{\mathsf{T}_j} \left(\ell og \frac{\mathsf{T}_j}{t}\right)^{\ltimes_{j-1}-1} f(t, z(t), I_{\mathsf{T}_{j-1}^+}^{\ltimes_{j-1}} z(t)) \frac{dt}{t}\right.$$

$$\left. + \frac{b}{\Gamma(\ltimes_n)} \int_{\mathsf{T}_n}^{\mathsf{T}} \left(\ell og \frac{\mathsf{T}}{t}\right)^{\ltimes_n - 1} f(t, z(t), I_{\mathsf{T}_n^+}^{\ltimes_n} z(t)) \frac{dt}{t} + b\sum_{j=1}^{n} L_j(z(\mathsf{T}_j^-)) - c\right]$$

$$- \sum_{0<\mathsf{T}_j<s} \frac{1}{\Gamma(\ltimes_{j-1})} \int_{\mathsf{T}_{j-1}}^{\mathsf{T}_j} \left(\ell og \frac{\mathsf{T}_j}{t}\right)^{\ltimes_{j-1}-1} f(t, z(t), I_{\mathsf{T}_{j-1}^+}^{\ltimes_{j-1}} z(t)) \frac{dt}{t}$$

$$- \frac{1}{\Gamma(\ltimes_i)} \int_{\mathsf{T}_i}^{s} \left(\ell og \frac{s}{t}\right)^{\ltimes_i - 1} f(t, z(t), I_{\mathsf{T}_i^+}^{\ltimes_i} z(t)) \frac{dt}{t} - \sum_{0<\mathsf{T}_j<s} L(z(\mathsf{T}_t^-)) \leq \epsilon \frac{\mathsf{T}^{\ltimes^*}}{2(\sqrt{2}-1)}$$

holds.

According to Proposition 9, for $s \in (\mathsf{T}_i, \mathsf{T}_{i+1}]$, $i = 1, ..., n$ the unique solution $\varkappa$ of the ITVP (6.1)–(6.3) is given by

$$\begin{aligned}
\varkappa(s) &= \frac{-1}{a+b}\Bigg[b\sum_{j=1}^{n}\frac{1}{\Gamma(\ltimes_{j-1})}\int_{\mathsf{T}_{j-1}}^{\mathsf{T}_j}\left(\ell og\frac{\mathsf{T}_j}{t}\right)^{\ltimes_{j-1}-1} f(t,\varkappa(t), I^{\ltimes_{j-1}}_{\mathsf{T}_{j-1}^+}\varkappa(t))\frac{dt}{t}\\
&+\frac{b}{\Gamma(\ltimes_n)}\int_{\mathsf{T}_n}^{\mathsf{T}}\left(\ell og\frac{\mathsf{T}}{t}\right)^{\ltimes_n-1} f(t,\varkappa(t), I^{\ltimes_n}_{\mathsf{T}_n^+}\varkappa(t))\frac{dt}{t} + b\sum_{j=1}^{n}L_j(z(\mathsf{T}_j^-)) - c\Bigg]\\
&+\sum_{0<\mathsf{T}_j<s}\frac{1}{\Gamma(\ltimes_{j-1})}\int_{\mathsf{T}_{j-1}}^{\mathsf{T}_j}\left(\ell og\frac{\mathsf{T}_j}{t}\right)^{\ltimes_{j-1}-1} f(t,\varkappa(t), I^{\ltimes_{j-1}}_{\mathsf{T}_{j-1}^+}\varkappa(t))\frac{dt}{t}\\
&+\frac{1}{\Gamma(\ltimes_i)}\int_{\mathsf{T}_i}^{s}\left(\ell og\frac{s}{t}\right)^{\ltimes_i-1} f(t,\varkappa(t), I^{\ltimes_i}_{\mathsf{T}_i^+}\varkappa(t))\frac{dt}{t} + \sum_{0<\mathsf{T}_j<s} L(\varkappa(\mathsf{T}_t^-)).
\end{aligned}$$

Let $s \in (\mathsf{T}_i, \mathsf{T}_{i+1}]$, $i = 1, ..., n$. Then, we get

$$\begin{aligned}
&|z(s) - \varkappa(s)|\\
= &\quad |z(s) + \frac{1}{a+b}\Bigg[b\sum_{j=1}^{n}\frac{1}{\Gamma(\ltimes_{j-1})}\int_{\mathsf{T}_{j-1}}^{\mathsf{T}_j}\left(\ell og\frac{\mathsf{T}_j}{t}\right)^{\ltimes_{j-1}-1} f(t,\varkappa(t), I^{\ltimes_{j-1}}_{\mathsf{T}_{j-1}^+}\varkappa(t))\frac{dt}{t}\\
+ &\quad \frac{b}{\Gamma(\ltimes_n)}\int_{\mathsf{T}_n}^{\mathsf{T}}\left(\ell og\frac{\mathsf{T}}{t}\right)^{\ltimes_n-1} f(t,\varkappa(t), I^{\ltimes_n}_{\mathsf{T}_n^+}\varkappa(t))\frac{dt}{t} + b\sum_{j=1}^{n}L_j(\varkappa(\mathsf{T}_j^-)) - c\Bigg]\\
- &\quad \sum_{0<\mathsf{T}_j<s}\frac{1}{\Gamma(\ltimes_{j-1})}\int_{\mathsf{T}j-1}^{\mathsf{T}_j}\left(\ell og\frac{\mathsf{T}_j}{t}\right)^{\ltimes_{j-1}-1} f(t,\varkappa(t), I^{\ltimes_{j-1}}_{\mathsf{T}_{j-1}^+}\varkappa(t))\frac{dt}{t}\\
- &\quad \frac{1}{\Gamma(\ltimes_i)}\int_{\mathsf{T}_i}^{s}\left(\ell og\frac{s}{t}\right)^{\ltimes_i-1} f(t,\varkappa(t), I^{\ltimes_i}_{\mathsf{T}_i^+}\varkappa(t))\frac{dt}{t} - \sum_{0<\mathsf{T}_j<s} L(\varkappa(\mathsf{T}_t^-))|\\
\leq &\quad |z(s) + \frac{1}{a+b}\Bigg[b\sum_{j=1}^{n}\frac{1}{\Gamma(\ltimes_{j-1})}\int_{\mathsf{T}_{j-1}}^{\mathsf{T}_j}\left(\ell og\frac{\mathsf{T}_j}{t}\right)^{\ltimes_{j-1}-1} f(t,z(t), I^{\ltimes_{j-1}}_{\mathsf{T}_{j-1}^+}z(t))\frac{dt}{t}\\
+ &\quad \frac{b}{\Gamma(\ltimes_n)}\int_{\mathsf{T}_n}^{\mathsf{T}}\left(\ell og\frac{\mathsf{T}}{t}\right)^{\ltimes_n-1} f(t,z(t), I^{\ltimes_n}_{\mathsf{T}_n^+}z(t))\frac{dt}{t} + b\sum_{j=1}^{n}L_j(z(\mathsf{T}_j^-)) - c\Bigg]\\
- &\quad \sum_{0<\mathsf{T}_j<s}\frac{1}{\Gamma(\ltimes_{j-1})}\int_{\mathsf{T}j-1}^{\mathsf{T}_j}\left(\ell og\frac{\mathsf{T}_j}{t}\right)^{\ltimes_{j-1}-1} f(t,z(t), I^{\ltimes_{j-1}}_{\mathsf{T}_{j-1}^+}z(t))\frac{dt}{t}\\
- &\quad \frac{1}{\Gamma(\ltimes_i)}\int_{\mathsf{T}_i}^{s}\left(\ell og\frac{s}{t}\right)^{\ltimes_i-1} f(t,z(t), I^{\ltimes_i}_{\mathsf{T}_i^+}z(t))\frac{dt}{t} - \sum_{0<\mathsf{T}_j<s} L(z(\mathsf{T}_t^-))|
\end{aligned}$$

$$
\begin{aligned}
&+ \frac{1}{|a+b|}\Bigg[|b|\sum_{j=1}^{n}\frac{1}{\Gamma(\ltimes_{j-1})}\int_{\mathsf{T}_{j-1}}^{\mathsf{T}_j}\left(\ell og\frac{\mathsf{T}_j}{t}\right)^{\ltimes_{j-1}-1} \\
&|f(t,z(t),I_{\mathsf{T}_{j-1}^+}^{\ltimes_{j-1}}z(t)) - f(t,\varkappa(t),I_{\mathsf{T}_{j-1}^+}^{\ltimes_{j-1}}\varkappa(t))|\frac{dt}{t} \\
&+ \frac{|b|}{\Gamma(\ltimes_n)}\int_{\mathsf{T}_n}^{\mathsf{T}}\left(\ell og\frac{\mathsf{T}}{t}\right)^{\ltimes_n-1}|f(t,z(t),I_{\mathsf{T}_n^+}^{\ltimes_n}z(t)) - f(t,\varkappa(t),I_{\mathsf{T}_n^+}^{\ltimes_n}\varkappa(t))|\frac{dt}{t} \\
&+ |b|\sum_{j=1}^{n}|L_j(z(\mathsf{T}_j^-)) - L_j(\varkappa(\mathsf{T}_j^-))|\Bigg] \\
&+ \sum_{0<\mathsf{T}_j<s}\frac{1}{\Gamma(\ltimes_{j-1})}\int_{\mathsf{T}_{j-1}}^{\mathsf{T}_j}\left(\ell og\frac{\mathsf{T}_j}{t}\right)^{\ltimes_{j-1}-1} \\
&|f(t,z(t),I_{\mathsf{T}_{j-1}^+}^{\ltimes_{j-1}}z(t)) - f(t,\varkappa(t),I_{\mathsf{T}_{j-1}^+}^{\ltimes_{j-1}}\varkappa(t))|\frac{dt}{t} + \frac{1}{\Gamma(\ltimes_i)}\int_{\mathsf{T}_i}^{s}\left(\ell og\frac{s}{t}\right)^{\ltimes_i-1} \\
&|f(t,z(t),I_{\mathsf{T}_i^+}^{\ltimes_i}z(t)) - f(t,\varkappa(t),I_{\mathsf{T}_i^+}^{\ltimes_i}\varkappa(t))|\frac{dt}{t} + \sum_{0<\mathsf{T}_j<s}|L_j(z(\mathsf{T}_j^-)) - L_j(\varkappa(\mathsf{T}_j^-))| \\
\leq\ & \epsilon\frac{\mathsf{T}^{\ltimes^*}}{2(\sqrt{2}-1)} + \frac{1}{|a+b|}|b|\sum_{j=1}^{n}\frac{1}{\Gamma(\ltimes_{j-1})}\int_{\mathsf{T}_{j-1}}^{\mathsf{T}_j}\left(\ell og\frac{\mathsf{T}_j}{t}\right)^{\ltimes_{j-1}-1} \\
&\left(F_1|z(t)-\varkappa(t)| + F_2|I_{\mathsf{T}_{j-1}^+}^{\ltimes_{j-1}}(z(t)-\varkappa(t))|\right)\frac{dt}{t} + \frac{|b|}{\Gamma(\ltimes_n)}\int_{\mathsf{T}_n}^{\mathsf{T}}\left(\ell og\frac{\mathsf{T}}{t}\right)^{\ltimes_n-1} \\
&\left(F_1|z(t)-\varkappa(t)| + F_2|I_{\mathsf{T}_n^+}^{\ltimes_n}(z(t)-\varkappa(t))|\right)\frac{dt}{t} \\
&+ \sum_{j=1}^{n}\frac{1}{\Gamma(\ltimes_{j-1})}\int_{\mathsf{T}_{j-1}}^{\mathsf{T}_j}\left(\ell og\frac{\mathsf{T}_j}{t}\right)^{\ltimes_{j-1}-1} \\
&\left(F_1|z(t)-\varkappa(t)| + F_2|I_{\mathsf{T}_{j-1}^+}^{\ltimes_{j-1}}(z(t)-\varkappa(t))|\right)\frac{dt}{t} + \frac{1}{\Gamma(\ltimes_i)}\int_{\mathsf{T}_i}^{s}\left(\ell og\frac{s}{t}\right)^{\ltimes_i-1} \\
&\left(F_1|z(t)-\varkappa(t)| + F_2|I_{\mathsf{T}_i^+}^{\ltimes_i}(z(t)-\varkappa(t)|\right)\frac{dt}{t} \\
&+ \left(\frac{|b|}{|a+b|}+1\right)\sum_{j=1}^{n}F_4|z(\mathsf{T}_j^-) - \varkappa(\mathsf{T}_j^-)| \\
\leq\ & \epsilon\frac{\mathsf{T}^{\ltimes^*}}{2(\sqrt{2}-1)}\frac{1}{|a+b|}\left[|b|\frac{n(\ell og\mathsf{T})^{\ltimes_{j-1}}}{\Gamma(\ltimes_{j-1}+1)}\left(F_1 + F_2\frac{(\ell og\mathsf{T})^{\ltimes_{j-1}}}{\Gamma(\ltimes_{j-1}+1)}\right)\|z-\varkappa\|\right. \\
&\left.+ \frac{|b|(\ell og\mathsf{T})^{\ltimes_n}}{\Gamma(\ltimes_n+1)}\left(F_1 + F_2\frac{(\ell og\mathsf{T})^{\ltimes_n}}{\Gamma(\ltimes_n+1)}\right)\|z-\varkappa\|\right]
\end{aligned}
$$

$$+ \quad \frac{n(\ell og\mathsf{T})^{\ltimes_{j-1}}}{\Gamma(\ltimes_{j-1}+1)}\left(F_1+F_2\frac{(\ell og\mathsf{T})^{\ltimes_{j-1}}}{\Gamma(\ltimes_{j-1}+1)}\right)\|z-\varkappa\|$$

$$+ \quad \frac{(\ell og\mathsf{T})^{\ltimes_i}}{\Gamma(\ltimes_i+1)}\left(F_1+F_2\frac{(\ell og\mathsf{T})^{\ltimes_i}}{\Gamma(\ltimes_i+1)}\right)\|z-\varkappa\|+\left(\frac{|b|}{|a+b|}+1\right)\sum_{j=1}^{n}F_4\|z-\varkappa\|$$

$$\leq \quad \epsilon\frac{\mathsf{T}^{\ltimes^*}}{2(\sqrt{2}-1)}+\left(\frac{|b|}{|a+b|}+1\right)\left[\frac{(n+1)(\ell og\mathsf{T})^{\ltimes^*}}{2(\sqrt{2}-1)}\left(F_1+F_2\frac{(\ell og\mathsf{T})^{\ltimes^*}}{2(\sqrt{2}-1)}\right)\right.$$

$$+ \quad \left. nF_4\right]\|z-\varkappa\|.$$

Then,

$$\|z-\varkappa\|\left[1-\left(\frac{|b|}{|a+b|}+1\right)\left[\frac{(n+1)(\ell og\mathsf{T})^{\ltimes^*}}{2(\sqrt{2}-1)}\left(F_1+F_2\frac{(\ell og\mathsf{T})^{\ltimes^*}}{2(\sqrt{2}-1)}\right)+nF_4\right]\right]$$

$$\leq \epsilon\frac{\mathsf{T}^{\ltimes^*}}{2(\sqrt{2}-1)}.$$

Thus, we obtain,

$$|z(s)-\varkappa(s)|$$

$$\leq \frac{\mathsf{T}^{\ltimes^*}\epsilon}{2(\sqrt{2}-1)\left[1-\left(\frac{|b|}{|a+b|}+1\right)\left[\frac{(n+1)(\ell og\mathsf{T})^{\ltimes^*}}{2(\sqrt{2}-1)}\left(F_1+F_2\frac{(\ell og\mathsf{T})^{\ltimes^*}}{2(\sqrt{2}-1)}\right)+nF_4\right]\right]}$$
$$:= c_g\epsilon.$$

Consequently, the ITVP (6.1)–(6.3) is Ulam-Hyers stable.

6.1.4 Applied Example

As an example, consider the following impulsive ITVP:

$$^{CH}F^{\ltimes(s)}_{\mathsf{T}_i^+}\varkappa(s)=\frac{e^{-3(s-1)}}{(e^{e^{\frac{3(s-1)^2}{s}}}+5)(1+|\varkappa(s)|+|^{H}I^{\ltimes(s)}_{\mathsf{T}_i^+}\varkappa(s)|)},\quad s\in\mathcal{J}:=\mathcal{J}_0\cup\mathcal{J}_1, \tag{6.11}$$

$$\Delta\varkappa|_{s=\frac{3}{2}}=\frac{|(\varkappa(\frac{3}{2}^-))|}{10+|(\varkappa(\frac{3}{2}^-))|}, \tag{6.12}$$

$$3\varkappa(1)+2\varkappa(3)=5, \tag{6.13}$$

where

$$\mathsf{T}_0=1,\ \mathsf{T}_1=\frac{3}{2},\ \mathsf{T}_2=\mathsf{T}=3,\ n=1,\ \mathcal{J}=[1,3],\ \mathcal{J}_0=[1,\frac{3}{2}],\ \mathcal{J}_1=[\frac{3}{2},3],$$

and

$$\ltimes(s) = \begin{cases} \frac{1}{2}, & s \in \mathcal{J}_0, \\ \frac{3}{4}, & s \in \mathcal{J}_1. \end{cases} \tag{6.14}$$

Let

$$f(s, \varkappa, y) = \frac{e^{-3(s-1)}}{(e^{e^{\frac{3(s-1)^2}{s}}} + 5)(1 + |\varkappa| + |y|)}, \quad (s, \varkappa, y) \in \mathcal{J} \times \mathbb{R} \times \mathbb{R}.$$

For each $s \in \mathcal{J}$, and $\varkappa_1, \varkappa_2, y_1, y_2 \in \mathbb{R}$, we have

$$(\ell ogs)^{\frac{1}{3}} |f(s, \varkappa_1, y_1) - f(s, \varkappa_2, y_2)| \leq \frac{1}{e+5}(|\varkappa_1 - \varkappa_2| + |y_1 - y_2|).$$

Thus, assumption (H2) is satisfied with $\digamma_1 = \digamma_2 = \frac{1}{e+5}$.

Let

$$|L_1(\varkappa) - L_1(y)| \leq \frac{1}{7}|\varkappa - y|.$$

Then, the assumption (H4) holds with $\digamma_4 = \frac{1}{7}$.

We will also verify that assumption (6.10) is fulfilled with $M = 3$, $n = 1$, $\digamma_1 = \digamma_2 = \frac{1}{e+5}$, $\digamma_4 = \frac{1}{7}$ and $\ltimes^* = \frac{3}{4}$. Indeed,

$$\left(\frac{|b|}{|a+b|} + 1\right) \left[\frac{(n+1)(\ell og\mathsf{T})^{\ltimes^*}}{2(\sqrt{2}-1)} \left(\digamma_1 + \digamma_2 \frac{(\ell og\mathsf{T})^{\ltimes^*}}{2(\sqrt{2}-1)}\right) + n\digamma_4\right] \simeq 0.6256 < 1.$$

Hence, assumption (6.10) is satisfied. By Theorem 53 the ITVP (6.11)–(6.13) has a unique solution on $PC(\mathcal{J}, \mathbb{R})$.

In addition, according to Theorem 54, the ITVPs (6.11)–(6.13) is Ulam-Hyers stable.

6.2 On the Caputo-Hadamard Fractional IVP with Variable Order using the Upper–Lower Solutions Technique

We shall investigate whether the following fractional nonlinear initial value problem (IVP), involving the Caputo-Hadamard fractional derivative of variable order, admits solutions:

$$\begin{cases} {}^{CH}\digamma_{p^+}^{\ltimes(s)} \varkappa(s) = f(s, \varkappa(s)), \quad s \in \mathbf{J} := [p, \mathsf{T}], \\ \varkappa(p) = \varkappa_p, \end{cases} \tag{6.15}$$

where $0 < p < \mathsf{T} < \infty$, $\varkappa_p \in \mathbb{R}$, and $0 < \ltimes(s) \leq 1$ is the variable order. The function $f : \mathbf{J} \times \mathbb{R} \to \mathbb{R}$ is given, and ${}^{CH}\digamma_{p^+}^{\ltimes(s)}$ denotes the Caputo-Hadamard fractional derivative of order $\ltimes(s)$.

6.2.1 Results on Existence

Let us state the underlying assumptions, which will serve as the basic foundational step in proving the results of this section.

(H1) For $n \in \mathbb{N}$, the finite sequence of points $\{\mathsf{T}_i\}_{i=0}^{n}$ such that $p = \mathsf{T}_0 < \mathsf{T}_i < \mathsf{T}_n = \mathsf{T}$ for $i = 1, \ldots, n-1$. Define $\mathbf{J}_i := (\mathsf{T}_{i-1}, \mathsf{T}_i]$ for $i = 1, 2, \ldots, n$. Consequently, $\mathcal{P} = \bigcup_{i=1}^{n} \mathbf{J}_i$ is a partition of $\mathbf{J}$.
The symbol $E_i = C(\mathbf{J}_i, \mathbb{R})$, for $i = 1, 2, \ldots, n$ denotes the Banach space of continuous functions $\varkappa : \mathbf{J}_i \to \mathbb{R}$, endowed with the norm $\|\varkappa\|_{E_i} = \sup_{s \in \mathbf{J}_i} |\varkappa(s)|$.
Suppose that $\ltimes(s) : \mathbf{J} \to (0, 1]$ is defined by $\ltimes(s) = \sum_{i=1}^{n} \ltimes_i I_i(s)$, where $0 < \ltimes_i \leq 1$ are constants, and I_i denotes the indicator function of $\mathbf{J}_i$, which is piecewise constant with respect to the partition $\mathcal{P}$, defined by

$$I_i(s) = \begin{cases} 1, & for\ s \in \mathbf{J}_i, \\ 0, & elsewhere. \end{cases}$$

Then, for any $s \in \mathbf{J}_i, i = 1, 2, ..., n$, the left Caputo-Hadamard derivative for the function $\varkappa \in C(\mathbf{J}, \mathbb{R})$ with variable order $\ltimes(t)$ may be expressed as the sum of the left Caputo-Hadamard derivatives of the constant orders $\ltimes_k$, $k = 1, 2, \ldots, i$, i.e.,

$$\begin{aligned} \digamma_{p^+}^{\ltimes(s)} \varkappa(s) &= \frac{s \ltimes'(s)}{\Gamma(2 - \ltimes(s))} \int_p^s (\ell og \frac{s}{t})^{1-\ltimes(s)} \varkappa'(t) \Big[\frac{1}{1 - \ltimes(s)} - \ell og \Big(\ell og \frac{s}{t} \Big) \Big] dt \\ &+ \frac{1}{\Gamma(1 - \ltimes(s))} \int_p^s (\ell og \frac{s}{t})^{-\ltimes(s)} \varkappa'(t) dt \\ &= \frac{1}{\Gamma(1 - \ltimes(s))} \Big(\sum_{k=1}^{i-1} \int_{\mathsf{T}_{k-1}}^{\mathsf{T}_k} (\ell og \frac{s}{t})^{-\ltimes_k} \varkappa'(t) dt + \int_{\mathsf{T}_{i-1}}^{s} (\ell og \frac{s}{t})^{-\ltimes_i} \varkappa'(t) dt \Big). \end{aligned}$$

Then, the Caputo-Hadamard derivative for the IVP (6.15) can be expressed in the following form:

$$\frac{1}{\Gamma(1 - \ltimes(s))} \Big(\sum_{k=1}^{i-1} \int_{\mathsf{T}_{k-1}}^{\mathsf{T}_k} (\ell og \frac{s}{t})^{-\ltimes_k} \varkappa'(t) dt + \int_{\mathsf{T}_{i-1}}^{s} (\ell og \frac{s}{t})^{-\ltimes_i} \varkappa'(t) dt \Big) = f(s, \varkappa(s)). \tag{6.16}$$

To solve the integral equation (6.16), let the function $\tilde{\varkappa} \in C(\mathbf{J}_i, \mathbb{R})$ be such that $\tilde{\varkappa}(s) \equiv 0$ on $s \in [p, \mathsf{T}_{i-1}]$. Then, equation (6.16) is transformed into

$$ {}^{CH}F^{\ltimes_i}_{\mathsf{T}_{i-1}^+} \tilde{\varkappa}(s) = f(s, \tilde{\varkappa}(s)), \quad s \in \mathbf{J}_i. $$

For the obtained Caputo-Hadamard constant order fractional differential equations, we consider the following auxiliary initial value problem (IVP) of constant order:

$$ \begin{cases} {}^{CH}F^{\ltimes_i}_{\mathsf{T}_{i-1}^+} \varkappa_i(s) = f(s, \varkappa_i(s)), \ s \in \mathbf{J}_i, \\ \varkappa_i(\mathsf{T}_{i-1}) = \varkappa_{\mathsf{T}_{i-1}}, \end{cases} \tag{6.17} $$

for each $i = 1, 2, \ldots, n$.

The main foundational theorem can now be stated.

Theorem 55 *Assume that $f : \mathbf{J}_i \times \mathbb{R} \to \mathbb{R}$ is a continuous function. The solution to the integral equation, i.e., $\varkappa_i \in C(\mathbf{J}_i; \mathbb{R})$, given by*

$$ \varkappa_i(s) = \varkappa_{\mathsf{T}_{i-1}} + \frac{1}{\Gamma(\ltimes_i)} \int_{\mathsf{T}_{i-1}}^{s} (\ell og\frac{s}{t})^{\ltimes_i - 1} \frac{f(t, \varkappa_i(t))}{t} dt \quad for \ s \in \mathbf{J}_i, \tag{6.18} $$

solves the auxiliary initial value problem of constant order (6.17).

Proof: Assume that $\varkappa_i \in C(\mathbf{J}_i; \mathbb{R})$ is a solution of (6.18). Naturally, we take $\varkappa(\mathsf{T}_{i-1}) = \varkappa_{\mathsf{T}_{i-1}}$ and the mapping $s \to^H I^{\ltimes_i}_{\mathsf{T}_{i-1}^+} \varkappa_i(s) \in C(\mathbf{J}_i; \mathbb{R})$. The definition of the Hadamard integral ${}^H I^{\ltimes_i}_{\mathsf{T}_{i-1}^+}$ and the continuity of f guarantee that the mapping $s \to f(s, \varkappa_i(s))$ is also continuous, and

$$ {}^H I^{\ltimes_i}_{\mathsf{T}_{i-1}^+} f(s, \varkappa_i(s))|_{s=\mathsf{T}_{i-1}} = 0. $$

Since the mapping $s \to^H I^{\ltimes_i}_{\mathsf{T}_{i-1}^+} f(s, \varkappa_i(s))$ is continuous, we can conclude that $\varkappa_i$ is differentiable for almost every $s \in (\mathsf{T}_{i-1}, \mathsf{T}_i)$; see (6.18). That is, $\varkappa_i \in AC(\mathbf{J}_i; \mathbb{R})$. From Lemma 5, we have

$$ {}^{CH}F^{\ltimes_i}_{\mathsf{T}_{i-1}^+} \cdot {}^H I^{\ltimes_i}_{\mathsf{T}_{i-1}^+} f(s, \varkappa_i(s)) = f(s, \varkappa_i(s)) \quad \text{for} \ s \in \mathbf{J}_i. $$

On the other hand, we have

$$ \begin{aligned} {}^{CH}F^{\ltimes_i}_{\mathsf{T}_{i-1}^+} [\varkappa_i(s) - \varkappa_{\mathsf{T}_{i-1}}] &= \frac{1}{\Gamma(1-\ltimes_i)} \int_{\mathsf{T}_{i-1}}^{s} (\ell og\frac{s}{t})^{-\ltimes_i} [\varkappa_i(t) - \varkappa_{\mathsf{T}_{i-1}}]' dt \\ &= \frac{1}{\Gamma(1-\ltimes_i)} \int_{\mathsf{T}_{i-1}}^{s} (\ell og\frac{s}{t})^{-\ltimes_i} \varkappa_i'(t) dt \\ &= {}^{CH}F^{\ltimes_i}{}_{\mathsf{T}_{i-1}^+} \varkappa_i(s), \end{aligned} $$

for each $s \in \mathbf{J}_i$. From all the above, we conclude that $\varkappa_i \in C(\mathbf{J}_i;\mathbb{R})$ is a solution of the auxiliary initial value problem of constant order (6.17).

Definition 56 *Let $(\underline{\varkappa_i}, \overline{\varkappa_i}) \in C(\mathbf{J}_i;\mathbb{R}) \times C(\mathbf{J}_i;\mathbb{R})$. A pair of functions $(\underline{\varkappa_i}, \overline{\varkappa_i})$ is called an upper-lower solution of the auxiliary IVP of constant order (6.17), respectively, if*

$$\underline{\varkappa_i}(s) \leq \varkappa_{\mathsf{T}_{i-1}} + \frac{1}{\Gamma(\ltimes_i)} \int_{\mathsf{T}_{i-1}}^{s} (\ell og\frac{s}{t})^{\ltimes_i - 1} \frac{f(t, \underline{\varkappa_i}(t))}{t} dt \quad \text{for all } s \in \mathbf{J}_i,$$

and

$$\overline{\varkappa_i}(s) \geq \varkappa_{\mathsf{T}_{i-1}} + \frac{1}{\Gamma(\ltimes_i)} \int_{\mathsf{T}_{i-1}}^{s} (\ell og\frac{s}{t})^{\ltimes_i - 1} \frac{f(t, \overline{\varkappa_i}(t))}{t} dt \quad \text{for all } s \in \mathbf{J}_i.$$

Assume that the upper-lower solution to the the auxiliary IVP of constant order (6.17) is $(\underline{\varkappa_i}, \overline{\varkappa_i})$. In the following, we define an acceptable set of solutions for the auxiliary IVP of constant order (6.17), which is controlled by two upper-lower solutions $(\underline{\varkappa_i}, \overline{\varkappa_i})$ as follows:

Let

$$\mathcal{S}_{(\underline{\varkappa_i}, \overline{\varkappa_i})} := \Big\{ \varkappa_i \in C(\mathbf{J}_i;\mathbb{R}) : \underline{\varkappa_i}(s) \leq \varkappa_i(s) \leq \overline{\varkappa_i}(s), s \in \mathbf{J}_i \text{ and } \varkappa_i \text{ is a solution of (6.17)} \Big\}.$$

Theorem 56 *Let $f \in C(\mathbf{J}_i \times \mathbb{R};\mathbb{R})$ and $(\underline{\varkappa_i}, \overline{\varkappa_i}) \in C(\mathbf{J}_i;\mathbb{R}) \times C(\mathbf{J}_i;\mathbb{R})$. The auxiliary IVP of constant order (6.17) admits a pair of upper-lower solutions satisfying $\underline{\varkappa_i}(s) \leq \overline{\varkappa_i}(s)$ and $s \in \mathbf{J}_i$. If the mapping $\varkappa_i \to f(s, \varkappa_i)$ is nondecreasing, that is,*

$$f(s, \varkappa_1) \leq f(s, \varkappa_2) \quad \text{for } \varkappa_1 \leq \varkappa_2,$$

then there exist minimum and maximum solutions $\varkappa_{i,m}, \varkappa_{L,m} \in \mathcal{S}_{(\underline{\varkappa_i}, \overline{\varkappa_i})}$ in $\mathcal{S}_{(\underline{\varkappa_i}, \overline{\varkappa_i})}$; such that for each $\varkappa_i \in \mathcal{S}_{(\underline{\varkappa_i}, \overline{\varkappa_i})}$,

$$\varkappa_{L,i}(s) \leq \varkappa_i(s) \leq \varkappa_{M,i}(s) \quad \text{for } s \in \mathbf{J}_i.$$

Proof: We provide two sequences $\{\vartheta_{n,i}\}$ and $\{\curlywedge_{n,i}\}$ defined as

$$\begin{cases} \vartheta_{0,i} = \underline{\varkappa_i}, \\ \vartheta_{n+1,i}(s) = \varkappa_{\mathsf{T}_{i-1}} + \frac{1}{\Gamma(\ltimes_i)} \int_{\mathsf{T}_{i-1}}^{s} (\ell og\frac{s}{t})^{\ltimes_i - 1} \frac{f(t, \vartheta_{n,i}(t))}{t} dt, s \in \mathbf{J}_i \text{ and } n = 0, 1, ..., \end{cases} \tag{6.19}$$

and

$$\begin{cases} \curlywedge_{0,i} = \overline{\varkappa_i}, \\ \curlywedge_{n+1,i}(s) = \varkappa_{\mathsf{T}_{i-1}} + \frac{1}{\Gamma(\ltimes_i)} \int_{\mathsf{T}_{i-1}}^{s} (\ell og\frac{s}{t})^{\ltimes_i - 1} \frac{f(t, \curlywedge_{n,i}(t))}{t} dt, s \in \mathbf{J}_i \text{ and } n = 0, 1, ... \end{cases} \tag{6.20}$$

The proof is now divided into three steps.

STEP 1: The sequences $\{\vartheta_{n,i}\}$ and $\{\leftthreetimes_{n,i}\}$ satisfy the following relation:

$$\underline{\varkappa_i}(s) = \vartheta_{0,i}(s) \leq \vartheta_{1,i}(s) \leq \vartheta_{2,i}(s) \leq ... \leq \vartheta_{n,i}(s) \leq ... \leq \leftthreetimes_{n,i}(s) \leq ... \leq \leftthreetimes_{1,i}(s) \leq \leftthreetimes_{0,i}(s) = \overline{\varkappa_i}(s) \tag{6.21}$$

for each $s \in \mathbf{J}_i$.

We will first demonstrate that the sequence $\{\vartheta_{n,i}\}$ is nondecreasing and

$$\vartheta_{n,i}(s) \leq \leftthreetimes_{0,i}(s), s \in \mathbf{J}_i \quad \text{for all} \quad n \in \mathbb{N}.$$

Therefore, by a recurrence relation, we prove

$$\vartheta_{n-1,i}(s) \leq \vartheta_{n,i}(s), \ \ \forall s \in \mathbf{J}_i. \tag{6.22}$$

By the definition of $\vartheta_{0,i}(s)$, we have $\vartheta_{0,i}(s) \leq \vartheta_{1,i}(s)$ for each $s \in \mathbf{J}_i$. We suppose that relation (6.22) holds for n, and we prove for $n+1 : \vartheta_{n,i}(s) \leq \vartheta_{n+1,i}(s), \ \ \forall s \in \mathbf{J}_i$.

We have

$$\vartheta_{n,i}(s) = \varkappa_{\intercal_{i-1}} + \frac{1}{\Gamma(\ltimes_i)} \int_{\intercal_{i-1}}^{s} (\ell og\frac{s}{t})^{\ltimes_i - 1} \frac{f(t, \vartheta_{n-1,i}(t))}{t} dt,$$

and

$$\vartheta_{n+1,i}(s) = \varkappa_{\intercal_{i-1}} + \frac{1}{\Gamma(\ltimes_i)} \int_{\intercal_{i-1}}^{s} (\ell og\frac{s}{t})^{\ltimes_i - 1} \frac{f(t, \vartheta_{n,i}(t))}{t} dt.$$

Using the monotonicity of f, we obtain

$$\vartheta_{n,i}(s) \leq \vartheta_{n+1,i}(s),$$

for all $s \in J_i$.

As $\vartheta_{n,i}(s)$ is noncreasing, it follows from the definition of $\leftthreetimes_{0,i}(s)$ that

$$\vartheta_{n,i}(s) \leq \vartheta_{n+1,i}(s) \leq \leftthreetimes_{0,i}(s).$$

Furthermore, we will show that

$$\vartheta_{n,i}(s) \leq \leftthreetimes_{n,i}(s) \quad \text{for all} \quad s \in \mathbf{J}_i \quad \text{and} \quad n \in \mathbb{N}.$$

Since $n = 0$, it is evident that $\underline{\varkappa_i}(s) = \vartheta_{0,i}(s) \leq \leftthreetimes_{0,i}(s) = \overline{\varkappa_i}(s)$ for all $s \in \mathbf{J}_i$. Now, we make the inductive assumption

$$\vartheta_{n,i}(s) \leq \leftthreetimes_{n,i}(s), s \in \mathbf{J}_i.$$

Accordingly, given that f is monotonic with respect to the second variable, it is straightforward to conclude that

$$\vartheta_{n+1,i}(s) \leq \leftthreetimes_{n+1,i}(s), s \in \mathbf{J}_i.$$

Moreover, the sequence $\{\leftthreetimes_{n,i}\}$ is nonincreasing.

STEP 2: Both sequences $\{\vartheta_{\mathbf{n,i}}\}$ and $\{\leftthreetimes_{\mathbf{n,i}}\}$ are relatively compact in $\mathbf{C}(\mathbf{J_i};\mathbb{R})$. Since f is continuous and $(\underline{\varkappa_i}, \overline{\varkappa_i}) \in C(\mathbf{J}_i;\mathbb{R})$, it follows from Step 1 that $\{\vartheta_{n,i}\}$ and $\{\leftthreetimes_{n,i}\}$ also belong to $C(\mathbf{J}_i;\mathbb{R})$. Moreover, by equation (6.21) both sequences are uniformly bounded. On the other hand, for any $s_1, s_2 \in \mathbf{J}_i$, without loss of generality, let $s_1 \leq s_2$. Then we have

$$|\vartheta_{n+1,i}(s_1) - \vartheta_{n+1,i}(s_2)|$$

$$= \frac{1}{\Gamma(\ltimes_i)}\Big|\int_{\mathsf{T}_{i-1}}^{s_2} (\ell og\frac{s_2}{t})^{\ltimes_i - 1}\frac{f(t,\vartheta_{n,i}(t))}{t}dt$$

$$- \int_{\mathsf{T}_{i-1}}^{s_1} (\ell og\frac{s_1}{t})^{\ltimes_i - 1}\frac{f(t,\vartheta_{n,i}(t))}{t}dt\Big|$$

$$= \frac{1}{\Gamma(\ltimes_i)}\Big|\int_{\mathsf{T}_{i-1}}^{s_1} \Big[(\ell og\frac{s_2}{t})^{\ltimes_i - 1} - (\ell og\frac{s_1}{t})^{\ltimes_i - 1}\Big]\frac{f(t,\vartheta_{n,i}(t))}{t}dt$$

$$+ \int_{s_1}^{s_2} (\ell og\frac{s_2}{t})^{\ltimes_i - 1}\frac{f(t,\vartheta_{n,i}(t))}{t}dt\Big|$$

$$\leq \frac{M}{\Gamma(\ltimes_i)}\Big|\int_{\mathsf{T}_{i-1}}^{s_1} \frac{1}{t}\Big[(\ell og\frac{s_2}{t})^{\ltimes_i - 1} - (\ell og\frac{s_1}{t})^{\ltimes_i - 1}\Big]dt + \int_{s_1}^{s_2} \frac{1}{t}(\ell og\frac{s_2}{t})^{\ltimes_i - 1}dt\Big|$$

$$= \frac{M}{\Gamma(\ltimes_i)}\Big| - \frac{1}{\ltimes_i}\Big[\Big(\ell og\frac{s_2}{t}\Big)^{\ltimes_i}\Big]_{\mathsf{T}_{i-1}}^{s_1} + \frac{1}{\ltimes_i}\Big[\Big(\ell og\frac{s_1}{t}\Big)^{\ltimes_i}\Big]_{\mathsf{T}_{i-1}}^{s_1} - \frac{1}{\ltimes_i}\Big[\Big(\ell og\frac{s_2}{t}\Big)^{\ltimes_i}\Big]_{s_1}^{s_2}\Big|$$

$$= \frac{M}{\Gamma(\ltimes_i)}\Big|\frac{1}{\ltimes_i}\Big(\Big(\ell og\frac{s_2}{\mathsf{T}_{i-1}}\Big)^{\ltimes_i} - \Big(\ell og\frac{s_2}{s_1}\Big)^{\ltimes_i}\Big)\Big) + \Big(\frac{1}{\ltimes_i}\Big(- \Big(\ell og\frac{s_1}{\mathsf{T}_{i-1}}\Big)^{\ltimes_i} + \Big(\ell og\frac{s_1}{s_1}\Big)^{\ltimes_i}\Big)\Big)$$

$$+ \Big(\frac{1}{\ltimes_i}\Big(\Big(\ell og\frac{s_2}{s_1}\Big)^{\ltimes_i} - \Big(\ell og\frac{s_2}{s_2}\Big)^{\ltimes_i}\Big)\Big|\Big)$$

$$= \frac{M}{\Gamma(\ltimes_i)}\Big|\frac{1}{\ltimes_i}\Big(\Big(\ell og\frac{s_2}{\mathsf{T}_{i-1}}\Big)^{\ltimes_i} - \Big(\ell og\frac{s_1}{\mathsf{T}_{i-1}}\Big)^{\ltimes_i}\Big)\Big|$$

$$= \frac{M}{\Gamma(\ltimes_i + 1)}\Big|\Big(\ell og\frac{s_2}{\mathsf{T}_{i-1}}\Big)^{\ltimes_i} - \Big(\ell og\frac{s_1}{\mathsf{T}_{i-1}}\Big)^{\ltimes_i}\Big|$$

$$\to 0, \quad as\ s_1 \to s_2,$$

where $M > 0$ is a constant independent of n, s_1, and s_2. This establishes that the sequence $\{\vartheta_{n,i}\}$ is equicontinuous in $C(\mathsf{T}_{i-1}, \mathsf{T}_i; \mathbb{R})$.

We conclude that the sequence $\{\vartheta_{n,i}\}$ is relatively compact in $C(\mathsf{T}_{i-1}, \mathsf{T}_i; \mathbb{R})$ by the Arzela-Ascoli Theorem. Similar, we find that the sequence $\{\lambda_{n,i}\}$ is also relatively compact in $C(\mathbf{J}_i; \mathbb{R})$.

STEP 3: In $\mathcal{S}_{(\underline{\varkappa_i}, \overline{\varkappa_i})}$, there exist minimum and maximum solutions.

The sequences $\{\vartheta_{n,i}\}$ and $\{\lambda_{n,i}\}$ are monotone and relatively compact in $C(\mathsf{T}_{i-1}, \mathsf{T}_i; \mathbb{R})$, as demonstrated in Steps 1 and 2. Evidently, there exist continuous functions ϑ_i and λ_i such that $\vartheta_{n,i}(s) \leq \vartheta_i(s) \leq \lambda_i(s) \leq \lambda_{n,i}(s)$ for all $s \in \mathbf{J}_i$ and $n \in \mathbb{N}$, and the sequences $\{\vartheta_{n,i}\}$ and $\{\lambda_{n,i}\}$ converge uniformly to ϑ_i and λ_i, respectively, in $C(\mathsf{T}_{i-1}, \mathsf{T}_i; \mathbb{R})$. Therefore, the solutions to the auxiliary IVP of constant order (6.17) are ϑ_i and λ_i; that is,

$$\begin{aligned} \vartheta_i(s) &= \varkappa_{\mathsf{T}_{i-1}} + \tfrac{1}{\Gamma(\Bbbk_i)} \int_{\mathsf{T}_{i-1}}^{s} (\ell og\tfrac{s}{t})^{\Bbbk_i - 1} \tfrac{f(t, \vartheta_i(t))}{t} dt, \\ \lambda_i(s) &= \varkappa_{\mathsf{T}_{i-1}} + \frac{1}{\Gamma(\Bbbk_i)} \int_{\mathsf{T}_{i-1}}^{s} (\ell og\frac{s}{t})^{\Bbbk_i - 1} \frac{f(t, \lambda_i(t))}{t} dt, \end{aligned}$$

for each $s \in \mathbf{J}_i$. Therefore,

$$\underline{\varkappa_i}(s) \leq \vartheta_i(s) \leq \lambda_i(s) \leq \overline{\varkappa_i}(s) \quad for\ s \in \mathbf{J}_i.$$

Finally, we will prove that ϑ_i and λ_i are the minimum and maximum solutions in $\mathcal{S}_{(\underline{\varkappa_i}, \overline{\varkappa_i})}$. If $\varkappa_i \in \mathcal{S}_{(\underline{\varkappa_i}, \overline{\varkappa_i})}$, then

$$\underline{\varkappa_i}(s) \leq \varkappa_i(s) \leq \overline{\varkappa_i}(s), \quad s \in \mathbf{J}_i.$$

Recalling that the second and third arguments do not induce a decrease in f, we introduce the following inequality:

$$\underline{\varkappa_i}(s) \leq \vartheta_{n,i}(s) \leq \varkappa_i(s) \leq \lambda_{n,i}(s) \leq \overline{\varkappa_i}(s) \quad \text{for} \quad s \in \mathbf{J}_i \quad \text{and} \quad n \in \mathbb{N}.$$

As $n \to \infty$ in the above inequality, it follows that

$$\underline{\varkappa_i}(s) \leq \vartheta_i(s) \leq \varkappa_i(s) \leq \lambda_i(s) \leq \overline{\varkappa_i}(s) \quad \text{for} \quad s \in \mathbf{J}_i.$$

This concludes the proof of theorem by identifying $\varkappa_{L,i} = \vartheta_i$ and $\varkappa_{i,i} = \lambda_i$, respectively, as the minimum and maximum solutions in $\mathcal{S}_{(\underline{\varkappa_i}, \overline{\varkappa_i})}$.

Theorem 57 *Assume that the hypotheses of Theorem 56 are satisfied. Then the auxiliary IVP of constant order (6.17) admits at least one solution in* $C(\mathbf{J}_i; \mathbb{R})$.

Proof: Based on Theorem 56, we obtain $\mathcal{S}_{(\underline{\varkappa_i}, \overline{\varkappa_i})} \neq \emptyset$, implying that the solution set associated with the auxiliary IVP of constant order (6.17) is not empty in $C(\mathsf{T}_{i-1}, \mathsf{T}_i; \mathbb{R})$. By establishing that the auxiliary IVP of constant order (6.17) admits at least one solution in $C(\mathsf{T}_{i-1}, \mathsf{T}_i; \mathbb{R})$, the proof of the theorem is complete.

We now proceed to investigate the existence result for the nonlinear IVP (6.15).

Theorem 58 *Let all $i \in \{1, 2, \ldots, n\}$ satisfy condition (H1). Then the nonlinear IVP (6.15) admits at least one solution in E.*

Proof: Based on the preceding proofs, we conclude that the IVP of constant order (6.17) admits at least one solution $\widetilde{\varkappa}_i \in E_i$, for each $i \in \{1, 2, \ldots, n\}$. This result is consistent with Theorems 56 and 57.

We define the solution function for each $i \in \{1, 2, \ldots, n\}$ as

$$\varkappa_i = \begin{cases} 0, & s \in [p, \mathsf{T}_{i-1}], \\ \widetilde{\varkappa}_i, & s \in \mathbf{J}_i. \end{cases} \tag{6.23}$$

Thus, the function $\varkappa_i \in C(\mathsf{T}_{i-1}, \mathsf{T}_i; \mathbb{R})$ is a solution to the Hadamard integral equation (6.16) for each $s \in \mathbf{J}_i$, satisfying $\varkappa_i(p) = 0$ and $\varkappa_i(\mathsf{T}_i) = \widetilde{\varkappa}_i(\mathsf{T}_i) = 0$. Therefore, the function

$$\varkappa(t) = \begin{cases} \varkappa_1(s), \quad s \in \mathbf{J}_1, \\ \varkappa_2(s) = \begin{cases} 0, & s \in \mathbf{J}_1, \\ \widetilde{\varkappa}_2, & s \in \mathbf{J}_2, \end{cases} \\ \vdots \\ \varkappa_n(s) = \begin{cases} 0, & s \in [p, \mathsf{T}_{n-1}], \\ \widetilde{\varkappa}_n, & s \in \mathbf{J}_n, \end{cases} \end{cases}$$

is a solution to the given nonlinear IVP (6.15) in E.

6.2.2 Numerical Representation

Let $\mathbf{J} := [1, e^2]$, and define $\mathsf{T}_0 = 1,\ \mathsf{T}_1 = e,\ \mathsf{T}_2 = e^2$. Consider the following nonlinear initial value problem (IVP):

$$\begin{cases} {}^{CH}\digamma_{1^+}^{\ltimes(s)} \varkappa(s) = \frac{1}{\pi}(\sqrt{\ell ogs} + (\ell ogs)^4) + \varkappa(s), & s \in \mathbf{J}, \\ \varkappa(1) = 0, \end{cases} \tag{6.24}$$

where the piecewise function $\ltimes(s)$ is defined by

$$\ltimes(s) = \begin{cases} \frac{1}{2}, & s \in \mathbf{J}_1 := [1, e], \\ \frac{3}{2}, & s \in \mathbf{J}_2 :=]e, e^2]. \end{cases} \tag{6.25}$$

Denote

$$f(s, \varkappa) = \frac{1}{\pi}(\sqrt{\ell ogs} + (\ell ogs)^4) + \varkappa(s), \ (s, \varkappa) \in [1, e^2] \times \mathbb{R}.$$

Using (6.25) and (6.17), we consider two auxiliary constant order initial value problems (IVPs) as follows:

$$\begin{cases} {}^{CH}\digamma_{1^+}^{\frac{1}{2}} \varkappa(s) = \frac{1}{\pi}(\sqrt{\ell ogs} + (\ell ogs)^4) + \varkappa(s), & s \in \mathbf{J}_1, \\ \varkappa(1) = 0, \end{cases} \tag{6.26}$$

and

$$\begin{cases} {}^{CH}\digamma_{e^+}^{\frac{3}{2}} \varkappa(s) = \frac{1}{\pi}(\sqrt{\ell ogs} + (\ell ogs)^4) + \varkappa(s), & s \in \mathbf{J}_2, \\ \varkappa(e) = 1. \end{cases} \tag{6.27}$$

For $\mathbf{i} = 1$: By Theorem 55, constant-order IVP (6.26) admits at least one solution $\widetilde{\varkappa}_1 \in E_1$ as

$$\varkappa_1(s) = I_{1^+}^{\frac{1}{2}}\Big(\frac{1}{\pi}(\sqrt{\ell ogs} + (\ell ogs)^4) + \varkappa_1(s)\Big) \quad \text{for } s \in \mathbf{J}_1. \tag{6.28}$$

In fact, as one can observe $(\underline{\varkappa_1(s)}, \overline{\varkappa_1(s)}) = (0, \ell ogs + (\ell ogs)^5)$ denotes the lower and upper bounds of the solution to (6.28). We can compute the sequences $\{\vartheta_{n,1}\}$ and $\{\curlywedge_{n,1}\}$ by

$$\begin{cases} \vartheta_{0,1} = \underline{\varkappa_1} \\ \vartheta_{n+1,1}(s) = I_{1^+}^{\frac{1}{2}} f(s, \vartheta_{n,1}(s)), n = 0, 1, ..., \end{cases}$$

and

$$\begin{cases} \curlywedge_{0,1} = \overline{\varkappa_1} \\ \curlywedge_{n+1,1}(s) = I_{1^+}^{\frac{1}{2}} f(s, \curlywedge_{n,1}(s)), n = 0, 1, ..., \end{cases}$$

for each $s \in \mathbf{J}_1$. We can now invoke Theorem 56 to determine that $\vartheta_{n,1} \to \vartheta_1 \in E_1$ and $\curlywedge_{n,1} \to \curlywedge_1 \in E_1$ as $n \to \infty$. Meanwhile, for $s \in \mathbf{J}_1$ we obtain $\curlywedge_1(s) = \vartheta_1(s) = \dfrac{\pi(\ell ogs)^2}{3}$.

We use Maple to calculate the sequences $\{\vartheta_{n,2}\}$ and $\{\curlywedge_{n,2}\}$ for each n, where these sequences are defined as integrals with distinct initial values.

Then, we evaluate the values of these sequences at selected instants and plot them using MATLAB®. In Table 6.1, we present the error—defined as the supremum of the absolute value of the difference—between the sequences $\{\vartheta_{n,1}\}$, $\{\leftthreetimes_{n,1}\}$ and the exact solution for $n = 5, 10, 15, 20$. In Figure 6.1, we plot the sequences $\{\vartheta_{n,1}\}$, $\{\leftthreetimes_{n,1}\}$, and the exact solution for $n = 0, 1, 2, 10, 30$.

TABLE 6.1: Error analysis for $i = 1$.

	$n = 5$	$n = 10$	$n = 15$	$n = 20$
$\sup_{s\in[1,e]} \lvert\vartheta_{n,1}(s) - \vartheta_1(s)\rvert$	4.7692×10^{-2}	6.3900×10^{-4}	$4. \times 10^{-6}$	$1. \times 10^{-16}$
$\sup_{s\in[1,e]} \lvert \leftthreetimes_{n,1} - \leftthreetimes_1 \rvert$	4.6818×10^{-2}	$7.8299 \times 10^{e-4}$	$5. \times 10^{-6}$	$9. \times 10^{-8}$

In Figure 6.1, We observe that as n increases, the sequences $\{\vartheta_{n,1}\}$ and $\{\leftthreetimes_{n,1}\}$ increasingly approximate the exact solution $\dfrac{\pi(\ell ogs)^2}{3}$. Moreover, Table 6.1, confirms this observation, as the error approaches to zero when $n \to +\infty$.

For i = 2: By Theorem 55, the auxiliary constant-order IVP (6.27) admits at least one solution $\widetilde{\varkappa}_2 \in E_2$, given by

$$\varkappa_2(s) = I^1_{e+}\Big(\frac{1}{\pi}(\sqrt{\ell ogs} + (\ell ogs)^4) + \varkappa_2(s)\Big) \quad \text{for} \quad s \in \mathbf{J}_2. \tag{6.29}$$

In fact, we observe that $(\underline{\varkappa_2(s)}, \overline{\varkappa_2(s)}) = (1, (\ell ogs)^2 + (\ell ogs)^5)$ serves as an upper-lower solution to equation (6.29). We can compute the sequences $\{\vartheta_{n,2}\}$ and $\{\leftthreetimes_{n,2}\}$ by numerical discretization.

We define the sequences $\{\vartheta_{n,2}\}$ and $\{\leftthreetimes_{n,2}\}$

$$\begin{cases} \vartheta_{0,2} = \underline{\varkappa_2} \\ \vartheta_{n+1,2}(s) = I^1_{e+} f(s, \vartheta_{n,2}(s)), n = 0, 1, ..., \end{cases}$$

and

$$\begin{cases} \leftthreetimes_{0,2} = \overline{\varkappa_2} \\ \leftthreetimes_{n+1,2}(s) = I^1_{e+} f(s, \leftthreetimes_{n,2}(s)), n = 0, 1, ..., \end{cases}$$

for each $s \in \mathbf{J}_2$. We can now invoke Theorem 56 to establish the convergence $\vartheta_{n,2} \to \vartheta_2 \in E_2$ and $\leftthreetimes_{n,2} \to \leftthreetimes_2 \in E_2$ as $n \to \infty$. Meanwhile, for some $s \in \mathbf{J}_2$, we may obtain $\leftthreetimes_2(s) = \vartheta_2(s) = \pi\frac{(\ell ogs)^5}{30}$.

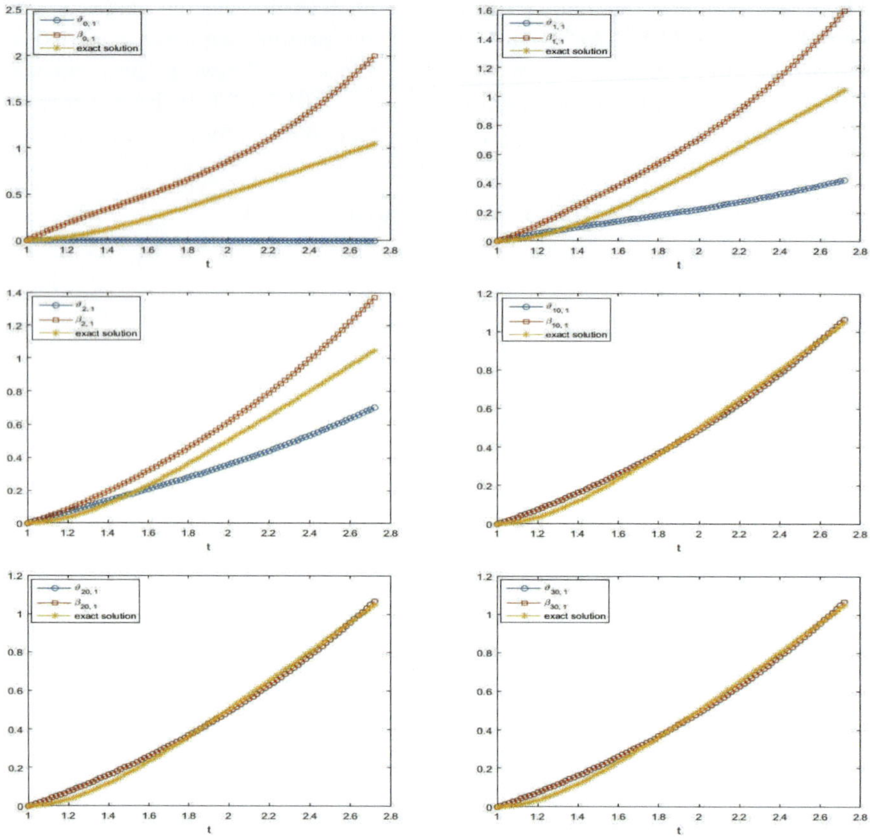

FIGURE 6.1: A plot of $\vartheta_{n,1}$, $\leftthreetimes_{n,1}$ and exact solution, for $n = 0, 1, 2, 10, 20, 30$.

We use Maple to compute the sequences $\{\vartheta_{n,2}\}$ and $\{\leftthreetimes_{n,2}\}$ for each n, where each sequence is defined via integral formulations with distinct initial *v*alues. Subsequently, we evaluate these sequences at each point $s \in \mathbf{J}_2$ and visualize them using MATLAB. In Table 6.2, we present the error—defined as the supremum of the absolute *v*alue of the difference—between the sequences $\{\vartheta_{n,2}\}$, $\{\leftthreetimes_{n,2}\}$ and the exact solution for $n = 5, 10, 15, 20$. In Figure 6.1, we plot the sequences $\{\vartheta_{n,2}\}$, $\{\leftthreetimes_{n,2}\}$ and the exact solution, for $n = 0, 1, 2, 10, 30$. From this figure, we observe that as n increases, both sequences $\{\vartheta_{n,2}\}$ and $\{\leftthreetimes_{n,2}\}$ converge toward the exact solution $\dfrac{\pi(\ell ogs)^5}{30}$.

In Table 6.2, we confirm our previous remark, because the error approaches to 0 when n converges to $+\infty$.

Consequently, in accordance with Theorem 58, the nonlinear initial value problem (IVP) (6.24) admits a solution of the form

$$\varkappa(s) = \begin{cases} \widetilde{\varkappa}_1(s), & s \in \mathbf{J}_1, \\ \varkappa_2(s), & s \in \mathbf{J}_2, \end{cases}$$

where

$$\varkappa_2(s) = \begin{cases} 0, & s \in \mathbf{J}_1, \\ \widetilde{\varkappa}_2(s), & s \in \mathbf{J}_2. \end{cases}$$

TABLE 6.2: Error analysis for $i = 2$.

	$n = 5$	$n = 10$	$n = 15$	$n = 20$
$\sup_{s\in]e,e^2]} \lvert\vartheta_{n,2}(s) - \vartheta_2(s)\rvert$	6.8509×10^{-3}	$1. \times 10^{-6}$	$6. \times 10^{-10}$	$1. \times 10^{-10}$
$\sup_{s\in]e,e^2]} \lvert\lambda_{n,2}(s) - \lambda_2(s)\rvert$	3.1123×10^{-2}	$1. \times 10^{-6}$	$1. \times 10^{-7}$	$1. \times 10^{-12}$

6.3 Stability of an Initial and Terminal Value Problem Involving the Riemann–Liouville Fractional Derivative in the Sense of Atangana–Baleanu of Variable Order

We study properties such as the existence and stability of solutions for a variable-order multiterm initial–terminal value problem (ITVP) with nonlocal conditions, given by

$$\begin{cases} {}^{ABR}\mathsf{F}_{0^+}^{\ltimes(s)} y(s) = f_y(s), \quad s \in J, \\ y(0) = 0, y(\mathsf{T}) = \sum_{k=1}^{m} a_k y(s_k), \quad s_k \in]0, \mathsf{T}], \end{cases} \tag{6.30}$$

where $f_y(s) := f(s, y(s), {}^{AB}I_{0^+}^{\ltimes(s)} y(s))$, with $2 < \ltimes(s) \leq 3$, and $f : \mathcal{J} \times \mathbb{R}^2 \to \mathbb{R}$ being a given function. Here, ${}^{AB}\mathsf{F}_{0^+}^{\ltimes(s)}$ and ${}^{AB}I_{0^+}^{\ltimes(s)}$ denote the Riemann–Liouville fractional derivative and integral of variable order $\ltimes(\cdot)$ in the sense of Atangana-Baleanu framework.

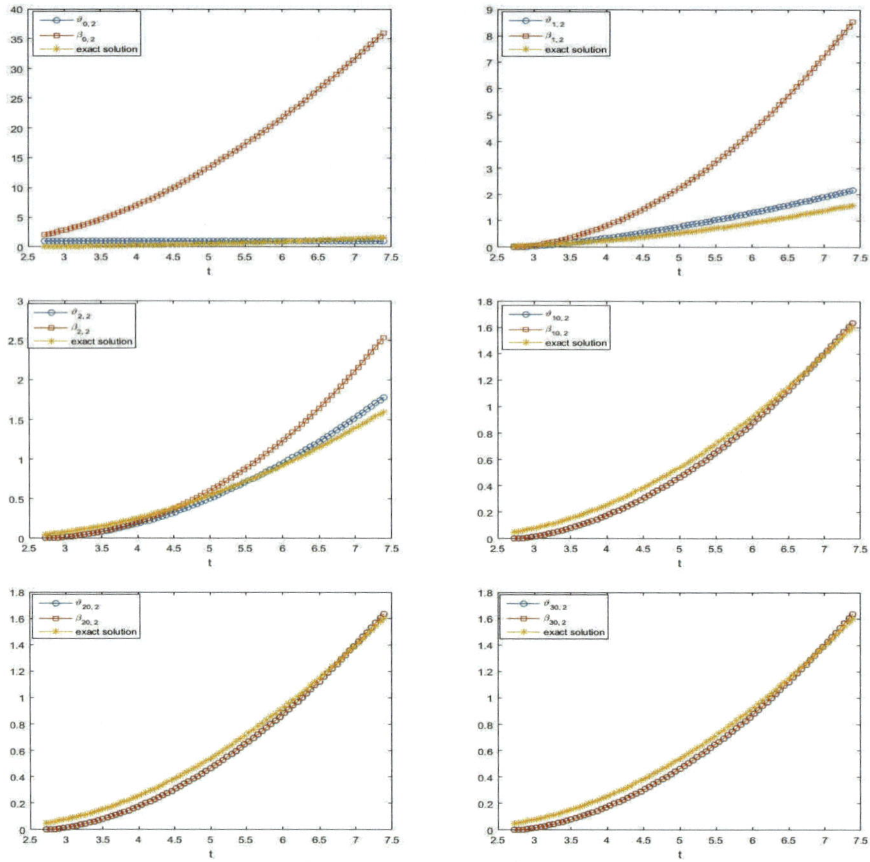

FIGURE 6.2: A plot of $\vartheta_{n,2}$, $\leftthreetimes_{n,2}$ and exact solution, for $n = 0, 1, 2, 10, 20, 30$.

6.3.1 Existence of Solutions

Let us prepare some required hypotheses:

(H1) Suppose $n \in \mathbb{N}$, and let $\{\mathsf{T}_k\}_{k=0}^{n}$ be a sequence of points satisfying $0 = \mathsf{T}_0 < \mathsf{T}_1 < \mathsf{T}_2 < \cdots < \mathsf{T}_{n-1} < \mathsf{T}_n = \mathsf{T}$, where $k = 1, \ldots, n-1$. Define $J_k := (\mathsf{T}_{k-1}, \mathsf{T}_k]$ for $k = 1, \ldots, n$. Consequently, the collection $\mathcal{P} = \{J_k : k = 1, 2, \ldots, n\}$ forms a partition of the interval J. Consider $\ltimes : J \to (2, 3]$ as a piecewise constant function with respect to the partition $\mathcal{P}$, defined by:

$$\ltimes(s) = \sum_{k=1}^{n} \ltimes_k I_k(s) = \begin{cases} \ltimes_1, & if\ s \in J_1, \\ \ltimes_2, & if\ s \in J_2, \\ \cdot \\ \cdot \\ \cdot \\ \ltimes_n, & if\ s \in J_n, \end{cases}$$

where the constants $\ltimes_k$ are such that $2 < \ltimes_k \leq 3$. Moreover, I_k is an indicator of J_k, for $k = 1, 2, ..., n$; that is,

$$I_k(s) = \begin{cases} 1, & for\ s \in J_k, \\ 0, & for\ elsewhere. \end{cases}$$

(H2) Let $f : J \times \mathbb{R}^2 \to \mathbb{R}$ be continuous. Then there exists a constant $K_f > 0$ such that

$$|f(s, y_1, z_1) - f(s, y_2, z_2)| \leq K_f \left(|y_1 - y_2| + |z_1 - z_2|\right),$$

for all $y_1, y_2,\ z_1, z_2 \in \mathbb{R}$ and $s \in J$.

For every $s \in J_k$ (with $k = 1, 2, \ldots, n$), and by Definition 26, the Riemann–Liouville fractional derivative in the Atangana-Baleanu sense of a variable order $\ltimes(\cdot)$ for $y \in C(J, \mathbb{R})$ can be decomposed into a sum of Riemann–Liouville fractional derivatives of constant orders $\ltimes_k$:

$$\begin{aligned} {}^{ABR}\digamma_{0+}^{\ltimes(s)} y(s) &= {}^{ABR}\digamma_{0+}^{\nu(s)} y^{(2)}(s) \\ &= \frac{N(\nu(s))}{1-\nu(s)} \frac{d}{dt} \int_0^s \mathcal{E}_{\nu(\tau)} \left(-\nu(\tau) \frac{(s-\tau)^{\nu(\tau)}}{(1-\nu(\tau))}\right) y^{(2)}(\tau) d\tau \\ &= \frac{N(\nu(s))}{1-\nu(s)} \frac{d}{dt} \bigg(\int_0^{\mathsf{T}_1} \mathcal{E}_{\nu_1} \left(-\nu_1 \frac{(s-\tau)^{\nu_1}}{(1-\nu_1)}\right) y^{(2)}(\tau) d\tau \\ &+ \cdots + \int_{\mathsf{T}_{k-1}}^s \mathcal{E}_{\nu_k} \left(-\nu_k \frac{(s-\tau)^{\nu_k}}{(1-\nu_k)}\right) y^{(2)}(\tau) d\tau \bigg). \end{aligned} \tag{6.31}$$

Thus, according to (6.31), the Riemann–Liouville fractional variable-order multiterm initial-terminal value problem (ITVP) given in (6.30), in the sense of Atangana-Baleanu is rewritten in the form

$$\begin{aligned} &\frac{N(\nu(s))}{1-\nu(s)} \frac{d}{dt} \bigg(\int_0^{\mathsf{T}_1} \mathcal{E}_{\nu_1} \left(-\nu_1 \frac{(s-\tau)^{\nu_1}}{(1-\nu_1)}\right) y^{(2)}(\tau) d\tau \\ &\quad + \cdots + \int_{\mathsf{T}_{k-1}}^s \mathcal{E}_{\nu_k} \left(-\nu_k \frac{(s-\tau)^{\nu_k}}{(1-\nu_k)}\right) y^{(2)}(\tau) d\tau \bigg) = f_y(s), \end{aligned} \tag{6.32}$$

for all $s \in J_k$.

Now, we define the solution of the Riemann–Liouville fractional variable order multiterm initial-terminal value problem (ITVP) (6.30) in the sense of Atangana-Baleanu, which is required in this chapter.

Definition 57 *The Riemann–Liouville fractional variable-order multiterm initial-terminal value problem (ITVP) (6.30), in the sense of Atangana–Baleanu, admits a solution whenever there exist functions* $y_k \in C(0, \mathsf{T}_k], \mathbb{R})$ *satisfying equation (6.32) such that* $y_k(0) = 0$ *and* $y(\mathsf{T}_k) = \sum_{s_l \in J_k} a_l y(s)$*, for each* $k = 1, 2, \ldots, n$.

From our previous analysis, for $k \in \{1, 2, ..., n\}$, the Riemann–Liouville fractional variable-order multiterm initial-terminal value problem (ITVP) (6.30), in the sense of Atangana-Baleanu, can be expressed using equation (6.32), on the subintervals J_k.

For $0 \leq s \leq \mathsf{T}_{k-1}$, by taking $y(s) \equiv 0$, and invoking Definition (2.17), equation (6.31) reduces to:

$$\begin{aligned} {}^{ABR}F_{0+}^{\ltimes(s)} y(s) &= \frac{N(\nu(s))}{1-\nu(s)} \frac{d}{dt}\Big(\int_{\mathsf{T}_{k-1}}^{s} \mathcal{E}_{\nu_k}\left(-\nu_k \frac{(s-\tau)^{\nu_k}}{(1-\nu_k)} \right) y^{(2)}(\tau) d\tau \Big) \\ &= {}^{ABR}F_{\mathsf{T}_{k-1}^{+}}^{\nu_k} y^{(2)}(s) \\ &= {}^{ABR}F_{\mathsf{T}_{k-1}^{+}}^{\nu_k+2} y(s). \end{aligned}$$

Hence,

$$ {}^{ABR}F_{0+}^{\ltimes(s)} y(s) = {}^{ABR}F_{\mathsf{T}_{k-1}^{+}}^{\ltimes_k} y(s). $$

Then (6.32) can be rewritten as follows:

$$ {}^{ABR}F_{\mathsf{T}_{k-1}^{+}}^{\ltimes_k} y(s) = f(s, y(s), {}^{AB}I_{\mathsf{T}_{k-1}^{+}}^{\ltimes_k} y(s)), \quad s \in J_k. $$

In this step, we consider the Riemann–Liouville fractional constant-order multiterm initial–terminal value problem (ITVP) in the sense of Atangana-Baleanu.

$$ \begin{cases} {}^{ABR}F_{\mathsf{T}_{k-1}^{+}}^{\ltimes_k} y(s) = f(s, y(s), {}^{AB}I_{\mathsf{T}_{k-1}^{+}}^{\ltimes_k} y(s)), \quad s \in J_k, \\ y(\mathsf{T}_{k-1}) = 0, \qquad y(\mathsf{T}_k) = \sum_{s_l \in J_k} a_l y(s_l). \end{cases} \tag{6.33} $$

In what follows, we assume that $\sum_{s_l \in J_k}$ and that $a_l y(s_l) \neq \mathsf{T}_k - \mathsf{T}_{k-1}$. Set

$$ A = \frac{1}{(\mathsf{T}_k - \mathsf{T}_{k-1}) - \sum_{s_l \in J_k} a_l (s_l - \mathsf{T}_{k-1})}. $$

An auxiliary lemma is introduced to initiate the main results concerning the existence property of the Riemann–Liouville fractional constant-order multiterm initial–terminal value problem (ITVP) (6.33) in the sense of Atangana-Baleanu.

Lemma 26 *The function $y \in C([\mathsf{T}_{k-1}, \mathsf{T}_k], \mathbb{R})$ is a solution of the Riemann–Liouville fractional constant order multiterm initial–terminal value problem (ITVP) (6.33) in the sense of Atangana-Baleanu if and only if y satisfies:*

$$ y(s) = A\left[\sum_{s_l \in J_k} a_l \, {}^{AB}I_{\mathsf{T}_{k-1}^{+}}^{\ltimes_k} f_y(s_l) - {}^{AB}I_{\mathsf{T}_{k-1}^{+}}^{\ltimes_k} f_y(\mathsf{T}_k) \right] (s - \mathsf{T}_{k-1}) + {}^{AB}I_{\mathsf{T}_{k-1}^{+}}^{\ltimes_k} f_y(s). \tag{6.34} $$

Proof: Suppose $y \in C([\mathsf{T}_{k-1}, \mathsf{T}_k], \mathbb{R})$ is a solution to the Riemann–Liouville fractional constant order multiterm initial–terminal value problem (ITVP) (6.33) within the Atangana-Baleanu framework. By applying the operator ${}^{AB}I^{\ltimes_k}_{\mathsf{T}^+_{k-1}}$ to both sides of equation (6.33) and invoking Lemma 6, we obtain:

$$y(s) = \eta_1 + \eta_2(s - \mathsf{T}_{k-1}) + {}^{AB}I^{\ltimes_k}_{\mathsf{T}^+_{k-1}} f_y(s), \quad s \in J_k, \ k \in \{1, 2, ..., n\}. \quad (6.35)$$

By $y(\mathsf{T}_{k-1}) = 0$, we obtain $\eta_1 = 0$.

Let $s = s_l$ in (6.35), and we obtain:

$$y(s_l) = \eta_2(s_l - \mathsf{T}_{k-1}) + {}^{AB}I^{\ltimes_k}_{\mathsf{T}^+_{k-1}} f_y(s_l).$$

Thus, we obtain:

$$\begin{aligned} \sum_{s_l \in J_k} a_l y(s_l) &= \sum_{s_l \in J_k} a_l \left(\eta_2(s_l - \mathsf{T}_{k-1}) + {}^{AB}I^{\ltimes_k}_{\mathsf{T}^+_{k-1}} f_y(s_l) \right) \\ &= \eta_2 \sum_{s_l \in J_k} a_l(s_l - \mathsf{T}_{k-1}) + \sum_{s_l \in J_k} a_l \ {}^{AB}I^{\ltimes_k}_{\mathsf{T}^+_{k-1}} f_y(s_l). \quad (6.36) \end{aligned}$$

On the other hand, we have:

$$\begin{aligned} \eta_2(\mathsf{T}_k - \mathsf{T}_{k-1}) + {}^{AB}I^{\ltimes_k}_{N^+_{k-1}} f_y(\mathsf{T}_k) &= y(\mathsf{T}_k) \\ &= \sum_{s_l \in J_k} a_l y(s_l). \quad (6.37) \end{aligned}$$

Hence,

$$\eta_2(\mathsf{T}_k - \mathsf{T}_{k-1}) + {}^{AB}I^{\ltimes_k}_{\mathsf{T}^+_{k-1}} f_y(\mathsf{T}_k) = \eta_2 \sum_{s_l \in J_k} a_l(s_l - \mathsf{T}_{k-1}) + \sum_{s_l \in J_k} a_l \ {}^{AB}I^{\ltimes_k}_{\mathsf{T}^+_{k-1}} f_y(s_l), \quad (6.38)$$

implying

$$\eta_2 = A \left[\sum_{s_l \in J_k} a_l \ {}^{AB}I^{\ltimes_k}_{\mathsf{T}^+_{k-1}} f_y(s_l) - {}^{AB}I^{\ltimes_k}_{\mathsf{T}^+_{k-1}} f_y(\mathsf{T}_k) \right]. \quad (6.39)$$

By substituting equation (6.39) into equation (6.35), we immediately obtain equation (6.34).

Now we prove the sufficiency. Let $y \in C([\mathsf{T}_{k-1}, \mathsf{T}_k], \mathbb{R})$ satisfy equation (6.34). Applying the operator ${}^{ABR}F^{\ltimes_k}_{\mathsf{T}^+_{k-1}}$ to both sides of (6.34), and using Lemma 6 along with the identity

$${}^{ABR}F^{\ltimes_k}_{\mathsf{T}^+_{k-1}}(t - \mathsf{T}_{k-1}) = 0$$

we obtain:

$$\begin{aligned}
{}^{ABR}F^{\ltimes(s)}_{\mathsf{T}^+_{k-1}} y(s) &= A\left[\sum_{k=1}^{m} a_k^{AB} I^{\ltimes_k}_{\mathsf{T}^+_{k-1}} f_y(s_k) -^{AB} I^{\ltimes(\mathsf{T}_k)}_{\mathsf{T}^+_{k-1}} f_y(\mathsf{T}_k)\right] \\
&\times \;{}^{ABR}F^{\ltimes_k}_{\mathsf{T}^+_{k-1}} (s-\mathsf{T}_{k-1}) +^{ABR} F^{\ltimes_k}_{\mathsf{T}^+_{k-1}} \; {}^{AB}I^{\ltimes_k}_{\mathsf{T}^+_{k-1}} f_y(s) \\
&= f_y(s).
\end{aligned}$$

Let $s = s_l$ in (6.34), and we obtain

$$y(s_l) = A\left[\sum_{s_l\in J_k} a_l \; {}^{AB}I^{\ltimes_k}_{\mathsf{T}^+_{k-1}} f_y(s_l) -^{AB} I^{\ltimes_k}_{\mathsf{T}^+_{k-1}} f_y(\mathsf{T}_k)\right](s_l-\mathsf{T}_{k-1}) +^{AB} I^{\ltimes_k}_{\mathsf{T}^+_{k-1}} f_y(s_l).$$

Then, we derive

$$\begin{aligned}
\sum_{s_l\in J_k} a_l y(s_l) &= A\left[\sum_{s_l\in J_k} a_l \; {}^{AB}I^{\ltimes_k}_{\mathsf{T}^+_{k-1}} f_y(s_k) -^{AB} I^{\ltimes_k}_{\mathsf{T}^+_{k-1}} f_y(\mathsf{T}_k)\right] \sum_{s_l\in J_k} a_l(s_l-\mathsf{T}_{k-1}) \\
&+ \sum_{s_l\in J_k} a_l^{AB} I^{\ltimes_k}_{\mathsf{T}^+_{k-1}} f_y(s_l) \\
&= A\left[\sum_{s_l\in J_k} a_l \; {}^{AB}I^{\ltimes_k}_{\mathsf{T}^+_{k-1}} f_y(s_l) -^{AB} I^{\ltimes_k}_{\mathsf{T}^+_{k-1}} f_y(\mathsf{T}_k)\right]\left((\mathsf{T}_k-\mathsf{T}_{k-1})-\frac{1}{A}\right) \\
&+ \sum_{s_l\in J_k} a_l^{AB} I^{\ltimes_k}_{\mathsf{T}^+_{k-1}} f_y(s_l) \\
&= A\left[\sum_{s_l\in J_k} a_l \; {}^{AB}I^{\ltimes_k}_{\mathsf{T}^+_{k-1}} f_y(s_l) -^{AB} I^{\ltimes_k}_{\mathsf{T}^+_{k-1}} f_y(\mathsf{T}_k)\right](\mathsf{T}_k-\mathsf{T}_{k-1}) \\
&- \sum_{s_l\in J_k} a_l \; {}^{AB}I^{\ltimes_k}_{\mathsf{T}^+_{k-1}} f_y(s_l) +^{AB} I^{\ltimes_k}_{\mathsf{T}^+_{k-1}} f_y(\mathsf{T}_k) \\
&+ \sum_{s_l\in J_k} a_l^{AB} I^{\ltimes_k}_{\mathsf{T}^+_{k-1}} f_y(s_l) \\
&= A\left[\sum_{s_l\in J_k} a_l \; {}^{AB}I^{\ltimes_k}_{\mathsf{T}^+_{k-1}} f_y(s_l) -^{AB} I^{\ltimes_k}_{\mathsf{T}^+_{k-1}} f_y(\mathsf{T}_k)\right](\mathsf{T}_k-\mathsf{T}_{k-1}) \\
&+ {}^{AB}I^{\ltimes_k}_{\mathsf{T}^+_{k-1}} f_y(\mathsf{T}_k) \\
&= \xi(\mathsf{T}_k).
\end{aligned}$$

Now, the proof is complete.

Lemma 27 *If λ is constant such that $\lambda \in (n, n+1]$, $n \in \mathbb{B}$ and $\curlywedge = \lambda - n$, then*

1. ${}^{AB}I^{\lambda}_{\mathsf{T}^{+}_{k-1}} y = \frac{1-\curlywedge}{N(\curlywedge)} I^{n}_{\mathsf{T}_{k-1}} y + \frac{\curlywedge}{N(\curlywedge)} I^{\lambda}_{\mathsf{T}_{k-1}} y.$
2. *For $n = 2$, we obtain*

$$ {}^{AB}I^{\lambda}_{\mathsf{T}^{+}_{k-1}} y(s) = \frac{3-\lambda}{N(\lambda-2)} \int_{\mathsf{T}_{k-1}}^{s} (s-\tau)\xi(\tau)d\tau + \frac{\lambda-2}{N(\lambda-2)\Gamma(\lambda)} \int_{\mathsf{T}_{k-1}}^{s} (s-\tau)^{\lambda-1} y(\tau)d\tau. $$

Proof: By (2.16) and (2.14), we obtain

$$ \begin{aligned} {}^{AB}I^{\lambda}_{\mathsf{T}^{+}_{k-1}} y(s) &= I^{n}_{\mathsf{T}^{+}_{k-1}} \; {}^{AB}I^{\curlywedge}_{\mathsf{T}^{+}_{k-1}} y(s) \\ &= I^{n}_{\mathsf{T}^{+}_{k-1}} \left(\frac{1-\curlywedge}{N(\curlywedge)} y(s) + \frac{\curlywedge}{N(\curlywedge)} \int_{\mathsf{T}_{k-1}}^{s} \frac{(s-\tau)^{\curlywedge-1}}{\Gamma(\curlywedge)} y(\tau)d\tau \right) \\ &= I^{n}_{\mathsf{T}^{+}_{k-1}} \left(\frac{1-\curlywedge}{N(\curlywedge)} \xi(s) + \frac{\curlywedge}{N(\curlywedge)} I^{\curlywedge}_{\mathsf{T}^{+}_{k-1}} y(s) \right) \\ &= \frac{1-\curlywedge}{N(\curlywedge)} I^{n}_{\mathsf{T}^{+}_{k-1}} y(s) + \frac{\curlywedge}{N(\curlywedge)} I^{n+\curlywedge}_{\mathsf{T}^{+}_{k-1}} y(s) \\ &= \frac{1-\curlywedge}{N(\curlywedge)} I^{n}_{\mathsf{T}^{+}_{k-1}} y(s) + \frac{\curlywedge}{N(\curlywedge)} I^{\lambda}_{\mathsf{T}^{+}_{k-1}} y(s). \end{aligned} $$

This completes the proof of the first part. Now, we continue the following computations:

$$ \begin{aligned} {}^{AB}I^{\lambda}_{\mathsf{T}^{+}_{k-1}} y(s) &= \frac{1-\curlywedge}{N(\curlywedge)} I^{2}_{\mathsf{T}_{k-1}} y(s) + \frac{\curlywedge}{N(\curlywedge)} I^{\lambda}_{\mathsf{T}_{k-1}} y(s) \qquad (6.40) \\ &= \frac{3-\lambda}{N(\lambda-2)} \int_{\mathsf{T}_{k-1}}^{s} (s-\tau) y(\tau)d\tau + \frac{\lambda-2}{N(\lambda-2)\Gamma(\lambda)} \int_{\mathsf{T}_{k-1}}^{s} (s-\tau)^{\lambda-1} y(\tau)d\tau, \end{aligned} $$

and the proof is completed.

Theorem 59 *The function* $y \in C([\mathsf{T}_{k-1}, \mathsf{T}_k], \mathbb{R})$ *is a solution of the Riemann–Liouville fractional constant-order multiterm initial-terminal value problem (ITVP) (6.33) in the sense of Atangana-Baleanu. Then, it is given by:*

$$\begin{aligned} y(s) &= q_1(s - \mathsf{T}_{k-1})\left(\sum_{s_l \in J_k} a_l \int_{\mathsf{T}_{k-1}}^{s_l} (s_l - \tau) f_y(\tau) d\tau - \int_{\mathsf{T}_{k-1}}^{\mathsf{T}_k} (\mathsf{T}_k - \tau) f_y(\tau) d\tau\right) \\ &+ \frac{q_2(s - \mathsf{T}_{k-1})}{\Gamma(\ltimes_k)}\left(\sum_{s_l \in J_k} a_l \int_{\mathsf{T}_{k-1}}^{s_l} (s_l - \tau)^{\ltimes_k - 1} f_y(\tau) d\tau - \int_{\mathsf{T}_{k-1}}^{\mathsf{T}_k} (\mathsf{T}_k - \tau)^{\ltimes_k - 1} f_y(\tau) d\tau\right) \\ &+ q_3 \int_{\mathsf{T}_{k-1}}^{s} (s - \tau) f_y(\tau) d\tau + \frac{q_4}{\Gamma(\ltimes_k)} \int_{\mathsf{T}_{k-1}}^{s} (s - \tau)^{\ltimes_k - 1} f_y(\tau) d\tau, \end{aligned}$$

where

$$q_1 := \frac{A(3 - \ltimes_k)}{N(\ltimes_k - 2)}, \quad q_2 := \frac{A(\ltimes_k - 2)}{N(\ltimes_k - 2)},$$

$$q_3 := \frac{3 - \ltimes_k}{N(\ltimes_k - 2)}, \quad q_4 := \frac{\ltimes_k - 2}{N(\ltimes_k - 2)}.$$

Proof: We have

$$\begin{aligned}
y(s) &= A\left[\sum_{s_l\in J_k} a_l\,{}^{AB}I^{\ltimes_k}_{\mathsf{T}^+_{k-1}} f_y(s_l) - {}^{AB}I^{\ltimes_k}_{\mathsf{T}^+_{k-1}} f_y(\mathsf{T}_k)\right](s-\mathsf{T}_{k-1}) + {}^{AB}I^{\ltimes_k}_{\mathsf{T}^+_{k-1}} f_y(s) \\
&= A\left[\sum_{s_l\in J_k} a_l\left(\frac{3-\ltimes_k}{N(\ltimes_k-2)}\int_{\mathsf{T}_{k-1}}^{s_l}(s_l-\tau)y(\tau)d\tau + \frac{\ltimes_k-2}{N(\ltimes_k-2)\Gamma(\ltimes_k)}\int_{\mathsf{T}_{k-1}}^{s_l}(s_l-\tau)^{\ltimes_k-1}y(\tau)d\tau\right)\right. \\
&\quad \left. -\left(\frac{3-\ltimes_k}{N(\ltimes_k-2)}\int_{\mathsf{T}_{k-1}}^{\mathsf{T}_k}(\mathsf{T}_k-\tau)y(\tau)d\tau + \frac{\ltimes_k-2}{N(\ltimes_k-2)\Gamma(\ltimes_k)}\int_{\mathsf{T}_{k-1}}^{\mathsf{T}_k}(\mathsf{T}_k-\tau)^{\ltimes_k-1}y(\tau)d\tau\right)\right](s-\mathsf{T}_{k-1}) \\
&+ \left(\frac{3-\ltimes_k}{N(\ltimes_k-2)}\int_{\mathsf{T}_{k-1}}^{s}(s-\tau)y(\tau)d\tau + \frac{\ltimes_k-2}{N(\ltimes_k-2)\Gamma(\ltimes_k)}\int_{\mathsf{T}_{k-1}}^{s}(s-\tau)^{\ltimes_k-1}y(\tau)d\tau\right) \\
&= \frac{A(3-\ltimes_k)}{N(\ltimes_k-2)}(s-\mathsf{T}_{k-1})\left(\sum_{s_l\in\mathcal{N}_k} a_l\int_{\mathsf{T}_{k-1}}^{s_l}(s_l-\tau)f_y(\tau)d\tau - \int_{\mathsf{T}_{k-1}}^{\mathsf{T}_k}(\mathsf{T}_k-\tau)f_y(\tau)d\tau\right) \\
&+ \frac{A(\ltimes_k-2)(s-\mathsf{T}_{k-1})}{N(\ltimes_k-2)\Gamma(\ltimes_k)}\left(\sum_{s_l\in J_k} a_l\int_{\mathsf{T}_{k-1}}^{s_l}(s_l-\tau)^{\ltimes_k-1}f_y(\tau)d\tau - \int_{\mathsf{T}_{k-1}}^{\mathsf{T}_k}(\mathsf{T}_k-\tau)^{\ltimes_k-1}f_y(\tau)d\tau\right) \\
&+ \frac{3-\ltimes_k}{N(\ltimes_k-2)}\int_{\mathsf{T}_{k-1}}^{s}(s-\tau)f_y(\tau)d\tau + \frac{\ltimes_k-2}{N(\ltimes_k-2)\Gamma(\ltimes_k)}\int_{\mathsf{T}_{k-1}}^{s}(s-\tau)^{\ltimes_k-1}f_y(\tau)d\tau \\
&= q_1(s-\mathsf{T}_{k-1})\left(\sum_{s_l\in J_k} a_l\int_{\mathsf{T}_{k-1}}^{s_l}(s_l-\tau)f_y(\tau)d\tau - \int_{\mathsf{T}_{k-1}}^{\mathsf{T}_k}(\mathsf{T}_k-\tau)f_\xi(\tau)d\tau\right) \\
&+ \frac{q_2(s-\mathsf{T}_{k-1})}{\Gamma(\ltimes_k)}\left(\sum_{s_l\in J_k} a_l\int_{\mathsf{T}_{k-1}}^{s_l}(s_l-\tau)^{\ltimes_k-1}f_y(\tau)d\tau - \int_{\mathsf{T}_{k-1}}^{\mathsf{T}_k}(\mathsf{T}_k-\tau)^{\ltimes_k-1}f_y(\tau)d\tau\right) \\
&+ q_3\int_{\mathsf{T}_{k-1}}^{s}(s-\tau)f_y(\tau)d\tau + \frac{q_4}{\Gamma(\ltimes_k)}\int_{\mathsf{T}_{k-1}}^{s}(s-\tau)^{\ltimes_k-1}f_y(\tau)d\tau.
\end{aligned}$$

This completes the proof.

We shall prove the existence results for the Riemann–Liouville fractional constant order multiterm initial-terminal value problem (ITVP) (6.33) in the sense of Atangana-Baleanu. The following result is derived using Krasnoselskii's fixed point theorem.

Theorem 60 *Assume both hypotheses **(H1)** and **(H2)** hold, and that*

$$L_k K_f (1+\mathcal{G}) < 1. \tag{6.41}$$

Then, the Riemann–Liouville fractional constant-order multiterm initial-terminal value problem (ITVP) (6.33) in the sense of Atangana-Baleanu possesses a solution $\overline{y_k} \in C([\mathsf{T}_{k-1}, \mathsf{T}_k], \mathbb{R})$.

Proof: Transform the Riemann–Liouville fractional constant order multi-term initial–terminal value problem (ITVP) (6.33) in the sense of Atangana-Baleanu into a fixed point problem. Define the operator

$$\mathcal{P} : C([\mathsf{T}_{k-1}, \mathsf{T}_k], \mathbb{R}) \to C([\mathsf{T}_{k-1}, \mathsf{T}_k], \mathbb{R}),$$

by

$$\begin{aligned}(\mathcal{P}y)(s) &= q_1(s-\mathsf{T}_{k-1})\left(\sum_{s_l \in J_k} a_l \int_{\mathsf{T}_{k-1}}^{s_l} (s_l-\tau) f_y(\tau)d\tau - \int_{\mathsf{T}_{k-1}}^{\mathsf{T}_k} (\mathsf{T}_k-\tau) f_y(\tau)d\tau\right)\\ &+ \frac{q_2(s-\mathsf{T}_{k-1})}{\Gamma(\ltimes_k)}\left(\sum_{s_l \in J_k} a_l \int_{\mathsf{T}_{k-1}}^{s_k} (s_k-\tau)^{\ltimes_k-1} f_y(\tau)d\tau - \int_{\mathsf{T}_{k-1}}^{\mathsf{T}_k} (\mathsf{T}_k-\tau)^{\ltimes_k-1} f_y(\tau)d\tau\right)\\ &+ q_3 \int_{\mathsf{T}_{k-1}}^{s} (s-\tau) f_y(\tau)d\tau + \frac{q_4}{\Gamma(\ltimes_k)} \int_{\mathsf{T}_{k-1}}^{s} (s-\tau)^{\ltimes_k-1} f_y(\tau)d\tau. \end{aligned} \tag{6.42}$$

Let

$$R_k \geq \frac{L_k f^\star}{1 - L_k K_f (1+\mathcal{G})}, \tag{6.43}$$

where $\mathcal{G} := \dfrac{(3-\ltimes_k)(\mathsf{T}_k-\mathsf{T}_{k-1})^2}{2N(\ltimes_k-2)} + \dfrac{(\ltimes_k-2)(\mathsf{T}_k-\mathsf{T}_{k-1})^{\ltimes_k}}{N(\ltimes_k-2)\Gamma(\ltimes_k+1)}$ and $f^\star := \sup_{s\in J} |f(s,0,0)|$. Define the set:

$$\Phi_{R_k} = \{y \in C([\mathsf{T}_{k-1}, \mathsf{T}_k], \mathbb{R}),\ \|y\|_\infty \leq R_k\}.$$

Clearly, $\Phi_{R_k} \neq \emptyset$ is a closed convex set possessing the boundedness property.

The properties of fractional integrals and assumption **(H2)** imply that the operator $\mathcal{P}$ defined in (6.42) is well-defined.

To show that $\mathcal{P}(\Phi_{R_k}) \subseteq (\Phi_{R_k})$, observe that for any $y \in \Phi_{R_k}$, the condition **(H2)** yields (for $s \in J_k$) that:

$$\begin{aligned}|\mathcal{P}y(s)| &\leq q_1(s-\mathsf{T}_{k-1})\left(\sum_{s_l \in J_k} a_l \int_{\mathsf{T}_{k-1}}^{s_l} (s_l-\tau)|f_y(\tau)|d\tau + \int_{\mathsf{T}_{k-1}}^{\mathsf{T}_k} (\mathsf{T}_k-\tau)|f_y(\tau)|d\tau\right)\\ &+ \frac{q_2(s-\mathsf{T}_{k-1})}{\Gamma(\ltimes_k)}\left(\sum_{s_l \in J_k} a_l \int_{\mathsf{T}_{k-1}}^{s_l} (s_l-\tau)^{\ltimes_k-1}|f_y(\tau)|d\tau + \int_{\mathsf{T}_{k-1}}^{\mathsf{T}_k} (\mathsf{T}_k-\tau)^{\ltimes_k-1}|f_y(\tau)|d\tau\right)\\ &+ q_3 \int_{\mathsf{T}_{k-1}}^{s} (s-\tau)|f_y(\tau)|d\tau + \frac{q_4}{\Gamma(\ltimes_k)} \int_{\mathsf{T}_{k-1}}^{s} (s-\tau)^{\ltimes_k-1}|f_y(\tau)|d\tau. \end{aligned} \tag{6.44}$$

Moreover, we have for every $\tau \in J_k$,

$$\begin{aligned}
|f_y(\tau)| &= |f(\tau, y(\tau), {}^{AB} I^{\ltimes_k}_{\mathsf{T}^+_{k-1}} y(\tau))| \\
&\leq |f(\tau, y(\tau), {}^{AB} I^{\ltimes_k}_{\mathsf{T}^+_{k-1}} y(\tau)) - f(\tau, 0, 0)| + |f(\tau, 0, 0)| \\
&= K_f \left(|y(\tau)| + |{}^{AB} I^{\ltimes_k}_{\mathsf{T}^+_{k-1}} y(\tau)| \right) + f^{\star}, \qquad (6.45)
\end{aligned}$$

and

$$\begin{aligned}
|{}^{AB} I^{\ltimes_k}_{\mathsf{T}^+_{k-1}} y(s)| &= |\frac{3-\ltimes_k}{N(\ltimes_k - 2)} \int_{\mathsf{T}_{k-1}}^{s} (s-\tau) y(\tau) d\tau + \frac{\ltimes_k - 2}{N(\ltimes_k - 2)\Gamma(\ltimes_k)} \int_{\mathsf{T}_{k-1}}^{s} (s-\tau)^{\ltimes_k - 1} y(\tau) d\tau| \\
&\leq \frac{3-\ltimes_k}{N(\ltimes_k - 2)} \int_{N_{k-1}}^{s} (s-\tau)|\xi(\tau)| d\tau + \frac{\ltimes_k - 2}{N(\ltimes_k - 2)\Gamma(\ltimes_k)} \int_{\mathsf{T}_{k-1}}^{s} (s-\tau)^{\ltimes_k - 1} |y(\tau)| d\tau \\
&\leq \frac{3-\ltimes_k}{N(\ltimes_k - 2)} \frac{(\mathsf{T}_k - \mathsf{T}_{k-1})^2}{2} \|y\| + \frac{(\ltimes_k - 2)(\mathsf{T}_k - \mathsf{T}_{k-1})^{\ltimes_k}}{N(\ltimes_k - 2)\Gamma(\ltimes_k + 1)} \|y\| \\
&= \left(\frac{3-\ltimes_k}{N(\ltimes_k - 2)} \frac{(\mathsf{T}_k - \mathsf{T}_{k-1})^2}{2} + \frac{(\ltimes_k - 2)(\mathsf{T}_k - \mathsf{T}_{k-1})^{\ltimes_k}}{N(\ltimes_k - 2)\Gamma(\ltimes_k + 1)} \right) \|y\| \\
&= \left(\frac{(3-\ltimes_k)(\mathsf{T}_k - \mathsf{T}_{k-1})^2}{2N(\ltimes_k - 2)} + \frac{(\ltimes_k - 2)(\mathsf{T}_k - \mathsf{T}_{k-1})^{\ltimes_k}}{N(\ltimes_k - 2)\Gamma(\ltimes_k + 1)} \right) \|y\| \\
&= \mathcal{G} \|y\|.
\end{aligned}$$

Hence,

$$\|f_y\| \leq K_f\,(1+\mathcal{G})\,\|y\| + f^{\star}, \tag{6.46}$$

and

$$\begin{aligned}
|\mathcal{P}y(s)| &\leq q_1(s-\mathsf{T}_{k-1})\left(\sum_{s_l\in J_k} a_l \int_{\mathsf{T}_{k-1}}^{s_l}(s_l-\tau)|f_y(\tau)|d\tau + \int_{\mathsf{T}_{k-1}}^{\mathsf{T}_k}(\mathsf{T}_k-\tau)|f_y(\tau)|d\tau\right)\\
&+ \frac{q_2(s-\mathsf{T}_{k-1})}{\Gamma(\ltimes_k)}\left(\sum_{s_l\in J_k} a_l \int_{\mathsf{T}_{k-1}}^{s_l}(s_l-\tau)^{\ltimes_k-1}|f_y(\tau)|d\tau + \int_{\mathsf{T}_{k-1}}^{\mathsf{T}_k}(\mathsf{T}_k-\tau)^{\ltimes_k-1}|f_y(\tau)|d\tau\right)\\
&+ q_3\int_{\mathsf{T}_{k-1}}^{s}(s-\tau)\left(K_f\,(1+\mathcal{G})\,\|y\| + f^{\star}\right)d\tau\\
&+ \frac{q_4}{\Gamma(\ltimes_k)}\int_{\mathsf{T}_{k-1}}^{s}(s-\tau)^{\ltimes_k-1}\left(K_f\,(1+\mathcal{G})\,\|y\| + f^{\star}\right)d\tau.\\
&= \left(K_f\,(1+\mathcal{G})\,\|y\| + f^{\star}\right)\Bigg[q_1(s-\mathsf{T}_{k-1})\left(\sum_{s_l\in J_k} a_l \int_{\mathsf{T}_{k-1}}^{s_l}(s_l-\tau)d\tau + \int_{\mathsf{T}_{k-1}}^{\mathsf{T}_k}(\mathsf{T}_k-\tau)d\tau\right)\\
&+\frac{q_2(s-\mathsf{T}_{k-1})}{\Gamma(\ltimes_k)}\left(\sum_{s_l\in J_k} a_l \int_{\mathsf{T}_{k-1}}^{s_l}(s_l-\tau)^{\ltimes_k-1} + \int_{\mathsf{T}_{k-1}}^{\mathsf{T}_k}(\mathsf{T}_k-\tau)^{\ltimes_k-1}d\tau\right)\\
&+q_3\int_{\mathsf{T}_{k-1}}^{s}(s-\tau)d\tau + \frac{q_4}{\Gamma(\ltimes_k)}\int_{\mathsf{T}_{k-1}}^{s}(s-\tau)^{\ltimes_k-1}d\tau\Bigg]\\
&\leq \left(K_f\,(1+\mathcal{G})\,\|y\| + f^{\star}\right)\Bigg[\frac{q_1}{2}(\mathsf{T}_k-\mathsf{T}_{k-1})\left(\sum_{s_l\in J_k} a_l(s_l-\mathsf{T}_{k-1})^2 + (\mathsf{T}_k-\mathsf{T}_{k-1})^2\right)\\
&+\frac{q_2(\mathsf{T}_k-\mathsf{T}_{k-1})}{\Gamma(\mu_k+1)}\left(\sum_{s_l\in J_k} a_l(s_l-\mathsf{T}_{k-1})^{\ltimes_k} + (\mathsf{T}_k-\mathsf{T}_{k-1})^{\ltimes_k}\right)\\
&+\frac{q_3}{2}(\mathsf{T}_k-\mathsf{T}_{k-1})^2 + \frac{q_4}{\Gamma(\ltimes_k+1)}(\mathsf{T}_k-\mathsf{T}_{k-1})^{\ltimes_k}\Bigg]\\
&= L_k\left(K_f\,(1+\mathcal{G})\,\|y\| + f^{\star}\right)\\
&= L_k K_f\,(1+\mathcal{G})\,\|y\| + L_k f^{\star},
\end{aligned} \tag{6.47}$$

where

$$\begin{aligned} L_k &= \frac{q_1}{2}(\mathsf{T}_k - \mathsf{T}_{k-1})\left(\sum_{s_l \in J_k} a_l(s_l - \mathsf{T}_{k-1})^2 + (\mathsf{T}_k - \mathsf{T}_{k-1})^2\right) \\ &+ \frac{q_2(\mathsf{T}_k - \mathsf{T}_{k-1})}{\Gamma(\ltimes_k + 1)}\left(\sum_{s_l \in J_k} a_l(s_l - \mathsf{T}_{k-1})^{\ltimes_k} + (\mathsf{T}_k - \mathsf{T}_{k-1})^{\ltimes_k}\right) \\ &+ \frac{q_3}{2}(\mathsf{T}_k - \mathsf{T}_{k-1})^2 + \frac{q_4}{\Gamma(\ltimes_k + 1)}(\mathsf{T}_k - \mathsf{T}_{k-1})^{\ltimes_k}. \end{aligned} \tag{6.48}$$

Thus, (6.43) and (6.47) imply that

$$\begin{aligned} \|\mathcal{P}y\|_\infty &\leq L_k K_f (1 + \mathcal{G}) \|y\| + L_k f^\star \\ &\leq L_k K_f (1 + \mathcal{G}) R_k + (1 - L_k K_f (1 + \mathcal{G})) R_k \\ &\leq R_k, \end{aligned}$$

which implies $\mathcal{P}(\Phi_{R_k}) \subseteq \Phi_{R_k}$.

We define the operators $\mathcal{P}_1$ and $\mathcal{P}_2$ on Φ_{R_k} by

$$\begin{aligned} (\mathcal{P}_1 y)(s) &= q_1(s - \mathsf{T}_{k-1})\left(\sum_{s_l \in J_k} a_l \int_{\mathsf{T}_{k-1}}^{s_l} (s_l - \tau) f_y(\tau) d\tau - \int_{\mathsf{T}_{k-1}}^{\mathsf{T}_k} (\mathsf{T}_k - \tau) f_y(\tau) d\tau\right) \qquad (6.49) \\ &+ \frac{q_2(s - \mathsf{T}_{k-1})}{\Gamma(\ltimes_k)}\left(\sum_{l=1}^{m} a_l \int_{\mathsf{T}_{k-1}}^{s_l} (s_l - \tau)^{\ltimes_k - 1} f_y(\tau) d\tau - \int_{\mathsf{T}_{k-1}}^{\mathsf{T}_k} (\mathsf{T}_k - \tau)^{\ltimes_k - 1} f_y(\tau) d\tau\right), \end{aligned}$$

and

$$(\mathcal{P}_2 y)(s) = q_3 \int_{\mathsf{T}_{k-1}}^{s} (s - \tau) f_y(\tau) d\tau + \frac{q_4}{\Gamma(\ltimes_k)} \int_{\mathsf{T}_{k-1}}^{s} (s - \tau)^{\ltimes_k - 1} f_y(\tau) d\tau. \tag{6.50}$$

It follows that $\mathcal{P} = \mathcal{P}_1 + \mathcal{P}_2$.

We shall investigate that the $\mathcal{P}_1$ and $\mathcal{P}_2$ satisfy the conditions of Krasnoselskii's fixed point theorem through several steps:

STEP 1: We aim to show that $\mathcal{P}_1 y + \mathcal{P}_2 \acute{y} \in \Phi_{R_k}$ for all $y,\ \acute{y} \in \Phi_{R_k}$. Let $y,\ \acute{y} \in \Phi_{R_k}$. Then, by assumption (H2) and for all $s \in J_k$, we have

$$\begin{aligned}
|(\mathcal{P}y+\mathcal{P}_2\acute{y})(s)| &\leq q_1(s-\mathsf{T}_{k-1})\left(\sum_{s_l\in J_k} a_l \int_{\mathsf{T}_{k-1}}^{s_l}(s_l-\tau)|f_y(\tau)|d\tau + \int_{\mathsf{T}_{k-1}}^{\mathsf{T}_k}(\mathsf{T}_k-\tau)|f_y(\tau)|d\tau\right)\\
&+ \frac{q_2(s-\mathsf{T}_{k-1})}{\Gamma(\ltimes_k)}\left(\sum_{s_l\in J_k} a_l \int_{\mathsf{T}_{k-1}}^{s_l}(s_l-\tau)^{\ltimes_k-1}|f_y(\tau)|d\tau\right.\\
&+ \left.\int_{\mathsf{T}_{k-1}}^{\mathsf{T}_k}(\mathsf{T}_k-\tau)^{\ltimes_k-1}|f_y(\tau)|d\tau\right)\\
&+ q_3\int_{\mathsf{T}_{k-1}}^{s}(s-\tau)|f_{\acute{y}}(\tau)|d\tau + \frac{q_4}{\Gamma(\ltimes_k)}\int_{\mathsf{T}_{k-1}}^{s}(s-\tau)^{\ltimes_k-1}|f_{\acute{y}}(\tau)|d\tau. \qquad (6.51)
\end{aligned}$$

By Equation 6.46 ,6.43 and $y,\ \acute{y} \in \Phi_{R_k}$, we have

$$\begin{aligned}
|(\mathcal{P}_1y+\mathcal{P}_2\acute{y})(s)| &\leq \|f_y\|\left[q_1(s-\mathsf{T}_{k-1})\left(\sum_{s_l\in J_k} a_l \int_{\mathsf{T}_{k-1}}^{s_l}(s_l-\tau)d\tau + \int_{\mathsf{T}_{k-1}}^{\mathsf{T}_k}(\mathsf{T}_k-\tau)d\tau\right)\right.\\
&+ \left.\frac{q_2(s-\mathsf{T}_{k-1})}{\Gamma(\ltimes_k)}\left(\sum_{s_l\in J_k} a_l \int_{\mathsf{T}_{k-1}}^{s_l}(s_l-\tau)^{\ltimes_k-1}d\tau + \int_{\mathsf{T}_{k-1}}^{\mathsf{T}_k}(\mathsf{T}_k-\tau)^{\ltimes_k-1}d\tau\right)\right]\\
&+ \|f_{\acute{y}}\|\left[q_3\int_{\mathsf{T}_{k-1}}^{s}(s-\tau)d\tau + \frac{q_4}{\Gamma(\ltimes_k)}\int_{\mathsf{T}_{k-1}}^{s}(s-\tau)^{\ltimes_k-1}d\tau.\right]\\
&= \left(K_f\left(1+\mathcal{G}\right)R_k+f^\star\right)\left[q_1(s-\mathsf{T}_{k-1})\left(\sum_{s_l\in J_k} a_l \int_{\mathsf{T}_{k-1}}^{s_l}(s_l-\tau)d\tau\right.\right.\\
&+ \left.\int_{\mathsf{T}_{k-1}}^{\mathsf{T}_k}(\mathsf{T}_k-\tau)d\tau\right)\\
&+ \frac{q_2(s-\mathsf{T}_{k-1})}{\Gamma(\ltimes_k)}\left(\sum_{s_l\in J_k} a_l \int_{\mathsf{T}_{k-1}}^{s_l}(s_l-\tau)^{\ltimes_k-1}d\tau + \int_{\mathsf{T}_{k-1}}^{\mathsf{T}_k}(\mathsf{T}_k-\tau)^{\ltimes_k-1}d\tau\right)\\
&+ \left.q_3\int_{\mathsf{T}_{k-1}}^{s}(s-\tau)d\tau + \frac{q_4}{\Gamma(\ltimes_k)}\int_{\mathsf{T}_{k-1}}^{s}(s-\tau)^{\ltimes_k-1}d\tau.\right]\\
&\leq \left(K_f\left(1+\mathcal{G}\right)R_k+f^\star\right)L_k\\
&\leq R_k.
\end{aligned}$$

Therefore,

$$\|\mathcal{P}_1 y + \mathcal{P}_2 \acute{y}\|_\infty \leq R_k.$$

Thus, $\mathcal{P}_1 y + \mathcal{P}_2 \acute{y} \in \Phi_{R_k}$.

STEP 2: $\mathcal{P}_2$ is a contraction.
Let $y, \tilde{y} \in C([\mathsf{T}_{k-1}, \mathsf{T}_k], \mathbb{R})$ and $s \in J_k$. We have

$$\begin{aligned}
|\mathcal{P}_2 y(s) - \mathcal{P}_2 \tilde{y}(s)| &\leq q_3 \int_{\mathsf{T}_{k-1}}^{s} (s-\tau)|f_y(\tau) - f_{\tilde{y}}(\tau)|d\tau \\
&+ \frac{q_4}{\Gamma(\ltimes_k)} \int_{\mathsf{T}_{k-1}}^{s} (s-\tau)^{\ltimes_k - 1}|f_y(\tau) - f_{\tilde{y}}(\tau)|d\tau. \\
&\leq q_1(s - \mathsf{T}_{k-1}) \left(\sum_{l=1}^{m} a_l \int_{\mathsf{T}_{k-1}}^{s_l} (s_l - \tau)|f_y(\tau) - f_{\tilde{y}}(\tau)|d\tau \right. \\
&+ \left. \int_{\mathsf{T}_{k-1}}^{\mathsf{T}_k} (\mathsf{T}_k - \tau)|f_y(\tau) - f_{\tilde{y}}(\tau)|d\tau \right) \\
&+ \frac{q_2(s - \mathsf{T}_{k-1})}{\Gamma(\ltimes_k)} \left(\sum_{l=1}^{m} a_l \int_{\mathsf{T}_{k-1}}^{s_l} (s_l - \tau)^{\ltimes_k - 1}|f_y(\tau) - f_{\tilde{y}}(\tau)|d\tau \right. \\
&+ \left. \int_{\mathsf{T}_{k-1}}^{\mathsf{T}_k} (\mathsf{T}_k - \tau)^{\ltimes_k - 1}|f_y(\tau) - f_{\tilde{y}}(\tau)|d\tau \right) \\
&+ q_3 \int_{\mathsf{T}_{k-1}}^{s} (s-\tau)|f_y(\tau) - f_{\tilde{y}}(\tau)|d\tau \\
&+ \frac{q_4}{\Gamma(\ltimes_k)} \int_{\mathsf{T}_{k-1}}^{s} (s-\tau)^{\ltimes_k - 1}|f_y(\tau) - f_{\tilde{y}}(\tau)|d\tau. \qquad (6.52)
\end{aligned}$$

On the other side, for every $\tau \in J_k$, we write

$$\begin{aligned}
|f_y(\tau) - f_{\tilde{y}}(\tau)| &= |f(\tau, y(\tau), {}^{AB}I^{\ltimes_k}_{\mathsf{T}^+_{k-1}} y(\tau)) - f(\tau, \tilde{y}(\tau), {}^{AB}I^{\ltimes_k}_{\mathsf{T}^+_{k-1}} \tilde{y}(\tau))| \\
&\leq K_f \left(|y(\tau) - \tilde{y}(\tau)| + |{}^{AB}I^{\ltimes_k}_{\mathsf{T}^+_{k-1}} y(\tau) - {}^{AB}I^{\ltimes_k}_{\mathsf{T}^+_{k-1}} \tilde{y}(\tau)| \right) \\
&= K_f \left(|(y - \tilde{y})(\tau)| + |{}^{AB}I^{\ltimes_k}_{\mathsf{T}^+_{k-1}} (y - \tilde{y})(\tau)| \right),
\end{aligned}$$

and

$$\begin{aligned}
|{}^{AB}I^{\ltimes_k}_{\intercal_{k-1}^+}(y-\tilde{y})(s)| &= |\frac{3-\ltimes_k}{N(\ltimes_k-2)}\int_{\intercal_{k-1}}^{s}(s-\tau)(y-\tilde{y})(\tau)d\tau \\
&+ \frac{\ltimes_k-2}{N(\ltimes_k-2)\Gamma(\ltimes_k)}\int_{\intercal_{k-1}}^{s}(s-\tau)^{\ltimes_k-1}(y-\tilde{y})(\tau)d\tau| \\
&\le \frac{3-\ltimes_k}{N(\ltimes_k-2)}\int_{\intercal_{k-1}}^{s}(s-\tau)|(y-\tilde{y})(\tau)|d\tau \\
&+ \frac{\ltimes_k-2}{N(\ltimes_k-2)\Gamma(\ltimes_k)}\int_{\intercal_{k-1}}^{s}(s-\tau)^{\ltimes_k-1}|(y-\tilde{y})(\tau)|d\tau \\
&\le \frac{3-\ltimes_k}{N(\ltimes_k-2)}\frac{(\intercal_k-\intercal_{k-1})^2}{2}\|y-\tilde{y}\| \\
&+ \frac{(\ltimes_k-2)(\intercal_k-\intercal_{k-1})^{\ltimes_k}}{N(\ltimes_k-2)\Gamma(\ltimes_k+1)}\|y-\tilde{y}\| \\
&= \left(\frac{3-\ltimes_k}{N(\ltimes_k-2)}\frac{(\intercal_k-\intercal_{k-1})^2}{2}+\frac{(\ltimes_k-2)(\intercal_k-\intercal_{k-1})^{\ltimes_k}}{N(\ltimes_k-2)\Gamma(\ltimes_k+1)}\right)\|y-\tilde{y}\| \\
&= \left(\frac{(3-\ltimes_k)(\intercal_k-\intercal_{k-1})^2}{2N(\ltimes_k-2)}+\frac{(\ltimes_k-2)(\intercal_k-\intercal_{k-1})^{\ltimes_k}}{N(\ltimes_k-2)\Gamma(\ltimes_k+1)}\right)\|y-\tilde{y}\| \\
&= \mathcal{G}\|y-\tilde{y}\|.
\end{aligned}$$

Hence,

$$\|f_y-f_{\tilde{y}}\| \le K_f\left(1+\mathcal{G}\right)\|y-\tilde{y}\|. \tag{6.53}$$

By replacing (6.53) in the inequality (6.52), we obtain

$$\begin{aligned}
|\mathcal{P}_2y(s)-\mathcal{P}_2\tilde{y}(s)| &\le K_f\left(1+\mathcal{G}\right)\|y-\tilde{y}\|\left[\frac{q_1}{2}(\intercal_k-\intercal_{k-1})\left(\sum_{s_l\in J_k}a_l(s_l-\intercal_{k-1})^2+(\intercal_k-\intercal_{k-1})^2\right)\right. \\
&+ \frac{q_2(\intercal_k-\intercal_{k-1})}{\Gamma(\ltimes_k+1)}\left(\sum_{s_l\in J_k}a_l(s_l-\intercal_{k-1})^{\ltimes_k}+(\intercal_k-\intercal_{k-1})^{\ltimes_k}\right) \\
&+ \left.\frac{q_3}{2}(\intercal_k-\intercal_{k-1})^2+\frac{q_4}{\Gamma(\ltimes_k+1)}(\intercal_k-\intercal_{k-1})^{\ltimes_k}\right] \\
&\le L_kK_f\left(1+\mathcal{G}\right)\|y-\tilde{y}\|. \qquad (6.54)
\end{aligned}$$

Consequently by (6.41), $\mathcal{P}_2$ is a contraction.

STEP 3: $\mathcal{P}_1$ is continuous.
The Continuity of f implies that $\mathcal{P}_1$ is continuous.

STEP 4: $\mathcal{P}_1(\Phi_{R_k})$ is bounded in Φ_{R_k}.
Similar to Step 1, we know that $\mathcal{P}_1(\Phi_{R_k}) \subset \Phi_{R_k}$. It implies that $\mathcal{P}_1(\Phi_{R_i})$ is a bounded set in Φ_{R_k}.

STEP 5: $\mathcal{P}_1(\Phi_{R_k})$ is equicontinuous.

For arbitrary $s_1, s_2 \in J_k$, with $s_1 < s_2$, let $\xi \in \Phi_{R_k}$. We write

$$
\begin{aligned}
|\mathcal{P}_1(y)(s_2) \quad &- \quad \mathcal{P}_1(y)(s_1)| = \Big| q_1(s_2 - \mathsf{T}_{k-1}) \left(\sum_{s_l \in J_k} a_l \int_{\mathsf{T}_{k-1}}^{s_l} (s_l - \tau) f_y(\tau) d\tau - \int_{\mathsf{T}_{k-1}}^{\mathsf{T}_k} (\mathsf{T}_k - \tau) f_y(\tau) d\tau \right) \\
&+ \quad \frac{q_2(s_2 - \mathsf{T}_{k-1})}{\Gamma(\ltimes_k)} \left(\sum_{s_l \in J_k} a_l \int_{\mathsf{T}_{k-1}}^{s_l} (s_l - \tau)^{\ltimes_k - 1} f_y(\tau) d\tau - \int_{\mathsf{T}_{k-1}}^{\mathsf{T}_k} (\mathsf{T}_k - \tau)^{\ltimes_k - 1} f_y(\tau) d\tau \right) \\
&- \quad q_1(s_1 - \mathsf{T}_{k-1}) \left(\sum_{s_l \in J_k} a_l \int_{\mathsf{T}_{k-1}}^{s_l} (s_l - \tau) f_y(\tau) d\tau - \int_{\mathsf{T}_{k-1}}^{\mathsf{T}_k} (\mathsf{T}_k - \tau) f_y(\tau) d\tau \right) \\
&- \quad \frac{q_2(s_1 - \mathsf{T}_{k-1})}{\Gamma(\ltimes_k)} \left(\sum_{s_l \in J_k} a_l \int_{\mathsf{T}_{k-1}}^{s_l} (s_l - \tau)^{\ltimes_k - 1} f_y(\tau) d\tau - \int_{N_{k-1}}^{N_k} (N_k - \tau)^{\ltimes_k - 1} f_\xi(\tau) d\tau \right) . \Big| \\
&\le \quad q_1(s_2 - s_1) \Big| \left(\sum_{s_l \in J_k} a_l \int_{\mathsf{T}_{k-1}}^{s_l} (s_l - \tau) f_y(\tau) d\tau - \int_{\mathsf{T}_{k-1}}^{\mathsf{T}_k} (\mathsf{T}_k - \tau) f_y(\tau) d\tau \right) \Big| \\
&+ \quad \frac{q_2(s_2 - s_1)}{\Gamma(\ltimes_k)} \Big| \left(\sum_{s_l \in J_k} a_l \int_{\mathsf{T}_{k-1}}^{s_l} (s_l - \tau)^{\ltimes_k - 1} f_y(\tau) d\tau - \int_{\mathsf{T}_{k-1}}^{\mathsf{T}_k} (\mathsf{T}_k - \tau)^{\ltimes_k - 1} f_y(\tau) d\tau \right) \Big| .
\end{aligned}
$$

Hence, $|\mathcal{P}_1(y)(s_2) - \mathcal{P}_1(y)(s_1)| \to 0$ as $s_2 - s_1 \to 0$. This implies that the set $\mathcal{P}_1(\Phi_{R_k})$ is equicontinuous. Therefore, all conditions of Krasnoselskii's fixed point theorem are satisfied, and we deduce that the operator $\mathcal{P}$ has a fixed point $\overline{y}_k \in \Phi_{R_k}$. Consequently, the R-Liouville fractional constant order multiterm initial–terminal value problem (ITVP) (6.33) in the sense of Atangana-Baleanu admits a solution $\overline{y}_k \in C([\mathsf{T}_{k-1}, \mathsf{T}_k], \mathbb{R})$.

Now, we present the final theorem concerning the existence result in connection with the Riemann–Liouville fractional variable-order multiterm initial–terminal value problem (ITVP) (6.30) in the sense of Atangana–Baleanu.

Theorem 61 *Given that the conditions* ***(H1)****,* ***(H2)****, and the inequality (6.41) are satisfied for all* $k \in \{1, \dots, n\}$*, the Riemann–Liouville fractional variable-order multiterm initial–terminal value problem (ITVP) (6.30), formulated in the sense of the Atangana-Baleanu approach, admits a solution* $y \in C([0, \mathsf{T}], \mathbb{R})$.

Proof: For each $k \in \{1, \dots, n\}$ and based on Theorem 60, we know that the Riemann–Liouville fractional constant-order multiterm initial–terminal value problem (6.33), as formulated in the context of Atangana-Baleanu and discussed in Problems for Caputo–Hadamard Fractional Differential Equations with variable order [p. 271], admits a a solution $\overline{y}_k \in C([\mathsf{T}_{k-1}, \mathsf{T}_k], \mathbb{R})$. We define

$$y_1(s) = \overline{y}_1(s), \quad s \in J_1,$$

and for any $k \in \{2, ..., n\}$, define

$$y_k(s) = \begin{cases} 0, & s \in [0, \mathsf{T}_{k-1}], \\ \overline{y}_k(s), & s \in J_k. \end{cases}$$

Therefore, for each $y_k \in C([0, \mathsf{T}_k], \mathbb{R})$ solves the integral equation (6.32) for $s \in J_k$, subject to the initial and terminal conditions $y_k(0) = 0$ and $y_k(\mathsf{T}_k) = \sum_{s_l \in J_k} a_l y_k(s_l) = \sum_{s_l \in J_k} a_l \overline{y}_k(s_l)$.

We define the global solution $y : [0, \mathsf{T}] \to \mathbb{R}$ piecewise by:

$$y(s) = \begin{cases} y_1(s), \quad s \in J_1, \\ y_2(s) = \begin{cases} 0, & s \in J_1, \\ \overline{y}_2(s), & s \in J_2, \end{cases} \\ \cdot \\ \cdot \\ \cdot \\ y_n(s) = \begin{cases} 0, & s \in [0, \mathsf{T}_{n-1}], \\ \overline{y}_n(s), & s \in J_n, \end{cases} \end{cases}.$$

This function $y(s)$ provides the solution to the Riemann–Liouville fractional variable-order multiterm initial–terminal value problem (6.30) in the sense of Atangana-Baleanu.

6.3.2 Stability Properties

Remark 17 *Let $z \in C([\mathsf{T}_{k-1}, \mathsf{T}_k], \mathbb{R})$. Then z is a solution of equation (2.33) if and only if there exists a function $h \in C([\mathsf{T}_{k-1}, \mathsf{T}_k], \mathbb{R})$ (depending on z) such that:*

(i) $|h(s)| \leq \epsilon$, *for all* $s \in [\mathsf{T}_{k-1}, \mathsf{T}_k]$,

(ii) $F^{\ltimes(s)}_{N^+_{k-1}} z(s) = f_z(s) + h(s), \, \forall \, s \in J_k$.

Lemma 28 *If $z \in C([\mathsf{T}_{k-1}, \mathsf{T}_k], \mathbb{R})$ is a solution of equation (2.33), then z satisfies the inequality*

$$|z(s) - A_z - q_3 \int_{\mathsf{T}_{k-1}}^{s} (s-\tau) f_z(\tau) d\tau - \frac{q_4}{\Gamma(\ltimes_k)} \int_{\mathsf{T}_{k-1}}^{s} (s-\tau)^{\ltimes_k - 1} f_z(\tau) d\tau| \leq \epsilon L_k, \tag{6.55}$$

where the constant term A_z is given by

$$\begin{aligned} A_z &= q_1(s - \mathsf{T}_{k-1}) \left(\sum_{s_l \in J_k} a_l \int_{\mathsf{T}_{k-1}}^{s_l} (s_l - \tau) f_z(\tau) d\tau - \int_{\mathsf{T}_{k-1}}^{\mathsf{T}_k} (\mathsf{T}_k - \tau) f_z(\tau) d\tau \right) \qquad (6.56) \\ &+ \frac{q_2(s - \mathsf{T}_{k-1})}{\Gamma(\ltimes_k)} \left(\sum_{s_l \in J_k} a_l \int_{\mathsf{T}_{k-1}}^{s_l} (s_l - \tau)^{\ltimes_k - 1} f_z(\tau) d\tau - \int_{\mathsf{T}_{k-1}}^{\mathsf{T}_k} (\mathsf{T}_k - \tau)^{\ltimes_k - 1} f_z(\tau) d\tau \right). \end{aligned}$$

Proof: By Remark 17, we have that

$$^{ABR}F^{\ltimes_k}_{\mathsf{T}^+_{k-1}} z(s) = f(t, z(s), {}^{AB}I^{\ltimes_k}_{\mathsf{T}^+_{k-1}} z(s)) + h(s) = f_z(s) + h(s),$$

and equivalently,

$$\begin{cases} ^{ABR}F^{\ltimes_k}_{\mathsf{T}^+_{k-1}} z(s) = f_z(s) + h(s), \quad s \in J_k, \\ z(\mathsf{T}_{k-1}) = 0, \qquad z(\mathsf{T}_k) = \sum_{s_l \in J_k} a_l z(s_l). \end{cases} \tag{6.57}$$

Then, by Theorem 59, we get

$$
\begin{aligned}
z(s) = \;& q_1(s-\mathsf{T}_{k-1})\left(\sum_{s_l\in J_k} a_l \int_{\mathsf{T}_{k-1}}^{s_l}(s_l-\tau)(f_z(\tau)+h(\tau))d\tau - \int_{\mathsf{T}_{k-1}}^{\mathsf{T}_k}(\mathsf{T}_k-\tau)(f_z(\tau)+h(\tau))d\tau\right)\\
+\;& \frac{q_2(s-\mathsf{T}_{k-1})}{\Gamma(\ltimes_k)}\left(\sum_{s_l\in J_k} a_l \int_{\mathsf{T}_{k-1}}^{s_l}(s_l-\tau)^{\ltimes_k-1}(f_z(\tau)+h(\tau))d\tau\right.\\
-\;& \left.\int_{\mathsf{T}_{k-1}}^{\mathsf{T}_k}(\mathsf{T}_k-\tau)^{\ltimes_k-1}(f_z(\tau)+h(\tau))d\tau\right)\\
+\;& q_3\int_{N_{k-1}}^{s}(s-\tau)(f_z(\tau)+h(\tau))d\tau + \frac{q_4}{\Gamma(\ltimes_k)}\int_{\mathsf{T}_{k-1}}^{s}(s-\tau)^{\ltimes_k-1}(f_z(\tau)+h(\tau))d\tau.\\
=\;& q_1(s-\mathsf{T}_{k-1})\left(\sum_{s_l\in J_k} a_l \int_{\mathsf{T}_{k-1}}^{s_l}(s_l-\tau)f_z(\tau)d\tau - \int_{\mathsf{T}_{k-1}}^{\mathsf{T}_k}(\mathsf{T}_k-\tau)f_z(\tau)d\tau\right)\\
+\;& \frac{q_2(s-\mathsf{T}_{k-1})}{\Gamma(\ltimes_k)}\left(\sum_{s_l\in J_k} a_l \int_{\mathsf{T}_{k-1}}^{s_l}(s_l-\tau)^{\ltimes_k-1}f_z(\tau)d\tau - \int_{\mathsf{T}_{k-1}}^{\mathsf{T}_k}(\mathsf{T}_k-\tau)^{\ltimes_k-1}f_z(\tau)d\tau\right)\\
+\;& q_1(s-\mathsf{T}_{k-1})\left(\sum_{s_l\in J_k} a_l \int_{\mathsf{T}_{k-1}}^{s_l}(s_l-\tau)h(\tau)d\tau - \int_{\mathsf{T}_{k-1}}^{\mathsf{T}_k}(\mathsf{T}_k-\tau)h(\tau)d\tau\right)\\
+\;& \frac{q_2(s-\mathsf{T}_{k-1})}{\Gamma(\ltimes_k)}\left(\sum_{s_l\in J_k} a_l \int_{\mathsf{T}_{k-1}}^{s_l}(s_l-\tau)^{\ltimes_k-1}h(\tau)d\tau - \int_{\mathsf{T}_{k-1}}^{\mathsf{T}_k}\mathsf{T}_k-\tau)^{\ltimes_k-1}h(\tau)d\tau\right)\\
+\;& q_3\int_{\mathsf{T}_{k-1}}^{s}(s-\tau)(f_z(\tau)+h(\tau))d\tau + \frac{q_4}{\Gamma(\ltimes_k)}\int_{\mathsf{T}_{k-1}}^{s}(s-\tau)^{\ltimes_k-1}(f_z(\tau)+h(\tau))d\tau.\\
=\;& A_z + q_3\int_{\mathsf{T}_{k-1}}^{s}(s-\tau)f_z(\tau)d\tau + \frac{q_4}{\Gamma(\ltimes_k)}\int_{\mathsf{T}_{k-1}}^{s}(s-\tau)^{\ltimes_k-1}f_z(\tau)d\tau\\
+\;& q_1(s-\mathsf{T}_{k-1})\left(\sum_{s_l\in J_k} a_l \int_{\mathsf{T}_{k-1}}^{s_l}(s_l-\tau)h(\tau)d\tau - \int_{\mathsf{T}_{k-1}}^{\mathsf{T}_k}(\mathsf{T}_k-\tau)h(\tau)d\tau\right)\\
+\;& \frac{q_2(s-\mathsf{T}_{k-1})}{\Gamma(\ltimes_k)}\left(\sum_{s_l\in J_k} a_l \int_{\mathsf{T}_{k-1}}^{s_l}(s_l-\tau)^{\ltimes_k-1}h(\tau)d\tau - \int_{\mathsf{T}_{k-1}}^{\mathsf{T}_k}(\mathsf{T}_k-\tau)^{\ltimes_k-1}h(\tau)d\tau\right)\\
+\;& q_3\int_{\mathsf{T}_{k-1}}^{s}(s-\tau)h(\tau)d\tau + \frac{q_4}{\Gamma(\ltimes_k)}\int_{\mathsf{T}_{k-1}}^{s}(s-\tau)^{\ltimes_k-1}h(\tau)d\tau. \qquad (6.58)
\end{aligned}
$$

It follows that

$$
\begin{aligned}
& |z(s) - A_z - q_3 \int_{\mathsf{T}_{k-1}}^{s} (s-\tau) f_z(\tau) d\tau - \frac{q_4}{\Gamma(\ltimes_k)} \int_{\mathsf{T}_{k-1}}^{s} (s-\tau)^{\ltimes_k - 1} f_z(\tau) d\tau| \\
\leq \ & \epsilon \left[\frac{q_1}{2} (\mathsf{T}_k - \mathsf{T}_{k-1}) \left(\sum_{s_l \in J_k} a_l (s_l - \mathsf{T}_{k-1})^2 + (\mathsf{T}_k - \mathsf{T}_{k-1})^2 \right) \right. \\
+ \ & \frac{q_2 (\mathsf{T}_k - \mathsf{T}_{k-1})}{\Gamma(\ltimes_k + 1)} \left(\sum_{s_l \in J_k} a_l (s_l - \mathsf{T}_{k-1})^{\ltimes_k} + (\mathsf{T}_k - \mathsf{T}_{k-1})^{\ltimes_k} \right) \\
+ \ & \left. \frac{q_3}{2} (\mathsf{T}_k - \mathsf{T}_{k-1})^2 + \frac{q_4}{\Gamma(\ltimes_k + 1)} (\mathsf{T}_k - \mathsf{T}_{k-1})^{\ltimes_k} \right] \\
\leq \ & \epsilon L_k.
\end{aligned}
$$

Now, the proof is complete.

Theorem 62 *Suppose that assumptions (H1), (H2), and condition (6.41) hold. Then the Riemann–Liouville fractional variable order multiterm initial–terminal value problem (ITVP) (6.30), in the sense of Atangana-Baleanu is Ulam-Hyers stable.*

Proof: For each $\epsilon > 0$, let $z \in C([\mathsf{T}_{k-1}, \mathsf{T}_k], \mathbb{R})$ be a solution of (2.33). According to Theorem 61, the Riemann–Liouville fractional variable order multiterm initial–terminal value problem (ITVP) (6.30), in the sense of Atangana-Baleanu admits a solution $y \in C([0, B], \mathbb{R})$ as $y(s) = y_k(s)$ for $s \in [0, \mathsf{T}_k]$, $k = 1, 2, ..., n$, in which

$$
y_1(s) = \overline{y}_1(s), \quad s \in J_1, \tag{6.59}
$$

and for any $k \in \{2, ..., n\}$

$$
y_k(s) = \begin{cases} 0, & s \in [0, \mathsf{T}_{k-1}], \\ \overline{y}_k(s), & s \in J_k, \end{cases} \tag{6.60}
$$

and $\overline{y}_k \in C([\mathsf{T}_{k-1}, \mathsf{T}_k], \mathbb{R})$ is a solution of the Riemann–Liouville fractional constant order multiterm ITVP in the sense of Atangana-Baleanu

$$
\begin{cases} {}^{ABR}F^{\ltimes_k}_{\mathsf{T}^+_{k-1}} y(s) = f(s, y(s), {}^{AB}I^{\ltimes_k}_{\mathsf{T}^+_{k-1}}(s)), \quad s \in J_k, \\ y(\mathsf{T}_{k-1}) = z(\mathsf{T}_{k-1}) = 0, \quad y(\mathsf{T}_k) = z(\mathsf{T}_k) = \sum_{s_l \in J_k} a_l y(s_l) = \sum_{s_l \in J_k} a_l z(s_l). \end{cases} \tag{6.61}
$$

Then, by Theorem 59, the solution of the constant-order initial–terminal value problem (ITVP) (6.61) takes the form:

$$
\overline{y}_k(s) = A_{\overline{y}} + q_3 \int_{\mathsf{T}_{k-1}}^{s} (s-\tau) f_{\overline{y}}(\tau) d\tau + \frac{q_4}{\Gamma(\ltimes_k)} \int_{\mathsf{T}_{k-1}}^{s} (s-\tau)^{\ltimes_k - 1} f_{\overline{y}}(\tau) d\tau.
$$

On the other hand, since $y(\mathsf{T}_{k-1}) = z(\mathsf{T}_{k-1}) = 0$ and $\sum_{s_l\in J_k} a_l y(s_l) = \sum_{s_l\in J_k} a_l z(s_l)$, it follows that $A_z = A_{\overline{y}}$. Then we have:

$$\overline{y}_k(s) = A_z + q_3 \int_{\mathsf{T}_{k-1}}^{s} (s-\tau) f_{\overline{y}}(\tau)d\tau + \frac{q_4}{\Gamma(\ltimes_k)} \int_{\mathsf{T}_{k-1}}^{s} (s-\tau)^{\ltimes_k - 1} f_{\overline{y}}(\tau)d\tau. \quad (6.62)$$

Let $s \in J_k$, where $k \in \{1, 2, ..., n\}$. According to Lemma 28 and equations (6.53), (6.59), (6.60), and (6.62), we obtain:

$$
\begin{aligned}
& |z(s) - y(s)| = |z(s) - y_k(s)| = |z(s) - \overline{y}_k(s)| \\
= & \left| z(s) - A_z - q_3 \int_{\mathsf{T}_{k-1}}^{s} (s-\tau) f_{\overline{y}}(\tau)d\tau - \frac{q_4}{\Gamma(\ltimes_k)} \int_{\mathsf{T}_{k-1}}^{s} (s-\tau)^{\ltimes_k - 1} f_{\overline{y}}(\tau)d\tau \right| \\
\leq & \left| z(s) - A_z - q_3 \int_{\mathsf{T}_{k-1}}^{s} (s-\tau) f_z(\tau)d\tau + \frac{q_4}{\Gamma(\ltimes_k)} \int_{\mathsf{T}_{k-1}}^{s} (s-\tau)^{\ltimes_k - 1} f_z(\tau)d\tau \right| \\
+ & q_3 \int_{\mathsf{T}_{k-1}}^{s} (s-\tau)|f_z(\tau) - f_{\overline{y}}(\tau)|d\tau + \frac{q_4}{\Gamma(\ltimes_k)} \int_{\mathsf{T}_{k-1}}^{s} (s-\tau)^{\ltimes_k - 1} |(f_z\tau) - f_{\overline{y}}(\tau)|d\tau \\
\leq & \epsilon L_k + K_f (1 + \mathcal{G}) \|z - y\| \left(q_3 \int_{\mathsf{T}_{k-1}}^{s} (s-\tau)d\tau + \frac{q_4}{\Gamma(\ltimes_k)} \int_{\mathsf{T}_{k-1}}^{s} (s-\tau)^{\ltimes_k - 1} d\tau \right) \\
\leq & \epsilon L_k + K_f (1 + \mathcal{G}) \left(\frac{q_3}{2} (\mathsf{T}_k - \mathsf{T}_{k-1})^2 + \frac{q_4}{\Gamma(\ltimes_k + 1)} (\mathsf{T}_k - \mathsf{T}_{k-1})^{\ltimes_k} \right) \|z - y\| \\
\leq & \epsilon L_k + K_f (1 + \mathcal{G}) L_k \|z - y\| \\
\leq & \epsilon L_k + \nu \|z - y\|,
\end{aligned}
$$

where

$$\nu = L_k K_f (1 + \mathcal{G}).$$

Then

$$\|z - y\|(1 - \nu) \leq \epsilon L_k,$$

and so by assuming $c_f := \frac{L_k}{(1-\nu)}$,

$$\|z - y\| \leq c_f \epsilon,$$

i.e.,

$$|z(s) - y(s)| \leq c_f \epsilon, \quad s \in [\mathsf{T}_{k-1}, \mathsf{T}_k].$$

Finally, Definition 43 implies that the Riemann–Liouville fractional variable order multiterm initial–terminal value problem (ITVP) (6.30), in the sense of Atangana–Baleanu, is Ulam–Hyers stable.

6.3.3 Numerical Approaches

Numerical Approach 1 Consider the Riemann–Liouville fractional variable order multiterm initial–terminal value problem (ITVP) in the sense of Atangana–Baleanu given by:

$$\begin{cases} {}^{ABR}F_{0+}^{\ltimes(s)} y(s) = \frac{1}{100(1+|y(s)|+|^{AB}I_{0+}^{\ltimes(s)} y(s))|}, & s \in J :=]0,2], \\ y(0)=0, \quad y(2) = \frac{1}{2}y(\frac{1}{2}) + \frac{2}{3}y(\frac{3}{2}), & \end{cases} \tag{6.63}$$

where the variable order $\ltimes(s)$ is defined as:

$$\ltimes(s) = \begin{cases} \frac{7}{3}, & s \in J_1 := [0,1], \\ \frac{8}{3}, & s \in J_2 :=]1,2]. \end{cases} \tag{6.64}$$

Let

$$f(s,y,z) = \frac{1}{100(1+|y|+|z|)}, \quad (s,y,z) \in [0,2] \times \mathbb{R} \times \mathbb{R}.$$

Let $y_1, y_2, z_1, z_2 \in \mathbb{R}$ and $s \in J$. Then, we have:

$$\begin{aligned} |f(s,y_1,z_1) - f(s,y_2,z_2)| &= \left| \left(\frac{1}{100(1+|y_1|+|z_1|)} - \frac{1}{100(1+|y_2|+|z_2|)} \right) \right| \\ &\leq \frac{||y_2|+|z_2|-|y_1|-|z_1||}{100(1+|y_1|+|z_1|)(1+|y_2|+|z_2|)} \\ &\leq \frac{1}{100}(|y_1-y_2|+|z_1-z_2|). \end{aligned}$$

Hence the condition **(H2)** holds with $K_f = \frac{1}{100}$.

By equation (6.64), and according to (6.33), we design two auxiliary Riemann–Liouville fractional constant-order multiterm initial–terminal value problems (ITVPs) in the sense of Atangana–Baleanu as follows:

$$\begin{cases} {}^{ABR}F_{0+}^{\frac{7}{3}} y(s) = \frac{1}{100(1+|\xi(s)|+|^{AB}I_{0+}^{\frac{7}{3}} \xi(s)|)}, & s \in J_1, \\ y(0)=0, \ \xi(1) = \frac{1}{2}y(\frac{1}{2}), & \end{cases} \tag{6.65}$$

and

$$\begin{cases} {}^{ABR}F_{1+}^{\frac{8}{3}} y(s) = \frac{1}{100(1+|\xi(s)|+|^{AB}I_{1+}^{\frac{8}{3}} \xi(s)|)}, & s \in J_2, \\ y(1)=0, \ y(2) = \frac{2}{3}\xi(\frac{3}{2}). & \end{cases} \tag{6.66}$$

We will now confirm that condition (6.41) is satisfied for $k = 1$. The necessary calculations are as follows:

$$A = \frac{1}{(1-0) - \frac{1}{2}(\frac{1}{2}-0)} = \frac{4}{3}, \quad N(\ltimes_1 - 2) = 1 - (\frac{7}{3} - 2) + \frac{\frac{7}{3} - 2}{\Gamma(\frac{7}{3} - 2)} = 0.79109,$$

$$q_1 := \frac{4}{3}\frac{(3 - \frac{7}{3})}{N(\ltimes_1 - 2)} = 1.12361, \quad q_2 := \frac{\frac{4}{3}(\frac{7}{3} - 2)}{N(\ltimes_1 - 2)} = 0.56180,$$

$$q_3 := \frac{3 - \frac{7}{3}}{N(\ltimes_1 - 2)} = 0.84271, \quad q_4 := \frac{\frac{7}{3} - 2}{N(\ltimes_1 - 2)} = 0.42135,$$

$$\mathcal{G} := \frac{(3 - \frac{7}{3})(1-0)^2}{2N(\ltimes_1 - 2)} + \frac{(\frac{7}{3} - 2)(1-0)^{\frac{7}{3}}}{N(\ltimes_1 - 2)\Gamma(\frac{7}{3} + 1)} = 0.29353,$$

and

$$\begin{aligned} L_1 &= \frac{q_1}{2}(1-0)\left(\frac{1}{2}(\frac{1}{2} - 0)^2 + (1-0)^2\right) \\ &+ \frac{q_2(1-0)}{\Gamma(\frac{7}{3}+1)}\left(\frac{1}{2}(\frac{1}{2} - 0)^{\frac{7}{3}} + (1-0)^{\frac{7}{3}}\right) \\ &+ \frac{q_3}{2}(1-0)^2 + \frac{q_4}{\Gamma(\frac{7}{3}+1)}(1-0)^{\frac{7}{3}} = 1.42733. \end{aligned} \tag{6.67}$$

Hence,

$$L_1 K_f\,(1 + \mathcal{G}) \simeq 0.018462 < 1.$$

Clearly, the condition (6.41) is satisfied. From Theorem 60, the problem (6.65) admits a solution $y_1 \in C([0,1], \mathbb{R})$, where

$$y_1(s) = \overline{y}_1(s), \quad s \in J_1.$$

In the next step, we investigate the satisfaction of condition (6.41) for $k = 2$. Indeed, we first compute the following values:

$$A = \frac{1}{(2-1) - \frac{2}{3}(\frac{3}{2}-1)} = \frac{3}{2}, \quad N(\ltimes_2 - 2) = 1 - (\frac{8}{3} - 2) + \frac{\frac{8}{3} - 2}{\Gamma(\frac{8}{3} - 2)} = 0.8256587410811,$$

$$q_1 := \frac{3}{2}\frac{(3 - \frac{8}{3})}{N(\ltimes_2 - 2)} 0.302788534246790, \quad q_2 := \frac{\frac{3}{2}(\frac{8}{3} - 2)}{N(\ltimes_2 - 2)} = 0.60557706849359,$$

$$q_3 := \frac{3 - \frac{8}{3}}{N(\ltimes_2 - 2)} = 0.40371804566239, \quad q_4 := \frac{\frac{8}{3} - 2}{N(\ltimes_2 - 2)} = 0.80743609132479,$$

$$\mathcal{G} := \frac{(3 - \frac{8}{3})(2-1)^2}{2N(\ltimes_2 - 2)} + \frac{(\frac{8}{3} - 2)(2-1)^{\frac{8}{3}}}{N(\ltimes_2 - 2)\Gamma(\frac{8}{3} + 1)} = 0.33885494976913,$$

and

$$\begin{aligned} L_2 &= \frac{q_1}{2}(2-1)\left(\frac{2}{3}(\frac{3}{2}-1)^2+(2-1)^2\right) \\ &+ \frac{q_2(2-1)}{\Gamma(\frac{8}{3}+1)}\left(\frac{2}{3}(\frac{3}{2}-1)^{\frac{8}{3}}+(2-1)^{\frac{8}{3}}\right) \\ &+ \frac{q_3}{2}(2-1)^2+\frac{q_4}{\Gamma(\frac{8}{3}+1)}(2-1)^{\frac{8}{3}} = 0.44418362357126. \quad (6.68) \end{aligned}$$

Hence,

$$L_2K_f\,(1+\mathcal{G}) \simeq 0.00594697443025 < 1.$$

Thus, the condition (6.41) is satisfied.

It follows from the conclusion of Theorem 60 that the auxiliary Riemann–Liouville fractional constant-order multiterm initial–terminal value problem (ITVP) (6.66), in the sense of Atangana–Baleanu, possesses a solution $y_2 \in C([0,2],\mathbb{R})$, where

$$y_2(s) = \begin{cases} 0, & s \in [0,1], \\ \overline{y}_2(s), & s \in J_2. \end{cases}$$

Therefore, by Theorem 61, the Riemann–Liouville fractional variable-order multiterm initial–terminal value problem (ITVP) (6.63), in the sense of Atangana–Baleanu, admits a solution given by:

$$y(s) = \begin{cases} y_1(s), & s \in J_1, \\ y_2(s) = \begin{cases} 0, & s \in J_1, \\ \overline{y}_2(s), & s \in J_2, \end{cases} \end{cases}$$

Moreover, according to Theorem 62, the Riemann–Liouville fractional variable-order multiterm initial–terminal value problem (ITVP) (6.63), in the sense of Atangana-Baleanu is Ulam-Hyers stable.

Numerical approach 2 Consider the Riemann–Liouville fractional variable-order multiterm initial–terminal value problem (ITVP), in the sense of Atangana-Baleanu as

$$\begin{cases} {}^{ABR}\mathcal{F}_{0+}^{\ltimes(s)}y(s) = \frac{1}{(s+4)(1+|y(s)|+|{}^{AB}I_{0+}^{\ltimes(s)}y(s)|)}, & s \in J :=]0,2], \\ y(0)=0, \quad y(2) = \frac{1}{4}y(\frac{1}{4})+\frac{1}{2}y(\frac{1}{2})+\frac{2}{3}y(\frac{3}{2}), \end{cases} \quad (6.69)$$

where

$$\ltimes(s) = \begin{cases} \frac{7}{3}, & s \in J_1 := [0,1], \\ \frac{8}{3}, & s \in J_2 :=]1,2]. \end{cases} \quad (6.70)$$

Let

$$f(s,y,z)=\frac{1}{(s+4)(1+|y|+|z|)},\ (s,y,z)\in[0,2]\times\mathbb{R}\times\mathbb{R}.$$

Let $y_1,y_2,z_1,z_2\in\mathbb{R}$ and $s\in J$. Then, we have

$$\begin{aligned}|f(s,y_1,z_1)-f(s,y_2,z_2)| &= \left|\left(\frac{1}{(s+4)(1+|y_1|+|z_1|)}-\frac{1}{(s+4)(1+|y_2|+|z_2|)}\right)\right|\\ &\le \frac{||y_2|+|z_2|-|y_1|-|z_1||}{(s+4)(1+|y_1|+|z_1|)(1+|y_2|+|z_2|)}\\ &\le \frac{1}{4}(|y_1-y_2|+|z_1-z_2|).\end{aligned}$$

Hence the condition **(H2)** holds with $K_f=\frac{1}{4}$.

By (6.70), and according to (6.33) we design two auxiliary Riemann–Liouville fractional constant-order multiterm initial–terminal value problems (ITVPs) in the sense of Atangana–Baleanu as follows:

$$\begin{cases} {}^{ABR}\mathcal{F}_{0+}^{\frac{7}{3}}y(s)=\frac{1}{(s+4)(1+|\xi(s)|+|{}^{AB}I_{0+}^{\frac{7}{3}}\xi(s)|)}, & s\in J_1,\\ y(0)=0,\ y(1)=\frac{1}{4}y(\frac{1}{4})+\frac{1}{2}y(\frac{1}{2}), \end{cases} \tag{6.71}$$

and

$$\begin{cases} {}^{ABR}\mathcal{F}_{1+}^{\frac{8}{3}}y(s)=\frac{1}{(s+4)(1+|\xi(s)|+|{}^{AB}I_{1+}^{\frac{8}{3}}\xi(s)|)}, & s\in J_2,\\ y(1)=0,\ y(2)=\frac{2}{3}\xi(\frac{3}{2}). \end{cases} \tag{6.72}$$

Next, we verify that the condition (6.41) holds for $k=1$ by calculating the following values:

$$A=\frac{1}{(1-0)-\frac{1}{4}(\frac{1}{4}-0)-\frac{1}{2}(\frac{1}{2}-0)}=\frac{16}{11},\ N(\ltimes_1-2)=1-(\frac{7}{3}-2)+\frac{\frac{7}{3}-2}{\Gamma(\frac{7}{3}-2)}=0.79109,$$

$$q_1:=\frac{4}{3}\frac{(3-\frac{7}{3})}{N(\ltimes_1-2)}=1.032982,\ q_2:=\frac{\frac{4}{3}(\frac{7}{3}-2)}{N(\ltimes_1-2)}=0.612886,$$

$$q_3:=\frac{3-\frac{7}{3}}{N(\ltimes_1-2)}=0.84271,\ q_4:=\frac{\frac{7}{3}-2}{N(\ltimes_1-2)}=0.42135,$$

$$\mathcal{G}:=\frac{(3-\frac{7}{3})(1-0)^2}{2N(\ltimes_1-2)}+\frac{(\frac{7}{3}-2)(1-0)^{\frac{7}{3}}}{N(\ltimes_1-2)\Gamma(\frac{7}{3}+1)}=0.29353,$$

and

$$\begin{aligned} L_1 &= \frac{q_1}{2}(1-0)\left(\frac{1}{4}(\frac{1}{4}-0)^2+\frac{1}{2}(\frac{1}{2}-0)^2+(1-0)^2\right) \\ &+ \frac{q_2(1-0)}{\Gamma(\frac{7}{3}+1)}\left(\frac{1}{4}(\frac{1}{4}-0)^{\frac{7}{3}}+\frac{1}{2}(\frac{1}{2}-0)^{\frac{7}{3}}+(1-0)^{\frac{7}{3}}\right) \\ &+ \frac{q_3}{2}(1-0)^2+\frac{q_4}{\Gamma(\frac{7}{3}+1)}(1-0)^{\frac{7}{3}} = 1.406810. \end{aligned} \tag{6.73}$$

Hence,

$$L_1K_f\,(1+\mathcal{G}) \simeq 0.49989239 < 1.$$

Evidently the condition (6.41) is satisfied. From Theorem 60, the problem (6.71) admits a solution $y_1 \in C([0,1],\mathbb{R})$, where

$$y_1(s) = \overline{y}_1(s), \quad s \in J_1.$$

We proceed by verifying the satisfaction of condition (6.41) for $k = 2$. Indeed, first, we first compute the following values:

$$A = \frac{1}{(2-1)-\frac{2}{3}(\frac{3}{2}-1)} = \frac{3}{2}, \quad N(\ltimes_2-2) = 1-(\frac{8}{3}-2)+\frac{\frac{8}{3}-2}{\Gamma(\frac{8}{3}-2)} = 0.8256587410811,$$

$$q_1 := \frac{3}{2}\frac{(3-\frac{8}{3})}{N(\ltimes_2-2)}0.302788534246790, \; q_2 := \frac{\frac{3}{2}(\frac{8}{3}-2)}{N(\ltimes_2-2)} = 0.60557706849359,$$

$$q_3 := \frac{3-\frac{8}{3}}{N(\ltimes_2-2)} = 0.40371804566239, \; q_4 := \frac{\frac{8}{3}-2}{N(\ltimes_2-2)} = 0.80743609132479,$$

$$\mathcal{G} := \frac{(3-\frac{8}{3})(2-1)^2}{2N(\ltimes_2-2)} + \frac{(\frac{8}{3}-2)(2-1)^{\frac{8}{3}}}{N(\ltimes_2-2)\Gamma(\frac{8}{3}+1)} = 0.33885494976913,$$

and

$$\begin{aligned} L_2 &= \frac{q_1}{2}(2-1)\left(\frac{2}{3}(\frac{3}{2}-1)^2+(2-1)^2\right) \\ &+ \frac{q_2(2-1)}{\Gamma(\frac{8}{3}+1)}\left(\frac{2}{3}(\frac{3}{2}-1)^{\frac{8}{3}}+(2-1)^{\frac{8}{3}}\right) \\ &+ \frac{q_3}{2}(2-1)^2+\frac{q_4}{\Gamma(\frac{8}{3}+1)}(2-1)^{\frac{8}{3}} = 0.44418362357126. \end{aligned} \tag{6.74}$$

Hence,

$$L_2K_f\,(1+\mathcal{G}) \simeq 0.148674 < 1.$$

Thus, the condition (6.41) is satisfied.

By the conclusion of Theorem 60, the auxiliary Riemann–Liouville fractional constant-order multiterm initial–terminal value problem (ITVP) (6.72), in the sense of Atangana-Baleanu, possesses a solution $y_2 \in C([0,2],\mathbb{R})$, where

$$y_2(s) = \begin{cases} 0, & s \in [0,1], \\ \overline{y}_2(s), & s \in J_2. \end{cases}$$

Therefore, by Theorem 61, the Riemann–Liouville fractional variable-order multiterm initial–terminal value problem (ITVP) (6.69), in the sense of Atangana-Baleanu, admits a solution as

$$y(s) = \begin{cases} y_1(s), & s \in J_1, \\ y_2(s) = \begin{cases} 0, & s \in J_1, \\ \overline{y}_2(s), & s \in J_2, \end{cases} \end{cases}$$

Moreover, according to Theorem 62, the Riemann–Liouville fractional variable-order multiterm initial–terminal value problem (ITVP) (6.69), in the sense of Atangana-Baleanu is Ulam-Hyers stable.

7

Conclusion

Recent advances in variable-order fractional calculus have attracted growing interest across scientific disciplines, leading to a surge in research exploring its unique modeling capabilities. While prior reviews have largely focused on mathematical foundations for specialized audiences, this review aims to serve the broader scientific community by highlighting its diverse applications in science and engineering.

This book focuses on applying variable-order fractional calculus to the numerical modeling of complex physical systems. Despite ongoing mathematical challenges limiting its broader adoption, we aim to support the development of tailored numerical methods that unlock its potential. A key step is establishing closed-form links between the operators' mathematical properties and the system's physical parameters—essential for making variable-order models a practical tool in simulation.

This book aims to generalize the study of stability, uniqueness, and existence of solutions for differential equations involving both constant and variable-order fractional derivatives. To achieve this, we introduced abstract variable-order fractional models and partitioned them using generalized intervals. With piecewise-constant functions and Green's functions, these were transformed into equivalent constant-order problems. Using fixed point theory, upper–lower solution methods, and measures of noncompactness, we established Ulam–Hyers and Ulam–Hyers–Rassias stability results. Approximate solutions were also illustrated through examples using MATLAB® and Euler's discretization.

DOI: 10.1201/9781003687900-7

The findings in this book highlight strong potential for multidisciplinary applications and may inspire future research in this evolving field. Moving forward, we aim to explore initial and terminal value problems under integral, integro-derivative, nonlocal, and infinite delay conditions. These results can be extended to other variable-order fractional models, including those used in simulating COVID-19 dynamics. We hope this book serves as a valuable resource for researchers, authors, and readers in the scientific community.

References

[1] S. Abbas and Benchohra. M. On the generalized Ulam-Hyers-Rassias stability for Darboux problem for partial fractional implicit differential equations. *Applied Mathematics E-Notes*, pages 20–28, 2014.

[2] T. Abdeljawad. A Lyapunov type inequality for fractional operators with nonsingular Mittag-Leffler kernel. *Journal of Inequalities and Applications*, pages 1–11, 2017.

[3] M. Abramowitz and I. Stegun. *Handbook of Mathematical Functions.* Dover, New York, 1964.

[4] R. A. Adams. *Sobolev Spaces.* Academic Press, 1975.

[5] R. A. Adams and J. Fournier. *Sobolev Spaces, Pure and Applied Mathematics* . Boston, MA: Academic Press, 2003.

[6] R.R. Akhmerov, M.I. Kamenskii, A.S. Patapov, A.E. Rodkina, and B.N. Sadovskii. *Birkhauser Verlag, Basel.* Measures of Noncompactness and Condensing Operators, trans. from the Russian by A. Iacob, 1992.

[7] A. Almeida and S. Samko. Fractional and hypersingular operators in variable exponent spaces on metric measure spaces. *Mediterr. J. Math*, 6:215–232, 2009.

[8] R. Almeida. Caputo-Hadamard fractional derivatives of variable order. *Numer. Funct. Anal. Optim.*, 38(1):1–19, 2017.

[9] R. Almeida, R. Bastos, and M.T.T. Monteiro. A fractional Malthusian growth model with variable order using an optimization approach. *Stat. Optim. Inf. Comput.*, 6(1):4–11, 2018.

[10] R. Almeida and D.F. M. Torres. Computing Hadamard type operators of variable fractional order. *Applied Mathematics and Computation*, 257:74–88, 2015.

[11] C. Alsina and R. Ger. On some inequalities and stability results related to the exponential function. *J. Inequal. Appl.*, 2:373–380, 1998.

[12] J.C. Alvarez. Measure of noncompactness and fixed points of nonexpansive condensing mappings in locally convex spaces. *Rev. Real. Acad. Cienc. Exact. Fis. Natur., Madrid*, 79:53–66, 1985.

[13] T. Aoki. On the stability of the linear transformation in Banach spaces. *J. Math. Soc. Japan*, 2:64–66, 1950.

[14] A. Atangana and D. Baleanu. New fractional derivatives with nonlocal and non-singular kernel: Theory and application to heat transfer model. *Therm. Sci.*, 20(2):763–769, 2016.

[15] A. Ayache and P.R. Bertrand. A process very similar to multifractional Brownian motion. *Recent Developments in Fractals and Related Fields*, pages 311–326, 2010.

[16] A. Ayache and J.L. Vehel. On the identification of the pointwise Holder exponent of the generalized multifractional Brownian motion. *Stoch. Process. their Appl.*, 111:119–156, 2004.

[17] J. Banas. Applications of measures of weak noncompactness and some classes of operators in the theory of functional equations in the Lebesgue space. *Nonlinear Anal.*, 30:3283–3293, 1997.

[18] J. Banas and K. Goebel. *Measures of Noncompactness in Banach Spaces.* Lecture Notes in Pure and Applied Mathematics, 60, Marcel Dekker, New York, 1980.

[19] J. Banas and L. Olszowy. Measures of noncompactness related to monotonicity. *Comment. Math. Prace. Mat.*, 41:13–23, 2001.

[20] J. Banas and B. Rzepka. An application of a measure of noncompactness in the study of asymptotique stability. *Appl. Math. Lett.*, 16:1–6, 2003.

[21] J. Banas and K. Sadarangani. On some measures of noncompactness in the space of continuous functions. *Nonlinear Anal.*, 68:377–383, 2006.

[22] M. Benchohra, S. Bouriah, and J. Henderson. Existence and stability results for nonlinear implicit neutral fractional differential equations with finite delay and impulses. *Communications on Applied Nonlinear Analysis*, 22(1):46–67, 2015.

[23] M. Benchohra, S. Bouriah, J.E. Lazreg, and J.J. Nieto. Nonlinear Implicit Hadamard's Fractional Differential Equations with Delay in Banach Space. *Acta Univ. Palacki. Olomuc. Fac. Rer. Nat. Mathematica*, 55(1):15–26, 2016.

[24] A. Benkerrouche, D. Baleanu, M. S. Souid, A. Hakem, and M. Inc. Boundary value problem for nonlinear fractional differential equations of variable order via Kuratowski MNC technique. *Advances in Difference Equations*, 365:1–19, 2021.

[25] A. Benkerrouche, S. Etemad, M. S. Souid, S. Rezapour, H. Ahmad, and T. Botmart. Fractional variable order differential equations with impulses: A study on the stability and existence properties. *AIMS Mathematics*, 8:775–791, 2022.

[26] A. Benkerrouche, M.S. Souid, S. Chandok, and A. Hakem. Existence and stability of a Caputo variable-order boundary value problem. *Journal of Mathematics*, 2021:1– 16, 2021.

[27] A. Benkerrouche, M.S. Souid, S. Etemad, A. Hakem, P. Agarwal, S. Rezapour, S.K. Ntouyas, and J. Tariboon. Qualitative study on solutions of a Hadamard variable order boundary problem via the Ulam-Hyers-Rassias stability. *Fractal and Fractional*, 108:1–20, 2021.

[28] A. Benkerrouche, M.S. Souid, Jarad F., and A. Hakem. On boundary value problems of Caputo fractional differential equation of variable order via Kuratowski MNC technique. *Advances in Continuous and Discrete Models: Theory and Applications*, 43:1–19, 2022.

[29] A. Benkerrouche, M.S. Souid, E. Karapinar, and A. Hakem. On the boundary value problems of Hadamard fractional differential equations of variable order. *Math. Meth. Appl. Sci.*, 46(3):1–17, 2022.

[30] A. Benkerrouche, M.S. Souid, K. Sitthithakerngkiet, and A. Hakem. Implicit nonlinear fractional differential equations of variable order. *Boundary Value Problems*, 64:1– 16, 2021.

[31] A. Benkerrouche, M.S. Souid, G. Stamov, and I. Stamova. Multiterm impulsive Caputo-Hadamard type differential equations of fractional variable order. *Axioms*, 634:1–18, 2022.

[32] A. Benkerrouche, M.S. Souid, G. Stamov, and I. Stamova. On the solutions of a quadratic integral equation of the Urysohn type of fractional variable order. *Entropy*, 886:1–14, 2022.

[33] Z. Bouazza, S Sabit, S. Etemad, M. S. Souid, A Akgul, S. Rezapour, and M De la Sen. On the Caputo-Hadamard Fractional IVP with Variable Order Using the Upper-lower Solutions Technique. *AIMS Mathematics*, 8:5484–5501, 2022.

[34] Z. Bouazza, M. S. Souid, C. H. C. Hussin, A. Mandangan, and S Sabit. Variable-order Implicit Fractional Differential Equations based on the Kuratowski MNC Technique. *Malaysian Journal of Mathematical Sciences*, 17:305–332, 2023.

[35] T. A. Burton. *Stability by Fixed Point Theory for Functional Differential Equations.* Dover Publictions, New York, 2006.

[36] A. Charef and D. Idiou. Design of analog variable fractional order differentiator and integrator. *Nonlinear Dyn.*, 69:1577–1588, 2012.

[37] W. Chen and S. Holm. Modified Szabo's wave equation models for lossy media obeying frequency power law. *J. Acoust. Soc. Am.*, 114:2570–2574, 2003.

[38] Y. J. Cho, Th. M. Rassias, and R. Saadati. *Science-Business Media, Springer.* Dover Publictions, New York, 2013.

[39] C. F. M. Coimbra. Mechanics with variable-order differential operators. *Ann. Phys.*, 12:692–703, 2003.

[40] A. Coronel-Escamilla, J.F. Gomez-Aguilar, L. Torres, and R. Escobar-Jiménez. A numerical solution for a variable-order reaction-diffusion model by using fractional derivatives with non-local and non-singular kernel. *Physica. A*, 491:406–424, 2018.

[41] G. Diaz and C. F. M. Coimbra. Nonlinear dynamics and control of a variable order oscillator with application to the van der Pol equation. *Nonlinear Dynam.*, 56:145–157, 2009.

[42] P. Gavruta. A generalization of the Hyers-Ulam-Rassias stability of approximately additive mappings. *J. Math. Anal. Appl.*, 184:431–436, 1994.

[43] B. Ghanbari and J.F. Gomez-Aguilar. Modeling the dynamics of nutrient-phytoplanktonzooplankton system with variable-order fractional derivatives. *Chaos Solitons Fractals*, 116:114–120, 2018.

[44] J. F. Gomez-Aguilar. Analytical and numerical solutions of a nonlinear alcoholism model via variable-order fractional differential equations. *Physica. A*, 494:52–75, 2018.

[45] A. Granas and J. Dugundji. *Fixed point theory.* Springer, New York, 2003.

[46] D. Guo, V. Lakshmikantham, and X. Liu. *Nonlinear Integral Equations in Abstract Spaces, Mathematics and its Applications.* Kluwer Academic Publishers Group, Dordrecht, 1996.

[47] D. Guo and V. Lakshmikanthan. *Nonlinear Problems in Abstract Cones.* Academic Press, New York, 1988.

[48] J. P. Hollkamp, M. Semperlotti, and F. Sen. Model-order reduction of lumped parameter systems via fractional calculus. *Journal of Sound and Vibration*, 419:526– 543, 2018.

[49] Ch. Horvath. Measure of Non-compactness and multivalued mappings in complete metric topological spaces. *J. Math. Anal. Appl.*, 108:403–408, 1985.

[50] S. Hristova, A. Benkerrouche, M. S. Souid, and A. Hakem. Boundary value problems of Hadamard fractional differential equations of variable order. *Symmetry*, 13:1– 16, 2021.

[51] D. H. Hyers. On the stability of the linear functional equation. *Proc. Natl. Acad. Sci, USA*, 27:222–224, 1941.

[52] D. H. Hyers, G. Isac, and T. M. Rassias. *Stability of functional equations in several variables.* Birkhuser Boston. Inc, Boston, MA, 1998.

[53] W. R. Ibrahim. Stability for Univalent Solutions of Complex Fractional Differential Equations. *Proceedings of the Pakistan Academy of Sciences*, 49(3):227–232, 2012.

[54] P. Ivady. A note on a gamma function inequality. *J. Math. Inequal.*, 3:227–236, 2009.

[55] A. Jiahui and C. Pengyu. Uniqueness of solutions to initial value problem of fractional differential equations of variable-order. *Dynamic Systems and Applications*, 28:607–623, 2019.

[56] K.W. Jun and H.M. Kim. On the stability of an n-dimensional quadratic and additive functional equation. *Math. Inequal. Appl.*, 19:854–858, 2006.

[57] S. M. Jung. On the Hyers-Ulam stability of the functional equations that have the quadratic property. *J. Math. Anal. Appl.*, 222:126–137, 1998.

[58] S. M. Jung. Hyers-Ulam stability of linear differential equations of first order. *Appl. Math. Lett.*, 17:1135–1140, 2004.

[59] S.M. Jung. *Hyers-Ulam-Rassias stability of functional equations in mathematical analysis.* Hadronic Press, 2001.

[60] S.M. Jung. Hyers-Ulam stability of linear differential equations of first order, II. *Appl. Math. Lett.*, 19:854–858, 2006.

[61] S.M. Jung and K.S. Lee. Hyers-Ulam stability of first order linear partial differential equations with constant coefficients. *Math. Inequal. Appl.*, 10:261–266, 2007.

[62] K. A. M. Kaabar, A. Refice, M. S. Souid, F. Martinez, S. Etemad, Z. Siri, and S. Rezapour. Existence and U-H-R stability of solutions to the implicit nonlinear FBVP in the variable order settings. *Mathematics*, 9:1–17, 2021.

[63] A.A. Kilbas, H. M. Srivastava, and J.J. Trujillo. *Theory and Applications of Fractional Differenatial Equations. North-Holland Mathematics Studies.* Elsevier Science B.V., Amsterdam, 2006.

[64] G. H. Kim. On the stability of functional equations with square-symmetric operation. *Math. Inequal. Appl.*, 17:257–266, 2001.

[65] W.A. Kirk and B. Sims. *Handbook of Metric Fixed Point Theory, Springer-Science, Business Media.* B.V, Dordrecht, 2001.

[66] K. Lan and J. Webb. Positive solutions of semilinear differential equations with singularities. *J. Differ. Equ.*, 148(2):407–421, 1998.

[67] C.F. Lorenzo and T.T. Hartley. Variabe order and distributed order fractional operators. *Nonlinear Dynamics*, 29:57–98, 2002.

[68] T. Miura, S. Miyajima, and S. E. Takahasi. A characterization of Hyers-Ulam stability of first order linear differential operators. *J. Math. Anal. Appl*, 286:136–146, 2003.

[69] J. P. Neto, R. M. Coelho, D. Valério, S. Vinga, D. Sierociuk, W. Malesza, M. Macias, and A. Dzielinski. Variable order differential models of bone remodelling. *IFAC*, 50:8066–8071, 2017.

[70] J. P. Neto, R. M. Coelho, D. Valério, S. Vinga, D. Sierociuk, W. Malesza, M. Macias, and A. Dzielinski. Simplifying biochemical tumorous bone remodeling models through variable order derivatives. *Comput. Math. Appl.*, 75:3147–3157, 2018.

[71] A. D. Obembe, M. E. Hossain, and S. A. Abu-Khamsin. Variable-order derivative time fractional diffusion model for heterogeneous porous media. *J. Pet. Sci. Eng.*, 152:391–405, 2017.

[72] M. Obloza. Connections between Hyers and Lyapunov stability of the ordinary differential equations. *Rocznik Nauk.-Dydakt. Prace. Mat.*, 14:141–146, 1997.

[73] T. Odzijewicz, A. B. Malinowska, and D.F.M. Torres. Fractional variational calculus of variable order . *Advances in Harmonic Analysis and Operator Theory*, 14:1–12, 2012.

[74] S. Patnaik, J. P. Hollkamp, and F. Semperlotti. Applications of variable-order fractional operators: A review. *Proc. R. Soc. A. 476. 20190498*, pages 1–32, 2020.

[75] H.T.C. Pedro, M.H. Kobayashi, J.M.C. Pereira, and C.F.M. Coimbra. Variable order modeling of diffusive-convective effects on the oscillatory flow past a sphere. *J. Vib. Control*, 14:1569–1672, 2008.

[76] I. Podlubny. *Fractional Differential Equations.* Academic Press, New York, USA, 1999, Science B. V, 2006.

[77] Almeida R., D. Tavares, and D.F.M. Torres. *The Variable-Order Fractional Calculus of Variations.* Springer, 2019.

[78] L.E.S. Ramirez and C.F.M. Coimbra. On the selection and meaning of variable order operators for dynamic modeling. *Int. J. Differ. Equ.*, 2010:1–16, 2010.

[79] L.E.S. Ramirez and C.F.M. Coimbra. On the variable order dynamics of the nonlinear wake caused by a sedimenting particle. *Physica D: Nonlinear Phenomena*, 240:1111–1118, 2011.

[80] M. R. Rapaic and A. Pisano. Variable-order fractional operators for adaptive order and parameter estimation. *IEEE Trans. Autom. Control.*, 59:798– 803, 2013.

[81] J.M. Rassias. *Functional Equations, Difference Inequalities and Ulam Stability Notions (F.U.N.).* Nova Science Publishers, Inc, New York, 2010.

[82] T. M. Rassias and J. Brzdek. *Functional Equations in Mathematical Analysis.* pringer, 86, New York. Dordrecht. Heidelberg. London, 2012.

[83] T.M. Rassias. On the stability of the linear mapping in Banach spaces. *Proc. Amer. Math. Soc.*, 72:297–300, 1978.

[84] A. Refice, M. S. Souid, and I. Stamova. On the boundary value problems of Hadamard fractional differential equations of variable order via Kuratowski MNC technique. *Mathematics*, 1134:1–16, 2021.

[85] A. Refice, M. S. Souid, and A. Yakar. Some qualitative properties of nonlinear fractional integro-differential equations of variable order. *An International Journal of Optimization and Control: Theories, Applications*, 11:68–78, 2021.

[86] S. Rezapour, M. S. Souid, Z. Bouazza, A. Hussain, and S. Etemad. On the fractional variable order thermostat model: existence theory on cones via piece-wise constant functions. *Journal of Function Spaces*, 22:1–11, 2022.

[87] B. Ross and S.G. Samko. Fractional integration operator of a variable order in the Holder spaces $H^{\{}\lambda(x)\}(x)$. *Internat. J. Math. Math. Sci.*, 18:777–788, 1995.

[88] I.A. Rus. Ulam stabilities of ordinary differential equations in a Banach space. *Carpathian J. Math.*, 26:103–107, 2010.

[89] E. A. Saied. Anomalous diffusion on fractal objects, additional analytic solutions. *Chaos Solitons Fractals*, 11:1369–1376, 2000.

[90] S. Samko, A. A. Kilbas, and O. Marichev. *Fractional Integrals and Derivatives, Theory and Applications.* Gordon and Breach, Amsterdam, 1993.

[91] S.G. Samko. Fractional integration and differentiation of variable order. *Anal. Math.*, 21:213–236, 1995.

[92] S.G. Samko and B. Boss. Integration and differentiation to a variable fractional order. *Integr. Transforms. Spec. Funct.*, 1:277–300, 1993.

[93] F. Santamaria, S. Wils, E. De Schutter, and G. J. Augustine. Anomalous diffusion in Purkinje cell dendrites caused by spines. *Neuron*, 52:635–648, 2006.

[94] H. Sheng, H. G. Sun, C. Coopmans, Y.Q. Chen, and G.W. Bohannan. A Physical experimental study of variable-order fractional integrator and differentiator. *The European Physical Journal Special Topics*, 193:93–104, 2011.

[95] J. J. Shyu, S.C. Pei, and C. H. Chan. An iterative method for the design of variable fractional-order FIR differintegrators. *Signal Process.*, 89:320–327, 2009.

[96] D. R. Smart. *Fixed Point Theorems.* Cambridge University Press, 1980.

[97] M. S. Souid, Z. Bouazza, and A. Yakar. Existence, uniqueness, and stability of solutions to variable fractional order boundary value problems. *Journal of New Theory*, 39:100–111, 2022.

[98] H. G. Sun, Y.Q. Chen, and W. Chen. Variable-order fractional differential operators in anomalous diffusion modeling. *Physica. A*, 338:4586–4592, 2009.

[99] H.G. Sun, W. Chen, C. Li, and Y.Q. Chen. Fractional differential models for anomalous diffusion. *Physica. A*, 389:2719–2724, 2010.

[100] H.G. Sun, W. Chen, H. Sheng, and Y.Q. Chen. On mean square displacement behaviors of anomalous diffusions with variable and random orders. *Phys. Lett. A*, 374:906–910, 2010.

[101] H.G. Sun, W. Chen, H. Wei, and Y.Q. Chen. A comparative study of constant-order and variable-order fractional models in characterizing memory property of systems. *Eur. Phys. J. Special Topics*, 193:185–192, 2011.

[102] H.G. Sun, Y.Q. Chen, and W. Chen. Random-order fractional differential equation models. *Signal Process*, 91:525–530, 2011.

[103] H.G. Sun, Y. Zhang, W. Chen, and D.M. Reeves. Use of a variable-index fractional-derivative model to capture transient dispersion in heterogeneous media. *J. Contam. Hydrol.*, 157:47–58, 2014.

[104] N.H. Sweilam and S. M. Al-Mekhlafi. Numerical study for multi-strain tuberculosis (TB) model of variable-order fractional derivatives. *J. Adv. Res.*, 7:271–283, 2016.

[105] D. Tavares, R. Almeida, and D.F.M. Torres. Caputo derivatives of fractional variable order Numerical approximations. *Commun. Nonlinear Sci. Numer. Simul.*, 35:69–87, 2016.

[106] B. Telli, M. S. Souid, and I. Stamova. Boundary-Value Problem for Nonlinear Fractional Differential Equations of Variable Order with Finite Delay via Kuratowski Measure of Noncompact. *Axioms*, 35:69–87, 2016.

[107] C. C. Tseng. Series expansion design of variable fractional order integrator and differentiator using logarithm. *Signal Processing*, 88:2278–2292, 2008.

[108] C. C. Tseng. Design of variable and adaptive fractional order FIR differentiators. *Axioms*, 80:1–22, 2023.

[109] S. M. Ulam. *John Wiley and Sons, New York, USA*. Problems in Modern Mathematics, Chapter 6, 1940.

[110] S. M. Ulam. *A Collection of Mathematical Problems.* Interscience Publishers. Inc, New York, 1968.

[111] D. Valerio and J. S. Costa. Variable-order fractional derivatives and their numerical approximations. *Signal Processing*, 91:470–483, 2011.

[112] C. Vanterler da, J. Sousa, and E. Capelas de Oliverira. Two new fractional derivatives of variable order with non-singular kernel and fractional differential equation. *Comput. Appl. Math.*, 37:5375–5394, 2018.

[113] J. Wang, M. Fečkan, and Y. Zhou. Ulam's type stability of impulsive ordinary differential equations. *J. Math. Anal. Appl.*, 395:258–264, 2012.

[114] J. Wang, L. Lv, and Y. Zhou. Ulam stability and data dependence for fractional differential equations with Caputo derivative. *Electronic Journal of Qualitative Theory of Differential Equations*, 63:1–10, 2011.

[115] J. Wang and Y. Zhang. Existence and stability of solutions to nonlinear impulsive differential equations in β-normed spaces. *Electronic Journal of Differential Equations*, 83:1–10, 2014.

[116] G.C. Wu, D. Baleanu, H. P. Xie, and S. D. Zeng. Discrete fractional diffusion equation of chaotic order. *International Journal of Bifurcation and Chaos*, 26:1–6, 2016.

[117] Y. Xu, B. Telli, M. S. Souid, S. Etemad, J. Xu, and S. Rezapour. Stability on a boundary problem with RL-Fractional derivative in the senseof Atangana-Baleanu of variable-order. *AIMS Electronic Research Archive*, 32:134–159, 2023.

[118] J. Yang, H. Yao, and B. Wu. An efficient numberical method for variable order fractional functional differential equation. *Appl. Math. Lett.*, 76:221–226, 2018.

[119] L. Zhang, S. Hu. Unique existence result of approximate solution to initial value problem for fractional differential equation of variable order involving the derivative arguments on the half-axis. *Mathematics*, 286:1–23, 2019.

[120] S. Zhang. Existence of solutions for two point boundary value problems with singular differential equations of variable order. *Electronic Journal of Differential Equations*, 245:1–16, 2013.

[121] S. Zhang. The uniqueness result of solutions to initial value problems of differential equations of variable-order. *Rev. R. Acad. Cienc. Exactas Fs. Nat. Ser. A Math.*, 112:407–423, 2018.

[122] S. Zhang and L. Hu. The existeness and uniqueness result of solutions to initial value problems of nonlinear diffusion equations involving with the conformable variable. *Azerbaijan Journal of Mathematics*, 9:22–45, 2019.

[123] S. Zhang and L. Hu. The existence of solutions and generalized Lyapunov-type inequalities to boundary value problems of differential equations of variable order. *Aims Mathematics*, 5:2923–2943, 2020.

[124] S. Zhang and L. Sun, S. Hu. Approximate solutions to initial value problem for differential equation of variable order. *Journal of Fractional Calculus and Applications*, 9:93–112, 2018.

Index

G

H

I

K

M

N

O

P

R

S

T

U

V

About the Authors

Mohammed Said Souid is a Professor Doctor of Mathematics at the University of Tiaret, Algeria. He earned a magister degree in 2010 and a doctorate in mathematics in 2015 from Sidi Bel Abbes University, Algeria. Since 2010, he has been teaching at Ibn Khaldoun University in Tiaret. He specializes in nonlinear differential equations and fractional calculus, fields in which he has made significant research contributions. Throughout his academic career, Souid has developed a deep expertise in fractional differential equations, particularly focusing on variable-order operators and boundary value problems. His research has resulted in numerous publications that have advanced the understanding and applications of fractional calculus in mathematical analysis and applied sciences. Known for his dedication to education and research, Souid has mentored many students and contributed extensively to curriculum development in advanced mathematics. His work combines theoretical insights with practical applications, making him a respected figure in both the academic and research communities.

Mustafa Inc is a Professor of Applied Mathematics at Fırat University in Elazığ, Türkiye. He obtained a PhD in Applied Mathematics from Fırat University in 2002. Professor Inc has contributed extensively to the study of analytical and numerical methods for solving complex mathematical equations, optical soliton, Lie symmetry and conservation laws, including solitary wave solutions and generalized functions. He has published numerous scientific articles in international journals and is recognized for his expertise in applied numerical mathematics. His academic career involves teaching a variety of mathematics courses while actively engaging in research projects and supervising graduate students. Mustafa Inc is noted for his scholarly impact within the field of mathematical sciences, holding a notable position among top mathematicians in Turkey and globally. His continuous dedication to advancing applied mathematics has made significant contributions to both theoretical and practical aspects of the discipline.

Amar Benkerrouche is an Associate Professor at Djelfa University, Algeria, specializing in the applications of fractional calculus within the physical sciences. He has a PhD in mathematics, his work centers on modeling complex systems using advanced mathematical tools. Amar's research integrates fractional differential equations to better describe phenomena in diverse scientific fields, contributing valuable insights into the behavior of complex and nonlinear systems. His expertise extends beyond pure mathematics, encompassing interdisciplinary collaborations that connect mathematical theory with practical challenges in applied sciences. Throughout his career, he has participated in various projects that bridge the gap between abstract mathematical concepts and their real-world applications, demonstrating his commitment to advancing both theory and practice. Amar's work is marked by a strong focus on innovative modeling techniques, enabling more accurate and efficient representations of dynamic processes in physics and related disciplines.

For Product Safety Concerns and Information please contact our EU representative GPSR@taylorandfrancis.com Taylor & Francis Verlag GmbH, Kaufingerstraße 24, 80331 München, Germany